THE CONCISE ANIMAL ENCYCLOPEDIA

THE CONCISE ANIMAL ENCYCLOPEDIA

Conceived and produced by
Weldon Owen Pty Ltd
Ground Floor 42–44 Victoria Street, McMahons Point
Sydney NSW 2060, Australia
weldonowenpublishing.com

This edition first published in 2012

WELDON OWEN PTY LTD
Managing Director Kay Scarlett
Publisher Corinne Roberts
Creative Director Sue Burk
Senior Vice President, International Sales Stuart Laurence
Sales Manager, North America Ellen Towell
Administration Manager, International Sales Kristine Ravn

Managing Editor Averil Moffat
Senior Editor this edition Barbara McClenahan
Designer this edition Gabrielle Green
Design Assistant this edition Oliver Black
Images Manager Trucie Henderson
Production Director Todd Rechner
Production and Prepress Controller Mike Crowton

Project Editors Stephanie Goodwin, Angela Handley
Designers Clare Forte, Hilda Mendham, Heather Menzies, Helen Perks, Sue Rawkins, Jacqueline Richards, Karen Robertson
Jacket Design John Bull
Picture Research Annette Crueger
Copy Editors Janine Flew, Lynn Humphries
Editorial Administrator Jessica Cox
Editorial Coordinator Jennifer Losco
Text Jenni Bruce, Karen McGhee, Luba Vangelova, Richard Vogt

Species Gallery Illustrations MagicGroup s.r.o. (Czech Republic) —www.magicgroup.cz
Pavel Dvorský, Eva Göndörová, Petr Hloušek, Pavla Hochmanová, Jan Hošek, Jaromír a Libuše Knotkovi, Milada Kudrnová, Petr Liška, Jan Maget, Vlasta Matoušová, Jiří Moravec, Pavel Procházka, Petr Rob, Přemysl Vranovský, Lenka Vybíralová

ISBN: 978-1-74252-251-7

This edition first printed in 2011

10 9 8 7 6 5 4 3 2 1

Printed and bound in China by
1010 Printing International Ltd

A WELDON OWEN PRODUCTION

Consultants

Dr. Fred Cooke
President-Elect
American Ornithologists' Union
Norfolk, UK

Dr. Hugh Dingle
Professor Emeritus
University of California
Davis, USA

Dr. Stephen Hutchinson
Visiting Senior Fellow
Southampton Oceanography Centre
Southampton, UK

Dr. George McKay
Consultant in Conservation Biology
Sydney, Australia

Dr. Richard Schodde
Fellow
Australian National Wildlife Collection CSIRO
Canberra, Australia

Dr. Noel Tait
Consultant in Invertebrate Biology
Sydney, Australia

Dr. Richard Vogt
Curator of Herpetology and Professor
National Institute for Amazon Research
Manaus, Amazonas, Brazil

CONTENTS

HOW TO USE THIS BOOK

Each of the six taxonomic sections of the book (Mammals, Birds, Reptiles, Amphibians, Fishes, and Invertebrates) comprises an introductory feature discussing the group in general, then is broken down into chapters devoted to particular subgroups. The sections do not necessarily provide a similar amount of coverage; for example, there is a discrepancy between Mammals and Invertebrates, the latter profiling larger taxonomic groups to give sufficient coverage to Invertebrates as a whole. The book concludes with a glossary and index.

HABITAT ICONS

The 19 habitat icons below indicate at a glance the various habitats in which a species or group can be found. It should be noted that the icons are used in the same order throughout the book, rather than in their order of significance. A more detailed profile of each habitat can be found on pages 40–53.

- Tropical rain forest
- Tropical monsoon forest
- Temperate forest
- Coniferous forest
- Moorlands and heath
- Open habitat, including savanna, grassland, fields, pampas, and steppes
- Desert and semidesert
- Mountains and highlands
- Tundra
- Polar regions
- Seas and oceans
- Coral reefs
- Mangrove swamps
- Coastal areas, including beaches, oceanic cliffs, sand dunes, intertidal rock pools, and/or coastal waters (as applicable to group)
- Rivers and streams, including river and stream banks
- Wetlands, including swamps, marshes, fens, floodplains, deltas, and bogs
- Lakes and ponds
- Urban areas
- Parasitic

Section and chapter
This indicates the group of animals under discussion.

Classification box indicates the taxonomic groups to which the animals belong.

Group global distribution shows the distribution of the group being profiled.

98 | **MAMMALS** SEALS AND SEA LIONS

SEALS AND SEA LIONS

CLASS	Mammalia
ORDER	Carnivora
FAMILIES	3
GENERA	21
SPECIES	36

With flexible, torpedo-shaped bodies, limbs modified to become flippers, and insulating layers of blubber and hair, seals, sea lions, and walruses are superbly adapted to a life in water. They have not, however, completely severed their lin with land and must return to shore to breed. Collectively known as pinnipeds, these marine mammals were once placed in their own order, but are now considered to be part of Carnivora. Most feed on fish, squid, and crustaceans, but som also eat penguins and carrion and may attack the pups of othe seal species. They can dive to great depths in search of prey, with the elephant seal able to stay submerged for up to 2 hours

Cold-water creatures Although monk seals are found in warmer waters, most seals, sea lions, and walruses are restricted to the colder, highly productive seas of the world's polar and temperate regions. The fossil record shows that the three families all originated in the North Pacific.

THREE GROUPS
There are three pinniped families. The Phocidae are known as the true seals. They swim mainly with strokes of their hind flippers, whic cannot bend forward to act as feet making their movement on land particularly ungainly. Although the hearing, especially under water, is good, true seals lack external ears.

Sea lions and fur seals belong t the family Otariidae. These "eared seals" have small external ears. The rely mostly on their front flippers for swimming, and can bend their hind flippers forward when on lar allowing them to walk "four-foote and sit in a semi-upright position.

The third family, Odobenidae, contains a single species, the walru instantly recognizable by the long canine teeth that form tusks on both sexes. Like true seals, walruse use their hind flippers for swimmi and lack external ears. Like eared seals, however, walruses can bend their hind flippers forward.

Sea lion life Australian sea lions often hunt squid and fish together. They can swim a month or so after being born. When they dive, their heartbeat slows from about 100 beats a minute to as low as 10 beats a minute.

CONSERVATION INFORMATION

Within the fact files, each profiled species is allocated a conservation status, using IUCN and other conservation categories, as follows:

✝ **Indicates that a species is listed under the following categories:**

Extinct (IUCN) It is beyond reasonable doubt that the last individual of a given species has died.

Extinct in the wild (IUCN) Only known to survive in captivity or as a naturalized population outside its former range.

Indicates that a species is listed under the following categories:

Critically endangered (IUCN) Facing a very high and immediate risk of extinction in the wild.

Endangered (IUCN) Facing a very high risk of extinction in the wild in the near future.

The following categories are also used:

Vulnerable (IUCN) Facing a high risk of extinction in the wild in the foreseeable future.

Near threatened (IUCN) Likely to qualify for one of the above categories in the near future.

Conservation dependent (IUCN) Dependent upon species- or habitat-specific conservation programs to keep it out of one of the above threatened categories.

Data deficient (IUCN) Inadequate information available to make an assessment of its risk.

Not known Not evaluated or little studied.

Common Widespread and abundant.

Locally common Widespread and abundant within its range.

Uncommon Occurs widely in low numbers in preferred habitat(s).

Rare Occurs in only some of preferred habitat or in small restricted areas.

Name labels
Labels indicate the common and scientific names, as well as the species' family, order, or class, where appropriate.

Habitat icons
The icons indicate the various habitats in which the profiled animal(s) can be found.

Distribution map
This shows the species' or group's range (and former range, where appropriate).

FACT FILE STATISTICS

Length
- **Mammals:** head and body
- **Birds:** tip of bill to tip of tail
- **Reptiles:** snakes and lizards: snout to vent; other reptiles: head and body including tail
- **Turtles:** length of carapace
- **Amphibians:** head and body, including tail
- **Fishes:** entire fish

Height
- **Mammals:** shoulder height
- **Birds:** head and body height

Wingspan: Birds
- From wingtip to wingtip

Weight/Mass
- Body weight

Social Unit: Mammal
- Solitary
- Pair
- Small to large group
- Varies between the above

Plumage: Birds
- Sexes alike
- Sexes differ

Reproduction: Birds and Reptiles
- Number of eggs

Migration: Birds
- Migrant
- Partial migrant
- Sedentary
- Nomadic

Habit: Reptiles and Amphibians
- Terrestrial
- Aquatic
- Burrowing
- Arboreal
- Varies between the above

Breeding season: Amphibians
- When breeding occurs, e.g. spring

Breeding: Reptiles and Fishes
- Viviparous (producing live young)
- Oviparous (producing eggs that develop outside the maternal body)
- Ovoviviparous (producing eggs that develop in the female)

Sex: Reptiles and Fishes
- ♀♂ Reptiles: indicates whether a species is temperature sex determined (TSD) or genetically sex determined (GSD); Fishes: indicates separate male and female, hermaphrodite, or sequential hermaphrodite

Number of genera and species: Birds, Reptiles, and Invertebrates
The number of genera and species in the relevant taxonomic group

ANIMALS

Animalia is one of five kingdoms into which biologists usually divide the living world. Monera includes bacteria and blue-green algae; Protista contains mostly large, unicellular organisms such as amoeba and paramecium, which were once considered animals but are now assigned to a kingdom of their own; molds, mildews, and mushrooms fall within the kingdom Fungi; and, as its name suggests, Plantae contains the plants. With more than 1 million of the 1.75 million living species currently described on Earth, kingdom Animalia is by far the largest and most diverse. It contains all the organisms most of us easily recognize as animals, as well as some species whose status would confuse many non-scientists. The overwhelming majority of animals are invertebrates—animals without a backbone. Among those, the insects dominate in terms of both individual numbers and species diversity. It is, however, the vertebrates with which most of us are familiar—fishes, amphibians, lizards, birds, and mammals—the group to which our own species is most closely affiliated in the evolutionary sense.

DEFINING FEATURES

The kingdom Animalia is also often referred to as Metazoa, a term that indicates the multicellular nature of all its members. As in plants, the tissues of animals are always made up of eukaryotic cells. In animals, however, these lack cell walls. Instead, animal cells are held in place by an extracellular matrix that inevitably contains collagen and provides a mostly flexible framework within which the cells are organized.

Another key feature of all animals is that they are heterotrophs. Unlike plants, which are termed autotrophs, they cannot produce their own food and so, instead, must consume other organisms, either directly or indirectly, to get their nourishment. This has helped drive the extraordinary biodiversity seen among animals today and a vast range of ways to track, capture and consume food has evolved within the group.

Food requirements have also had an impact on animal body plans; most have some sort of centralized digestive system into which they take food and break it down.

The necessity for seeking out, or being near, food sources has also led to capabilities for movement. And, although, it is not always clear-cut, mobility separates most plants from most animals. Questions of mobility have caused confusion in the past among scientists about the status of some groups, in particular the sponges, the animals that have the longest evolutionary history. These are the only living animals that lack organization beyond the cell level so that they have no organs or tissues, a unique feature that has added to past confusions about their status. Their cells are, however, capable of small movements, most have a free-swimming larval stage, and today there is no doubt they are animals and not plants.

The necessity and ability for independent movement has also helped impel in most animals the development of a nervous system with associated sensory equipment to coordinate and guide movements. Most also have body tissues, such as muscle, that facilitate movement.

Sexual reproduction is another unifying feature. Almost all animal species create offspring at some stage in their life-cycle by sexual reproduction; some have the capabilities to reproduce by asexual methods.

Food for warmth The brown bear (*Ursus arctos*), one of the largest ursids, feeds on tubers, berries, fishes, and carrion. Before winter, the northern temperate species stores up body fat and retreats into a den. The bears go into a winter dormancy, a state distinguished from true hibernation because their body temperature does not drop, and live entirely off their body fat.

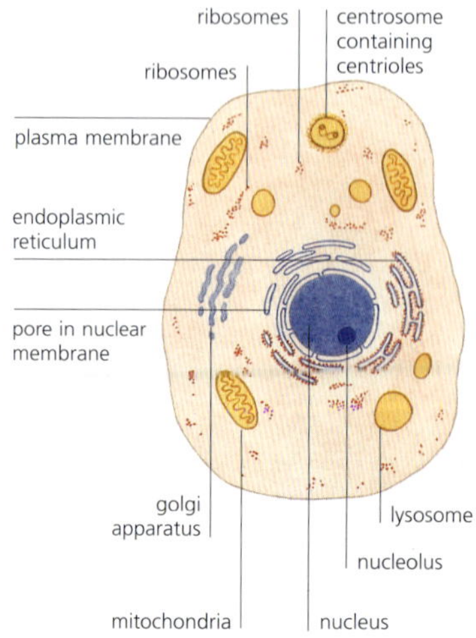

Building blocks All life on Earth is made of cells. The first were prokaryotes—simple assemblages of genetic material within a cell wall. This structure is seen in bacteria. The development from prokaryotes of the mostly larger, far more complex, eukaryotic cells underpinned the evolution of all plants and animals. The genetic material of the eukaryotes is contained within a membrane-bound nucleus and separate organelles that perform specific metabolic functions.

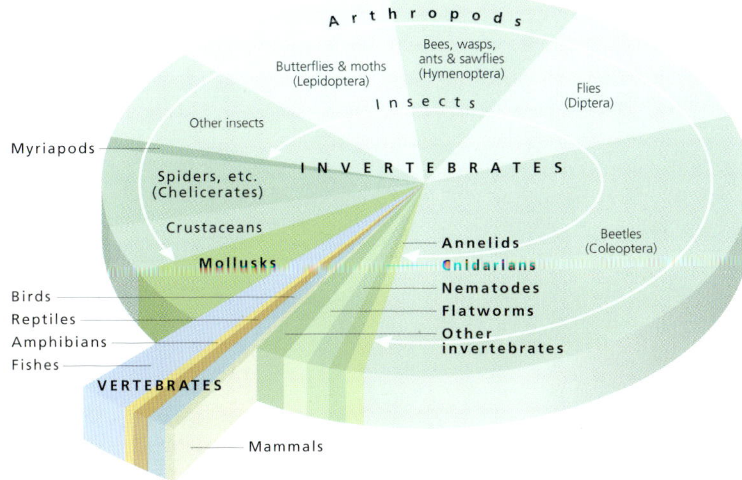

Counting the species

Scientists have described in the region of 1.75 million species of life-forms on Earth, but that is thought to be only a small proportion of the true figure. Estimates place this at between 5 and 100 million species. Even the figures for the number of known species are hard to pin down because new species are being discovered all the time. Vertebrates (see chart at left) are the best-described group but make up only about 5 percent of animal species. There are about 1 million known species of insects but the true number could be more in the region of 30 million.

CLASSIFYING ANIMALS

Humans have a natural tendency to sort and organize. And it seems that, ever since the days of the ancient Greek philosopher Aristotle, we have been attempting to do just that with the many different living organisms with which we share the planet. Biologists have so far discovered, described, and assigned names to about 1.7 million species of plant, animal, and microorganism that presently exist on Earth—a mere fraction of what the total number is thought to be. They have also named many species that lived in the past but have since gone extinct. Modern classification attempts, in part, to provide order and structure to the huge amount of information that has been gathered on different organisms. This is done by providing a unique name for each of these organisms and sorting them into hierarchies of increasingly exclusive groups, or "taxa," based on their evolutionary relationships. This allows for individual organisms to be unequivocally recognized while, at the same time, being associated with other organisms sharing a common ancestry. All classifications evolve and change with the gathering of more knowledge and new discoveries. In recent years, the capabilities to investigate and compare the DNA of organisms through genetic techniques have forced scientists to rethink the classification of many animals.

Looks can deceive Evolutionary relationships are not always obvious from superficial anatomical features. Consider, for example, the African and Asian family of mammals known as hyraxes (far left). All living species are roughly rabbit-sized and look very much like rodents. But these animals are actually ungulates, or hoofed animals, and their closest living relatives are elephants (left) and ocean dwelling manatees. Most species in one hyrax genus lead a solitary existence in the trees, making them the only living hoofed animals to have an arboreal lifestyle.

SPECIES: *catus* – domestic cat

GENUS: *Felis* – domestic cat, sand cat, jungle cat, black-footed cat, Chinese desert cat

FAMILY: Felidae – domestic cat, lion, tiger, leopard, panther, puma, lynx, *Smilodon*

ORDER: Carnivora – domestic cat, seal, wolf, dog, bear, thylacine

CLASS: Mammalia – domestic cat, human, lemur, dolphin, platypus, woolly mammoth

PHYLUM: Chordata – domestic cat, fish, salamander, dinosaur, albatross

KINGDOM: Animalia – domestic cat, stick insect, sea urchin, sponge

Linnaean classification Each grouping in this system of nested categories contains organisms with progressively similar characteristics. The domestic cat, for example, belongs to kingdom Animalia; phylum Chordata (animals with a centralized nerve chord); subphylum Vertebrata (the chord is within a bony vertebral column connected to the head); and class Mammalia (warm-blooded vertebrates with hair, milk glands, and a four-chambered heart). Categories continue through subclass Eutheria (placental mammals that bear live young); order Carnivora (with specialized teeth for eating meat); family Felicidae (all cats); and genus *Felis* (all small cats). Finally, no other organism shares the species scientific name, *Felis catus*.

SCIENTIFIC SORTING

Systematics is the name given to the science of discovering the diversity and evolutionary relationships of organisms. Aspects of systematics involved with naming and classifying organisms form the sub-discipline of taxonomy. The basic taxon, or group, for classification is the species. Ideally, all higher or more inclusive taxa comprise an ancestral species and all its descendants. Determination of this relationship is based on all members of a group sharing one or more derived features (evolutionary novelties).

NAMING RIGHTS

Animals are known by different vernacular or common names that vary from language to language, often even within the same country. And so, to avoid confusion, scientists use Latinized names for groups of organisms. This provides universality and stability and avoids the necessity of translating names into many different languages. In this way, no matter what one's native tongue, the Latinized scientific name is immediately associated with the same group of organisms.

The basic classification category is the species, which are populations of organisms that share one or more similarities not found in related organisms. Species also form closed genetic systems. This means that individuals can only reproduce with another individual of the same species, although closely related species may occasionally hybridize.

AN ENDURING SYSTEM

In the early 18th century, the Swedish naturalist Carl Linnaeus developed a scheme for naming, ranking, and classifying different organisms according to the presence or absence of observable similarities. His earliest version of this system was published in 1735 under the name *Systema Naturae*. Linnaeus refined it many times in his lifetime, and it has been improved many times since. It still, however, provides foundations for the approach to classification now used by biologists around the world. As a result, Linnaeus is often referred to as the "father of taxonomy."

The principal feature of the Linnaean system is that it assigns a unique two-part name to each and every different organism. In this binomial system, the first part of the name indicates the genus to which the organism belongs. Known as the generic name, this indicates other organisms, both living and dead, with which the organism shares its closest evolutionary relationships.

The second name is the specific name. Only one organism within any genus will ever be assigned this name. Scientists throughout the world know that when they use this name they are invariably talking about the same organism.

The generic name always begins with a capital letter, while the specific name is always written in the lower case. Both names are always printed in italics.

When populations of a species are separated geographically and have relatively consistent differences, they may be recognized as a subspecies and assigned a third name. Like the species name, this is written in the lower case and also printed in italics.

Leaping alike All frogs and toads have one adaptation that sets them apart from other amphibians: the ankle bones are greatly elongated to form an extra segment in the hind leg, providing greater leverage for leaping. This is a distinguishing condition of the order Anura, within which all frogs and toads are placed. Frogs and toads also possess another adaptation to leaping; they have a short vertebral column with no more than 10 free vertebrae followed by a bony rod (the coccyx, representing fused tail vertebra).

MAMMALS

The 26 orders of mammals are divided into three major groups based on the structure of their reproductive tracts. The most primitive are the egg-laying mammals with a single order, the monotremes. Marsupials, which give birth to young in a very early stage of development, are now considered to consist of seven orders. The other 18 orders are made up of placental mammals. The most recent evidence from DNA sequence analysis has revealed that whales and even-toed ungulates are more closely related to each other than to any other group. DNA also indicates that there have been three major radiations of placental mammals: in Africa, South America, and the Northern Hemisphere.

Class Mammalia

EGG-LAYING MAMMALS
Order Monotremata
Monotremes

MARSUPIALS
Order Didelphimorphia
American opossums

Order Paucituberculata
Shrew opossums

Order Microbiotheria
Monito del monte

Order Dasyuromorphia
Quolls, dunnarts, marsupial mice, numbat, and allies

Order Peramelemorphia
Bandicoots

Order Notoryctemorphia
Marsupial moles

Order Diprotodontia
Possums, kangaroos, koalas, wombats, and allies

PLACENTAL MAMMALS
Order Xenarthra
Sloths, anteaters, and armadillos

Order Pholidota
Pangolins

Order Insectivora
Insectivores

Order Dermoptera
Flying lemurs

Order Scandentia
Tree shrews

Order Chiroptera
Bats

Order Primates
Primates

Suborder Strepsirhini
Prosimians

Suborder Haplorhini
Monkeys and apes

Order Carnivora
Carnivores

Family Canidae
Dogs and foxes

Family Ursidae
Bear and pandas

Family Mustelidae
Mustelids

SEALS AND SEA LIONS
Family Phocidae
True seals

Family Otariidae
Sea lions and fur seals

Family Odobenidae
Walrus

Family Procyonidae
Raccoons

Family Hyaenidae
Hyenas and aardwolf

CIVETS AND MONGOOSES
Family Viverridae
Civets, genets, and linsangs

Family Herpestidae
Mongooses

Family Felidae
Cats

Order Proboscidea
Elephants

Order Sirenia
Dugong and manatees

Family Trichechidae
Manatees

Family Dugongidae
Dugong

Order Perissodactyla
Odd-toed ungulates

Family Equidae
Horses, zebras, and asses

Family Tapiridae
Tapirs

Family Rhinocerotidae
Rhinoceroses

Order Hyracoidea
Hyraxes

Order Tubulidentata
Aardvark

Order Artiodactyla
Even-toed ungulates

Family Bovidae
Cattle, antelopes, and sheep

DEER
Family Cervidae
Deer

Family Tragulidae
Chevrotains

Family Moschidae
Musk deer

Family Antilocapridae
Pronghorn

Family Giraffidae
Giraffe and okapi

Family Camelidae
Camels and llamas

Family Suidae
Pigs

Family Tayassuidae
Peccaries

Family Hippopotamidae
Hippopotamuses

Order Cetacea
Cetaceans

Suborder Odontoceti
Toothed whales

Suborder Mysticeti
Baleen whales

Order Rodentia
Rodents

Suborder Sciurognathi
Squirrel-like rodents, mouse-like rodents, and gundis

Suborder Hystricognathi
Cavy-like rodents

Order Lagomorpha
Hares, rabbits, and pikas

Order Macroscelidea
Elephant shrews

Family Cervidae, page 142

BIRDS

Since Charles Darwin, birds have been classified according to perceptions of their natural relationships. This approach clusters similar-looking, interbreeding populations in species—related or sister species in genera; sister genera in families; and sister families in orders—representing the "tree" of bird evolution.

We now identify species as similar-looking populations of organisms that can interbreed freely. Genera, families, and orders are based on structural similarities in limbs, skeletons, and feathers. Thus birds of the parrot family have a foot with two toes forward and two back, and a hooked bill with vertically twisted palate bones.

The accumulated data led the American Alexander Wetmore to devise a classification of the orders and families of birds in the 1930s. This epic work became the standard for bird classifications in the 20th century. Since then, DNA and other molecular studies have shown that many structural traits used for classifying birds are unreliable due to convergent evolution, especially within the families of passerines, or songbirds, which involve over half the world's bird species. In particular, they found that the wrens, flycatchers, robins, and warblers of Australasia were unrelated to Eurasian look-alikes. Modern research indicates that these Australasian groups are the old ancestral lineages of the world's songbirds, from Gondwana.

The bird classification used here takes account of these changes. At species, generic, and family levels, it is based on the most up-to-date and authoritative world checklist current, the Howard and Moore *Complete Checklist of Birds of the World*, published in 2003. That checklist did not consider orders; but here we have adapted its families to the familiar Wetmore arrangement.

Class Aves

Order Tinamiformes
Tinamous

Order Struthioniformes
Ostrich

Order Rheiformes
Rheas

Order Casuariiformes
Cassowaries and emus

Order Apterygiformes
Kiwis

Order Apterygiformes, page 199

Order Galliformes
Gamebirds

Order Anseriformes
Waterfowl

Order Sphenisciformes
Penguins

Order Gaviiformes
Divers

Order Podicipediformes
Grebes

Order Procellariiformes
Albatrosses and petrels

Order Phoenicopteriformes
Flamingos

Order Ciconiiformes
Herons and allies

Order Pelecaniformes
Pelicans and allies

Order Falconiformes
Birds of prey

Order Gruiformes
Cranes and allies

Order Charadriiformes
Waders and shorebirds

Order Pteroclidiformes
Sandgrouse

Order Columbiformes
Pigeons

Order Psittaciformes
Parrots

Order Cuculiformes
Cuckoos and turacos

Order Pelecaniformes, page 219

Order Strigiformes
Owls

Order Caprimulgiformes
Nightjars and allies

Order Apodiformes
Hummingbirds and swifts

Order Coliiformes
Mousebirds

Order Trogoniformes
Trogons

Order Coraciiformes
Kingfishers and allies

Order Piciformes
Woodpeckers and allies

Order Passeriformes
Passerines

Order Passeriformes, page 270

REPTILES

Reptilia—living reptiles—traditionally includes turtles, crocodilians, tuatara, and squamates (lizards, snakes, and worm lizards). Crocodilians are most closely related to birds, but as birds are treated separately this section is assumed to cover non-avian reptiles. The position of turtles is in controversy; some studies show that the lineage of turtles is so distant from the squamates that they belong in a separate class. Lizards lost their limbs to become snakes or limbless lizards or amphisbaenians, so the formation of suborders within Squamata is controversial. Only the status of the tuatara as an ancient order by itself within the reptiles is agreed upon. The traditional separation of groups has been used, but the reader should be aware of the artificial nature of these groupings.

Class Reptilia

Order Testudines
Tortoises and turtles

Order Crocodilia
Crocodilians

Order Rhyncocephalia
Tuatara

Order Squamata
Lizards and snakes

Order Testudines, page 300

Suborder Amphisbaenia
Worm lizards

AMPHIBIANS

Lissamphibia, the living amphibians, includes three orders: the frogs and toads (Anura), the salamanders, newts, and sirens (Urodela), and the caecilians (Gymnophiona). Amphibians are all derived from the same common ancestor (monophyletic). The main common feature all amphibians have is smooth skin without scales. The use of families, genera, species, and common names follows Amphibian Species of the World: an on-line reference (http://research.amnh.org/herpetology/amphibia/index.html). Other common names were sourced from regional guides or compendiums.

Class Amphibia

Order Caudata
Salamanders and newts

Order Gymnophiona
Caecilians

Order Anura
Frogs and toads

Order Anura, page 366

FISHES

Any study of the fishes shows that they are an immensely diverse array of animals, differing greatly in the range of habitats they occupy and their body forms and adaptations. As a consequence, most biologists regard the term "fishes," as a convenient name, rather than a closely defined taxonomic entity, that describes aquatic vertebrates such as hagfishes, lampreys, sharks, rays, lungfishes, sturgeons, gars, and the advanced ray-finned fishes. There are a number of classification schemes for the fishes but one of the most widely accepted recent ones recognizes five classes of living species and three classes that are now extinct. The five classes, whose classification is detailed below, are hagfishes, lampreys, cartilaginous fishes, lobe-finned fishes, and ray-finned fishes. These are grouped into two superclasses: jawless fishes and jawed fishes. The three extinct classes are the pteraspidomorphs—jawless armored fishes, the jawed placoderms that were encased in bony plates, and the acanthodians, small true bony fishes with two long dorsal spines.

JAWLESS FISHES
Superclass Agnatha
Lampreys and hagfishes

JAWED FISHES
Superclass Gnathostomata
(includes all the groups below)

CARTILAGINOUS FISHES "Chondrichthyes"
Class Chondrichthyes
Sharks, rays, and allies

Subclass Elasmobranchii
Sharks
Rays and allies

Subclass Holocephali
Chimaeras

BONY FISHES "Osteichthyes"

Class Sarcopterygii
Lungfishes and allies

Subclass Elasmobranchii, page 398

Class Actinopterygii

Subclass Chondrostei
Bichirs and allies

Subclass Neopterygii

Primitive Neopterygii
(gars and bowfin)

Division Teleostei

Subdivision Osteoglossomorpha
Bonytongues and allies

Subdivision Elopomorpha
Eels and allies

Subdivision Clupeomorpha
Sardines and allies

Subdivision Euteleostei
(includes all the groups below)

Superorder Ostariophysi
Catfish and allies

Superorder Protacanthopterygii
Salmons and allies

Superorder Stenopterygii
Dragonfishes and allies

Superorder Cyclosquamata
Lizardfishes and allies

Superorder Scopelomorpha
Lanternfishes

Superorder Polymixiomorpha
Beardfishes

Superorder Lampridiomorpha
Opahs and allies

Superorder Paracanthopterygii
Cod, anglerfishes, and allies

Superorder Acanthopterygii
Spiny-rayed fishes

Superorder Acanthopterygii, page 434

INVERTEBRATES

Over 95 percent of all animals are invertebrates. They are characterized by a structure that they all lack: a backbone or vertebral column. Invertebrates are divided into about 30 phyla, each displaying a distinct body form. Their evolutionary relationships can be inferred from their anatomy, their early development, and more recently from molecular analyses, particularly DNA, the genetic code. Features that define phyla include the organization of the body from a loose association of cells (Porifera), through tissue formation (Cnidaria) to the development of organs (Platyhelminthes). The acquisition of a fluid-filled body cavity was a defining point in animal evolution that allowed animals, such as Nematoda, Annelida, and many other phyla of worms, to move about by an hydraulic system driven by fluid pressure. The origin and form of these body cavities characterize different phyla. While these phyla are soft-bodied, others are protected and supported by various types of skeletons, such as shells in Mollusca and a jointed exoskeleton in Arthropoda. The division of the body into segments allowed for specialization of parts of the body. In arthropods, this has led to the development of segmental appendages that carry out specific functions, such as sensory perception, feeding and locomotion. Details of early embryonic development divide many advanced phyla into two lineages, one leading through the Echinodermata to the Chordata, the phylum to which vertebrates belong, the other containing the bulk of animal phyla. While molecular analyses have confirmed many of our ideas about the course of evolution based on anatomy and development, there are a number of instances where they are at variance. Hence, our classificatory system is undergoing revision. Furthermore, the continual identification of new species of invertebrates indicates that we are nowhere near their full inventory, and certainly far from understanding their vital roles in the sustainability of ecosystems.

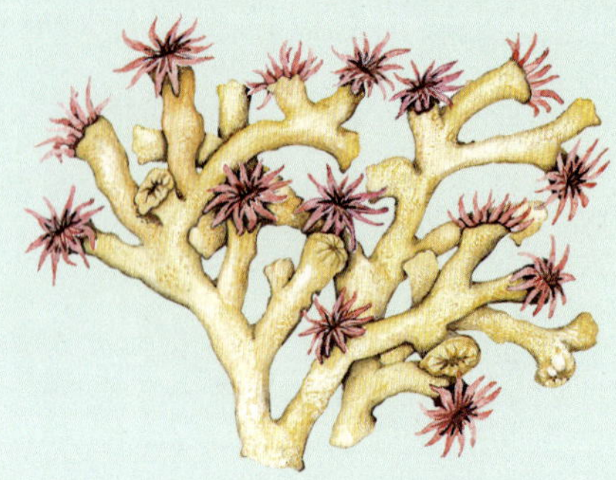

Phylum Cnidaria, page 454

Phylum Chordata
Invertebrate Chordates

Subphylum Urochordata
Sea squirts

Subphylum Cephalochordata
Lancelets

Phylum Porifera
Sponges

Phylum Cnidaria
Cnidarians (sea anemones, corals, jellyfishes, etc.)

Phylum Platyhelminthes
Flatworms

Phylum Nematoda
Roundworms

Phylum Mollusca
Mollusks (bivalves, snails, squids, etc.)

Phylum Annelida
Segmented worms

Phylum Arthropoda
Arthropods

Subphylum Chelicerata
Chelicerates

Class Arachnida
Arachnids

Class Merostomata
Horseshoe crabs

Class Pycnogonida
Sea spiders

Class Arachnida, page 470

Subphylum Myriapoda
Myriapods (centipedes, etc.)

Subphylum Crustacea
Crustaceans

Subphylum Hexapoda
Hexapods

Class Insecta
Insects

Order Odonata
Dragonflies and damselflies

Order Mantodea
Mantids

Order Blattodea
Cockroaches

Order Isoptera
Termites

Order Orthoptera
Crickets and grasshoppers

Order Hemiptera
Bugs

Order Coleoptera
Beetles

Order Diptera
Flies

Order Lepidoptera
Butterflies and moths

Order Hymenoptera
Bees, wasps, ants, and sawflies

Order Embioptera
Webspinners

Order Dermaptera
Earwigs

Order Thysanura
Silverfish

Order Phasmatodea
Stick and leaf insects

Order Archaeognatha
Bristletails

Order Raphidioptera
Snakeflies

Order Ephemeroptera
Mayflies

Order Megaloptera
Dobsonflies and alderflies

Order Phthiraptera
Parasitic lice

Order Psocoptera
Book lice and bark lice

Order Siphonaptera
Fleas

Order Thysanoptera
Thrips

Order Neuroptera
Lacewings, antlions, and allies

Order Mecoptera
Scorpionflies

Order Plecoptera
Stoneflies

Order Trichoptera
Caddisflies

Order Zoraptera
Angel insects

Order Grylloblattodea
Rock crawlers

Order Strepsiptera
Strepsipterans

Class Collembola
Springtails

Class Protura
Proturans

Class Diplura
Diplurans

Order Odonata, page 488

Order Coleoptera, page 497

Order Lepidoptera, page 504

Order Siphonaptera, page 514

Phylum Echinodermata
Echinoderms (sea stars, sea urchins, sea cucumbers, etc.)

Phylum Nemertea
Ribbon worms

Phylum Entoprocta
Goblet worms

Phylum Tardigrada
Water bears

Phylum Ctenophora
Comb jellies

Phylum Rotifera
Rotifers (wheel animals)

Phylum Hemichordata
Hemichordates (acorn worms)

Phylum Chaetognatha
Arrow worms

Phylum Gastrotricha
Gastrotrichs

Phylum Kinorhyncha
Spiny-crown worms

Phylum Phoronida
Horseshoe worms

Phylum Onychophora
Velvet worms

Phylum Brachiopoda
Brachiopods (lamp shells)

Phylum Bryozoa
Bryozoans (lace animals)

Phylum Sipuncula
Peanut worms

Phylum Echiura
Spoon worms

Phylum Loricifera
Brushheads

Phylum Priapulida
Phallus worms

Phylum Nematomorpha
Horsehair worms

Phylum Acanthocephala
Spiny-headed worms

Phylum Pogonophora
Beard worms

Phylum Gnathostomulida
Sand worms

Phylum Cycliophora
Cycliophorans

Phylum Placozoa
Placozoans

Phylum Orthonectida
Orthonectids

Phylum Rhombozoa
Rhombozoans

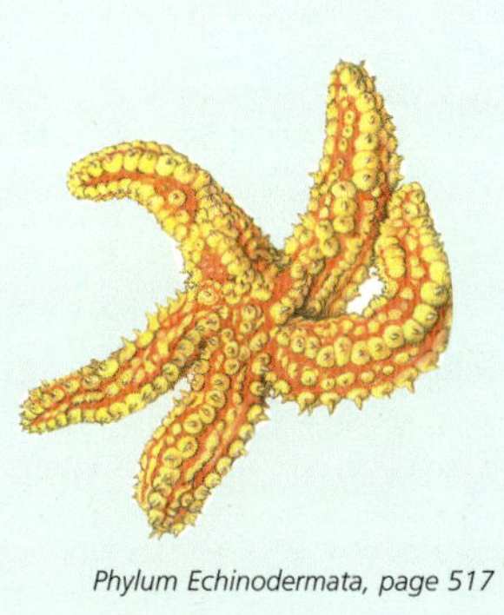
Phylum Echinodermata, page 517

Phylum Onychophora, page 521

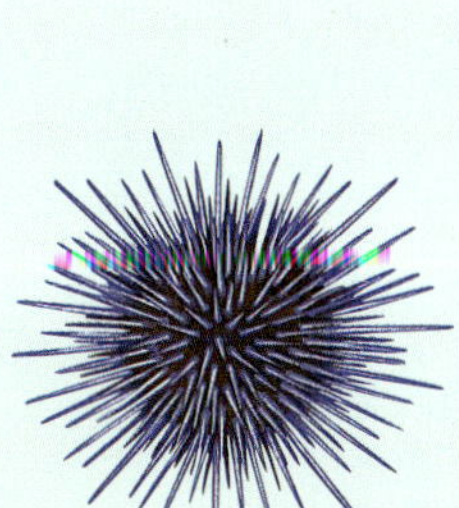
Phylum Echinodermata, page 517

Habitats and Adaptations

Habitats are the locations or surroundings in which living organisms survive. They are the providers of food, shelter, and other fundamental requirements necessary for plants and animals. Different habitats are most often characterized by a combination of their climate and geography but sometimes also by the communities of species that dominate them. Coral reefs, for example, occur in shallow and mostly tropical oceanic waters in which the coral polyps have laid down their calcium-carbonate skeletons. Different forest types, however, are defined by combinations of characteristics such as their latitude, total rainfall, whether the rainfall is received consistently throughout the year or in a marked season, and the dominant vegetation.

Natural vegetation This map shows the vegetation that would occur naturally without human interference. Climate and soil determine the vegetation of an area, and the vegetation, in turn, determines the extent and diversity of animal life. Together, the plants and animals of each zone form an ecological community known as a biome. The rainfall and warmth of the tropics have given rise to the greatest biodiversity in the world. In addition to the terrestrial zones indicated, freshwater and marine biomes contain abundant, diverse life-forms. The oceans, for example, are home to life-forms ranging from minuscule plankton to giant whales.

From high to low Altitude as well as latitude play a part in delineating habitats. Snow-covered Mt. Kilimanjaro contrasts with the semiarid grasslands below. Made up of three main cones, Mt Kilimanjaro is a dormant volcano that rises 19,340 feet (5,895 m) above the surrounding plain to form Africa's highest point.

Tree dweller The koala (*Phascolarctos cinereus*) builds neither nest nor den. For this marsupial, eucalypts provide both a home and a food source. The species of eucalypts that the koala eats have a high fiber and cellulose content as well as high concentrations of toxic compounds. The koala must chew on large quantities of leaves to extract sufficient nutrients.

BETWEEN WORLDS

While most species exist within one habitat, there are many that exploit or regularly move between more than one. Emperor penguins huddle together on ice in Antarctic breeding colonies during the bitter polar winters but feed on fishes in open oceanic waters long distances away. Migrating species, too, often move between habitats to exploit or avoid different or changing conditions. Within habitats, animals function in ecological niches. This defines the roles that animals play and includes what they eat, what eats them, where they find shelter, where they find food, and how they respond to and interact with other organisms. Some species are adapted to a very specific niche with a narrow range of parameters. Koalas, for example, survive only where their food trees—specific eucalypt species—grow.

Other animals such as some cockroach pest species have more generalist adaptations and are able to operate within wide niches or across several niches.

ENVIRONMENT CHANGE

Habitats are not fixed or static, and it is natural that they change over time. Many are shaped by sudden, dramatic geological events such as earthquakes and volcanoes, while some shift more slowly in response to a persistent natural force such as erosion. Changing weather patterns often create seasonal, recurring, and reversible change. In addition, some habitats change in response to the animals and plants that are living and evolving within them.

However, many changes being witnessed within Earth's habitats are not the result of natural forces or events. Rather, it is human activity that now drives the most significant transformations. Most habitats, for example, are shrinking in area. Major exceptions include the ever-expanding desert, semi-desert, and urban habitats. It is also significant that the diversity of life within most habitats is seen to be on the decline.

The more complex and species-rich habitats, such as the rain forests and coral reefs, were once thought to be the most resilient to change. In recent times, however, they have exhibited signs of being among the most fragile.

DEMARCATION ISSUES

The line where one habitat begins and another ends is not necessarily abrupt or clearly defined. Many forest habitats, for example, give way gradually at the outer margins to open habitats such as grasslands. In some cases, however, it is very clear where a habitat begins and ends. Caves are habitats that are frequently well delineated by their geology. Life within is defined by factors such as low light levels and the availability of food. Conditions at a cave's entrance are often very different to those further into the cave. Similarly, the deep-sea hydrothermal vent habitats discovered in recent years occur only within very well-defined and highly productive oceanic areas.

MAMMALS

CLASS Mammalia
ORDERS 26
FAMILIES 137
GENERA 1,142
SPECIES 4,785

The astonishing diversity of the class Mammalia ranges from tiny field mice no bigger than a thimble, to the massive blue whale, which weighs 1,750 times more than a man. Their flexibility, adaptability, and intelligence have allowed mammals to occupy all continents and almost all habitats, exploiting niches on land, below ground, in trees, in the air, and in fresh and salt water. Despite their diverse body forms, mammals share some unique characteristics. Their defining feature is that the lower jaw bone, the dentary, attaches directly to the skull. Mammals are endothermic ("warm-blooded"), nurse their young with milk produced by mammary glands, and usually have hair on the body.

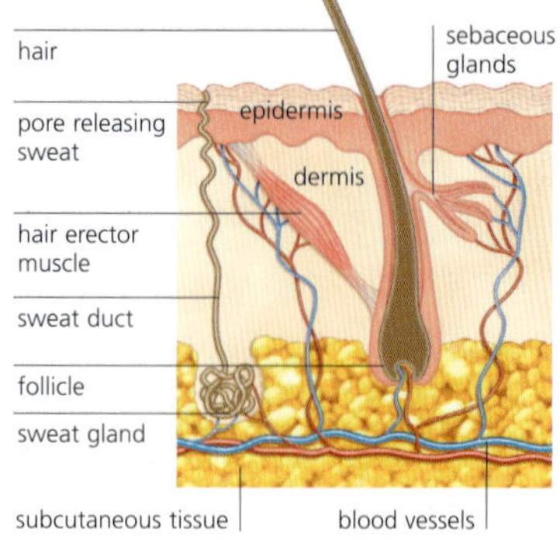

In the skin Mammalian skin is made up of two layers: the epidermis, an outer layer of dead cells, and the dermis, which contains blood vessels, nerve-endings, and glands. Muscles can raise or lower hairs to trap or release an insulating layer of air.

ORIGINS AND ANATOMY

Mammals evolved from a group of mammal-like reptiles that had a powerful bite and complex teeth. The first true mammals, with the distinctive jaw hinge between the dentary and the skull, emerged some 195 million years ago. These morganucodontids were nocturnal, shrew-like creatures, only about an inch (2.5 cm) long, that fed on insects. For the next 130 million years, Earth was dominated by dinosaurs, and mammals remained small. About 65 million years ago, however, a cataclysmic change in global climate led to the extinction of 70 percent of all animal species, including the dinosaurs. Mammals survived and began filling the newly empty niches, the start of a process that led to the remarkable variety of species that we see today.

The key to the survival of mammals during a period of severe climate change may have been the ability to regulate their internal body temperature through such methods as adjusting their metabolic rate or blood flow, and shivering, sweating, or panting. Endothermy allowed mammals to remain active regardless of extreme external temperatures and thus to colonize a great range of habitats.

One feature unique to mammals is the presence of hair on the body. Many mammals have a double coat of soft underfur and coarser guard hairs, which insulates against heat and cold. Specialized hairs form whiskers, or vibrissae, which are extremely sensitive to touch. Coat patterns can provide camouflage or communicate information, while spines and quills are modified hairs that create a defensive shield.

The skin of mammals contains a number of different types of glands. In females, mammary glands secrete milk for nourishing their young. Sebaceous glands produce an oily lubricant to protect and waterproof the fur. Sweat glands are important in temperature control, allowing many mammals to release sweat and cool down through its evaporation. Scent glands produce complex odors that convey a mammal's status, sexual condition, and other information. In some species, such as skunks, scent is also used in defense.

The brains of mammals are large (relative to body size) and complex. The sense of smell is important in communication. Color vision has arisen independently in a number of mammals, and binocular vision allows primates to judge distances. All mammals have three bones in the middle ear, and most have an external ear to collect sound.

Fur and whiskers Almost all mammals, from kangaroos to aquatic common seals, rely on a fur coat for insulation. Made of cells strengthened by keratin, hairs can also take the form of vibrissae, touch-sensitive whiskers that provide important tactile information about the creature's environment.

In water Streamlined for sleek movement through water, cetaceans such as dolphins have lost their body hair and rely instead on a layer of blubber as insulation.

Leaf-eating monkey The douc langur lives in the rain-forest canopy, gripping branches with its grasping hands and feet and leaping from tree to tree. Its potbelly contains a complex digestive system to break down large quantities of leaves.

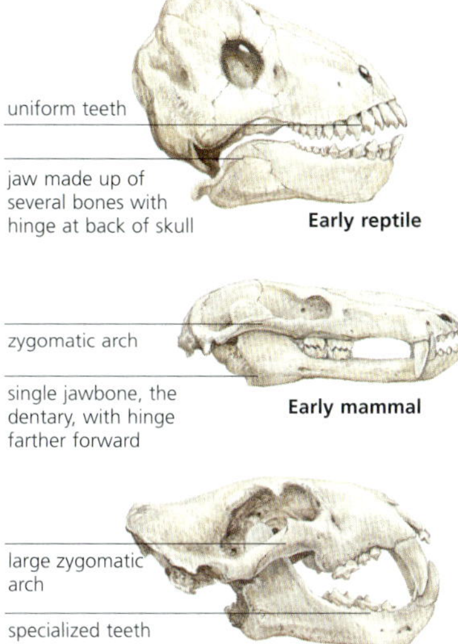

Jaws and teeth Mammals are the only animals that chew their food. They differ from their reptilian ancestors by having a single jawbone, powerful jaw muscles in a zygomatic arch, and complex teeth.

DIVERSE LIFESTYLES

As mammals moved into different habitats, they adapted their anatomy, locomotion, diet, and social habits to suit their particular niche. Most terrestrial mammals walk on four feet, but some move on two. Some, such as bears, plant the entire sole on the ground, while others walk on just the digits (cats) or tips of the toes (deer). Many shelter below ground, with moles and some other efficient diggers spending almost their entire lives in their burrows. Trees are home for many primates, rodents, and other mammals. In some, grasping hands and feet provide a secure grip on branches and a prehensile tail acts as a fifth limb. A number of species have a gliding membrane that allows them to glide from tree to tree. Bats have developed true flight, with flapping wings allowing them to traverse great distances through the air. Many mammals are good swimmers, but aquatic mammals such as whales, dolphins, seals, and manatees have fully adapted to a life spent in water. Their shape has become streamlined and their limbs have been modified into flippers.

To fuel the constant regulation of their body temperature, mammals must consume a rich or plentiful diet. The structure of the jaw bones and muscles provides mammals with a powerful bite and the ability to cut and chew their food. Teeth are highly specialized and reflect the diet of the animal. The cheek teeth of carnivores, for example, are sharp for cutting flesh and bones; those of herbivores are broad for grinding plant matter; and omnivores have multicusped cheek teeth that enable them to chew both animals and plants. While many mammals are opportunistic feeders, others are specialists. Some carnivores concentrate on insects and other invertebrates, or on small or large vertebrates. Herbivores may focus on fruit, leaves, or grasses. Many plant-eaters tend to have a complex digestive system and depend on microorganisms in the gut to break down plant cellulose.

MAMMAL REPRODUCTION

In all mammals, the female's eggs are fertilized internally by the male's sperm. Monotremes lay eggs and incubate them until they hatch. Marsupials and placental mammals both give birth to live young. The young of marsupials have short gestations and are poorly developed at birth, but latch onto a teat and are nourished by the mother's milk. Placental mammals grow for longer inside the uterus, nourished by a placenta. At birth, they tend to be in a more advanced state of development.

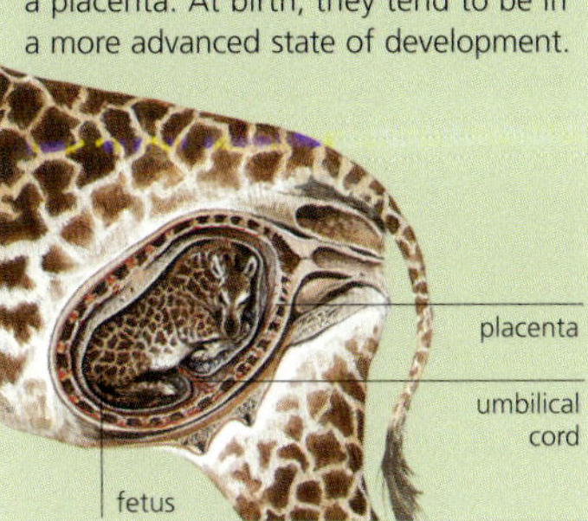

The adaptability of mammals is reflected in the diversity of their social organization. Adult mammals may be solitary, live in pairs or small family groups, or form harems, herds, or colonies. These groupings may be flexible and temporary, or highly stable and permanent. In all mammals, the fact that young are nourished by their mother's milk develops some kind of social bond. This may be rudimentary—the tree shrew, for example, visits its young only every couple of days for a brief nursing session, and some rodents are weaned after a few weeks. Many young mammals, however, remain dependent on the mother for many months or even years, as is the case with primates, elephants, and whales.

Mammals communicate with one another through scent, touch, sound, posture, and gesture. Both competition and cooperation are highly developed in many mammal species. Individuals of the same species will compete for territory, food resources, the right to mate, and social dominance. Social species may rely on fellow group members to help raise the young, to hunt cooperatively or share information about food sources, to warn of danger, and to deter predators.

MONOTREMES

CLASS	Mammalia
ORDER	Monotremata
FAMILIES	2
GENERA	3
SPECIES	3

Like other mammals, monotremes are covered in fur, lactate to feed their young, and have a four-chambered heart, a single bone in the lower jaw, and three bones in the middle ear. They are unusual, however, because they lay eggs rather than give birth to live young, and also have some anatomical similarities to reptiles, such as extra bones in their shoulders. The two families in this order are Tachyglossidae, which includes two species of echidna, and Ornithorhynchidae, which contains just the platypus. With its duck-like bill, webbed feet, furred body, and beaver-like tail, the platypus has fascinated scientists since the first specimen was sent to Britain in 1799.

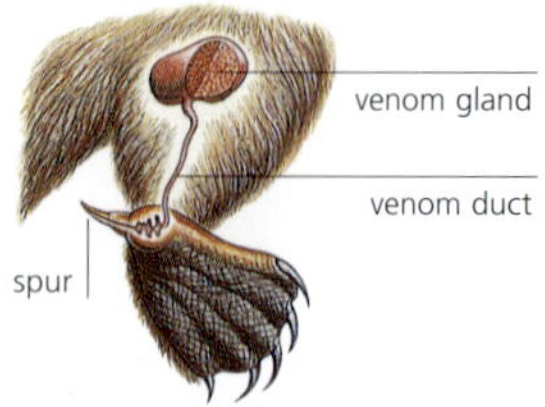

Poisonous spur The male platypus can use the spur on its rear ankle to inflict a paralyzing sting, making it one of the world's few venomous mammals. This feature may help it fight other males for territory and mates in the breeding season.

Spiny defense To protect itself from predators such as dingoes, the short-nosed echidna can speedily burrow vertically down into the earth until only the tips of its spines are visible. If threatened on hard ground, the echidna curls into a spiky ball. Its narrow snout and long tongue can search for ants and termites in narrow cavities.

EGG-LAYING MAMMALS

Monotreme eggs are soft-shelled and hatch after about 10 days. Once hatched, the young rely on their mother's milk for several months, a dependence typical of mammals.

After mating in spring, female platypuses lay up to three eggs in a bankside nesting burrow, and curl up to incubate them between the tail and body. The hatched young stay in the burrow for 3–4 months, feeding on milk that oozes from two nipple-like patches onto the mother's fur. After they emerge from the burrow, the young are gradually weaned and eventually lead a largely solitary life. Wild platypuses have been known to live for 15 years.

During their winter mating season, several male short-nosed echidnas may follow a female for up to 14 days, conducting digging and pushing contests until one wins the right to mate. The female then lays a single egg into her pouch. After hatching, the young stays in the pouch until it develops spines and moves to a burrow. The mother lacks teats but her mammary glands open into patches of skin in her pouch. She may continue to feed the young for up to 7 months.

Little is known of the long-nosed echidna's breeding cycle; it is thought to be like the short-nosed echidna's.

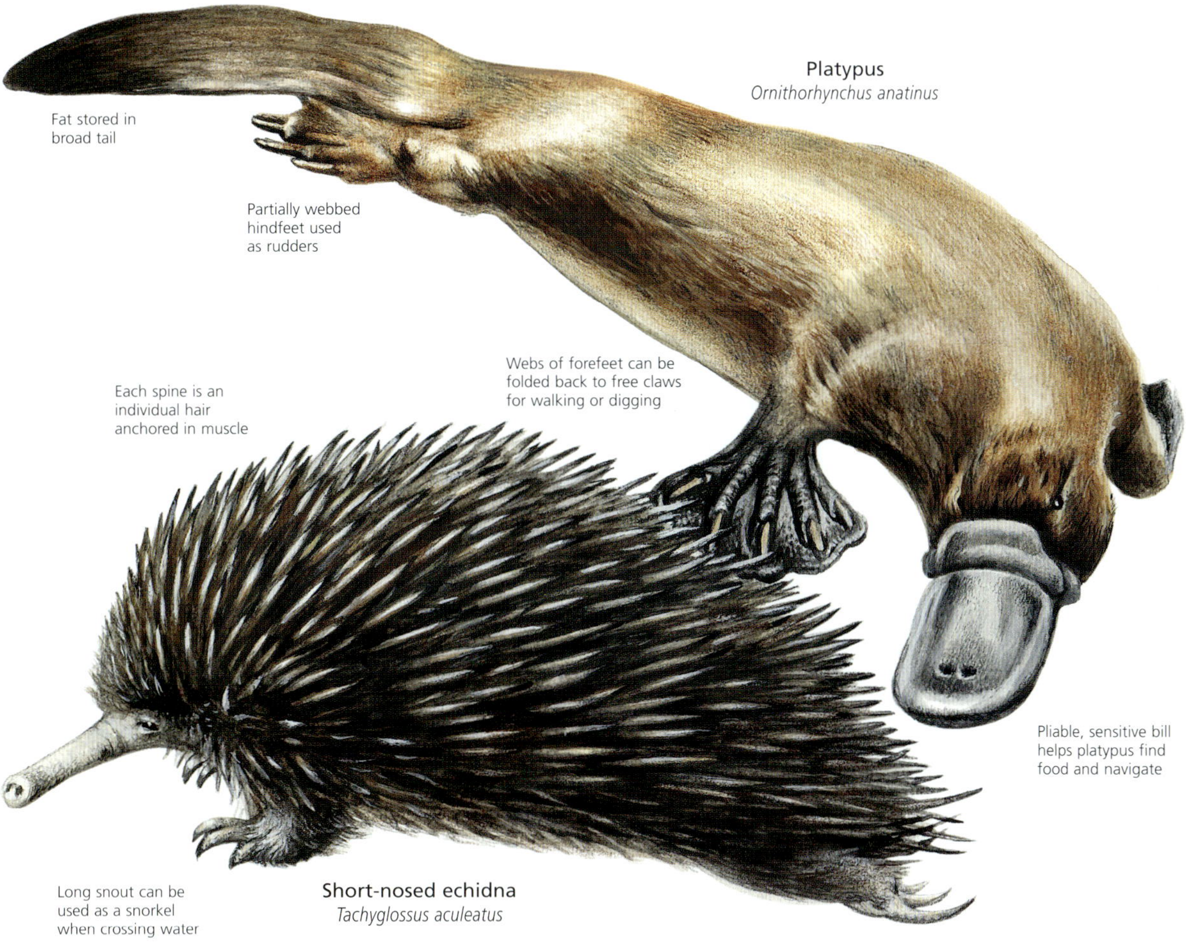
Platypus
Ornithorhynchus anatinus
Fat stored in broad tail
Partially webbed hindfeet used as rudders
Webs of forefeet can be folded back to free claws for walking or digging
Each spine is an individual hair anchored in muscle
Pliable, sensitive bill helps platypus find food and navigate
Long snout can be used as a snorkel when crossing water
Short-nosed echidna
Tachyglossus aculeatus

Long-nosed echidna
Zaglossus bruijni
Mouth at end of snout
Walks with a rolling gait

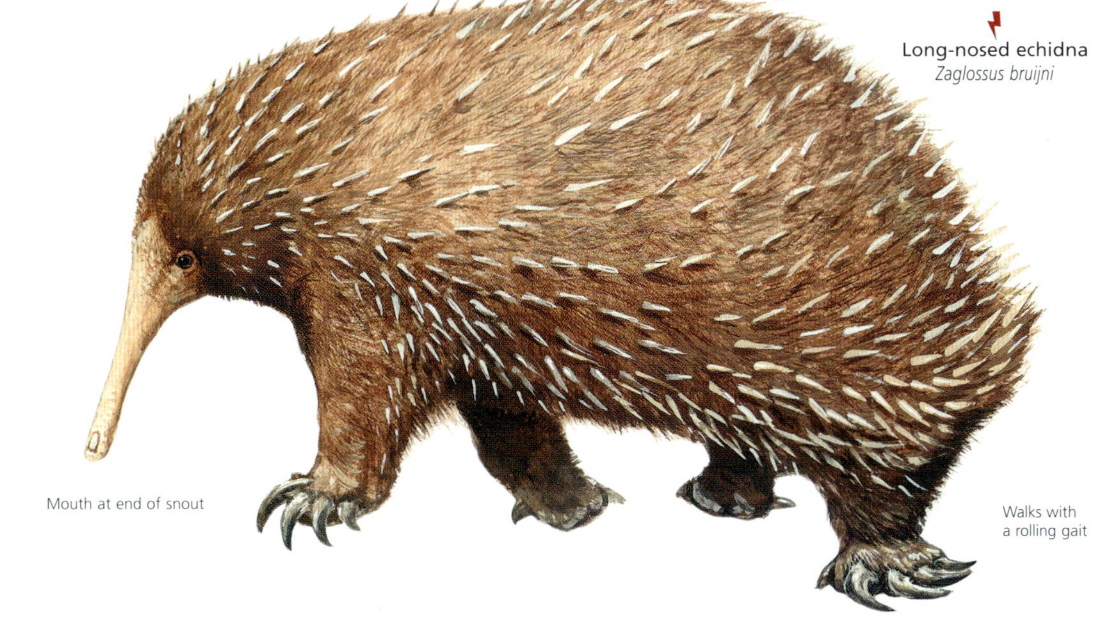

Platypus Perhaps one of the most unusual of all animals, this amphibious mammal has a pliable duck-like bill, thick fur, and webbed feet. It lives in riverbank burrows and feeds on insect larvae and other invertebrates.

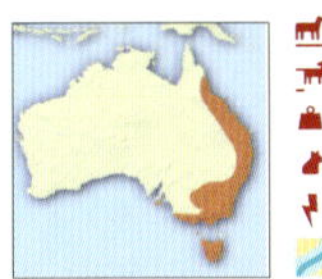

Up to 16 in (40 cm)
Up to 6 in (15 cm)
Up to 5½ lb (2.4 kg)
Solitary
Locally common

E. Australia, Tasmania, Kangaroo I., King I.

Short-nosed echidna Its stout body covered in long spines and shorter fur, this echidna walks with a rolling gait. It survives in a wide range of habitats, from semiarid to alpine, and feeds mainly on ants and termites.

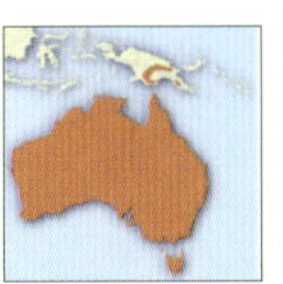

Up to 14 in (35 cm)
Up to 4 in (10 cm)
Up to 15½ lb (7 kg)
Solitary
Locally common

Australia, Tasmania, New Guinea

Long-nosed echidna This echidna has more hair and fewer spines than the short-nosed species. Tiny spines on its tongue help capture the earthworms that make up the bulk of its diet.

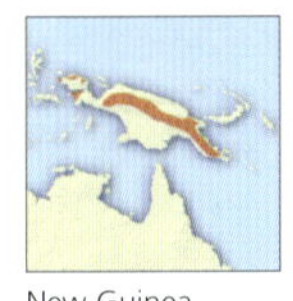

Up to 31½ in (80 cm)
None
Up to 22 lb (10 kg)
Solitary
Endangered

New Guinea

MARSUPIALS

CLASS	Mammalia
ORDERS	7
FAMILIES	19
GENERA	83
SPECIES	295

Commonly known as the pouched mammals, marsupials are virtually embryonic when born and must immediately drag themselves to their mother's teats, which are usually enclosed in a pouch of some kind. They latch firmly onto a teat for some weeks or months and come off only after reaching a level of development comparable to that of newborn mammals that have been nourished by a placenta in the womb. Most marsupials differ from placental mammals in other ways as well, with more incisors, a relatively smaller brain, an opposable toe on the hindfoot, and a slightly lower body temperature and metabolism.

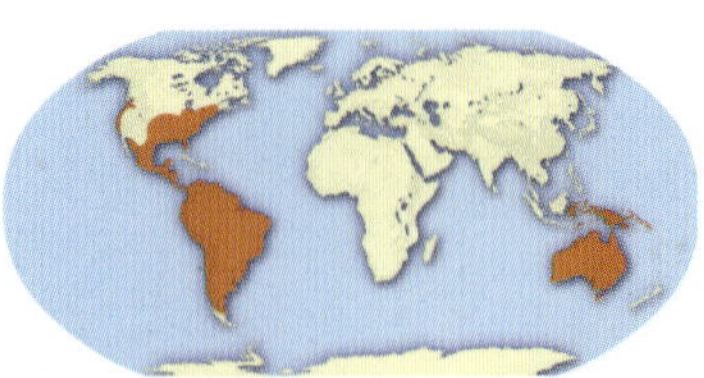

Marsupial success While some have flourished in the Americas, marsupials are most diverse in Australia and New Guinea, where there were no placental mammals. Marsupials have been introduced in New Zealand, Hawaii, and Britain.

FINDING A NICHE

Once considered a single order, marsupials are much more diverse than any order of placental mammals and are now classified into seven orders. Didelphimorphia (containing American opossums), Paucituberculata (shrew opossums), and Microbiotheria (monito del monte) are found in the Americas, while the Australia–New Guinea region hosts Dasyuromorphia (quolls, dunnarts, marsupial mice, numbat), Peramelemorphia (bandicoots), Notoryctemorphia (marsupial moles), and Diprodontia (possums, kangaroos, koala, and wombats).

The fossil record suggests that marsupials and placental mammals diverged more than 100 million years ago. In North America and Europe, marsupials died out as placental mammals diversified.

South America, on the other hand, was isolated from North America for about 60 million years, and marsupials evolved to fill a variety of ecological niches. When the Americas joined up, 2–5 million years ago, northern carnivores such as the jaguar soon displaced South America's large carnivorous marsupials, but small omnivorous marsupials persisted, with some, including the common opossum, recolonizing North America. It was in the region of Australia and New Guinea, however, that marsupials remained free of competition for the longest period and attained their greatest diversity.

Koala The koala looks sleepy because it lives on a diet of highly toxic eucalypt leaves. It has the ability to detoxify the poisons but the effort required leaves little energy to spare. Koalas do not make nests but simply sleep in the fork of a tree.

Black-shouldered opossum
Caluromysiops irrupta

Naked, prehensile tail

Bare-tailed woolly opossum
Caluromys philander

Forelimbs longer than hindlimbs

Gray four-eyed opossum
Philander opossum

Common opossum
Didelphis marsupialis

Carries young on back until they are fully weaned at 3–4 months of age

Robinson's mouse opossum
Marmosa robinsoni

Opposable thumbs on all four paws for gripping thin branches and vines

SIMILAR SOLUTIONS

Australasian marsupials often display similar adaptations to placental mammals elsewhere, an observed phenomenon known as convergent evolution. The striped possum, a marsupial of Australia and New Guinea, and the aye-aye, a placental mammal of Madagascar, are both tree-dwelling insectivores with a very long finger that allows it to dig grubs out of wood.

Striped possum

Aye-aye

Black-shouldered opossum Like many tree-dwelling mammals, this opossum has a prehensile tail for clinging to branches, and large, protruding eyes. It feeds on the abundant fruit and nectar of its tropical forest home.

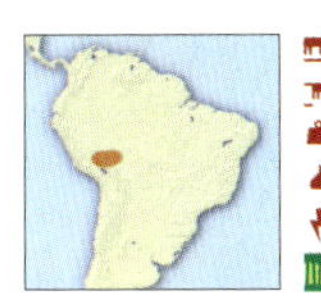
N.W. South America

Up to 12 in (30 cm)
Up to 16 in (40 cm)
Not known
Solitary
Vulnerable

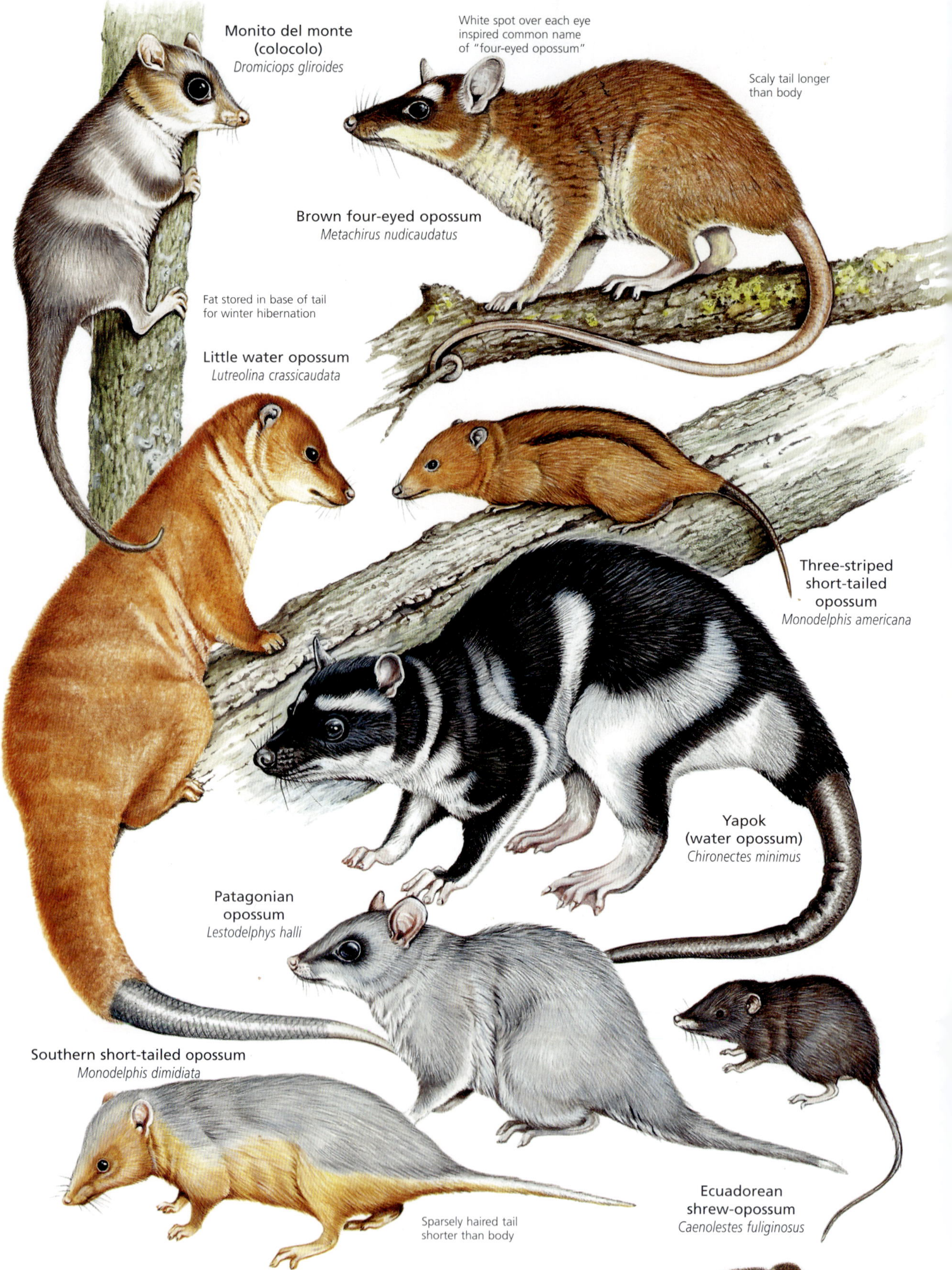

Monito del monte
(colocolo)
Dromiciops gliroides
Fat stored in base of tail
for winter hibernation
White spot over each eye
inspired common name
of "four-eyed opossum"
Scaly tail longer
than body
Brown four-eyed opossum
Metachirus nudicaudatus
Little water opossum
Lutreolina crassicaudata
Three-striped
short-tailed
opossum
Monodelphis americana
Yapok
(water opossum)
Chironectes minimus
Patagonian
opossum
Lestodelphys halli
Southern short-tailed opossum
Monodelphis dimidiata
Sparsely haired tail
shorter than body
Ecuadorean
shrew-opossum
Caenolestes fuliginosus

MARSUPIAL ORIGINS

DNA tests confirm that the monito del monte of Argentina and Chile is the only living member of the Microbiotheriidae, a family of South American marsupials. Together with fossil marsupials found in Antarctica, the monito del monte studies support the theory that marsupials spread from South America to Australia via Antarctica between 100 and 65 million years ago, when these continents formed a single landmass.

North America
South America
Antarctica
Australia

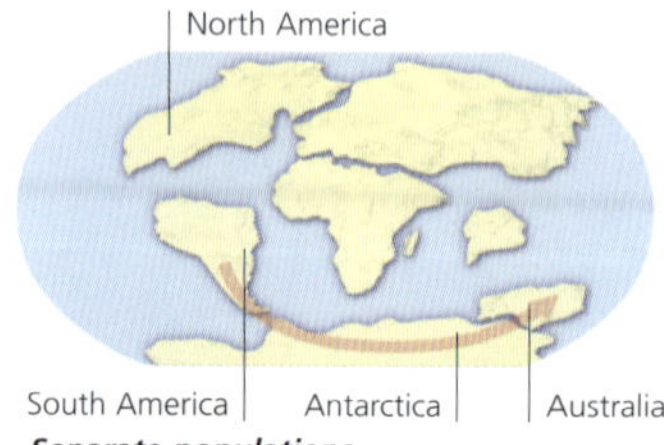

Separate populations
Marsupials radiated from South America to Antarctica and Australia. They flourished in Australia but vanished from Antarctica.

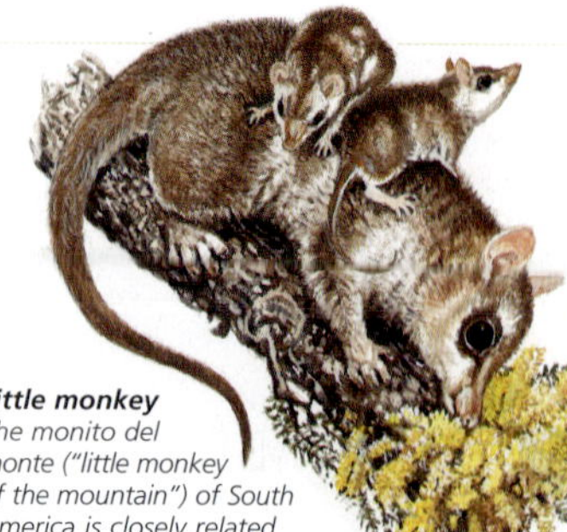

Little monkey
The monito del monte ("little monkey of the mountain") of South America is closely related to Australian marsupials.

Thylacine (Tasmanian tiger)
Thylacinus cynocephalus

13–19 dark vertical stripes along back

Stiff tail thick at base but tapers to a point

Padded, five-toed paws

Massive head with strong jaw and bone-crushing molar teeth

Tasmanian devil
Sarcophilus harrisii

Numbat
Myrmecobius fasciatus

Sticky cylindrical tongue can extend up to 4 inches (10 cm) from small mouth

Vertical white bands along back

Long, bushy tail

Lacks external ears; tiny, non-functional eyes are hidden in fur; horny shield on nose

Southern marsupial mole
Notoryctes typhlops

Spade-like claws for digging through sandy soil

Silky, iridescent fur stained pinkish or reddish by iron-rich soil

Thylacine The largest carnivorous marsupial to survive into historical times, the extinct thylacine resembled a wolf, but had a distinctive striped coat. Competition with the dingo led to its disappearance from mainland Australia 3,000 years ago.

Up to 51 in (130 cm)
Up to 27 in (68 cm)
Up to 77 lb (35 kg)
Solitary, small group
Extinct

Tasmania (until 1930s) Range prior to extinction

Tasmanian devil About the size of a terrier and now the largest carnivorous marsupial, the Tasmanian devil will hunt live prey such as possums and wallabies, but prefers to scavenge for food. It screeches or barks when threatened.

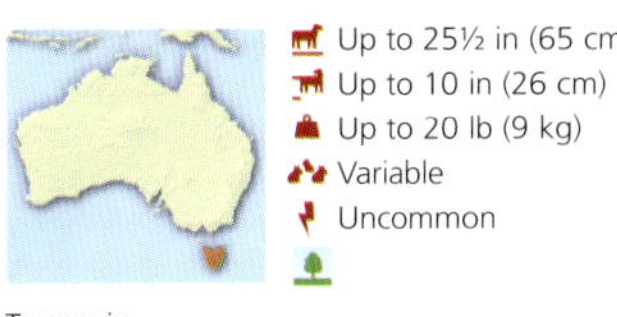

Up to 25½ in (65 cm)
Up to 10 in (26 cm)
Up to 20 lb (9 kg)
Variable
Uncommon

Tasmania

Numbat The only marsupial that is fully active by day, the numbat spends most of its time looking for termites, which make up almost its entire diet. It digs termites from loose earth with its front claws and captures them with its long, sticky tongue.

Up to 11 in (27.5 cm)
Up to 8½ in (21 cm)
Up to 24½ oz (700 g)
Solitary
Vulnerable

S.W. Australia Former range

Northern quoll
Dasyurus hallucatus

Grooved pads on hindfeet provide friction when climbing trees or rocks

Spotted-tailed quoll
Dasyurus maculatus

Only quoll wit[h] spotted tail

Brush-tailed phascogale (tuan)
Phascogale tapoatafa

Mulgara
Dasycercus cristicauda

Yellow-footed antechinus
Antechinus flavipes

Kowari
Dasycercus byrnei

Fat-tailed dunnart
Sminthopsis crassicaudata

Stores fat in tail

Common planigale
Planigale maculata

SLOWING DOWN

Dunnarts and other small insectivorous marsupials sometimes enter periods of torpor, during which the metabolism decreases and the heart and respiratory system slows down. By conserving energy, this state reduces the need to eat, a great advantage during the winter months when food can be scarce. A torpid period may last from a few hours up to several days.

A winter's tail *In winter, the fat-tailed dunnart may enter a torpid state and live off the fat that has been stored in its tail during more plentiful times.*

Kowari This little marsupial carnivore preys on insects and small birds, reptiles, and mammals. To survive in the arid climate, the kowari shelters in burrows and obtains all its moisture from its food, eliminating the need to drink water.

C. Australia

- Up to 7 in (18 cm)
- Up to 5½ in (14 cm)
- Up to 5 oz (140 g)
- Solitary
- Vulnerable

Long ears

Bilby
Macrotis lagotis

Long, pointed snout

Bicolored tail

Striped bandicoot
Microperoryctes longicauda

Southern brown bandicoot
Isoodon obesulus

Long-nosed bandicoot
Perameles nasuta

Stiff, spiny fur

Elongated hindfoot for running and hopping

Strong claws for digging

Raffray's bandicoot
Peroryctes raffrayana

Spiny bandicoot
Echymipera kalubu

Bilby The bilby is a powerful digger and may build up to 12 burrows on its home range. It can be distinguished from other bandicoots by its long ears. It feeds on insects, seeds, and fruits.

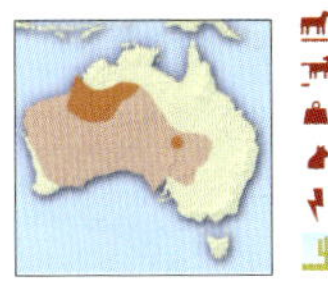

- Up to 22 in (55 cm)
- Up to 11½ in (29 cm)
- Up to 5½ lb (2.5 kg)
- Solitary
- Vulnerable

C. Australia ● Former range

BANDICOOTS

The relationship of the omnivorous bandicoots to other marsupials is unclear. They have an arrangement of teeth similar to that of carnivorous marsupials, but they also have fused toes on the hindfoot, a feature known as syndactyly that is shared with herbivorous kangaroos and wombats. Bandicoots are divided into the mainly Australian Peramelidae (comprising the *Perameles* and *Isoodon* species) and the mainly New Guinean Peroryctidae.

Rapid reproduction
After a short gestation period lasting 12 days or so, bandicoot young develop quickly and may reach sexual maturity in about 90 days.

Bobuk (mountain brushtail possum)
Trichosurus caninus

Gray cuscus
Phalanger orientalis

Spotted cuscus
Spilocuscus maculatus

Brushtail possum
Trichosurus vulpecula

Feathertail glider
Acrobates pygmaeus

Gliding membrane extends from wrist to knee

Feather-like arrangement of fur on tail unique among mammals

Prehensile tail covered in thick scales

Scaly-tailed possum
Wyulda squamicaudata

Mountain pygmy-possum
Burramys parvus

MOUNTAIN MEALS

The only Australian marsupial to live above the snowline, the mountain pygmy possum adjusts its eating habits according to the season. During the warmer months, it feeds mostly on Bogong moths and also eats small amounts of other insects. As the moths become less abundant around January, it switches to seeds and berries, storing caches of these for the cold winter months ahead.

Moth mouthful
During the warmer months, Bogong moths are between a third and all of the mountain pygmy-possum's diet.

Spotted cuscus This rain-forest marsupial spends most of its time in trees, sleeping by day in a small bed of leaves and feeding at night on a diet of fruit, flowers, and leaves. The spotted cuscus is threatened by habitat loss from logging and farming.

- Up to 23 in (58 cm)
- Up to 17½ in (45 cm)
- Up to 11 lb (4.9 kg)
- Solitary
- Vulnerable

N. Australia, New Guinea, some islands

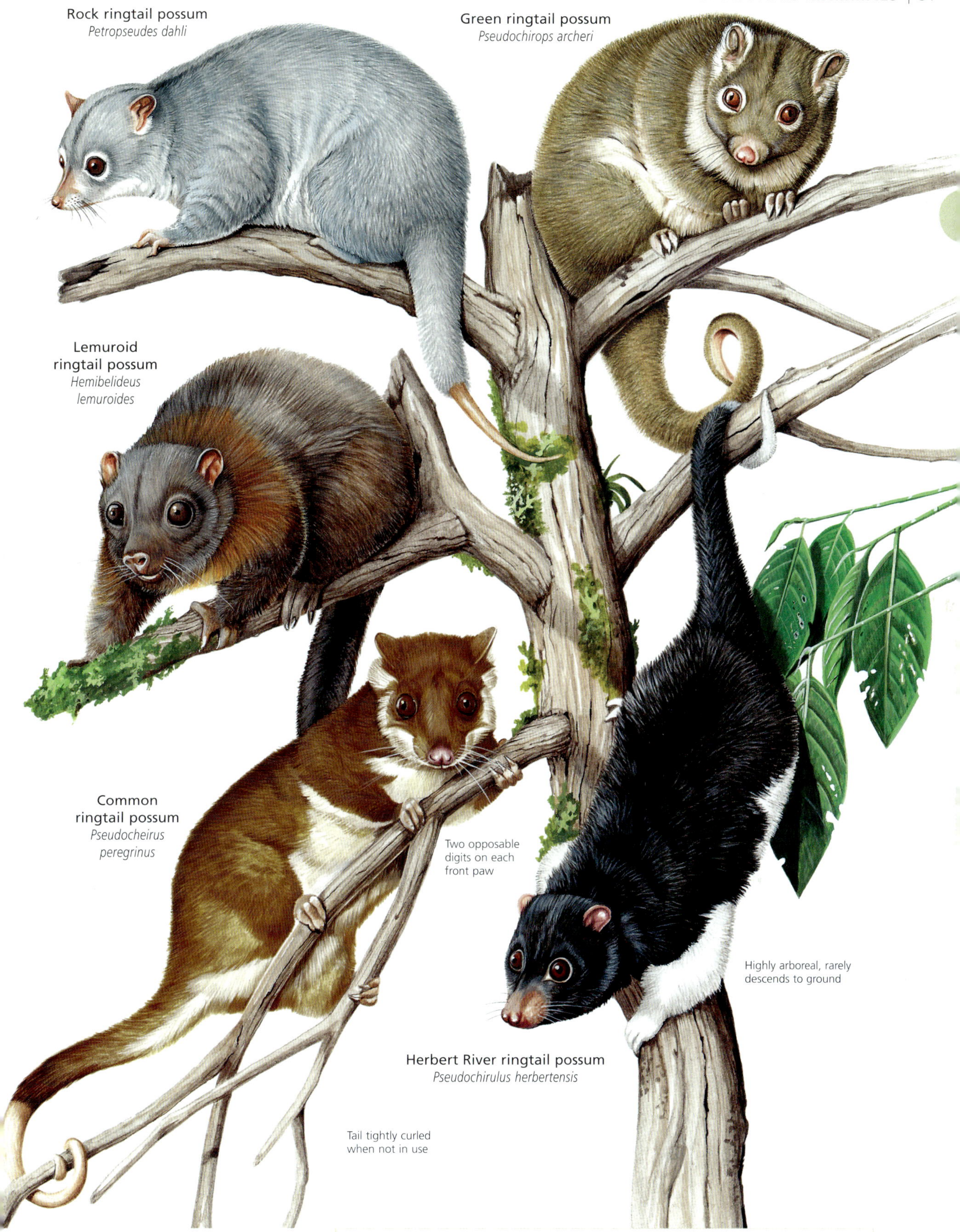

Rock ringtail possum
Petropseudes dahli
Green ringtail possum
Pseudochirops archeri
Lemuroid
ringtail possum
Hemibelideus
lemuroides
Common
ringtail possum
Pseudocheirus
peregrinus
Two opposable
digits on each
front paw
Highly arboreal, rarely
descends to ground
Herbert River ringtail possum
Pseudochirulus herbertensis
Tail tightly curled
when not in use

Rock ringtail possum By day, this possum stays cool and safe in rock crevices, emerging at night to feed in trees. Males and females share care of the young equally, a level of male participation that is rare in mammals and unknown in other marsupials.

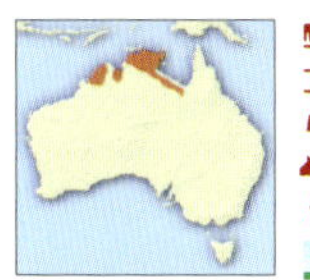

- Up to 15½ in (39 cm)
- Up to 10½ in (27 cm)
- Up to 4½ lb (2 kg)
- Pair
- Locally common

N. Australia

A POISONOUS DIET

The common ringtail primarily eats eucalyptus leaves, which are usually toxic and lacking in nutritional value. A specialized digestive system featuring an enlarged cecum (a pouch in the large intestine) detoxifies the leaves and then releases soft fecal pellets that the possum can eat. Any undigested material is then expelled as hard pellets. To cope with this low-energy diet, the ringtail has a slow metabolism.

small intestine
distal
colon
esophagus
cecum
stomach
rectum

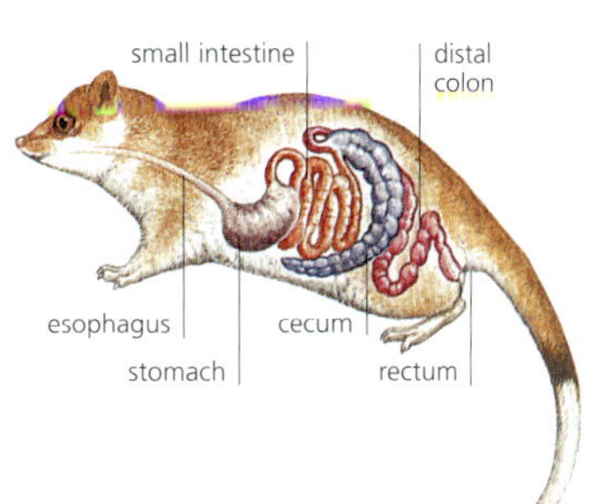

Greater glider
Petauroides volans

Yellow-bellied glider
Petaurus australis

Gliding membrane extends from elbows to ankles

Belly fur can be whitish, yellow, or orange

Sugar glider
Petaurus breviceps

Each hindfoot has an opposable big toe as well as two partially fused toes used for grooming

Leadbeater's possum
Gymnobelideus leadbeateri

Long, bushy tail acts as a rudder during glides

SAP SUCKERS

A membrane extending from the wrists to the ankles allows gliders to travel substantial distances through the air from one feeding tree to another. Once landed, these arboreal, nocturnal marsupials cut notches in the bark with their teeth and lap up the sap and gum. The yellow-bellied glider targets a number of eucalypt species, while the sugar glider prefers wattles and *Eucalyptus resinifera* trees.

Sticky feeders
Yellow-bellied gliders will vigorously defend their sap-feeding sites.

Leadbeater's possum This possum has exploited a niche created by the wildfire ecology of its highland forest home. When fires sweep through an area, they may kill some old trees and clear the way for new growth of wattle trees.

S.E. Australia

- Up to 6½ in (17 cm)
- Up to 7 in (18 cm)
- Up to 5½ oz (160 g)
- Pair, family group
- Endangered

Koala
Phascolarctos cinereus

Feeds almost solely on eucalyptus leaves

Needle-sharp claws can firmly grip smooth tree trunks

Pointed snout probes flowers for nectar, and long, bristly tongue collects pollen

Honey possum
Tarsipes rostratus

Elongated fourth finger with hooked nail used to extract wood-boring grubs

Striped possum
Dactylopsila trivirgata

Common wombat
Vombatus ursinus

Muzzle covered in fur

Powerful forelimbs with massive paws and long, sturdy claws used for digging burrows

Southern hairy-nosed wombat
Lasiorhinus latifrons

Koala Spending virtually their entire lives in trees, koalas sleep for 18 hours a day and feed at night, preferring the leaves of five particular eucalypt species. During the breeding season, competing males bellow throughout the night.

Up to 32½ in (82 cm)
None
Up to 33 lb (15 kg)
Solitary
Near threatened

S. & E. Australia

Southern hairy-nosed wombat Nicknamed "the bulldozer of the bush," this wombat can run at high speeds in play or fright. It shelters from the heat of the day in group warrens, emerging at night to feed on grasses, roots, bark, and fungi.

Up to 37 in (94 cm)
None
Up to 70½ lb (32 kg)
Solitary
Locally common

S. Australia

GOING UNDERGROUND

The common wombat relies on its powerful front limbs and claws to dig an extensive warren of burrows, which can be up to 20 inches (50 cm) wide and 100 feet (30 m) long.

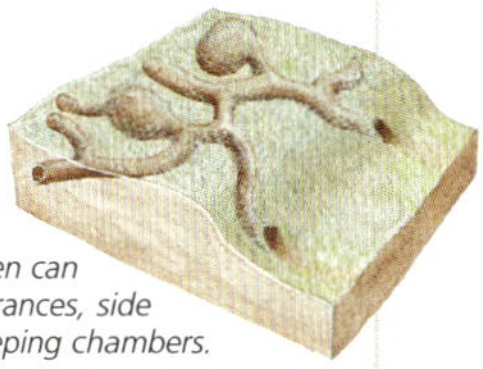

Layout
A wombat warren can have several entrances, side tunnels, and sleeping chambers.

Joey carried in pouch until it is a subadult

Bennett's tree kangaroo
Dendrolagus bennettianus

Goodfellow's tree kangaroo
Dendrolagus goodfellowi

Four limbs of roughly equal length

Long-nosed potoroo
Potorous tridactylus

Northern nail-tailed wallaby
Onychogalea unguifera

Brush-tailed bettong
Bettongia penicillata

Red-legged pademelon
Thylogale stigmatica

Rufous bettong
Aepyprymnus rufescens

Musky rat-kangaroo
Hypsiprymnodon moschatus

Goodfellow's tree kangaroo Limbs of equal length and sharp claws help this marsupial climb through rain-forest trees, where it shelters in small groups and feeds on leaves and fruit.

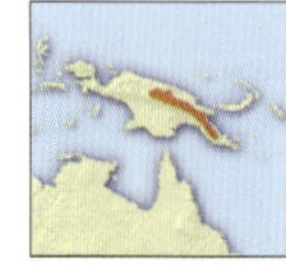

- Up to 25 in (63 cm)
- Up to 30 in (76 cm)
- Up to 18½ lb (8.5 kg)
- Solitary
- Endangered

New Guinea

Northern nail-tailed wallaby Named after the horny spur on its tail, this animal spends the day in a shallow nest beneath a bush. At night, it feeds on the roots of the grasses that grow in its savanna or woodland home.

- Up to 27½ in (70 cm)
- Up to 29 in (74 cm)
- Up to 20 lb (9 kg)
- Solitary
- Locally common

N. Australia

Red-legged pademelon The only ground-dwelling wallaby to live in wet tropical forests, this nocturnal creature searches the dense understory for food such as leaves, fruit, bark, and cicadas.

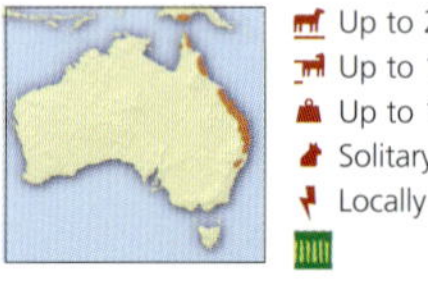

- Up to 21 in (54 cm)
- Up to 18½ in (47 cm)
- Up to 14 lb (6.5 kg)
- Solitary
- Locally common

E. Australia, New Guinea

MAKING MARSUPIALS

The unique nature of marsupial reproduction begins with the anatomy of the parents. On the outside, the female system seems simpler than in placental mammals, with a single opening, called a cloaca, for the digestive and reproductive tracts. Inside, however, there is a double reproductive tract involving two uteri, each with its own vagina. Many male marsupials have a forked penis, which directs semen into both vaginas. A pregnant female develops a third vagina as a birth canal. After a short gestation, ranging from 12 days in some bandicoot species to 38 days in the eastern gray kangaroo, almost embryonic young are born that crawl to a teat, which is usually protected by a pouch. Once the young are fully formed, they leave the teat, but are weaned gradually.

Diverse habitats Marsupials live in a variety of habitats. Trees provide shelter and food for possums, gliders, and the koala; rocky slopes are home to rock wallabies and wallaroos; and grassy nests hide the brown bandicoot. The elusive marsupial mole ploughs through soft sand without making a permanent burrow.

MOB MEMBERSHIP

Large kangaroos often congregate in groups, or mobs, of 50 or more, a strategy that helps deter predators such as dingoes. While males are not territorial, their access to mates depends on their position in the mob's hierarchy, which tends to be based on size. A dominant male Eastern gray kangaroo may father up to 30 young in a season, but most males never get the chance to mate.

Kickboxing *Male kangaroos sometimes fight to establish dominance, kicking with their powerful back legs.*

Red kangaroo The largest of the marsupials, the red kangaroo usually hops slowly, but can reach speeds of 35–45 miles per hour (55–70 km/h). Males of this species have a reddish coat, but females are bluish gray.

- Up to 55 in (140 cm)
- Up to 39 in (99 cm)
- Up to 187 lb (85 kg)
- Herd
- Common

Australia

Xenarthrans

CLASS Mammalia
ORDER Xenarthra
FAMILIES 4
GENERA 13
SPECIES 29

Some of the world's most bizarre animals—anteaters, sloths, and armadillos—make up Xenarthra, an ancient order that was once much more diverse and included ground sloths larger than elephants, and armored mammals bigger than polar bears. Originating in and confined to the Americas, xenarthrans all have extra joints, known as xenarthrales, in the lower spine, which limit twisting and turning but strengthen the lower back and hips, a particular advantage for the burrowing armadillos. The brains of xenarthrans are small and their teeth are rudimentary or, in the case of anteaters, absent. A slow metabolism has allowed these species to take advantage of narrow niches.

Built-in blanket The giant anteater spends up to 15 hours a day resting. It digs a shallow depression in the ground and lies down, curling its bushy, fan-like tail around itself. As well as providing warmth, this arrangement helps to disguise the anteater when it is at its most vulnerable.

Digging deep An anteater's sharp front talons can rip into concrete-like termite mounds, allowing its long, sticky tongue to collect the insects. An attack causes little permanent damage to a nest, however, because it lasts just a few minutes and only a small number of termites are eaten. The mound is repaired by the surviving termites.

SLOW AND STEADY

The low metabolic rate and body temperature of anteaters and sloths have enabled them to become highly specialized feeders, exploiting food sources that are abundant but low in energy content. Armadillos, on the other hand, have a varied diet, but live in deep underground tunnels, where their slow metabolism helps them to avoid overheating.

Ranging from the terrestrial giant anteater to the tree-dwelling silky anteater and tamanduas, anteaters use their keen sense of smell to detect ants and termites. From an especially long, tubular snout, an even longer tongue darts out, covered in tiny spines and sticky saliva to capture the prey.

Remarkably sluggish, sloths spend almost all their waking hours feeding on forest leaves, consuming such quantities that a full stomach can account for nearly a third of their body weight. Inside a sloth's multichambered stomach, toxins in the leaves are neutralized.

With a carapace of horn-covered plates, armadillos are shielded from predators. They use their powerful limbs and sharp claws to excavate up to 20 burrows in their home range.

Algae growing in fur provides camouflage in the tree canopy

Maned three-toed sloth
Bradypus torquatus

Shaggy fur harbors moths, beetles, and other insects

Pale-throated three-toed sloth
Bradypus tridactylus

Nine-banded armadillo
Dasypus novemcinctus

3–4 inch (8–10 cm) long curved claws hook onto branches

Southern naked-tailed armadillo
Cabassous unicinctus

Walks on soles of hindfeet and on tips of foreclaws

Silky anteater
Cyclopes didactylus

Can stretch out horizontally from branch, supported by prehensile tail

Larger hairy armadillo
Chaetophractus villosus

Giant anteater
Myrmecophaga tridactyla

Sticky tongue can extend up to 24 inches (61 cm) from long snout

Offspring may ride on mother's back for up to a year

IDENTICAL QUADRUPLETS

The only xenarthran that lives in the United States, the nine-banded armadillo has rapidly expanded its range in the past 150 years, inhabiting from central United States through most of northern South America. Along with the other species in the genus *Dasypus,* it is unique among vertebrates because the female produces a single fertilized egg that divides into a number of genetically identical embryos.

Family likeness
The nine-banded armadillo usually gives birth to a litter of four identical pups.

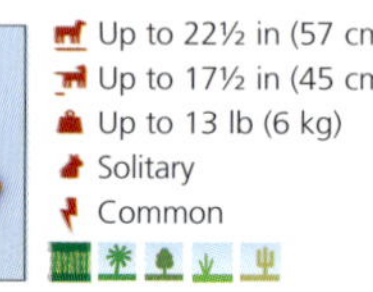

Up to 22½ in (57 cm)
Up to 17½ in (45 cm)
Up to 13 lb (6 kg)
Solitary
Common

North America & South America

Maned three-toed sloth Apart from a black mane on the shoulders, this animal's coarse, shaggy fur is grizzled tan, often tinged green by algae. The green helps to camouflage the slow-moving sloth in its treetop home.

Up to 19½ in (50 cm)
Up to 2 in (5 cm)
Up to 9½ lb (4.2 kg)
Solitary
Endangered

N.E. South America

Giant armadillo
Priodontes maximus

Protected from predators and thorny plants by shield of bony plates covered by horny skin

Large, sharp claws for digging

Southern three-banded armadillo
Tolypeutes matacus

Southern three-banded armadillo in defensive ball

Lesser fairy armadillo
Chlamyphorus truncatus

PANGOLINS

CLASS	Mammalia
ORDER	Pholidota
FAMILY	Manidae
GENUS	*Manis*
SPECIES	7

A covering of horny body scales growing from a thick underlying skin distinguishes pangolins from all other mammals. An extraordinary tongue, longer than the animal's head and body, is coiled in the animal's mouth when at rest, but can be extended and flicked into ant nests and termite mounds. Pangolins lack teeth, relying instead on powerful muscles and small pebbles in their stomachs to grind up their food. Terrestrial species, such as the giant ground pangolin, dig out underground burrows for shelter during the day. The tree-dwelling species have a prehensile tail and curl up into balls in tree hollows when resting.

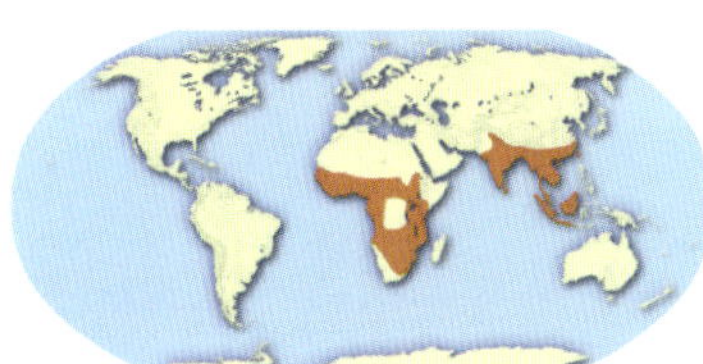

Asia and Africa Pangolins are found in much of Southeast Asia and subtropical Africa. While the Asian species have external ears and grow hair at the base of their scales, the African species lack external ears and have no scales on the tail's underside.

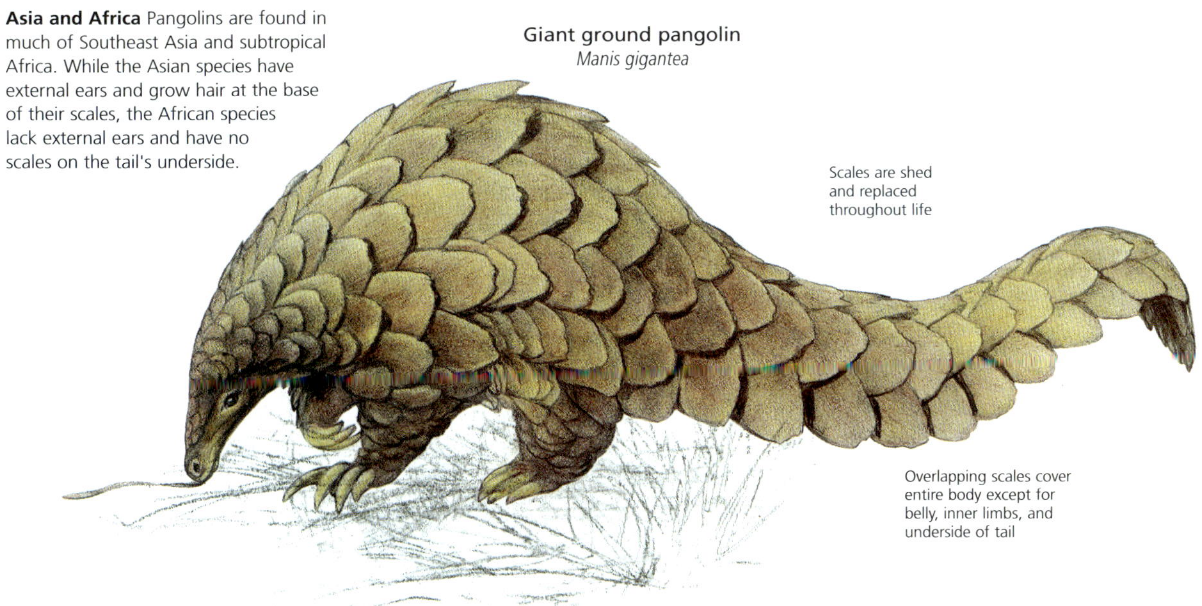

Giant ground pangolin
Manis gigantea

INSECTIVORES

CLASS	Mammalia
ORDER	Insectivora
FAMILIES	7
GENERA	68
SPECIES	428

Small, quick creatures with long, narrow snouts, shrews, moles, hedgehogs, and other insectivores make up a diverse order whose classification is much debated. While they share primitive features such as a small, smooth brain, simple bones in the ear, and rudimentary teeth, many also display specializations such as burrowing adaptations, defensive spines, or poisonous saliva. Insectivores were named after their tendency to eat insects, but many will take advantage of any available food source and readily consume plants and other animals. Usually shy and nocturnal, they rely on acute senses of smell and touch rather than vision and have very small or even minute eyes.

Worldwide spread While three insectivore families—hedgehogs and moonrats; moles and desmans; and shrews—are found in much of the world, solenodons, tenrecs, and otter shrews are highly localized.

Opportunity knocks Insectivores are often opportunistic feeders, prepared to eat a wide variety of prey and plants. European moles tunnel almost continuously as they chase prey, mainly earthworms. Their shallow feeding tunnels leave telltale dirt mounds, while permanent passageways are deeper down.

CONVERGING SPECIES

The order Insectivora contains numerous examples of convergent evolution, with animals in similar habitats displaying similar behaviors or physical adaptations, even though they are not closely related.

Some insectivores have exploited an aquatic niche. The desmans of Europe and the web-footed tenrec (*Limnogale mergulus*) of Madagascar evolved in total isolation from each other, but they share a dense, waterproof coat, a streamlined body, partially webbed feet, a long tail that acts as a rudder, and specialized mechanisms for breathing and detecting underwater prey.

European moles and African golden moles are very distantly related, with true moles evolving from a shrew-like animal, and golden moles more closely related to tenrecs. Nevertheless, they look very much alike and both display adaptations to a burrowing lifestyle. Their limbs are short and powerful, with large digging claws on the forefeet, and their eyes are minute and hidden by fur or skin.

The hedgehogs of Europe and the tenrecs of Africa employ similar means of self-defense. Both have a thick coat of spines and will curl up when threatened, becoming a spiky ball to discourage predators.

Raises stiff hairs along neck into a crest when threatened

Covered in coarse hair and sharp spines

Tail-less tenrec (common tenrec)
Tenrec ecaudatus

Large-eared tenrec
Geogale aurita

Thick, scaly tail

Strong claws for digging insects, worms, and small lizards out of leaf litter

Cuban solenodon
Solenodon cubanus

Flattened head allows nostrils, eyes, and ears to stay above water's surface while body is submerged

Dense fur traps insulating layer of air when swimming

Giant otter shrew
Potamogale velox

Fused toes on hindfoot used for grooming

Locates prey using sensitive whiskers

Tail used for propulsion and steering in water

Ruwenzori otter shrew
Micropotamogale ruwenzorii

Webbed feet

Cuban solenodon Like *Solenodon paradoxus* of Hispaniola, the Cuban solenodon releases venomous saliva through a groove in one of its lower incisors. It may use the venom to paralyze larger prey such as frogs.

E. Cuba

- Up to 15 in (39 cm)
- Up to 9½ in (24 cm)
- Up to 2 lb (1 kg)
- Solitary
- Endangered

Giant otter shrew The giant otter shrew swims powerfully through the water with thrusts of its laterally flattened tail. It hunts at night, searching out prey such as crabs, frogs, fish, and insects by scent and touch.

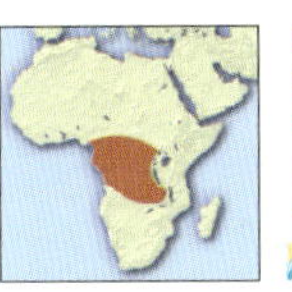

C. Africa

- Up to 14 in (35 cm)
- Up to 11½ in (29 cm)
- Up to 14 oz (400 g)
- Solitary
- Endangered

TENRECS OF MADAGASCAR

Tenrecs fill a variety of ecological niches, with species adapting to life in water, on land, in trees, and underground.

Bony tail
The lesser long-tailed shrew tenrec (Microgale longicaudata) has 47 vertebrae in its tail.

Coarse, spiky hair

Moonrat
Echinosorex gymnura

Long canine teeth

Shrew-hedgehog
Neotetracus sinensis

Lesser moonrat (lesser gymnure)
Hylomys suillus

Eats invertebrates such as worms and insects

Mindanao moonrat
Podogymnura truei

Sharp spines erected when threatened

Horny pad on nose

Hottentot golden mole
Amblysomus hottentotus

Belly covered in soft fur

Western European hedgehog
Erinaceus europaeus

Four-toed hedgehog
Atelerix albiventris

Powerful forelimbs and claws used for digging burrows and unearthing prey

HEDGEHOG HIBERNATION

To cope with the lack of food during cold winters, hedgehogs will go into hibernation, lowering their body temperature and heart and respiration rates to conserve energy. Their metabolic rate can become 100 times slower and they may stop breathing for up to 2 hours. Hibernating hedgehogs live off the extra body fat accumulated in summer. They rouse themselves for a day or so every couple of weeks to forage and defecate.

Getting settled *While the Western European hedgehog makes a hasty nest in summer, it takes more care in winter, choosing a well-insulated site.*

Moonrat The smell of rotting onions, stale sweat, or ammonia may indicate the presence of a moonrat, which produces a strong scent from two glands near the anus. The animal uses this powerful odor to mark its territory.

- Up to 18 in (46 cm)
- Up to 12 in (30 cm)
- Up to 4½ lb (2 kg)
- Solitary
- Locally common

Malay Peninsula, Sumatra, Borneo

Good sense of smell and sharp hearing

Small eyes with poor vision

House shrew
Suncus murinus

Piebald shrew
Diplomesodon pulchellum

Feet fringed with hairs to help shrew run across sand

Vocalizations include screams and twitters

Eurasian common shrew
Sorex araneus

Northern short-tailed shrew
Blarina brevicauda

Himalayan water shrew
Chimarrogale himalayica

Webbed feet for swimming

Elegant water shrew
Nectogale elegans

Western European hedgehog Like other hedgehogs, this species can raise or lower its spines and will protect its head and belly by rolling into a spiky ball. At night, it searches for insects, worms, eggs, and carrion.

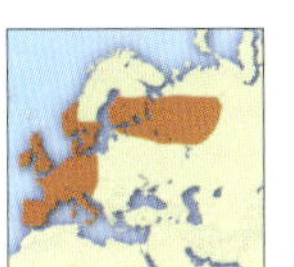

- Up to 10 in (26 cm)
- Up to 1 in (3 cm)
- Up to 3½ lb (1.6 kg)
- Solitary
- Common

N. & W. Europe

House shrew Like the house mouse, the house shrew has flourished living alongside humans. It is abundant through south Asia. Also known as the money shrew, it makes a jangling sound as it forages for insects in houses.

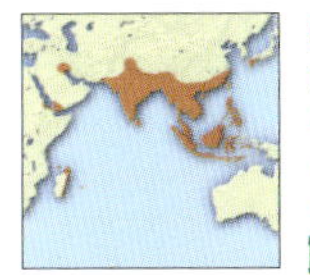

- Up to 6½ in (16 cm)
- Up to 3½ in (9 cm)
- Up to 3 oz (90 g)
- Solitary, commensal
- Common

S. & S.E. Asia

Northern short-tailed shrew This shrew has adapted to a burrowing lifestyle, with a body shape similar to that of the true moles. It immobilizes larger prey such as voles and mice by injecting venomous saliva when it bites.

- Up to 3 in (8 cm)
- Up to 1 in (3 cm)
- Up to 1 oz (30 g)
- Solitary
- Common

E. North America

Pyrenean desman
Galemys pyrenaicus

Nostrils closed by valves when underwater

Uses sensitive snout to probe riverbed for aquatic prey

Steers in water using long, flat tail broadened by a fringe of hairs

Russian desman
Desmana moschata

Star-nosed mole
Condylura cristata

Powerful forelimbs turned outward for digging

Hairy-tailed mole
Parascalops breweri

Tiny eyes hidden by fur can detect changes in light

American shrew mole
Neurotrichus gibbsii

Active throughout day and night, sleeping for only 1–8 minutes at a time and awake for periods of 2–18 minutes

European mole
Talpa europaea

SENSITIVE STARS

Fleshy tentacles around its nose helps the star-nosed mole sense small fish, leeches, snails, and other aquatic prey. It is an agile swimmer and lives in a network of burrowed tunnels.

Perceptive rays
Each of the star-nosed mole's 22 rays bears thousands of sensory organs.

Russian desman Once found across Europe, this gregarious species is now restricted to a handful of river basins in Eastern Europe. Its soft coat made it a target of trappers, but it is now legally protected.

E. Europe

- Up to 8½ in (22 cm)
- Up to 8½ in (22 cm)
- Up to 8 oz (220 g)
- Solitary
- Vulnerable

American shrew mole About the size of a shrew and without the large forelimbs of other moles, this animal is North America's smallest mole. It is also the only mole able to climb bushes, and it sleeps in quick spans.

W. North America

- Up to 3 in (8 cm)
- Up to 1½ in (4 cm)
- Up to ½ oz (11 g)
- Solitary
- Locally common

LIFE BELOW THE SURFACE

Often the only sign of moles in an area is the presence of molehills, small mounds of dirt created when a mole digs a vertical shaft to the surface. Spending almost their entire lives underground, moles dwell in a network of tunnels, sleeping and raising young in a subterranean nest and foraging in the tunnels for earthworms, insect larvae, slugs, and other soil invertebrates. A mole can create up to 65 feet (20 m) of tunnels in a day. It usually comes to the surface only to collect grasses and leaf litter to line a nest, or if a stronger animal has evicted it from its homeand it needs to find a new territory.

Raised in the dark Moles mate in the female's burrow during a frantic breeding season of just 24–48 hours. An average of three young are born a month later, and will be nursed in the nest for a further month. After initially exploring the tunnel system with the mother, young moles must soon leave and build their own tunnels elsewhere.

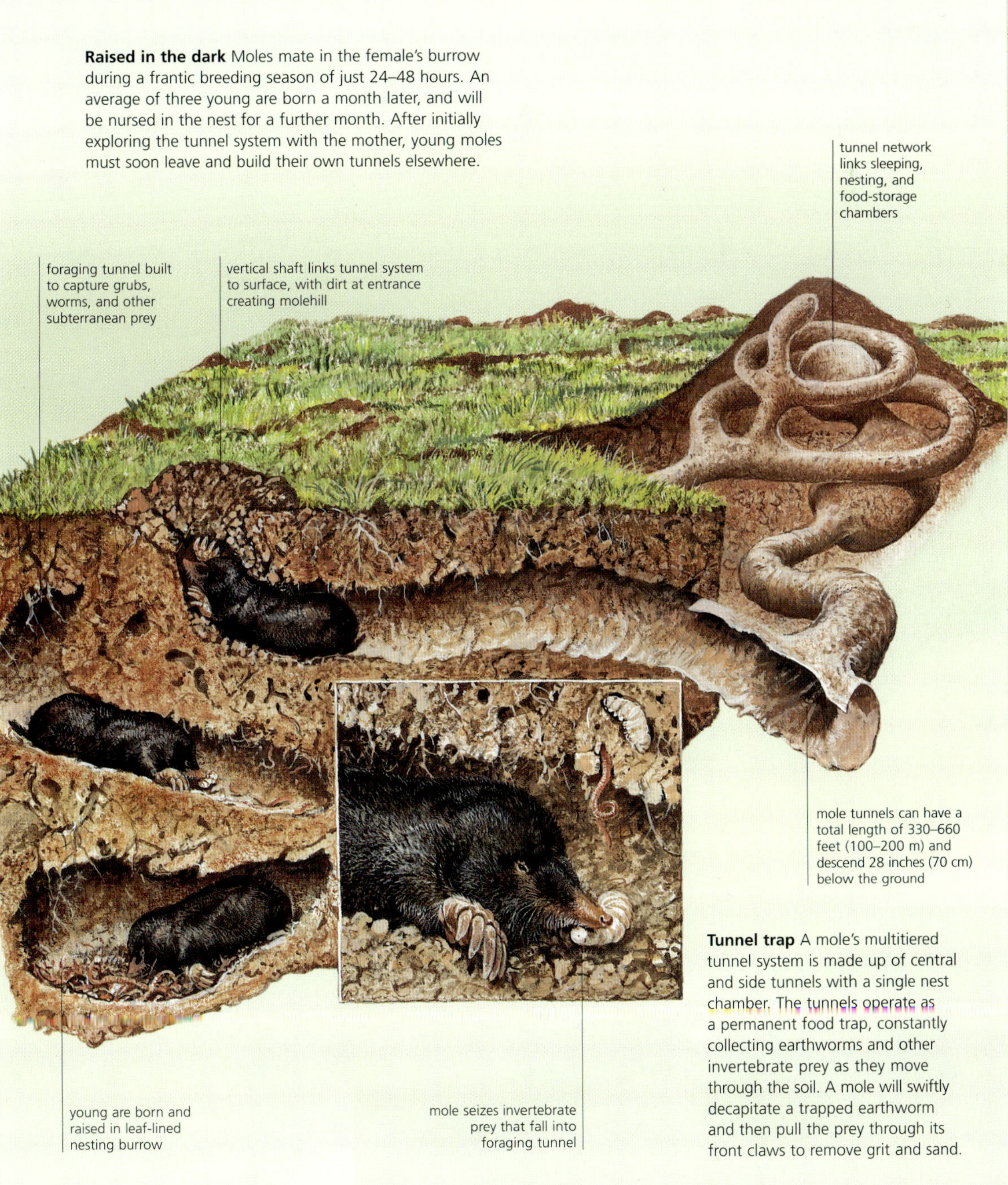

Tunnel trap A mole's multitiered tunnel system is made up of central and side tunnels with a single nest chamber. The tunnels operate as a permanent food trap, constantly collecting earthworms and other invertebrate prey as they move through the soil. A mole will swiftly decapitate a trapped earthworm and then pull the prey through its front claws to remove grit and sand.

FLYING LEMURS

CCLASS	Mammalia
ORDER	Dermoptera
FAMILY	Cynocephalidae
GENUS	Cynocephalus
SPECIES	2

Also known as colugos, flying lemurs do not really fly and are not true lemurs. Instead, they glide through the air, and are placed in their own small order, Dermoptera ("skin wing"). These cat-sized animals are helpless on the ground and climb awkwardly in a lurching fashion, but travel easily between the tall trees of their rain-forest home. To minimize the risk of being picked off by a swift bird of prey during a glide, flying lemurs are nocturnal. They spend the day resting, either in tree hollows or clinging to a trunk.

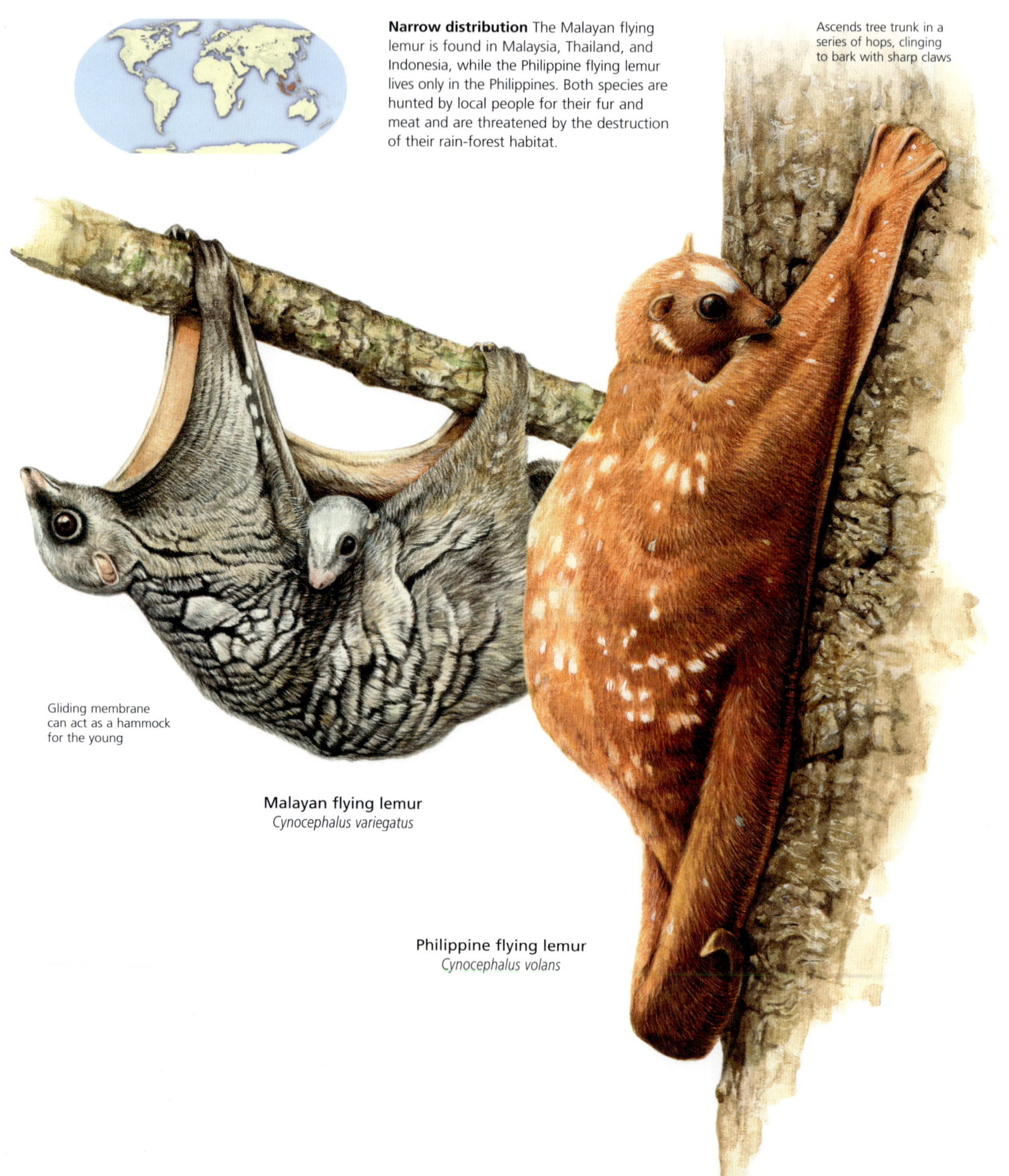

Narrow distribution The Malayan flying lemur is found in Malaysia, Thailand, and Indonesia, while the Philippine flying lemur lives only in the Philippines. Both species are hunted by local people for their fur and meat and are threatened by the destruction of their rain-forest habitat.

Malayan flying lemur
Cynocephalus variegatus

Philippine flying lemur
Cynocephalus volans

TREE SHREWS

CLASS	Mammalia
ORDER	Scandentia
FAMILY	Tupaiidae
GENERA	5
SPECIES	19

In some Asian tropical forests, small, squirrel-like mammals known as tree shrews scurry along the ground and up trees, foraging for insects, worms, small vertebrates, and fruit. Their sharp claws and splayed toes keep a firm grip on bark and rock alike, while a long tail helps with balance. When eating, they hold food in their hands and may sit on their haunches, alert for predators such as birds of prey, snakes, and mongooses. An average of three young are born in a nest of leaves, which is often made by the father in a tree hollow. Maternal care tends to be limited.

Divided family Tree shrews live in the tropical rain forests of south and Southeast Asia. Initially thought to be insectivores, then grouped with primates, they are now classified in their own order (Scandentia) in a single family (Tupaiidae). The subfamily Ptilocercinae contains just one species, the pen-tailed tree shrew; the subfamily Tupaiinae contains the other 18 species. The majority of these make their home on the island of Borneo, while the remainder are in eastern India and Southeast Asia.

Philippine tree shrew
Urogale everetti

Pen-tailed tree shrew
Ptilocercus lowii

Only fully nocturnal tree shrew

Common tree shrew
Tupaia glis

Scaly tail twitches continuously

BATS

CLASS	Mammalia
ORDER	Chiroptera
FAMILY	18
GENERA	177
SPECIES	993

The only mammals with flapping wings and therefore the only ones capable of true flight, bats can travel through the air at speeds of up to 30 miles per hour (50 km/h). This ability has enabled them to cover great distances, exploiting food sources over a wide range and colonizing most parts of the globe, including far-flung islands such as New Zealand and Hawaii, where they are the only native land mammals. Almost 1,000 species of bat form the order Chiroptera, which is the second-largest mammal order. Chiroptera is split into two suborders: the Old World fruit bats (Megachiroptera) and the mostly insect-eating New World bats (Microchiroptera).

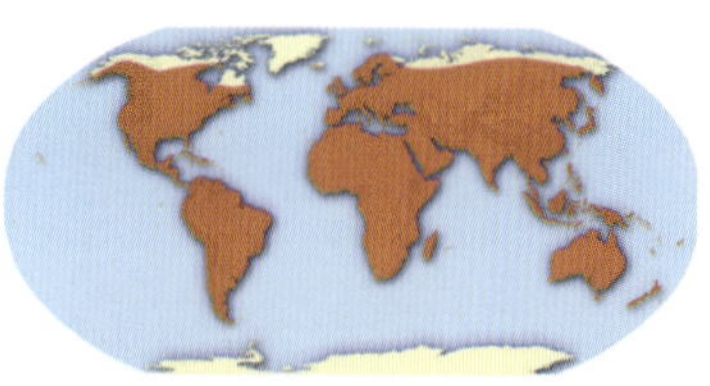

All over the world Bats make up nearly one-quarter of all mammal species. While most numerous in warmer regions, bat species are found worldwide except for the polar zones and a few isolated islands. Most bats are gregarious creatures, with some living in colonies of thousands or even millions of individuals.

FLYING FEEDERS

Bats are popularly portrayed as blood-sucking fiends, but only three species of bat drink blood and even these vampire bats display genuine altruism, sharing their food with hungry companions. Most bats are gregarious creatures, with some living in colonies of thousands or even millions of individuals.

More than 70 percent of bats consume insects that fly at night, a source of food exploited by few other animals. In addition to their flying ability, these hunters rely on echolocation, a means of locating both obstacles and prey by emitting high-pitched sounds and detecting the echoes. In many ecosystems, bats play a crucial role in keeping insect populations under control.

Most other bats are herbivores, using a keen sense of smell and effective night vision to locate fruit, flowers, nectar, and pollen. Such bats can be vital for pollination and seed dispersal, and some plants have adapted specifically to attract them, producing large fruits and strongly scented night-blooming flowers.

To minimize their consumption of energy, many bats regulate their body temperature, lowering it when roosting during the day. To cope with the scarcity of food in winter, some temperate species enter longer periods of hibernation, living off body fat deposited in autumn. Others migrate to warmer climes, with one species, the European noctule, flying up to 1,200 miles (2,000 km).

Hanging around Most bats hang upside down to sleep through the day, a position that allows for a fast takeoff. Bats' legs are designed to lock into place under the force of gravity when hanging upside down. Many species roost in caves, mines, or buildings, while others, such as the gray-headed flying fox (*Pteropus poliocephalus*), prefer trees.

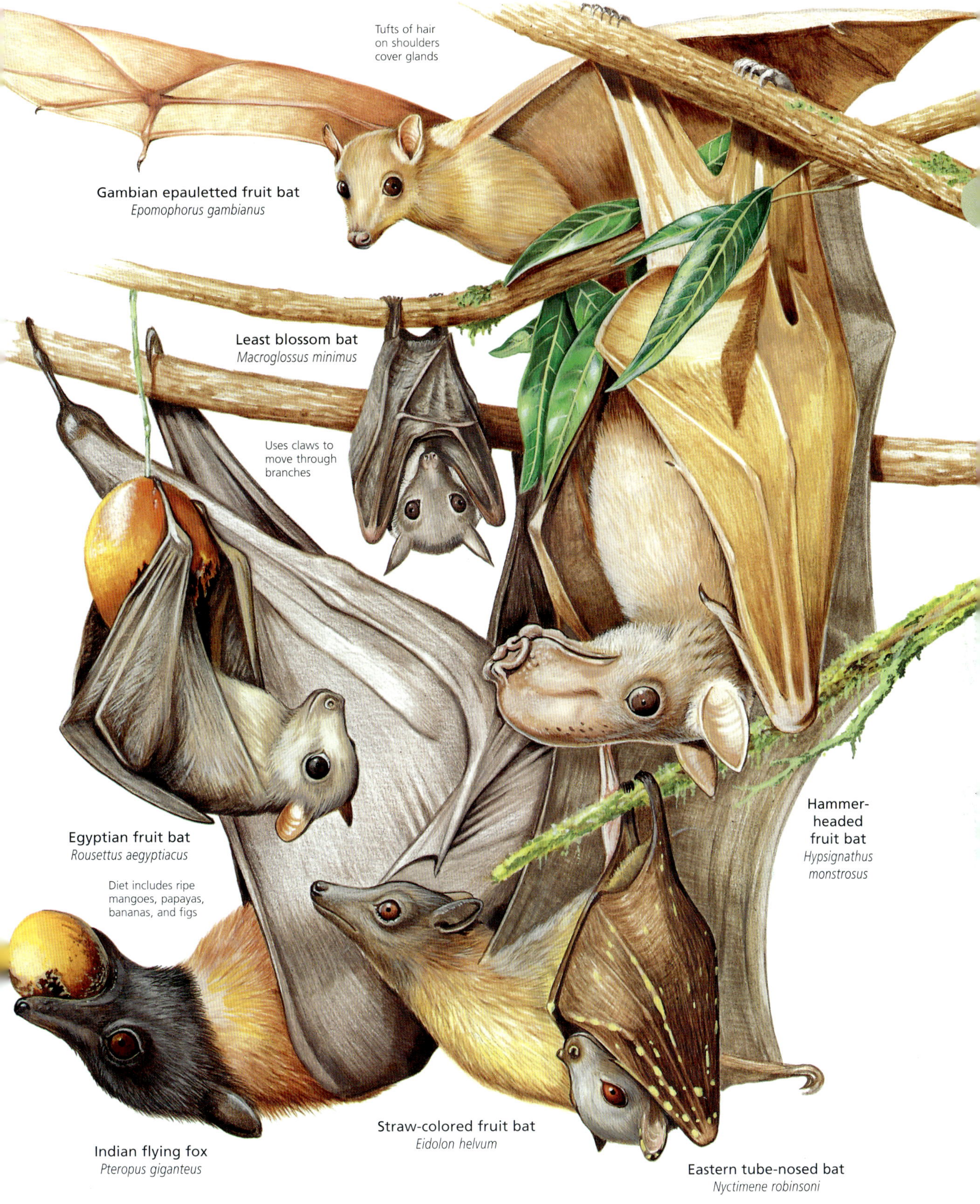

Tufts of hair on shoulders cover glands

Gambian epauletted fruit bat
Epomophorus gambianus

Least blossom bat
Macroglossus minimus

Uses claws to move through branches

Egyptian fruit bat
Rousettus aegyptiacus

Diet includes ripe mangoes, papayas, bananas, and figs

Hammer-headed fruit bat
Hypsignathus monstrosus

Indian flying fox
Pteropus giganteus

Straw-colored fruit bat
Eidolon helvum

Eastern tube-nosed bat
Nyctimene robinsoni

Least blossom bat This small bat uses a very long tongue to extract nectar and pollen from the blooms of banana, coconut, and mangrove trees. As the bat travels from feeding site to feeding site, it helps to pollinate the plants.

- Up to 3 in (7 cm)
- Up to ½ in (1 cm)
- Up to ½ oz (18 g)
- Solitary, pairs
- Locally common

N. Australia, New Guinea, some islands

Egyptian fruit bat While this species relies mostly on vision, it is one of the few Old World fruit bats with a crude echolocation mechanism, which comes in handy in its dim roosting caves.

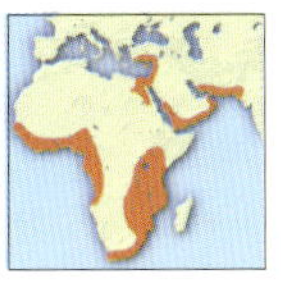

- Up to 5½ in (14 cm)
- Up to 1 in (2 cm)
- Up to 5½ oz (160 g)
- Colonial
- Common

Africa & Middle East

Hammer-headed fruit bat Widely distributed through western and central Africa, males of this large, canopy-roosting species have an enlarged snout and larynx, enabling them to produce loud calls.

- Up to 12 in (30 cm)
- None
- Up to 15 oz (420 g)
- Colonial, lekking
- Locally common

W. & C. Africa

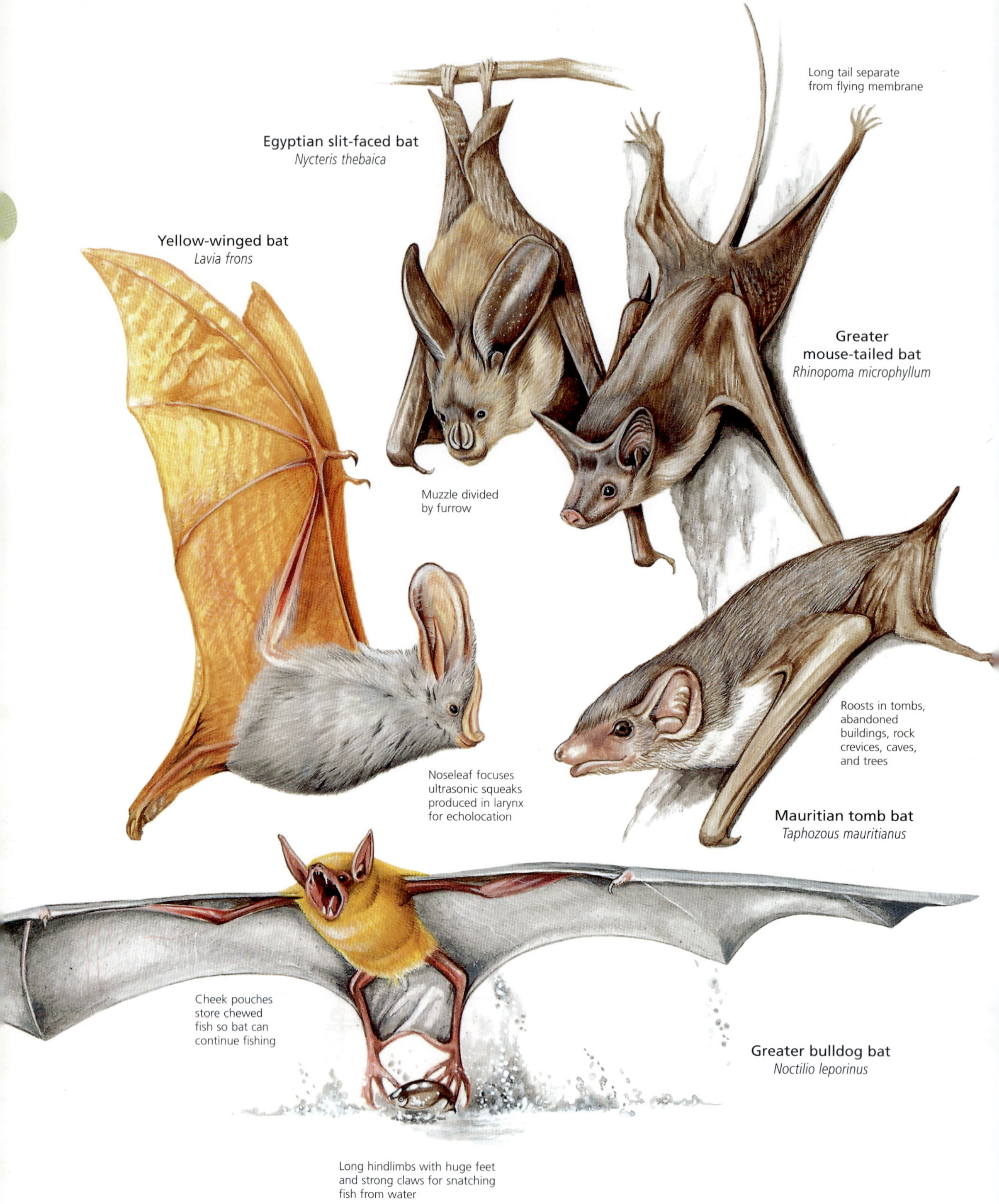

Egyptian slit-faced bat
Nycteris thebaica

Muzzle divided by furrow

Long tail separate from flying membrane

Greater mouse-tailed bat
Rhinopoma microphyllum

Yellow-winged bat
Lavia frons

Noseleaf focuses ultrasonic squeaks produced in larynx for echolocation

Roosts in tombs, abandoned buildings, rock crevices, caves, and trees

Mauritian tomb bat
Taphozous mauritianus

Cheek pouches store chewed fish so bat can continue fishing

Greater bulldog bat
Noctilio leporinus

Long hindlimbs with huge feet and strong claws for snatching fish from water

MAKING A TENT

The little Honduran tent bat is one of several fruit bats that create their own shelter. It curls a palm frond by chewing through the connection between the leaf's midrib and edges.

Yellow-winged bat Umbrella thorn trees are among this species' favorite roosting sites. When the trees flower, they attract swarms of insects, which are kept in check by the bats.

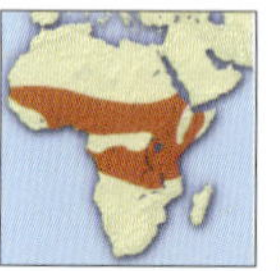

Sub-Saharan Africa

Up to 3 in (8 cm)
None
Up to 1¼ oz (36 g)
Pair, colonial
Locally common

Greater bulldog bat Common locally in Central and South America, this species zig-zags over ponds, rivers, and coastal waters, using echolocation and large, taloned feet to find and capture fish.

Central & South America, Caribbean

Up to 5 in (13 cm)
Up to 1½ in (4 cm)
Up to 3 oz (90 g)
Colonial
Locally common

American false vampire bat
Vampyrum spectrum

Greater horseshoe bat
Rhinolophus ferrumequinum

Diadem leaf-nosed bat
Hipposideros diadema

Bottom part of noseleaf is horseshoe-shaped

Eats small vertebrates but does not target animals for their blood

Hovers at flowers as it collects pollen and nectar with its long tongue

Pallas's long-tongued bat
Glossophaga soricina

Complex folds of skin around face

White bat
Ectophylla alba

Wrinkle-faced bat
Centurio senex

Modified thumbs and strong hindlimbs allow bat to walk, run, and hop on all fours as it hunts terrestrial prey

Orange leaf-nosed bat
Rhinonicteris aurantia

Common vampire bat
Desmodus rotundus

False vampire bat The American false vampire hunts birds, other bats, small rodents, reptiles, fish, and amphibians, but does not drink blood. This bat is a powerful predator.

Up to 6½ in (16 cm)
None
Up to 6½ oz (190 g)
Small group
Near threatened

Central & South America

Common vampire bat With razor-sharp teeth, this bat will slice away a small piece of skin from a cow, horse, deer, or other large mammal and lap up the blood. Most young vampires manage to find a meal only two nights out of three. It takes just a few days for a bat to starve to death, so a hungry vampire bat will beg for food from a roost mate, who will probably oblige by regurgitating blood. Vampire bats are among the few mammal species to display such a degree of generosity.

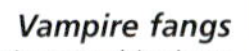

Vampire fangs
The enlarged upper canines and incisors of vampire bats are razor sharp.

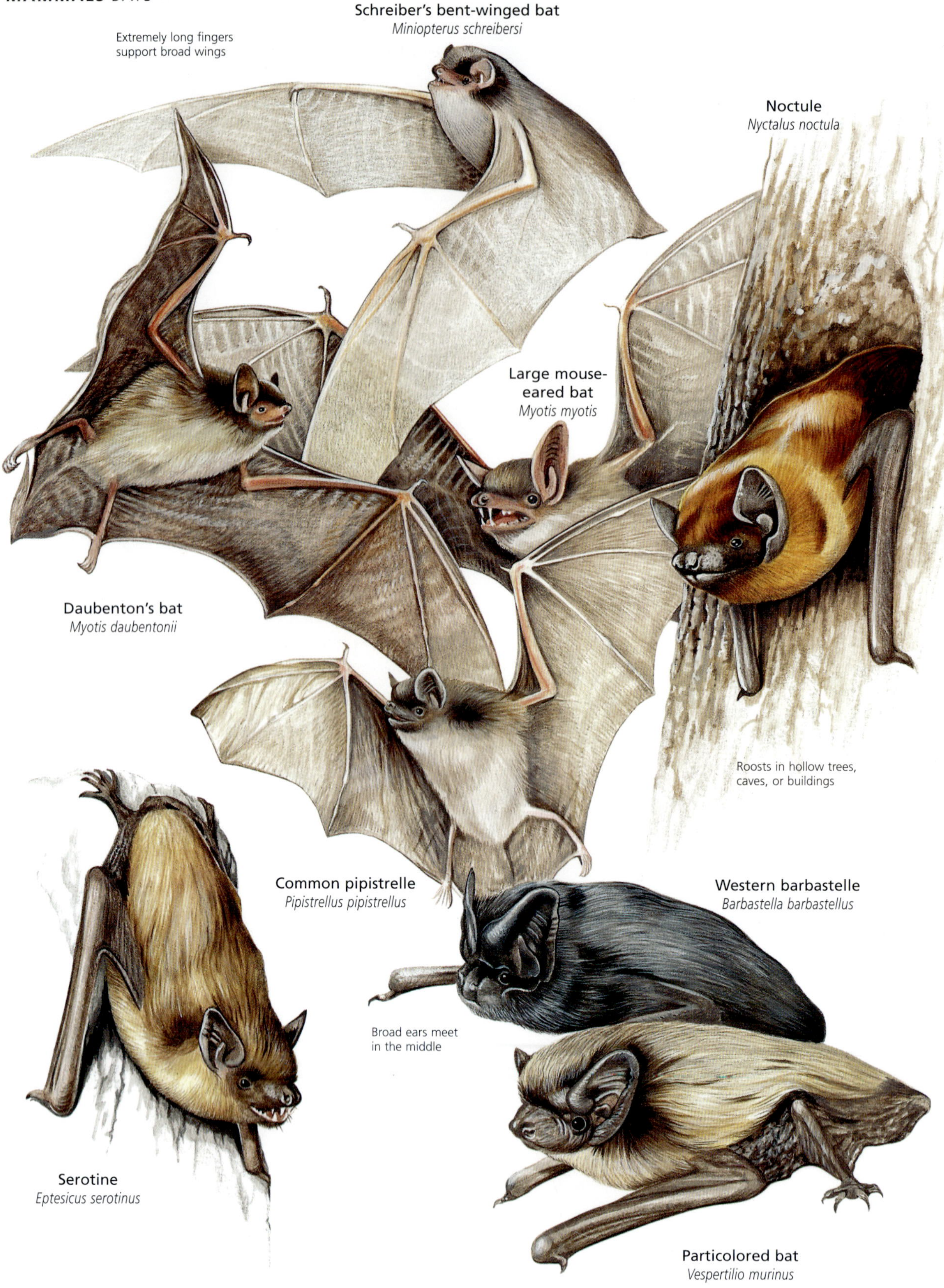

Schreiber's bent-winged bat
Miniopterus schreibersi

Extremely long fingers support broad wings

Noctule
Nyctalus noctula

Large mouse-eared bat
Myotis myotis

Daubenton's bat
Myotis daubentonii

Roosts in hollow trees, caves, or buildings

Common pipistrelle
Pipistrellus pipistrellus

Western barbastelle
Barbastella barbastellus

Broad ears meet in the middle

Serotine
Eptesicus serotinus

Particolored bat
Vespertilio murinus

Western barbastelle This medium-sized bat roosts in caves, mines, cellars, hollow trees, or under loose bark. It emerges at dusk to search for moths. Although widespread, it appears to be rare almost everywhere in its range.

Up to 2½ in (6 cm)
Up to 1½ in (4 cm)
Up to ¼ oz (10 g)
Small group
Vulnerable

W. Europe, Morocco, Canary Islands

Schreiber's bent-winged bat In some regions, this species migrates to warmer locations in winter. It roosts by day in caves or buildings, with the young placed in a communal group separate from the adults.

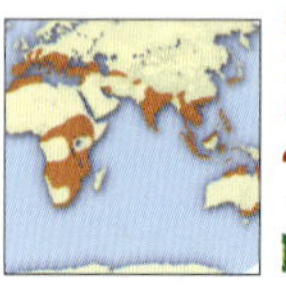

Up to 2½ in (6 cm)
Up to 2½ in (6 cm)
Up to ½ oz (20 g)
Large group
Near threatened

Europe, Africa, S. Asia, Australasia

Large mouse-eared bat Eating up to half its body weight each night, this bat preys on beetles and moths. Groups of 10–100 individuals share roosts in caves and attics. Young are born in April to June and must build up fat to survive the winter hibernation.

Up to 3 in (8 cm)
Up to 2½ in (6 cm)
Up to 1½ oz (45 g)
Small to large group
Near threatened

Europe & Israel

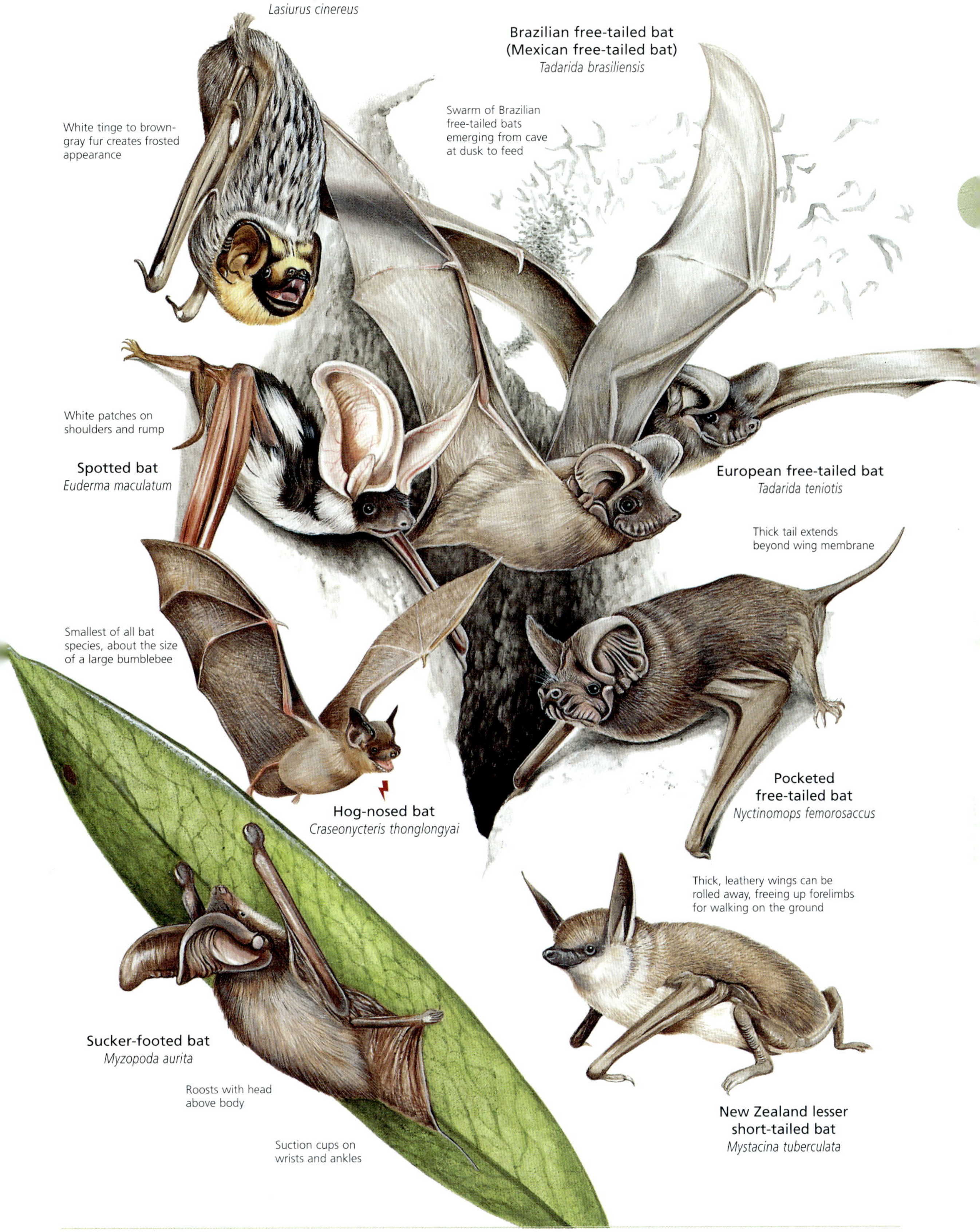

Hoary bat
Lasiurus cinereus

White tinge to brown-gray fur creates frosted appearance

Brazilian free-tailed bat (Mexican free-tailed bat)
Tadarida brasiliensis

Swarm of Brazilian free-tailed bats emerging from cave at dusk to feed

White patches on shoulders and rump

Spotted bat
Euderma maculatum

European free-tailed bat
Tadarida teniotis

Thick tail extends beyond wing membrane

Smallest of all bat species, about the size of a large bumblebee

Hog-nosed bat
Craseonycteris thonglongyai

Pocketed free-tailed bat
Nyctinomops femorosaccus

Thick, leathery wings can be rolled away, freeing up forelimbs for walking on the ground

Sucker-footed bat
Myzopoda aurita

Roosts with head above body

Suction cups on wrists and ankles

New Zealand lesser short-tailed bat
Mystacina tuberculata

Brazilian free-tailed bat With more than 20 million roosting in Bracken Cave, Texas, United States, Brazilian free-tailed bats hold the record for the largest gathering. It is estimated that they consume more than 250 tons (225 t) of insects per night.

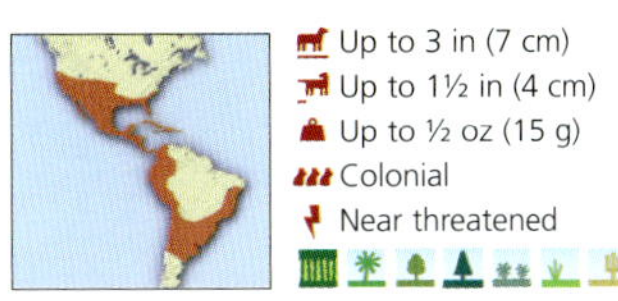

Up to 3 in (7 cm)
Up to 1½ in (4 cm)
Up to ½ oz (15 g)
Colonial
Near threatened

North & South America, Caribbean

Sucker-footed bat Now found only on Madagascar, this little bat uses the suction cups on its wrists and ankles to cling to leaves, supported by its tail. This adaptation appears to have evolved independently in the Old and New World species.

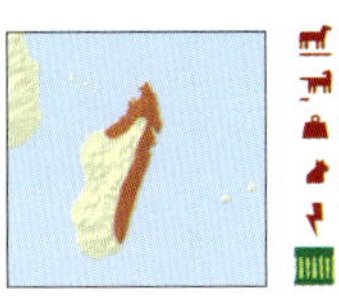

Up to 2½ in (6 cm)
Up to 2 in (5 cm)
Up to ¼ oz (10 g)
Not known
Vulnerable

E. & N. Madagascar

New Zealand lesser short-tailed bat Along with a larger bat species that became extinct early last century, these bats are New Zealand's only native land mammals. They forage on the ground and on tree trunks, flying only when necessary.

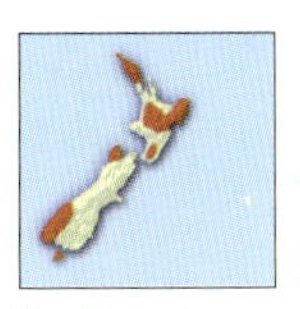

Up to 3 in (7 cm)
Up to ½ in (1 cm)
Up to ½ oz (15 g)
Colonial
Vulnerable

New Zealand

PRIMATES

CLASS	Mammalia
ORDER	Primates
FAMILIES	13
GENERA	60
SPECIES	295

Charismatic and intelligent, lemurs, monkeys, apes, and their close relatives make up the order known as Primates. Early primates were tree-dwellers and developed adaptations for an arboreal lifestyle: forward-facing eyes with stereoscopic vision to help judge distances when traveling through the trees; dextrous hands and feet to firmly grasp branches; and long, flexible limbs to enhance agility when foraging. Most primates are still largely arboreal, but even those that have opted for life on the ground retain at least some of these adaptations. Perhaps the most fascinating aspect of this order is the complex social behavior displayed by many species.

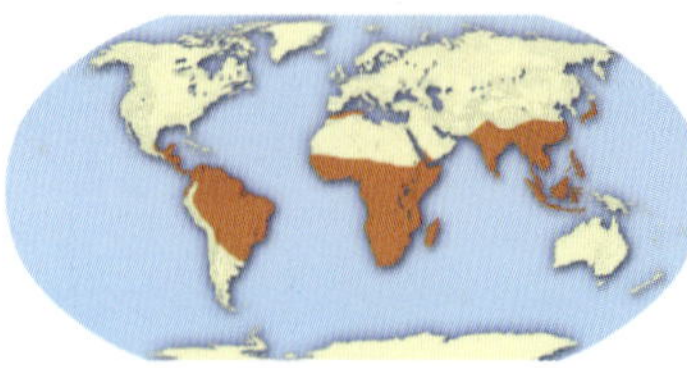

Tropical distribution Although most primates live in the tropical rain forests between 25°N and 30°S, a handful of species are found farther afield, in North Africa, China, and Japan.

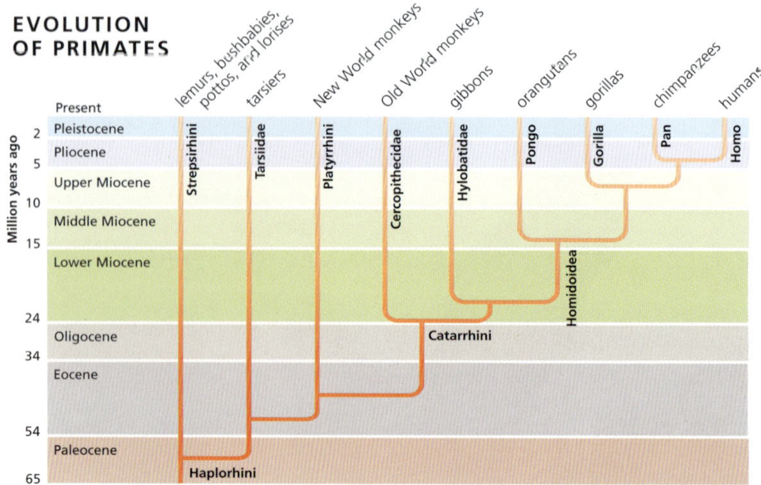

Vertical posture Gorillas, chimpanzees and other apes sit and sometimes walk upright, supported by a shorter back, broader rib-cage, and stronger pelvis than those of monkeys and lemurs. The arms, often used in locomotion, are longer than the legs, and the wrists are highly flexible. Along with monkeys and humans, the gorillas and other apes make up the suborder Haplorhini. Gorillas are social animals, living in troops of one or two silverback males, a few younger males, and several females and young.

PRIMATE DIVERSITY

While some small primates are solitary foragers, relying on hiding and nocturnal habits to escape predators, many larger primates are active by day and form groups as protection. A group offers many pairs of eyes to watch for predators. Even when a predator does attack, there is a good chance that it will take another member of the group. Some primates will fight off an attack together—baboons have been known to kill an attacking leopard.

The size and organization of primate groups vary enormously. Some species live in monogamous pairs, while others form troops of several females and one or more males. Stable troops of 150 geladas sometimes combine into herds of 600 individuals. The most common structure is based on related females and their offspring, often headed by a single male. Group living leads to

Temperate life Japanese macaques are among the few primates to live outside the tropics and subtropics. During the cold, snowy winters, their furry coats thicken and they feed on bark, buds, and stored food reserves, and warm up in thermal pools.

greater competition for food and mates, which is negotiated through complex hierarchies and alliances. These elaborate social networks rely on precise communication, with many species employing a range of nuanced visual and vocal signals. Relative to body size, primates have much larger brains than most other mammals, a feature that may be linked to their complex social lives.

The life of a primate moves slowly. Gestation periods are long; birth rates are low, with just one or two young in a litter; growth rates are slow, with long periods of dependency as infants and juveniles; and lifespans are long. This may be the cost of the large primate brain, which uses energy that would otherwise be available for growth and reproduction.

Primates range from the pygmy mouse lemur (*Microcebus myoxinus*), at 4 inches (10 cm) long and 1 ounce (30 g) in weight, to the gorilla, standing more than 5 feet (1.5 m) high and weighing 400 pounds (180 kg). Many small primates feed primarily on insects, a quickly processed food source to fuel their fast metabolisms. Larger species need greater volumes of food and often concentrate on leaves, shoots, and fruit, which are slow to digest but are in abundant supply. The general reliance on fruit, shoots, leaves, and insects has probably helped restrict primates largely to the tropics, where these foods are available year round.

The order Primates is split into two suborders: Strepsirhini, made up of lemurs and their relatives, and Haplorhini, comprised of tarsiers, monkeys, apes, and humans. Tarsiers and strepsirhines share a number of primitive features and have been collectively referred to as prosimians.

Primate survival About a third of all primate species are at risk of extinction, victims of habitat loss and hunting. Large species such as the orangutan (above) are especially vulnerable as they are easy for hunters to find.

Busy social lives The large brains of primates such as galadas may be needed to manage the complex social relationships associated with living in a hierarchical group. While group living can cause greater competition for food, it reduces the risk of being attacked by predators.

Prosimians

CLASS Mammalia
ORDER Primates
FAMILIES 8
GENERA 22
SPECIES 63

The name prosimians means "before monkeys," a reference to the fact that these creatures retain many features of the early primates. Absent from the Americas, prosimians include the lemurs of Madagascar, the bushbabies and pottos of Africa, and the lorises of Asia—all members of the suborder Strepsirhini. These species all possess a moist, pointed snout and most have a light-reflecting disk in the eye, a long toilet claw, and compressed lower teeth forming a dental comb. Tarsiers, now classified in the suborder Haplorhini, are also often referred to as prosimians because their appearance and solitary, nocturnal behavior is like that of many strepsirhines.

Red ruffed lemur
Varecia variegata rubra

SPECIALIZED SENSES

Prosimians tend to be relatively small, nocturnal tree-dwellers who forage alone but may occasionally form limited groups. Most are largely insectivorous but will also consume fruit, leaves, flowers, nectar, and gum. They display two specializations for grooming: a long claw on the second toe of the foot, known as the toilet claw; and the dental comb, a compressed row of protruding lower teeth. The dental comb also seems to be used for scraping resin off trees.

The moist, dog-like snout shared by lemurs, lorises, and the other strepsirhines supplies them with a wealth of information through smell. Sight is also important, but they do not have the full color vision of monkeys and apes. Sophisticated color vision would be of limited use to creatures that forage in the low-light conditions of night. Instead, most prosimians have a tapetum lucidum, a crystalline layer behind the retina that reflects light and produces the characteristic eyeshine of many nocturnal mammals.

While sound is important in prosimian communication, with many species employing alarm and territorial calls, scent plays a major role. Territories are often marked with urine, feces, or secretions from special scent glands, which can convey an individual's sex, identity, and breeding status.

Lemurs of Madagascar For millions of years, lemurs have been isolated on the island of Madagascar. They range from the size of a mouse to that of a medium domestic dog. Most lemurs are arboreal and nocturnal. Unusually among prosimians, ring-tailed lemurs are active during the day and spend most of their time on the ground. Their troops of 3–20 animals are dominated by female lemurs, a matriarchal structure found in few other primates.

Broad-nosed gentle lemur
Hapalemur simus

Mongoose lemur
Eulemur mongoz

Ring-tailed lemur
Lemur catta

Large forward-facing eyes

Black lemur
Eulemur macaco

Brown lemur
Eulemur fulvus

Coquerel's dwarf lemur
Microcebus coquereli

Long tail can store fat reserves

Fork-marked lemur
Phaner furcifer

Characteristic forked stripe on head

Broad-nosed gentle lemur More than 95 percent of this lemur's diet is made up of bamboo shoots, leaves, and stalks. One of the world's rarest mammals, it was not seen for a century and was considered extinct until its rediscovery in 1972.

S.E. Madagascar

- Up to 17½ in (45 cm)
- Up to 22 in (56 cm)
- Up to 5½ lb (2.5 kg)
- Family band
- Critically endangered

Mongoose lemur Usually nocturnal during the dry season, mongoose lemurs switch to daytime activity for the colder wet season. They live in small groups made up of a male and female pair and their offspring.

N.W. Madagascar (S. of Narinda Bay), Comoro I.

- Up to 17½ in (45 cm)
- Up to 25 in (64 cm)
- Up to 6½ lb (3 kg)
- Family band
- Vulnerable

Black lemur For years, male and female black lemurs were considered different species because they look so unalike. Males are black, but females are chestnut-brown with a white belly and ear tufts. This arboreal species feeds on fruit, flowers, and nectar.

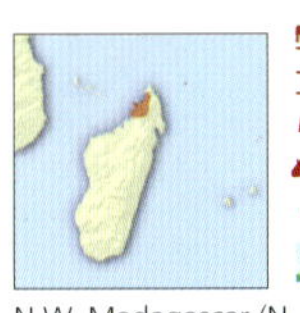

N.W. Madagascar (N. of Narinda Bay)

- Up to 17½ in (45 cm)
- Up to 25 in (64 cm)
- Up to 6½ lb (3 kg)
- Family band
- Vulnerable

Red-tailed sportive lemur
Lepilemur ruficaudatus

Rests in vertical position and moves between trees in short leaps

Aye-aye
Daubentonia madagascariensis

Weasel sportive lemur
Lepilemur mustelinus

Long, powerful hindlimbs power great leaps

Long middle finger for extracting grubs

Coat of a ruffed lemur can be black and red, or black and white

Black face, hands, feet, and tail regardless of coat color

Ruffed lemur
Varecia variegata

HANDS AND FEET

The contrasting lifestyles of prosimians are evident in the shape of their hands and feet. The indri and the tarsiers cling vertically and leap from tree to tree, while the aye-aye climbs through the branches.

Clinging indri
The indri's stout thumb and big toe help it to cling firmly to tree trunks.

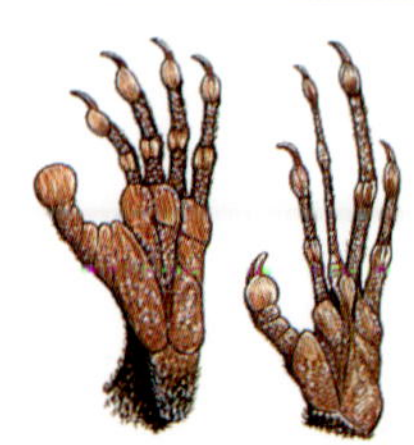

Digging aye-aye
Rather than cling, the aye-aye climbs by digging its long claws into the bark.

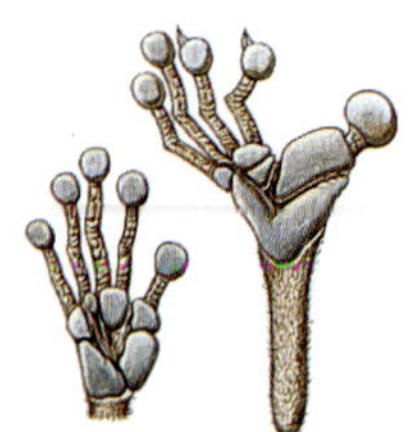

Tarsier friction
A tarsier's grip is strengthened by the friction of the disk-like pads on its fingers and toes.

Woolly lemur
Avahi laniger

Thick, woolly coat

Indri
Indri indri

Large, black, tufted ears

Furless black face

Verreaux's sifaka
Propithecus verreauxi

Only lemur with a very short tail

Diadem sifaka
Propithecus diadema

Almost completely arboreal

Woolly lemur The ascending alarm call of this species sounds like "Ava Hy," inspiring the genus name of *Avahi*. Family groups, made up of a breeding pair and their young, spend the day sleeping together among vines.

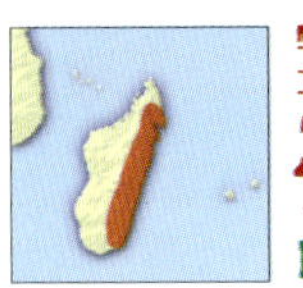

E. Madagascar

- Up to 17½ in (45 cm)
- Up to 16 in (40 cm)
- Up to 2½ lb (1.2 kg)
- Pair
- Near threatened

Indri The largest surviving lemur, the indri is active by day, leaping from tree to tree in search of fruit and flowers. It announces its presence with loud wailing calls and marks territory using scent glands in its cheeks.

N.E. Madagascar

- Up to 35½ in (90 cm)
- Up to 2 in (5 cm)
- Up to 22 lb (10 kg)
- Family band
- Endangered

LEAVING A TRACE

Lemurs scentmark to indicate their territories with secretions from glands on the head, hands, or rear. The indri's scent glands are in its cheeks, while the woolly lemur has them in its neck.

Lesser bushbaby
Galago senegalensis

Large, bat-like ears help to detect insects at night

Eastern needle-clawed bushbaby
Euoticus inustus

Large hands and feet with nails forming claws

Demidoff's galago
Galagoides demidoff

Needle-like claws on digits to grip branches

Western needle-clawed bushbaby
Euoticus elegantulus

Allen's bushbaby
Galago alleni

Bushy tail longer than body acts as stabilizer during leaps

CONTRASTING FAMILIES

Bushbabies, or galagos, belong to the family Galagonidae. Using their long hindlimbs for propulsion and long, bushy tail for balance, they leap from tree to tree with great agility. In constrast, the lorises, pottos, and angwantibos of the family Loridae creep slowly through the branches. As with other primates that move on all fours, their limbs are of roughly equal length and their tails are short.

Moving around
Extremely agile, arboreal primates, bushbabies cling onto and leap between tree branches.

Lesser bushbaby This nocturnal animal prefers to feed on grasshoppers and other insect prey. When insects become rare during times of drought, it will feed solely on acacia gum, allowing it to survive in drier habitats.

Up to 8 in (20 cm)
Up to 12 in (30 cm)
Up to 10½ oz (300 g)
Family band
Common

C. & S. Africa

Naked tail tipped with tuft of hair

Eyes cannot move but head can rotate almost a full circle

Western tarsier
Tarsius bancanus

Slender loris
Loris tardigradus

Climbs on all fours with slender limbs of equal length

Potto
Perodicticus potto

Angwantibo (golden potto)
Arctocebus calabarensis

Moves slowly through branches on all fours

Slow loris
Nycticebus coucang

Spectral tarsier
Tarsius spectrum

Very long, skinny digits for grasping branches

Western tarsier Like other tarsiers, this nocturnal species lacks a light-reflecting disk to enhance its night vision, but it does have incredibly large eyes. Each eye is bigger than its brain.

- Up to 6 in (15 cm)
- Up to 10½ in (27 cm)
- Up to 6 oz (165 g)
- Solitary
- Data deficient

Sumatra, Borneo, Bangka, Belitung, Serasan

Spectral tarsier Pushing off with its very long legs, this small primate can leap up to 20 feet (6 m) between trees. On the ground, it moves about by hopping on its back legs.

- Up to 6 in (15 cm)
- Up to 10½ in (27 cm)
- Up to 6 oz (165 g)
- Solitary
- Near threatened

Sulawesi, Sangihe, Peleng, Salayar

Defense position
A threatened potto tucks its head down and presents its neck, which has a shield of spiny vertebrae covered by horny skin.

MONKEYS

CLASS	Mammalia
ORDER	Primates
FAMILIES	3
GENERA	33
SPECIES	214

Geography has determined two separate lineages of monkeys: the New World monkeys of the Americas, classified within the suborder Haplorhini as platyrrhines; and the Old World monkeys of Africa and Asia, grouped with apes and humans as catarrhines. New and Old World monkeys are most easily distinguished by the shape of their noses and by the arrangement of their teeth. All New World monkeys live in trees and many have a strong prehensile tail to grip branches. Most Old World monkeys are also arboreal, but none has a prehensile tail and some species are semi-terrestrial. Some Old World monkeys have callous pads on their rumps, a feature found on no New World species.

Old and new *Old World monkeys (far left) have prominent noses with narrow, forward-facing nostrils. The noses of New World monkeys (left) are flattened with nostrils facing sideways.*

SOCIAL SIMIANS

Monkeys range from the pygmy marmoset, with a length of 6 inches (15 cm) and a weight of 5 ounces (140 g), to the mandrill, measuring 30 inches (76 cm) long and weighing in at 55 pounds (25 kg). Most live in social groups, are active by day, and eat mainly fruit and leaves. All Old World monkeys and many New World monkeys have developed full color vision, which allows them to easily spot fruit among the foliage.

Like apes, monkeys differ from lemurs and other strepsirhines by having a dry, slightly hairy muzzle, a greater dependence on sight over smell, and a larger brain relative to body size. Not only are the brains of monkeys and apes larger, but the neocortex, the brain's outer sheath, is especially well developed. The neocortex is associated with creative thinking, an advantage when dealing with the machinations of group life. Monkeys are known to deliberately deceive other members of their group, raising false alarm calls, for example, in order to distract them from a source of food.

The social arrangements of monkeys display great variety: small family groups may contain just one monogamous breeding pair and their offspring; harems of several females may be dominated by a single adult male; and large troops may include several adult males and many females. While large groups often involve fierce competition to establish rank, they also feature extensive cooperation. Relationships among monkeys tend to be close and enduring, cemented by regular episodes of mutual grooming.

Temperature life Japanese macaques are among the few primates to live outside the tropics and subtropics. During the cold, snowy winters, their furry coats thicken and they feed on bark, buds, and stored food reserves, and warm up in thermal pools.

Golden lion tamarin
Leontopithecus rosalia

Striking reddish-gold coat with long mane framing black face

Golden-headed lion tamarin
Leontopithecus chrysomelas

Pygmy marmoset
Callithrix pygmaea

Claws on all digits except big toe, which has a flat nail

White ear tufts on adults and juveniles, absent from infants

Cotton-top tamarin
Saguinus oedipus

Geoffroy's tamarin
Saguinus geoffroyi

Common marmoset
Callithrix jacchus

Golden lion tamarin There are only about 800 golden lion tamarins left in the wild. Their striking appearance made them popular pets and zoo exhibits, with many falling victim to the live-animal trade until it became illegal in the 1970s.

Up to 11 in (28 cm)
Up to 16 in (40 cm)
Up to 1½ lb (650 g)
Family band
Endangered

Coastal forest in Brazil

Pygmy marmoset The smallest monkey, this species gouges out holes in trees to release its favorite food of sap and gum. It runs on all fours along branches, leaping from tree to tree. Group members use high-pitched trillings to communicate.

Up to 6 in (15 cm)
Up to 8½ in (22 cm)
Up to 5 oz (140 g)
Family band
Locally common

W. Amazon basin

Cotton-top tamarin After spending the night resting in the forks of their sleeping tree, a group of 3–9 cotton-top tamarins will move through the trees of the canopy searching for insects, fruit, and gum.

Up to 10 in (25 cm)
Up to 16 in (40 cm)
Up to 1 lb (500 g)
Family band
Endangered

N. Colombia

Northern night monkey
Aotus trivirgatus

Large eyes for better night vision

Dusky titi
Callicebus moloch

White-faced saki
Pithecia pithecia

White-nosed saki
Chiropotes albinasus

Bald uakari (red uakari)
Cacajao calvus

Black-headed uakari
Cacajao melanocephalus

Northern night monkey The world's only nocturnal monkeys, night monkeys use their acute sense of smell and large eyes to find insects, fruit, nectar, and leaves in low light. They live in monogamous pairs, with the male responsible for the bulk of infant care.

Up to 18½ in (47 cm)
Up to 16 in (41 cm)
Up to 2½ lb (1.2 kg)
Pair
Common

S.W. Venezuela, N.W. Brazil

White-faced saki The long hindlimbs of these active tree-dwellers enable them to leap up to 30 feet (10 m) between trees. At night, they sleep curled up on branches like cats.

Up to 19 in (48 cm)
Up to 17½ in (45 cm)
Up to 5½ lb (2.4 kg)
Family band
Uncommon

Guianas, Venezuela, N. Brazil

Black-headed uakari This species lives in groups of up to 50 individuals, which include more than one adult male. Adult females and young engage in social grooming. For a tree-dweller, the uakari has an unusually short tail.

Up to 19½ in (50 cm)
Up to 8½ in (21 cm)
Up to 9 lb (4 kg)
Large troop
Rare

Upper Amazon basin

Black howler
Alouatta caraya

Males are black, females are brown or olive, and infants are golden

Prehensile tail grips branches

Common squirrel monkey
Saimiri sciureus

Mantled howler
Alouatta palliata

Red howler
Alouatta seniculus

White-fronted capuchin
Cebus albifrons

Brown capuchin
Cebus apella

Weeping capuchin
Cebus olivaceus

White-throated capuchin
Cebus capucinus

Brown capuchin Troops of a dozen or so brown capuchins forage together by day, led by a dominant male who has first pick of the food. These monkeys sometimes use simple tools, cracking open a nut with a stone, for example.

N.E. South America

Up to 19 in (48 cm)
Up to 19 in (48 cm)
Up to 10 lb (4.5 kg)
Small troop
Locally common

HOWLING

One of the loudest calls in the animal world is made by howler monkeys. At dawn, howler troops announce their presence with a deafening conversation of howls that resonate through the forest, traveling up to 3 miles (5 km) away. By helping troops avoid one another, the howls prevent territorial skirmishes that would waste time and energy that could be better spent eating or resting.

Thumbless hand acts as hook when swinging

Very long prehensile tail strong enough to support monkey's weight

Common woolly monkey
Lagothrix lagotricha

Woolly spider monkey
Brachyteles arachnoides

Coat can be reddish, dark to light brown, or dark to light gray

Long-haired spider monkey
Ateles belzebuth

Face color ranges from pink to black

Black spider monkey
Ateles paniscus

Black-handed spider monkey
Ateles geoffroyi

Common woolly monkey This heavy monkey spends most of its time in trees, but often descends to the forest floor, where it can walk upright on its back legs. Large, multi-male groups of up to 70 individuals sleep together at night.

Up to 23 in (58 cm)
Up to 31½ in (80 cm)
Up to 22 lb (10 kg)
Variable
Uncommon

Upper Amazon basin

Woolly spider monkey Unusually for primates, male woolly spider monkeys stay with their birth troop all their lives, while females must leave and join another troop when they reach adulthood.

Up to 25 in (63 cm)
Up to 31½ in (80 cm)
Up to 33 lb (15 kg)
Variable
Endangered

S.E. Brazil

Black spider monkey Groups of about 20 black spider monkeys will cooperatively defend their territory or mob a predator, but they will split into subgroups of up to six for foraging.

Up to 24½ in (62 cm)
Up to 35½ in (90 cm)
Up to 28½ lb (13 kg)
Variable
Locally common

N. of Amazon & E. of Rio Negro

Capped leaf monkey
Trachypithecus pileatus

Hanuman langur
Semnopithecus entellus

Coat can be dark brown, fawn, or gray

Douc langur
Pygathrix nemaeus

Female has brown face, male has blue face

Male's pendulous nose adds resonance to calls

Chinese snub-nosed monkey
Rhinopithecus roxellana

Slight webbing between digits helps to make proboscis monkey an excellent swimmer

Proboscis monkey
Nasalis larvatus

Proboscis monkey This monkey is named after the pendulous nose of the male. It lives in stable harems of one adult male and several females. Sexual contact is initiated by the female, who indicates her interest in a male by pursing her lips.

Lowland Borneo

- Up to 30 in (76 cm)
- Up to 30 in (76 cm)
- Up to 51 lb (23 kg)
- Harem
- Endangered

INFANTICIDE

Although most thoroughly studied among Hanuman langurs, infanticide is practiced by many primates. It occurs when a new male becomes dominant in a troop, often after a bachelor band has invaded. The new male kills all the troop's unweaned infants, presumably because lactation prevents the mothers from conceiving. Although the mothers often try to defend their young, they are usually unsuccessful.

Rival males *When young male Hanuman langurs are expelled from their birth troop, they form bachelor bands, which may invade a breeding troop and kill the nursing infants.*

COLOBUS MONKEYS

Colobus monkeys demonstrate great agility as they gallop along branches and take flying leaps to neighboring trees, using their thumbless hands as hooks to secure themselves. Most colobus monkeys live in groups of 10 or so individuals, with a fixed core of related females. Females in the group often "babysit" young that are not their own and may even suckle them.

Lift off *Colobus monkeys can leap spectacularly from one tree to another, either to reach a new source of food or to escape a pursuing predator.*

Mixed alliances
Colobus monkeys often form temporary or even stable associations with other monkey species. Red colobus monkeys and vervets, for example, may cooperate when drinking from a water hole, taking turns to watch for predators.

Rhesus macaque Up to 200 of these gregarious monkeys may live together in a group. Adapted to a wide range of habitats, they vary their diet according to season and location, with some in urban areas raiding gardens and trash cans.

Afghanistan & India to China

SNOW MONKEYS

Japanese macaques live farther north than any other primate (apart from humans). During the cold, snowy winters, they live off tree buds and bark, as well as stores of fat. Troops of 20–30 Japanese macaques are led by a dominant male. Their social lives tend to be harmonious, with much time spent grooming each other and sharing the care of the young.

Making snowballs *Entire troops of Japanese macaques make snowballs, packing a small ball in their hands then rolling it along the ground so it grows in size.*

Hamadryas baboon (savanna baboon)
Papio hamadryas

Male is grayish brown with shaggy mantle of silver hair on head

Female is olive-brown

Bright red callous pads on rump

Mane of hair on head of male

Heart-shaped patch of hairless skin on chest

Chacma baboon
Papio ursinus

Gelada
Theropithecus gelada

Highly opposable thumbs provide dexterity to select the best grass blades, rhizomes, and seeds

GRASS-EATING GELADAS

The only surviving species of a genus once widespread throughout Africa, geladas are now restricted to the highlands of northwest Ethiopia. They sleep on rocky cliffs beyond the reach of most predators, and forage in nearby grasslands by day. They survive almost entirely on grass, a degree of specialization that makes the species especially vulnerable as the local human population burgeons and requires ever larger areas of pasture for grazing livestock.

On alert *The dominant male in a troop is always on the alert for challenges from bachelor bands.*

Social creatures The basic unit of gelada society comprises one male, several females, and their offspring, but several families form foraging bands of around 70 animals. At times, a number of bands may congregate in vast herds of 600 or more individuals.

Mandrill The largest of the world's monkeys, Africa's mandrills are instantly recognizable by their striking red and blue faces. During the day, they come down from their sleeping trees to search for fruit, seeds, insects, and small vertebrates on the rain-forest floor. Troops of up to 250 mandrills are made up of smaller multi-male groups, each with about 20 animals led by a dominant male who fathers most of the young.

Big yawn *To threaten a predator or rival, a male will spread its arms wide and display its daunting teeth.*

Attractive colors *In addition to their vibrant faces, male mandrills have a yellow beard, mauve rump, red penis, and lilac scrotum. The skin colors are most intense on the dominant male—they seem to be linked to testosterone levels and may be advertising his virility.*

Sykes' monkey (blue monkey)
Cercopithecus mitis

Coat can be blue, reddish brown, or grayish brown

Vervet monkey
Chlorocebus aethiops

Male has turquoise-blue scrotum

White stripe across forehead thought to resemble shape of goddess Diana's bow, inspiring this monkey's common name

Diana monkey
Cercopithecus diana

Long white tufts on ears

Will freeze in position if it senses danger

Mona monkey
Cercopithecus mona

Sykes' monkey Among this species, a single adult male dominates troops of 10–40 females and their offspring. The females in the group help raise one another's young.

C., E. & S. Africa

- Up to 26½ in (67 cm)
- Up to 33½ in (85 cm)
- Up to 26½ lb (12 kg)
- Family band, variable
- Locally common

Vervet monkey Although it prefers to live in woodland along rivers, this adaptable monkey is found in many habitats, including human settlements. It spends most of its time in trees, eating, grooming, and sleeping.

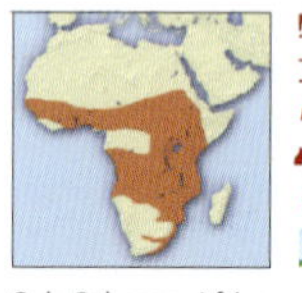

Sub-Saharan Africa

- Up to 24½ in (62 cm)
- Up to 28½ in (72 cm)
- Up to 20 lb (9 kg)
- Herd, troop
- Declining

Mona monkey Like many other Old World monkey species, this small primate stores fruit and insects in its cheek pouches while it forages. It is common in west African tropical forests, living in small family groups.

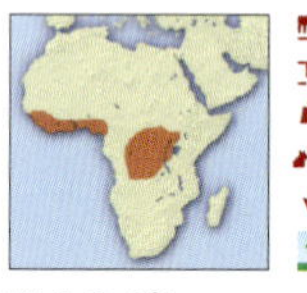

W. & C. Africa

- Up to 27½ in (70 cm)
- Up to 27½ in (70 cm)
- Up to 15½ lb (7 kg)
- Herd, troop
- Locally common

Redtail monkey
Cercopithecus ascanius

Grasping hands for gripping branches and collecting fruit

Chestnut fur on underside of tail gave rise to common name

CONSERVATION WATCH

Patas vulnerability This terrestrial monkey lives on the savannas of central Africa. Already vulnerable to rainfall fluctuations in this drought-prone area, patas monkeys are being killed by hunters for their meat and by farmers because they feed on crops.

Allen's swamp monkey
Allenopithecus nigroviridis

Lives in swamp forests and forages on ground or in shallow water

Webbing between digits helps with swimming

Patas monkey
Erythrocebus patas

Slender legs of equal length allow monkey to run at speeds of up to 35 mph (55 km/h)

Red-eared monkey
Cercopithecus erythrotis

MONKEY SIGNALS

Sometimes referred to as guenons, the monkeys in the genus *Cercopithecus* all use a range of signals to communicate with other members of their species. While many of these signals are vocal, encompassing barks, grunts, screams, booms, and chirps, others are tactile or visual. Nose-to-nose contact is a form of friendly greeting. Staring, head-bobbing, and yawning are often used as a threat display to intimidate.

On the nose *A nose-to-nose greeting between two redtail monkeys is often followed by grooming or play.*

Confident tail *Among vervets, tail position indicates if the animal is fearful or not. When on all fours, a tail arched over the body conveys confidence.*

APES

CLASS Mammalia	
ORDER Primates	
FAMILIES 2	
GENERA 5	
SPECIES 18	

Like humans, apes are intelligent and quick to learn, form complex social groups, and spend years on the care of each young. They are divided into two families: the gibbons of Hylobatidae, and the great apes—orangutans, chimpanzees, gorillas—of Hominidae, which includes humans. Although apes and Old World monkeys have a similar nose shape and dental structure and are grouped together as catarrhine primates, they differ in many ways. Apes have skeletons suited to sitting or standing upright. They have no tail, their lowest vertebrae fused to form the coccyx instead. Their spines are shorter, chests are barrel-shaped, and shoulders and wrists are very mobile.

Bonobo
Pan paniscus

Moving apes While all apes other than humans have arms longer than their legs, orangutans and gibbons are the only ones in which the arms are genuinely elongated compared to their trunk. The orangutan's head and body measure about 5 feet (1.5 m), while their arms have a spread of more than 7 feet (2.2 m). Gibbons move by brachiation, using their arms to swing from one branch to another. Orangutans do not brachiate, but climb slowly through the trees using some combination of all four limbs. Chimpanzees spend up to three-quarters of their time on the ground, but will brachiate when in trees. Gorillas are largely terrestrial and rarely climb trees.

Energy conservation
Long arms allow the orangutan to reach fruit with minimum exertion.

A flexible grip
The orangutan's strong hands and feet are hook-like, the thumb and big toe are short, and the other digits are long.

Hanging around
Although it does hang under branches, the orangutan uses all limbs to move through trees.

CLEVER APES

Ape societies are organized in various ways. Monogamous gibbons live in pairs with their offspring, making groups of up to six animals. Orangutans with overlapping ranges form loose associations and meet occasionally, but the male tends to be a solitary forager, and the female usually lives just with her single young. Chimpanzee communities involve 40–80 individuals, but these are rarely together at once and tend to forage in smaller groups. Gorillas live in harems, with one dominant male, possibly one or two lower-ranking adult males, several females, and their young.

The gibbons and the great apes evolved into distinct families at least 20 million years ago. Chimpanzees are believed to be the closest relatives of humans, sharing a common ancestor until about 6 million years ago. The great apes appear to work through problems much as humans do. Chimpanzees and orangutans are known to fashion tools in the wild, while in research centers all the great apes have been taught to use implements. By recognizing themselves in mirrors, great apes show a concept of self, and some have been taught to recognize and use symbols such as sign language.

Kloss's gibbon
Hylobates klossii

Hoolock
Hylobates hoolock

White face rings, hands, and feet with reddish or black coat

Lar gibbon
Hylobates lar

Throat sac bigger than head

Males are black; adult females are golden or buff, sometimes with black patches

Elongated arm with hook-like, grasping hand

Black gibbon
Hylobates concolor

Siamang
Hylobates syndactylus

Hoolock Found farther north and east than any other gibbon, this large species avoids competition with other primates by preferring especially ripe fruit. Its numbers are now decreasing, however, as hunting and habitat fragmentation take their toll.

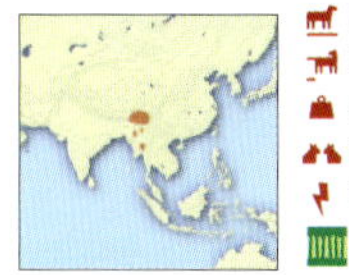

Up to 25½ in (65 cm)
None
Up to 17½ lb (8 kg)
Pair
Endangered

N.E. India, Bangladesh, S.W. China, Myanmar

Black gibbon This gibbon is born with a golden or buff coat, which turns black at about 6 months of age. Males remain black, but when females mature they turn golden or buff again, sometimes retaining patches of black.

Up to 25½ in (65 cm)
None
Up to 17½ lb (8 kg)
Pair
Endangered

S. China, N. Vietnam

Siamang The largest gibbon, the siamang spends 5 hours a day eating, often hanging by one arm as it feasts. While it consumes a lot of fruit and some insects and small vertebrates, up to half of its diet is made up of leaves.

Up to 35½ in (90 cm)
None
Up to 28½ lb (13 kg)
Pair
Near threatened

Malaya, Sumatra

Powerful grip

Flexible arms and legs can swing in most directions

Orangutan
Pongo pygmaeus

Male has large cheek pads and throat pouch with beard and mustache

Chimpanzee
Pan troglodytes

Arms longer than legs, with fingers longer than those of humans

Bonobo
Pan paniscus

Western gorilla
Gorilla gorilla

Walks on soles of feet and knuckles of hands

Slimmer body and more slender limbs than those of common chimpanzee

TOOL TIME

The ingenuity and manual dexterity of chimpanzees are demonstrated by their tool use. They strip twigs and grass stems to make wands for probing ant and termite nests. Specially selected stones are used to open nuts and hard-shelled fruit. During displays of strength or hunting, some chimps use sticks and rocks as missiles. Tool use varies between chimp populations.

Orangutan Asia's only great apes, orangutans are also the world's largest arboreal mammals. Most hardly ever descend to the ground, moving through the forest by swinging a tree back and forth until they can grasp the next one.

Borneo, Sumatra

- Up to 5 ft (1.5 m)
- None
- Up to 200 lb (90 kg)
- Solitary, pair
- Endangered

CONSERVING PRIMATES

According to the International Union for Conservation of Nature, about 200 primate species—nearly half of all primate types—are at risk of extinction within the next few decades. More than half of all colobus monkeys are threatened, while in the family Hominidae, humans are the only secure species, with all of the great apes considered endangered. The trade in live animals as pets or for biomedical research has contributed to the declining numbers, as has the hunting of primates as food or pests. The greatest threat, however, is from habitat destruction through logging, land clearing, and collection of wood for fuel. Because primates reproduce slowly, their populations take a long time to recover. Almost all are tropical animals and live in poorer countries, so conservation efforts are complicated by the pressing needs of the growing human population.

Habitat destruction When an area of forest is cut down, several primate species can be affected at once. Once the tree cover is lost, the soil is washed away by rain and the area may become barren. Gorillas, which move through the mountain forests of eastern and central Africa in family groups, often build tree nests for sleeping.

CARNIVORES

CCLASS	Mammalia
ORDER	Carnivora
FAMILIES	11
GENERA	131
SPECIES	278

Adaptable animals Carnivores are found in almost all of Earth's habitats. Arctic foxes, gray wolves, and these polar bears survive in the icy landscape of the Arctic; otters and seals spend most of their time in the water; big cats prowl both jungles and savanna; and jackals live in deserts.

From massive polar bears to little weasels, speedy cheetahs to lumbering elephant seals, pack-living wolves to solitary tigers, members of the order Carnivora display remarkable diversity. Although they are commonly referred to as carnivores, a term also used for any meat-eating animal, some of them eat meat only rarely, if at all. The one thing members of Carnivora have in common is a predatory ancestor with four carnassial teeth—scissor-like molars that can shear through meat. Most carnivores retain the carnassial teeth, which sets them apart from other meat-eating mammals. In mainly insectivorous or herbivorous carnivores, the carnassial teeth have been modified for grinding.

PRIME HUNTERS

As the dominant land predators on all continents except Antarctica, carnivores are built for hunting. They rely on their acute senses of sight, hearing, and smell to help them detect prey. The complex ear region, which often features more than one inner chamber, increases sensitivity to the frequencies that are produced by their prey species.

Intelligence, agility, and speed help carnivores stalk, chase, and catch their prey efficiently. Even apparently cumbersome species such as bears are capable of impressive sprints, while the cheetah is the fastest land animal in the world. All carnivores have fused bones in the forefeet, a feature that helps absorb the shock of running. A reduced collarbone increases the mobility of the shoulder muscles, allowing a longer stride and greater speed.

To kill their prey, carnivores generally use their strong jaws and sharp teeth. Weasels smash the prey's skull by biting into the back of its head. Cats strike small prey at the neck to snap the spinal cord, while dogs dislocate the neck by violently shaking the animal in their jaws. Cooperative hunting allows wolves, lions, and other pack animals to tackle much larger prey.

Almost all the larger meat-eating carnivores hunt vertebrates. Smaller carnivores mostly eat invertebrates, which are easier to catch but can rarely sustain a large animal. Various carnivores concentrate on termites, worms, fish, and crustaceans, and some are largely vegetarian, with a preference for berries, other fruit, nuts, seeds, nectar, or bamboo, but all tend to be opportunistic feeders and will take advantage of any easy meal that presents itself.

About 50 million years ago, the order Carnivora split into two lineages. The cat-like carnivores include civets (family Viverridae), cats (Felidae), hyenas (Hyaenidae), and mongooses (Herpestidae). Dog-like carnivores include dogs (Canidae), bears (Ursidae), raccoons (Procyonidae), weasels (Mustelidae), and seals (Otariidae and Phocidae). Until recent years, seals were often treated as a separate order known as Pinnipedia, but genetic studies indicate that they share a common ancestry with the other carnivores.

INTRODUCING CARNIVORES

The exceptional hunting abilities of carnivores have inspired many misguided attempts to use them for pest control in regions where they are not native. The consequences of such introductions have usually been disastrous. In the Caribbean and Hawaii, the small Indian mongooses (*Herpestes javanicus*) imported to control rodents and snakes ended up spreading rabies instead. On a number of the world's far-flung islands, feral cats, which were supposed to get rid of rats, preferred the easier prey of flightless birds and destroyed their populations.

Pack mentality Spotted hyenas (below) depend on hunting for food and pursue larger prey, such as zebra, wildebeest, gemsbok, and impala. When the prey is substantially larger than themselves, hyenas rely on cooperation to bring it down. Even when the prey is a young, smaller animal and a single hyena could complete the task alone, the kill supplies enough food to feed several hyenas, so pack hunting avoids waste.

Vegetarian carnivore While the giant panda (right) may eat small mammals, fish, and insects if they are available, bamboo makes up more than 99 percent of its diet. A plentiful, year-round food source, bamboo is low in nutrition, so giant pandas must spend 10–12 hours a day feeding in order to satisfy their energy requirements.

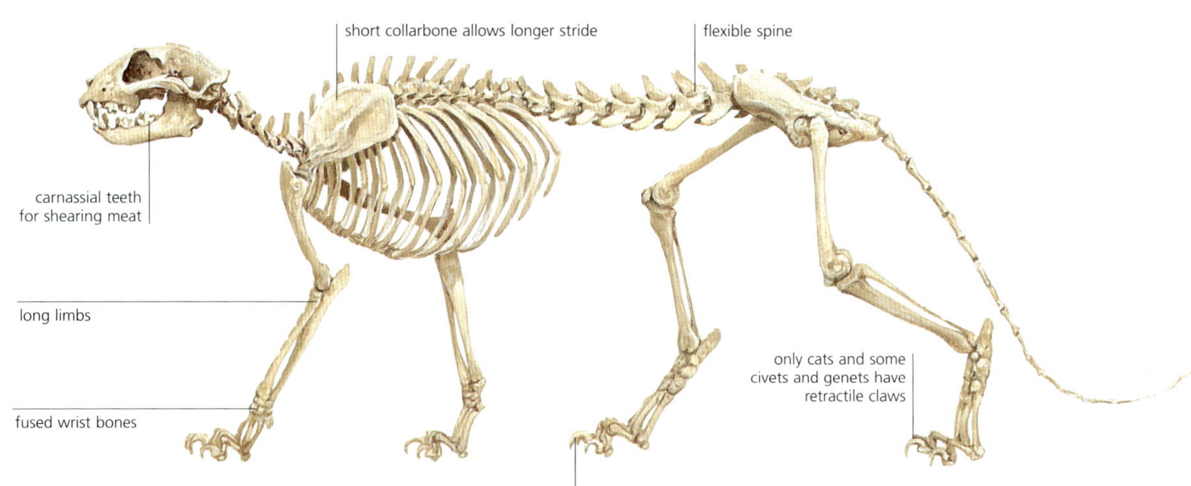

Designed to hunt The skeleton of a cat displays many of the anatomical features that have made carnivores such effective predators. A flexible spine, long limbs, fused wrist bones, and reduced collarbone all contribute to the cat's speed and agility.

THE DOG FAMILY

CLASS	Mammalia
ORDER	Carnivora
FAMILY	Canidae
GENERA	14
SPECIES	34

Perhaps no creatures have such an ambivalent relationship with humans as the dogs, wolves, coyotes, jackals, and foxes that make up the carnivore family Canidae. Domesticated at least 14,000 years ago, dogs were the first animals to enter a partnership with humans and have been used extensively for hunting, guarding, and companionship. At the same time, wild canids have been relentlessly persecuted, blamed for the loss of livestock and the spread of rabies, and hunted for sport and fashion. While some species such as red foxes and coyotes have adapted and thrived in the midst of urban development, others such as red wolves are on the verge of extinction.

A wide distribution Originating in North America 34–55 million years ago, wild canids now occur on every continent except Antarctica, but are absent from some islands. They were introduced to New Guinea and Australia in prehistoric times. The domestic dog is now worldwide.

Raccoon resemblance *The raccoon dog's black face mask, thick body, bushy tail, and ability to climb trees give it a great resemblance to the raccoon.*

Cooperative hunters *Gray wolves usually live in family units of 5–12 dogs. The pack cooperates to snare larger prey such as deer. The howling chorus of a wolf pack can be heard for about 6 miles (10 km) and tells other wolves to stay away.*

GRASSLAND CARNIVORES

Most canids live in open grasslands, where they capture their prey either by sudden pouncing or by extended pursuit. With their slender build, muscular, deep-chested bodies, and long, sturdy legs, they are capable of great endurance. In addition to the fused foot bones common to other carnivores, canid forelimb bones are locked to prevent them rotating as they run. A pointed muzzle houses the large scent organs that allow canids to track prey over long distances, while large, erect ears contribute to their acute hearing.

Canids prefer to eat freshly killed meat but will take advantage of whatever food is locally available and may eat fish, carrion, berries, and human garbage. Their social organization varies between and within species, often reflecting their diet. Smaller species such as jackals and foxes mostly eat small animals and often live alone or in pairs. Larger species such as gray wolves and African wild dogs usually live and hunt in hierarchical social groups, cooperating to bring down prey larger than themselves.

Group living has benefits other than for hunting. Some canids hunt alone but live communally, working together to care for the pack's young and defend their territory.

COYOTE COMMUNICATION

While many coyotes are solitary, others live and hunt in pairs or larger packs. Like other canids, coyotes use scent marking, facial expressions, body and tail posture, and a range of squeaks, yelps, and howls to communicate. Signs of submissiveness include crouching and holding the ears back and down. A snarl to show sharp teeth may be a sign of dominance or aggression, or part of a self-defence tactic.

Side-striped jackal
Canis adustus

Shorter legs and ears than other jackals, with white and black side stripe on drab coat

Color can vary from brown-tipped yellow in rainy season to pale gold in dry season

Golden jackal
Canis aureus

Black fur runs from back of neck to tail

Ethiopian wolf (Simien jackal)
Canis simensis

Long, pointed muzzle for catching small mammals

Black-backed jackal
Canis mesomelas

Dingo
Canis lupus dingo

Golden jackal The most widely distributed jackal, ranging from eastern Europe and northern Africa to Southeast Asia, this species has lived on the edge of human settlements since ancient times, when it featured prominently in Egyptian mythology.

- Up to 39½ in (100 cm)
- Up to 12 in (30 cm)
- Up to 33 lb (15 kg)
- Pair
- Common

N. Africa, S.E. Europe to Thailand, Sri Lanka

Black-backed jackal Near villages, this jackal is nocturnal, but elsewhere it may be active day or night. About half of its diet consists of insects; small mammals and fruit make up the rest. Males and females form lasting pairs and share the care of their pups.

- Up to 35½ in (90 cm)
- Up to 16 in (40 cm)
- Up to 26½ lb (12 kg)
- Pair
- Locally common

E. & S. Africa

Dingo Probably brought to Australia by Asian traders at least 3,500 years ago, the dingo became the dominant predator in many areas, cooperatively hunting large marsupials such as kangaroos and wallabies.

- Up to 39½ in (100 cm)
- Up to 14 in (36 cm)
- Up to 53 lb (24 kg)
- Solitary, small group
- Locally common

Mainland Australia

Arctic fox For winter, this fox grows a white coat to match its snowy habitat. As summer approaches, its coat turns brown or black to blend in with the tundra vegetation.

Winter coat

Up to 27½ in (70 cm)
Up to 16 in (40 cm)
Up to 20 lb (9 kg)
Solitary
Common

N. North America, N. Eurasia, Iceland, Greenland

Cape wild dog This canid is entirely carnivorous. After a kill, the healthy adult dogs will regurgitate food for any young, sick, or wounded pack members. Each Cape wild dog has a unique coat pattern.

Up to 35½ in (90 cm)
Up to 16 in (40 cm)
Up to 79½ lb (36 kg)
Solitary, small group
Endangered

E. & S. Africa

Tibetan fox
Vulpes ferrilata

Broad ears listen for rustling rodents

Corsac fox
Vulpes corsac

Blanford's fox
Vulpes cana

Displays cat-like movements

Kit fox
Vulpes macrotis

Pale fox
Vulpes pallida

Red fox
North American form
Vulpes vulpes fulva

Short reddish summer coat replaced by longer, thicker gray coat for winter

Swift fox
Vulpes velox

Bengal fox
Vulpes bengalensis

Red fox
Central European form
Vulpes vulpes crucigera

THE SUCCESSFUL RED FOX

One of the most widely distributed fox species—with a strong presence in North America, Europe, and Asia—the red fox's versatile feeding habits have allowed it to thrive in forest, prairie, farmland, and suburban areas. Hunted for sport and bred for fur, it has been blamed for killing poultry and spreading rabies. Where it has been introduced, as in Australia, the red fox poses a threat to native fauna.

Fox feast *Red foxes will eat almost anything, from rodents and rabbits to fruit and garbage.*

Kit fox The kit fox avoids the heat of the day by resting in its underground den, coming out at night to hunt. This fox relies on the moisture in its prey for fluids and thus kills more animals than are needed to meet its energy requirements alone.

S.W. USA & N. Mexico

- Up to 20½ in (52 cm)
- Up to 12½ in (32 cm)
- Up to 6 lb (2.7 kg)
- Solitary
- Conserv. dependent

Gray fox
Urocyon cinereoargenteus

Strong, hooked claws used for climbing trees

Black-tipped tail

Mane of erect black hairs

Crab-eating fox (common zorro)
Cerdocyon thous

Maned wolf
Chrysocyon brachyurus

Long legs help maned wolf see above tall grass of pampas

Argentine gray fox
Pseudalopex griseus

Small-eared zorro
Atelocynus microtis

Culpeo fox
Pseudalopex culpaeus

Bush dog
Speothos venaticus

Webbed feet for swimming after aquatic prey

Coat colors most vivid in northern part of range

Pampas fox
Pseudalopex gymnocercus

Maned wolf This omnivorous canid eats large amounts of bananas, guavas, and other fruit as well as animals such as armadillos, rabbits, rodents, snails, and birds. It hunts at night in a fox-like manner, suddenly pouncing on its prey.

- Up to 39½ in (100 cm)
- Up to 16 in (40 cm)
- Up to 53 lb (24 kg)
- Solitary, pair
- Near threatened

Brazil to Paraguay & Argentina

Bush dog With its squat stature and short face, the bush dog looks less like a dog than any other canid. Packs of up to 10 animals forage together in forest undergrowth, keeping in touch via high-pitched peeps and whines.

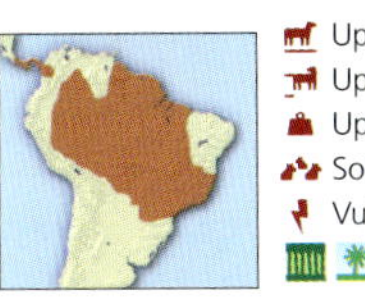

- Up to 29½ in (75 cm)
- Up to 5 in (13 cm)
- Up to 15½ lb (7 kg)
- Solitary to large group
- Vulnerable

W. Panama to Paraguay & N. Argentina

Pampas fox When threatened by humans, the pampas fox freezes and may remain motionless if handled. Pairs live together only during the mating season and share the care of the young.

- Up to 28½ in (72 cm)
- Up to 15 in (38 cm)
- Up to 17½ lb (7.9 kg)
- Solitary, pair
- Locally common

E. Bolivia & S. Brazil to N. Argentina

THE BEAR FAMILY

CLASS	Mammalia
ORDER	Carnivora
FAMILY	Ursidae
GENERA	6
SPECIES	9

In spite of their fearsome reputation, bears tend to be the most herbivorous of the carnivores. Only one species, the polar bear, is primarily a meat-eater, while berries, nuts, and tubers make up the bulk of the American black bear's diet, the sloth bear feeds mostly on insects, and the giant panda eats almost nothing but bamboo. The first bear species emerged from the canid family 20–25 million years ago in Eurasia. About the size of a raccoon, these early creatures had a long tail and the shearing carnassial teeth common to most Carnivora members. Over time, most bears became much bigger, their tails shortened, and their carnassials were flattened for grinding vegetable matter.

Northern diversity Bears are most abundant in the Northern Hemisphere, living in Europe, Asia, and North and South America. The brown bear was also found in North Africa until the 1800s. Today, all but two bear species are in danger, victims of habitat loss and overhunting.

STRENGTH AND BULK

Although the red panda may weigh only about 6 pounds (3 kg), most bears classified in the Ursidae family are substantial animals, with the polar bear and brown bear vying for the title of the largest terrestrial carnivore. As they often spend more of their time foraging than hunting, bears are built for strength rather than speed, with a stocky, muscular body, thick legs, and a massive skull. An enlarged snout reflects their keen sense of smell. Vision and sound are less important, and their eyes and ears are relatively small.

Bears are found in the tropics, but it is in the cold northern lands that they are most numerous. Here, their great size allows them to put on fat deposits through spring and summer when food is abundant. When the weather cools, the bears retreat to a den or cave and enter a long sleep that can last half the year. During this dormant period, they live entirely on their body fat; they do not eat, urinate, or defecate, and their heart and respiration rates drop. Unlike in true hibernation, however, their body temperature hardly falls. Remarkably, females give birth to young during the winter dormancy, maximizing the cubs' chance to grow and accumulate fat deposits before the next winter.

Most bears are solitary, although cubs often stay with the mother for 2–3 years. Rival males can be aggressive in the breeding season, with fights resulting in injury or sometimes even death.

Fish feeder In the northwest coastal regions of North America, brown bears will wait at waterfalls to catch spawning salmon as they swim upstream. This annual glut of fish provides important protein before winter begins.

Himalayan brown bear
Ursus arctos isabellinus

Brown bear's coat can be brown, blonde, silver-tipped, or nearly black

Pronounced shoulder hump

American black bear
Ursus americanus

Black bear's coat can be black or brown

European brown bear
Ursus arctos arctos

Polar bear
Ursus maritimus

Massive bulk helps polar bear withstand cold and store fat for times of scarcity

Largest species of bear

Large, paddle-like paws for swimming

Largest of all brown bear subspecies

Males twice as large as females

Kodiak bear
Ursus arctos middendorffi

Asiatic black bear
Ursus thibetanus

Brown bear This species once lived throughout Eurasia and North America and south to North Africa and Mexico. Its subspecies include the European brown bear (*Ursus arctos arctos*), the North American brown bear or grizzly (*U. a. horribilis*), the Kodiak bear (*U. a. middendorffi*), and the Himalayan brown bear (*U. a. isabellinus*).

Pockets in N.W. North America, Wyoming, W. & N. Europe, Himalayas, Japan

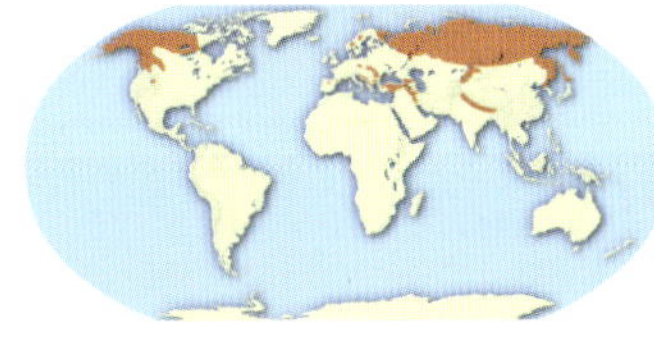

- Up to 9 ft (2.8 m)
- Up to 8½ in (21 cm)
- Up to 1,320 lb (600 kg)
- Solitary
- Locally common

Asiatic black bear Primarily herbivorous, this species readily climbs trees to gather fruit and nuts. To startle predators such as tigers, it will stand up and display its white chest mark.

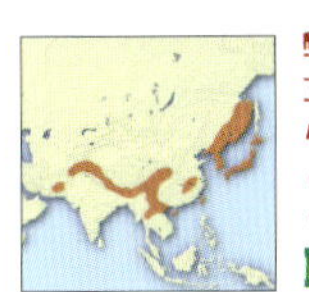

- Up to 6 ft (1.9 m)
- Up to 4 in (10 cm)
- Up to 375 lb (170 kg)
- Solitary
- Vulnerable

Afghanistan & Pakistan to China, Korea & Japan

Distinctive black and white markings make the giant panda one of the most widely recognized of all animal species

Spectacled bear
Tremarctos ornatus

Only bear species in South America

Giant panda
Ailuropoda melanoleuca

Giant panda's front paws have "pseudo-thumb," an extra opposable digit for grasping bamboo

Sloth bear
Melursus ursinus

Mobile snout with long tongue for capturing termites and ants

Shaggy coat may insulate from heat in tropical environment

Long tongue licks up larvae, insects, and honey

Red panda
Ailurus fulgens

Smallest bear apart from red panda

Sun bear
Helarctos malayanus

Long, curved claws help this highly arboreal bear to climb trees

Coat made up of long, coarse hairs and dense undercoat to insulate against the cold weather of its high-altitude habitat

Only bear with a long tail

BAMBOO BEAR

As China's human population has grown, most of the giant panda's habitat has been destroyed. Only about 1,000 pandas remain in the wild, in isolated mountain pockets of bamboo forest. Recognized as a symbol of conservation, the giant panda is still poached for its skin.

Red panda This species was once placed in the raccoon family due to a superficial resemblance, but genetic studies place it with the giant panda. It sleeps in trees by day and eats bamboo and fruit on the ground at night.

Up to 25½ in (65 cm)
Up to 19 in (48 cm)
Up to 6 lb (3 kg)
Pair
Endangered

Nepal to Myanmar & W. China

Sloth bear After ripping open termite mounds and ant nests with its long, curved claws, this bear uses its long tongue to suck up the insects. It also climbs trees to extract honey from hives.

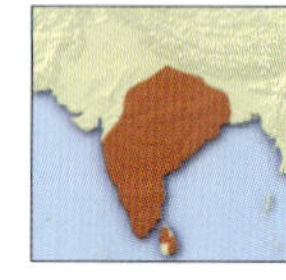

Up to 6 ft (1.8 m)
Up to 5 in (12 cm)
Up to 320 lb (145 kg)
Solitary
Vulnerable

Sri Lanka, India, Nepal

A YEAR AS A POLAR BEAR

Huge and powerful, polar bears are the top predators in the Arctic and the world's largest land-living carnivores. Adapted to the harsh climate of the Arctic, the polar bear lives near the ice-covered water that contains its main prey, the ringed seal. Unlike other bears in cold environments, this semi-aquatic carnivore tends to remain active through winter, but can enter a dormant state and live off its fat deposits at any time of the year if food becomes scarce. Its solitary existence is interrupted only during the breeding season or when adult males fast together in groups known as sloths.

Cycle of life In the fall, a pregnant polar bear digs her maternity den. Inside, she gives birth to a litter of cubs. They all remain in the den until the following spring. Male polar bears sometimes kill cubs, presumably to free up the female for breeding.

April–July feeding *Polar bears spend summer preying on the abundant and unwary pups of ringed seals. When the sea ice melts in late July, the bears come ashore and fast until it freezes once more.*

April–May mating *Female polar bears spend so long raising their young that they are available for mating only once every 3 years. This leads to intense competition among males for mates.*

February–April emergence *When the cubs are large enough to venture onto the sea ice, the mother leads them out of the den. Cubs stay with their mother for about 2½ years, learning the hunting skills that are crucial to their survival.*

November–January birth *While other polar bears remain active through winter, pregnant females retreat to a snow den, where they give birth to their young. Most litters contain two cubs, who are nursed until around late March.*

MUSTELIDS

CLASS	Mammalia
ORDER	Carnivora
FAMILY	Mustelidae
GENERA	25
SPECIES	65

As a group, the weasels, otters, skunks, and badgers of the family Mustelidae are the most successful and diverse carnivores, with many more species than any other family. Mustelids are found in almost every type of habitat, including forests, deserts, tundra, and fresh and salt water, and may be arboreal, terrestrial, burrowing, semiaquatic, or fully aquatic. While a few species, such as sea otters and wolverines, can weigh more than 55 pounds (25 kg), most members of the family are medium-sized, with the smallest, the least weasel, weighing as little as 1 ounce (30 g). Highly carnivorous, mustelids are voracious hunters and will often tackle and kill prey much larger than themselves.

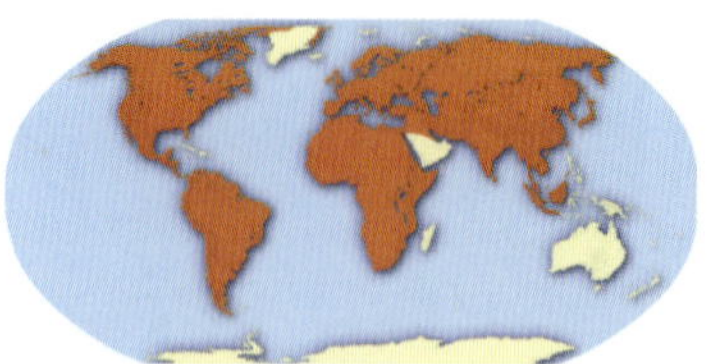

Widespread family Absent only from Australia and Antarctica, mustelids are found throughout Europe, Asia, Africa, and the Americas. Mustelids have been introduced in many places, either accidentally when they have escaped from fur farms or deliberately to control rodents and rabbits.

Badger There are nine species of badger globally, but only the Eurasian badger (*Meles meles*) occurs in the wild in Europe. The badger prefers to live in woodlands and grassy fields. While it resides in complex burrows, called setts, often in areas that experience cold winters and deep snowfalls, it does not hibernate. Animals enter into a state of torpor that can last several weeks.

RELENTLESS HUNTERS

With an elongated body and short legs, mustelids can pursue rodents and rabbits down burrows. Weasels tend to be slender and lithe, with a flexible spine that allows them to scamper and leap. Badgers, on the other hand, have squat bodies and shuffle along with a rolling gait. Many mustelids are good swimmers and climbers and readily exploit aquatic and tree-dwelling prey as well as terrestrial species.

The mustelid head has a low, flat skull and short face with small ears and eyes. Smell is usually the most important of the senses and is used to locate and track prey and for communication, with territories marked by scent. Most mustelids have long, curved, non-retractile claws that they use for digging. Aquatic and semiaquatic species often have webbing between the toes to help with swimming.

Mustelids have a double coat, with a layer of soft, dense underfur interspersed with longer guard hairs. This warm, water-repellent coat enables mustelids to hunt in water and stay active throughout cold winters, but it has also made many species the target of the fur trade.

Glands at base of tail produce musk scent

Enters and leaves the water at fixed locations within its home territory

European otter
Lutra lutra

Smooth-coated otter
Lutra perspicillata

Neotropical otter
Lontra longicaudis

Fingers lack webbing and claws but have opposable thumb and are highly dextrous and sensitive

Cape clawless otter
Aonyx capensis

Sensitive whiskers help locate prey

Giant otter
Pteronura brasiliensis

Spotted-necked otter
Lutra maculicollis

Sea otter
Enhydra lutris

Broad, flipper-like back paws with webbing to the tips of the toes

Uses stone as tool to crack open sea urchin

Insulating layer of air trapped between long guard hairs and dense underfur

Neotropical otter With a range that extends from Mexico through most of northern South America, much of this solitary otter's day is spent diving for fish. It eats small prey in the water but takes larger prey to the shore.

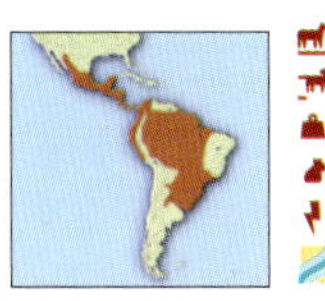

Up to 32 in (81 cm)
Up to 22½ in (57 cm)
Up to 33 lb (15 kg)
Solitary
Data deficient

Mexico to Uruguay

Giant otter This is the world's longest river otter, and overhunting has made this species the rarest otter. Family groups of several giant otters share a streamside burrow and may forage together.

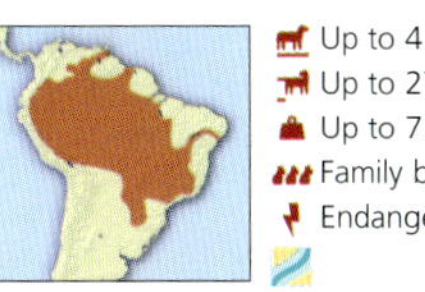

Up to 4 ft (1.2 m)
Up to 27½ in (70 cm)
Up to 75 lb (34 kg)
Family band
Endangered

S. Venezuela & Colombia to N. Argentina

Sea otter This often solitary species can spend its entire life in the ocean, but it may rest on shore in large sexually segregated groups. The only non-primate mammal known to employ tools, it will use a stone to dislodge abalone or crack open shellfish.

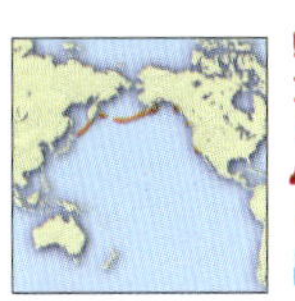

Up to 4 ft (1.2 m)
Up to 14 in (36 cm)
Up to 99 lb (45 kg)
Solitary, rests in group
Endangered

North Pacific

Sunda stink badger
Mydaus javanensis

May squirt foul secretion from anal glands when threatened

Hog badger
Arctonyx collaris

Elongated snout with pig-like nostrils

Eurasian badger
Meles meles

Powerful forelimbs and claws for digging

Chinese ferret badger
Melogale moschata

Long, bushy tail

American badger
Taxidea taxus

Burmese ferret badger
Melogale personata

CLIMBING BADGER

The smallest badger, the Chinese ferret badger forages at night for worms, insects, frogs, small rodents, and fruit. By day, it takes refuge in a burrow or rock crevice, or may use its long claws to climb a tree and then nest in the branches.

Hog badger Although the hog badger sometimes falls prey to leopards and tigers, it will put up a fight. When threatened, it arches its back, bristles its hair, and growls. It may also emit a noxious fluid from its anal glands.

- Up to 27½ in (70 cm)
- Up to 6½ in (17 cm)
- Up to 31 lb (14 kg)
- Not known
- Not known

N.E. India to N.E. China & S.E. Asia

American badger This solitary species spends much of its time busily digging for rodents such as prairie dogs and ground squirrels. It will rapidly burrow its way out of danger if threatened by a predator on the surface.

- Up to 28½ in (72 cm)
- Up to 6 in (15 cm)
- Up to 26½ lb (12 kg)
- Solitary
- Locally common

N. Canada to Mexico

Hooded skunk
Mephitis macroura

Longer, softer fur than striped skunk

Striped skunk
Mephitis mephitis

Eastern hog-nosed skunk
Conepatus leuconotus

Long claws on forefeet for digging

Spotted skunk
Spilogale putorius

Striped hog-nosed skunk
Conepatus semistriatus

No white stripe down center of face

Andean hog-nosed skunk
Conepatus chinga

Bare, projecting nose

Patagonian hog-nosed skunk
Conepatus humboldtii

Striped skunk This nocturnal creature is an opportunistic feeder and tends to consume a wide range of foods, from small mammals, insects, and fish to fruit, nuts, grains, and grasses. During winter, it becomes inactive and rarely emerges from its den.

Up to 31½ in (80 cm)
Up to 15½ in (39 cm)
Up to 14 lb (6.5 kg)
Solitary
Common

N. Canada to Mexico

Spotted skunk The only skunk that can climb, this species also tends to be more alert and active than other skunks. During the mating season in March and April, males may succumb to "mating madness," spraying any large animal they come across.

Up to 13 in (33 cm)
Up to 11 in (28 cm)
Up to 2 lb (900 g)
Solitary
Common

USA (E. of Rocky Mts)

Andean hog-nosed skunk This species dens in rocky crevices, hollow logs, or burrows abandoned by other animals. It preys on insects and small vertebrates such as rodents, lizards, and snakes. It has some resistance to the venom of pit vipers.

Up to 13 in (33 cm)
Up to 8 in (20 cm)
Up to 6½ lb (3 kg)
Solitary
Common

South America

American marten
Martes americana

Relatively large eyes and cat-like ears

Yellow-throated marten
Martes flavigula

Fisher
Martes pennanti

Long tail aids balance when climbing trees

Sable
Martes zibellina

Color varies from yellowish brown to dark brown

Japanese marten
Martes melampus

Pine marten
Martes martes

Soles of feet covered in fur in winter

COVETED FUR

A creature of the dense taiga forest of northern Asia, the sable hunts and dens on the forest floor. A careful, secretive predator with acute senses of smell and hearing, it feeds on birds, small mammals, and fish. This small carnivore was once found as far west as Scandinavia, but so many animals have been trapped for the fur trade that its range and numbers have been substantially reduced.

Valuable fur *The sable's long, thick, silky winter coat has made its pelt one of the most prized by the fur trade.*

Fisher While other predators are put off by a porcupine's quills, the fisher—the largest marten—is the perfect height to bite the unprotected face. Once repeated bites send the porcupine into shock, the fisher will flip it over and feed on its soft belly.

Up to 31 in (79 cm)
Up to 16 in (41 cm)
Up to 12 lb (5.5 kg)
Solitary
Uncommon

Alaska & Canada to N. California

Semiretractile claws used for climbing

Beech marten
Martes foina

Great variation in size, from 1–9 oz (35–250 g), depending on location

Coat becomes white in winter in northern part of range

Long-tailed weasel
Mustela frenata

Stoat (ermine) winter coat
Mustela erminea

Least weasel
Mustela nivalis

White winter coat helps stoat blend in with snow

Head paler than body

Malaysian weasel
Mustela nudipes

Stoat (ermine) summer coat
Mustela erminea

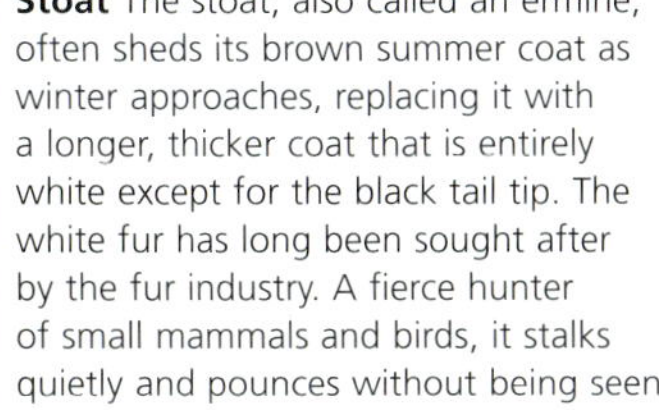

Stoat The stoat, also called an ermine, often sheds its brown summer coat as winter approaches, replacing it with a longer, thicker coat that is entirely white except for the black tail tip. The white fur has long been sought after by the fur industry. A fierce hunter of small mammals and birds, it stalks quietly and pounces without being seen.

Northern Hemisphere; introd. New Zealand

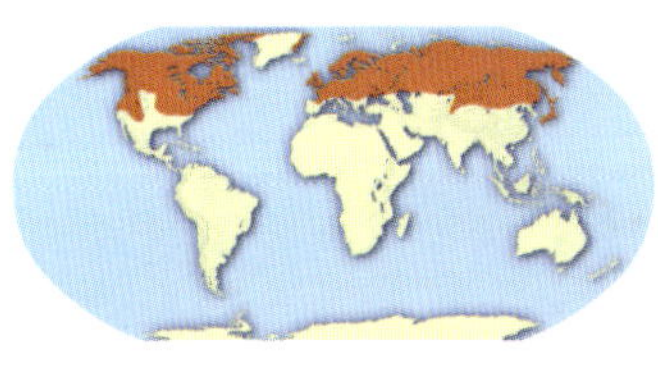

- Up to 12½ in (32 cm)
- Up to 5 in (13 cm)
- Up to 13 oz (365 g)
- Solitary
- Common

RELENTLESS CARNIVORES

Weasels will pursue prey underground or under snow and can carry up to half their own weight in meat as they run. While smaller weasels prefer mice and voles, and larger weasels favor rabbits, they all avail themselves of whatever animals they come across.

Siberian weasel
Mustela sibirica

American mink
Mustela vison

Partially webbed toes for swimming

Black face mask

†
Black-footed ferret
Mustela nigripes

Polecat
Mustela putorius

Males can weigh twice as much as females

Steppe polecat
Mustela eversmannii

Always has white patch on upper lip

European mink
Mustela lutreola

BATTLE OF THE MINKS

Both the European and American minks are versatile feeders that hunt in or near water. The European mink has been in decline ever since the American mink escaped from European fur farms and became a direct competitor in the wild.

Mink variation
While most American minks are brown, about 10 percent have blue-gray fur.

Black-footed ferret While most mustelids are opportunistic feeders, the black-footed ferret preys almost solely on prairie dogs and uses their burrows for shelter.

- Up to 18 in (46 cm)
- Up to 5½ in (14 cm)
- Up to 2½ lb (1.1 kg)
- Solitary
- Extinct in the wild

S. Canada to N.W. Texas (until late 1980s); reintrod. Montana, Dakota, & Wyoming

● Former range

Foul-smelling liquid secreted from anal glands

Patagonian weasel
Lyncodon patagonicus

Tayra
Eira barbara

Ratel (honey badger)
Mellivora capensis

Large hindfeet with long claws

Grison
Galictis vittata

Stands on hind legs to search for prey

Saharan striped weasel
Ictonyx libyca

Striped polecat (zorilla)
Ictonyx striatus

African weasel
Poecilogale albinucha

Marbled polecat
Vormela peregusna

Ratel This mainly terrestrial mustelid will climb trees to reach honey. The ratel's omnivorous diet also includes insects and both large and small vertebrates.

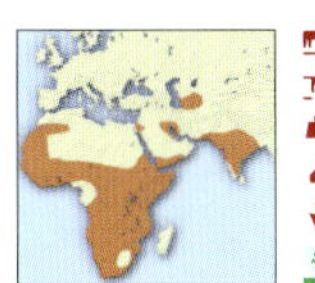

Up to 30½ in (77 cm)
Up to 12 in (30 cm)
Up to 28½ lb (13 kg)
Solitary
Uncommon

W. Africa, sub-Saharan Africa, Arabia, Iraq, Turkmenistan, Pakistan, India

PLAYING DEAD

When threatened, a striped polecat will fluff up its long tail and growl or scream. If that does not work, it then squirts the attacker with foul-smelling fluid from its anal glands. As a final resort, the polecat will pretend to be dead. Despite these various defense strategies, striped polecats do fall victim to predators such as domestic dogs and wild cats. More commonly, however, they are killed by cars.

Repulsive meal
Although faking death makes the striped polecat easier to attack, the unpleasant anal gland secretions on its fur may convince a predator to abandon the meal.

Seals and sea lions

CLASS	Mammalia
ORDER	Carnivora
FAMILIES	3
GENERA	21
SPECIES	36

With flexible, torpedo-shaped bodies, limbs modified to become flippers, and insulating layers of blubber and hair, seals, sea lions, and walruses are superbly adapted to a life in water. They have not, however, completely severed their link with land and must return to shore to breed. Collectively known as pinnipeds, these marine mammals were once placed in their own order, but are now considered to be part of Carnivora. Most feed on fish, squid, and crustaceans, but some also eat penguins and carrion and may attack the pups of other seal species. They can dive to great depths in search of prey, with the elephant seal able to stay submerged for up to 2 hours.

Cold-water creatures Although monk seals are found in warmer waters, most seals, sea lions, and walruses are restricted to the colder, highly productive seas of the world's polar and temperate regions. The fossil record shows that the three families all originated in the North Pacific.

THREE GROUPS

There are three pinniped families. The Phocidae are known as the true seals. They swim mainly with strokes of their hind flippers, which cannot bend forward to act as feet, making their movement on land particularly ungainly. Although their hearing, especially under water, is good, true seals lack external ears.

Sea lions and fur seals belong to the family Otariidae. These "eared seals" have small external ears. They rely mostly on their front flippers for swimming, and can bend their hind flippers forward when on land, allowing them to walk "four-footed" and sit in a semi-upright position.

The third family, Odobenidae, contains a single species, the walrus, instantly recognizable by the long canine teeth that form tusks on both sexes. Like true seals, walruses use their hind flippers for swimming and lack external ears. Like eared seals, however, walruses can bend their hind flippers forward.

Sea lion life Australian sea lions often hunt squid and fish together. They can swim a month or so after being born. When they dive, their heartbeat slows from about 100 beats a minute to as low as 10 beats a minute.

New Zealand fur seal
Arctocephalus forsteri

Length of male up to 7¼ ft (2.2 m), female up to 5½ ft (1.7 m)

Pronounced mane on male

Males can be up to three times larger than females

South American sea lion
Otaria byronia

South African fur seal
Arctocephalus pusillus

Length of male up to 7 ft (2.1 m), female up to 5 ft (1.5 m)

Northern fur seal
Callorhinus ursinus

Length of male up to 8 ft (2.5 m), female up to 6 ft (1.8 m)

Male has massive, maned neck

Californian sea lion
Zalophus californianus

Largest of the eared seals

Short stubble on black flippers

Steller's sea lion
Eumetopias jubatus

New Zealand fur seal In late spring, male fur seals establish territories on rocky shorelines, where they are joined by females for breeding. After the pups are born, the females visit the ocean to forage but the males stay put until the end of the breeding season. The range for these seals, once widely hunted, now extends to the waters around New Zealand and along the southern coast of Australia.

Male up to 795 lb (360 kg), female up to 245 lb (110 kg)
Harem
Common

S.W. Australia to New Zealand

TERRITORIAL MALES

Eared seals tend to be highly social and will gather in large numbers, in colonies of up to thousands of members, during the breeding season. Males defend their patch of shore and harem of females against other males, using aggressive postures and loud barks before resorting to fighting.

Bearded seal
Erignathus barbatus

Long, sensitive whiskers used to locate clams, snails, crabs, and shrimp

Baikal seal
Phoca sibirica

Common seal (harbor seal)
Phoca vitulina

Ribbon seal
Phoca fasciata

Length of male up to 6¼ ft (1.9 m), female up to 5½ ft (1.7 m)

Harp seal
Phoca groenlandica

White coats of pups shed by 3 weeks of age

Common name inspired by harp-shaped marking on back and sides of adult

Tusks on both males and females

Walrus
Odobenus rosmarus

Ringed seal
Phoca hispida

Spots surrounded by ring of lighter fur

SNOW LAIRS

The ringed seal lives in waters that are covered in ice for part of the year. A pregnant female digs a cave into the snow above her breathing hole, which provides her pup with protection from extreme Arctic weather and predators such as polar bears.

Walrus Instantly recognizable by the long canine teeth that form tusks on both sexes, walruses rely on a mustache of sensitive whiskers to locate their main food of clams and mussels, which they dig out of the sand with their snout. Most pinnipeds are gregarious animals and tend to live in large colonies. Walrus herds can number in the thousands and may be single sex or mixed. The largest animal with the longest tusks usually dominates the herd.

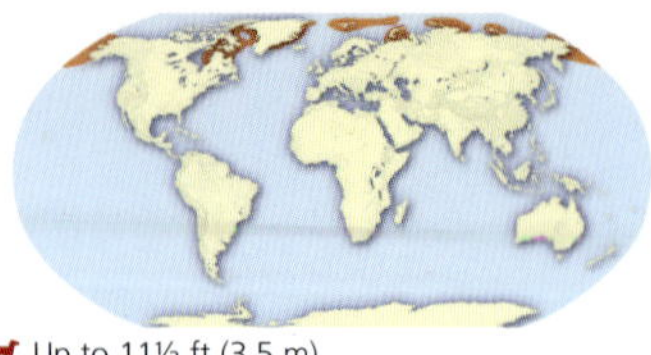

Up to 11½ ft (3.5 m)
Up to 3,640 lb (1,650 kg)
Large herd
Locally common
Shallow Arctic seas

Hooded seal
Cystophora cristata

Lining of left nostril inflated as part of mating display

Crabeater seal
Lobodon carcinophagus

Dark gray or brown back after January molt, becomes almost entirely blonde later in year

Gray seal
Halichoerus grypus

Swims with strokes of large flippers, while most other pinnipeds use tail

Weddell seal
Leptonychotes weddellii

Leopard seal
Hydrurga leptonyx

Mediterranean monk seal
Monachus monachus

Can be colored brown, gray, or black

Southern elephant seal
Mirounga leonina

Largest of all pinnipeds: length of male up to 20 ft (6 m), female up to 10 ft (3 m)

Mediterranean monk seal This species was once common throughout the warm Mediterranean coastal waters, but an ever-growing human presence has seen its numbers dwindle. Today, its main refuges are small, barren islands.

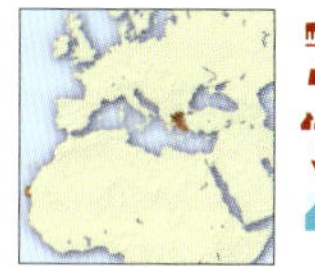

Up to 9 ft (2.8 m)
Up to 660 lb (300 kg)
Harem
Critically endangered

Coastal W. Africa, Aegean Sea

INFLATED NOSE

When it is trying to attract a mate, or is threatened or excited, mature male hooded seals have the unusual ability to blow up their black hood, which is an extension of the nasal cavity. They can also make the lining of the left nostril inflate into a red bladder. Found in the chilly waters of the Arctic and Northern Atlantic oceans, these seals have a dark face and a greyish coat with dark patches.

Blowing up *As part of its mating display, the male hooded seal may inflate either its red nostril bladder or its entire black hood.*

THE RACCOON FAMILY

CLASS	Mammalia
ORDER	Carnivora
FAMILY	Procyonidae
GENERA	6
SPECIES	19

Restricted to the New World, the family Procyonidae includes raccoons, coatis, kinkajous, ringtails, and olingos. All are medium-sized with a long body and tail, broad faces, and erect ears. Apart from the kinkajou, they also share mask-like markings on the face and alternating light and dark rings on the tail. Omnivorous feeding habits have enabled members of this family to thrive in habitats as varied as coniferous forest, rain forest, wetlands, desert, and urban areas. Procyonids tend to be highly vocal, barking and squeaking to maintain complex social structures. Raccoons often sleep in communal dens, and females form "consortships" with one to four males. Male coatis are solitary, but bands of about 15 females groom one another and share the care of young.

Opportunistic omnivores Although their preferred habitat is woodland near water, raccoons have adapted to live alongside humans in rural and urban environments. They often visit North American backyards to raid trash cans for food scraps, and may den in old buildings, cellars, or attics.

Ringtail
Bassariscus astutus

White-nosed coati
Nasua narica

Raccoon
Procyon lotor

HYENAS AND AARDWOLF

CLASS	Mammalia
ORDER	Carnivora
FAMILY	Hyaenidae
GENERA	3
SPECIES	4

The four species in the family Hyaenidae—the aardwolf and the brown, striped, and spotted hyenas—look rather like dogs but are classified as cat-like carnivores because they are more closely related to the cats and civets. Their spine distinctively slopes downward to the tail. A relatively massive head with a broad muzzle houses powerful, bone-crunching jaws and teeth. Unlike most other mammals, hyenas can digest skin and bone. They often scavenge the kills of lions and other predators, but sometimes capture their own prey. Spotted hyenas will cooperate to bring down large prey such as wildebeest. The aardwolf is primarily an insect-eater, using its long, sticky tongue and peg-like teeth to harvest up to 200,000 termites per night.

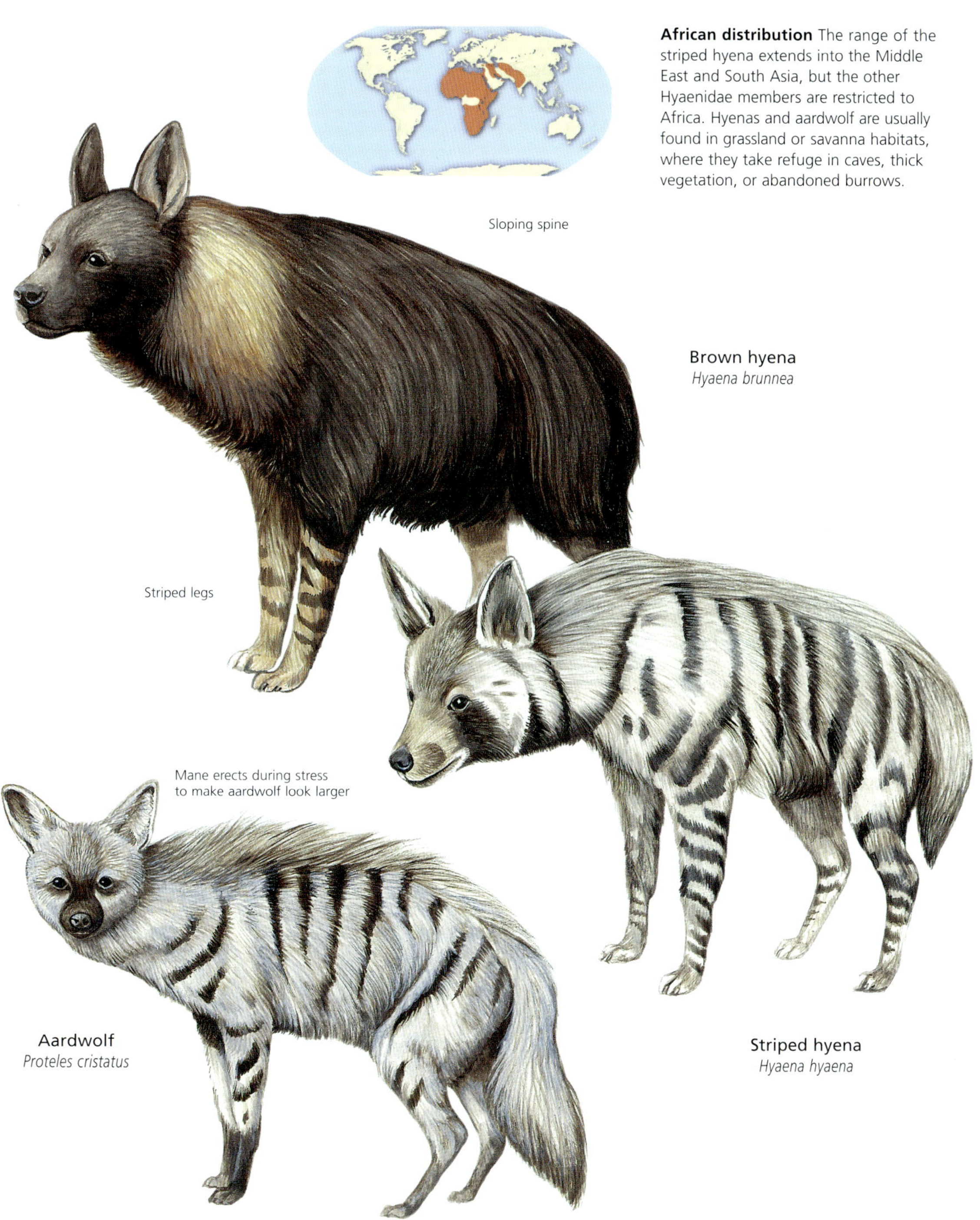

African distribution The range of the striped hyena extends into the Middle East and South Asia, but the other Hyaenidae members are restricted to Africa. Hyenas and aardwolf are usually found in grassland or savanna habitats, where they take refuge in caves, thick vegetation, or abandoned burrows.

Brown hyena
Hyaena brunnea

Striped hyena
Hyaena hyaena

Aardwolf
Proteles cristatus

CIVETS AND MONGOOSES

CLASS	Mammalia
ORDER	Carnivora
FAMILIES	2
GENERA	38
SPECIES	75

The family Viverridae includes civets, genets, and linsangs. It also once included mongooses, but these have now been placed in their own family Herpestidae. Related to cats and hyenas, viverrids and herpestids tend to be medium-sized with a long neck and head, a long, slender body, and short legs. Their skeletal structure and teeth closely resemble those of the earliest carnivores, but the inner-ear region is highly developed. Viverrids are generally nocturnal, arboreal forest-dwellers with long tails, retractile claws, and erect, pointed ears. Many have scent glands near the genitals that in some species produce civet oil, once widely used as a perfume base. Mongooses tend to be found in more open country and are usually terrestrial and often diurnal.

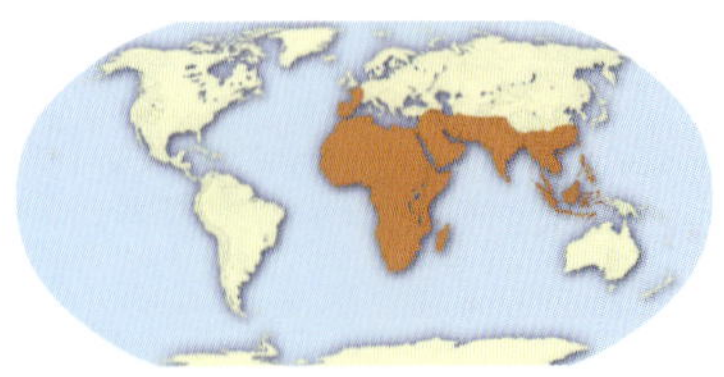

Old World families The civets, genets, and linsangs of Viverridae and the mongooses of Herpestidae are native to much of the Old World. Famed as agile ratters, some mongooses have also been introduced on many islands in the New World, often with disastrous consequences to native fauna.

Social life While all viverrids and many herpestids usually live alone or in pairs, some mongoose species are gregarious and live in colonies. Suricates (meerkats) live in troops of up to 30 animals, which cooperate to care for the young and watch for danger. Nonbreeding members will babysit, and sentinel duties are rotated.

Banded linsang
Prionodon linsang

Coat pattern provides camouflage in dappled light of forest

Fanaloka
(Malagasy civet)
Fossa fossana

Fat reserves stored in tail for winter

Hose's palm civet
Diplogale hosei

Falanouc
Eupleres goudotii

Owston's palm civet
Chrotogale owstoni

Differs from Owston's palm civet by lacking spots on the body

Banded palm civet
Hemigalus derbyanus

Black stripe along mane

Banded collar

Malayan civet
Viverra tangalunga

Banded linsang Sometimes called the tiger civet, this secretive forest animal sleeps in a vegetation-lined nest under a tree root or log. It mainly eats insects and small vertebrates such as squirrels, birds, and lizards.

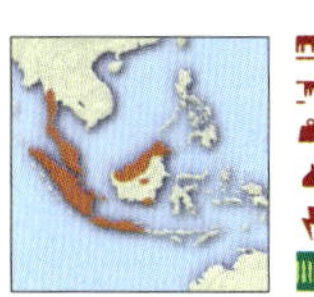

Up to 17½ in (45 cm)
Up to 16 in (40 cm)
Up to 28 oz (800 g)
Solitary
Uncommon

Thailand, Malaya, Sumatra, Java, Borneo

Fanaloka The young of the fanaloka are born highly developed, with a full coat of fur and open eyes. They walk within a few days, eat meat after a month, and are weaned by 10 weeks.

Up to 17½ in (45 cm)
Up to 8½ in (21 cm)
Up to 4½ lb (2 kg)
Pair
Vulnerable

N. & E. Madagascar

Owston's palm civet This little-studied civet appears to be largely terrestrial, with earthworms featuring in its diet. Its striking coat may serve to warn predators of the foul-smelling scent produced by its anal glands.

Up to 28½ in (72 cm)
Up to 18½ in (47 cm)
Up to 9 lb (4 kg)
Solitary
Vulnerable

N. Vietnam, N. Laos, S. China

Only viverrid with prehensile tail

Jerdon's palm civet
Paradoxurus jerdoni

Binturong
Arctictis binturong

Common palm civet (toddy cat)
Paradoxurus hermaphroditus

Brown palm civet
Macrogalidia musschenbroekii

African palm civet
Nandinia binotata

Masked palm civet
Paguma larvata

Three-striped palm civet
Arctogalidia trivirgata

Squirrels form part of civet's omnivorous diet, which also includes frogs, birds, insects, and fruit

MASKED MAMMAL

The masked palm civet of China and Southeast Asia plays a key role in its ecosystem. Through an omnivorous diet, it keeps insect and small vertebrate populations under control and helps disperse fruit seeds. In turn, it is a prey species of tigers, hawks, and leopards. To discourage predation, this civet relies on the powerful odor produced by its anal glands. Its distinctive facial markings may serve to warn predators.

THE FOSSA

The dominant predator on the island of Madagascar is the fossa (*Crytoprocta ferox*). This agile civet pursues lemurs through the trees, but will also hunt snakes, tenrecs, and guinea fowl on the ground.

Balancing act
The fossa's tail is about as long as its body and helps it to balance as it rushes through trees in pursuit of prey.

Narrow-striped mongoose
Mungotictis decemlineata

If threatened, bristles fur and arches back to increase apparent size

Ring-tailed mongoose
Galidia elegans

Large gray mongoose
Herpestes ichneumon

White-tailed mongoose
Ichneumia albicauda

Suricate (meerkat)
Suricata suricatta

Stands upright to survey terrain for danger

Indian gray mongoose
Herpestes edwardsii

Yellow mongoose
Cynictis penicillata

Yellow coat in south of range, grayish in north

Banded mongoose
Mungos mungo

Ring-tailed mongoose Usually living in small groups of a pair and their young, this dainty mongoose spends the night in burrows. By day, it is equally at ease on the ground or in trees as it searches for lemurs, other small vertebrates, insects, and fruit.

Madagascar

- Up to 15 in (38 cm)
- Up to 12 in (30 cm)
- Up to 32 oz (900 g)
- Solitary, pair
- Vulnerable

Suricate A sentinel suricate, or meerkat, will give an alarm bark if it senses danger. The troop may attempt to outrun terrestrial predators such as jackals, but eagles, hawks, and other birds of prey send them fleeing to their burrow system.

S. Africa

- Up to 12¼ in (31 cm)
- Up to 9½ in (24 cm)
- Up to 33½ oz (950 g)
- Family band
- Locally common

Yellow mongoose This gregarious diurnal species lives in complex social groups of 8–20 animals based on a breeding pair, their young, and others. It often shares a burrow system with suricates and ground squirrels.

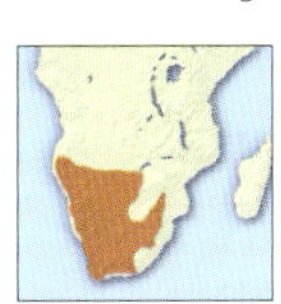
S. Africa

- Up to 14 in (35 cm)
- Up to 10 in (25 cm)
- Up to 32 oz (900 g)
- Family band
- Locally common

CATS

CLASS	Mammalia
ORDER	Carnivora
FAMILY	Felidae
GENERA	18
SPECIES	36

The ultimate hunters, cats eat very little other than meat, making them the most carnivorous of the carnivores. Their predatory expertise has placed them at the top of many food chains on all continents except Australia and Antarctica and in habitats ranging from deserts to Arctic regions. Within the family Felidae, there are considerable size differences but few variations in general form. All species have a strong, muscular body; a blunt face with large, forward-facing eyes; sharp teeth and claws; and acute senses and quick reflexes. They generally rely on stealth to capture prey, which they stalk or ambush. Largely terrestrial, cats are also agile climbers and often good swimmers.

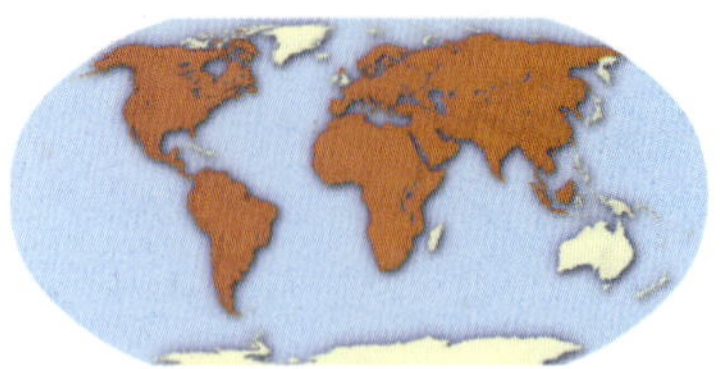

Worldwide spread Wild felids are the dominant predators on most continents, and are absent only from Australasia, Madagascar, Greenland, and Antarctica. First domesticated in Egypt thousands of years ago, the domestic cat has since spread everywhere except Antarctica.

JACKKNIFE CLAWS

Apart from the cheetah, all cats have retractile claws, which remain sharp because they are unsheathed only for capturing prey or climbing trees.

Surprise attack Small cats such as bobcats, wildcats, and lynx tend to hunt smaller mammals such as rodents, lizards, and birds. They creep up, then suddenly pounce with a jack-in-the-box movement, killing the prey with a bite to the neck.

PRIME PREDATORS

Cats are split into three subfamilies: the big cats of Pantherinae include tigers, lions, leopards, and jaguars; the small cats of Felinae include pumas (which can be bigger than some "big cats"), lynx, bobcats, and ocelots; and cheetahs are on their own in Acinonychinae. The chief difference between the big and small cats lies in the flexibility of the larynx, which allows big cats to roar. Cheetahs are distinguished by their non-retractile claws and by their exceptional speed, allowing them to outrun fast prey such as gazelles over short distances.

The sense of smell is used for communication among cats, which scentmark their territory. When hunting, felids depend more on vision and sound. Forward-facing eyes provide binocular vision that helps judge distances. A light-reflecting disk in the eye and rapidly adjusting irises enhance their night vision, which is up to six times sharper than that of humans. Large, mobile ears funnel sound to the sensitive inner ear, which can pick up the faint, high-pitched sounds of small prey such as mice.

Cats were domesticated in the Middle East 7,000 years or so ago, a development that saw them spread to almost every part of the globe as companion animals.

Lion
Panthera leo

Male is 30–50 percent heavier than female

Some males have dark mane, but most have golden mane

Lionesses do most of the hunting, but the male will usually eat first

"King" cheetah

In some cheetahs, a recessive gene produces a blotchy coat pattern with stripes down the spine

Cheetah
Acinonyx jubatus

Cheetah cubs stay with mother until 13–20 months old

Lion While most felids tend to be solitary, lions are famed for their close, enduring group relationships. Prides of 4–20 lionesses occupy a home range, cooperate to kill large prey, and share care of cubs. Males live alone or in coalitions with other males.

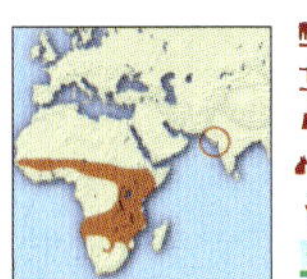

- Up to 7½ ft (2.3 m)
- Up to 39½ in (1 m)
- Up to 495 lb (225 kg)
- Family band
- Vulnerable

Sub-Saharan Africa, India

FAST AS A FLASH

Reaching speeds of 60 miles per hour (95 km/h) in pursuit of prey, the cheetah is the fastest mammal on land. With a burst of speed, it attempts to bring down hoofed animals such as Thomson's gazelle and wildebeest calves. Such sprints can last for only 20–60 seconds before the cheetah overheats and must rest. Sometimes it is too out of breath to defend its kill from scavengers.

Bengal tiger
Panthera tigris tigris

Stripe pattern unique to individual

Largest of all felids

Coat becomes lighter in winter

Siberian tiger (Amur tiger)
Panthera tigris altaica

VANISHING TIGERS

At the beginning of the 20th century, there were about 100,000 tigers roaming the tropical jungles, savanna, grasslands, mangrove swamps, deciduous woodlands, and snow-covered forests of Asia, from Turkey to far eastern Russia. Perhaps fewer than 2,500 breeding adults remain in the wild today. The Bali, Caspian, and Javan tigers—three of the eight tiger subspecies—are now extinct. Of the remaining subspecies, South China tigers exist only as a remnant population of 20–30 animals, and merely 500 or so Siberian or Amur tigers and 500 Sumatran tigers survive. The largest populations are of the Bengal tiger and the Indochinese tiger, but even these are in serious danger of becoming extinct. For years, tigers have been killed because they are considered pests; because their skins as well as body parts fetch good prices; and because they are prized as trophies by sport hunters. A continuing increase in local human populations has led to the degradation of much tiger habitat, as well as the decimation of hoofed mammals, the tiger's main prey.

Often rests in trees to avoid daytime heat or other predators

Leopard
Panthera pardus

"Black panthers" are in fact leopards with melanism, an excess of dark pigment

Clouded leopard
Neofelis nebulosa

Snow leopard
Uncia uncia

Jaguar
Panthera onca

Stockier build and larger head and jaw than leopard

Largest cat in the New World

Clouded leopard This largely arboreal cat will wait in a tree to ambush prey such as deer and pigs below. It will also seize primates and birds in the branches. The species is the smallest of the big cats and has only a soft roar.

- Up to 3½ ft (1.1 m)
- Up to 35½ in (90 cm)
- Up to 51 lb (23 kg)
- Solitary
- Vulnerable

Nepal to China, S.E. Asia

Jaguar This New World cat looks like the leopard, but fills a similar ecological niche to the tiger. It tends to live in dense vegetation near water and stalk large prey such as deer and peccaries. It also takes fish and other aquatic prey.

- Up to 6¼ ft (1.9 m)
- Up to 23½ in (60 cm)
- Up to 355 lb (160 kg)
- Solitary
- Near threatened

Mexico to Argentina

SNOWED IN

Adapted to high altitudes, the snow leopard lives in the remote mountains of Central Asia. It has a dense coat with very large, furry paws that act as snowshoes. Up to five cubs are born in a rocky den lined with the mother's fur.

Canadian lynx
Lynx canadensis

Eurasian lynx
Lynx lynx

Bobcat
Lynx rufus

Coat can be mainly striped, mainly spotted, or plain

In winter, coat thickens and becomes paler, and furry feet help lynx to travel over soft snow

Largest of the small cats

Puma
(cougar, mountain lion)
Puma concolor

Tufts of black fur on long, slender ears

Iberian lynx
Lynx pardinus

Caracal
Caracal caracal

Bobcat Rare in some parts of its range but more common in others, the bobcat usually hunts small prey such as rabbits and rodents by night but will also eat carrion. It rests by day, often in a cave.

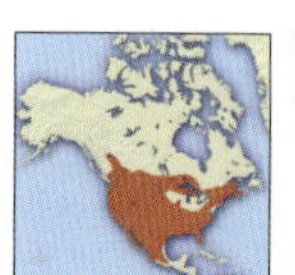

- Up to 41 in (105 cm)
- Up to 8 in (20 cm)
- Up to 68½ lb (31 kg)
- Solitary
- Locally common

Temperate North America to Mexico

Puma This cat's once vast range is now largely restricted to remote mountains, where it hunts white-tailed deer, moose, and caribou. The puma hisses, growls, whistles, and purrs, but cannot roar.

- Up to 5 ft (1.5 m)
- Up to 38 in (96 cm)
- Up to 265 lb (120 kg)
- Solitary
- Near threatened

Canada to S. Argentina & Chile

Caracal The swiftest of the small cats, the caracal can leap 10 feet (3 m) to snatch birds from the air. It also pounces on rodents and antelopes.

- Up to 36 in (92 cm)
- Up to 12¼ in (31 cm)
- Up to 42 lb (19 kg)
- Solitary
- Uncommon

Africa, Middle East, India & N.W. Pakistan

Coat color and pattern vary according to location; most commonly reddish brown or grayish with spots on lower flanks or all over

African golden cat
Profelis aurata

Jungle cat
Felis chaus

Long legs for chasing prey

Pallas's cat
Otocolobus manul

Chinese desert cat
Felis bieti

Black-footed cat
Felis nigripes

European wildcats tend to have darker coats than those in Africa

Soles of feet are black and covered in hair that protects them from hot sand

Sand cat
Felis margarita

Wildcat
Felis silvestris

Sand cat This desert creature can survive in extremely arid conditions, gaining enough water from its prey of rodents, hares, birds, and reptiles so that it does not need to drink.

- Up to 21 in (54 cm)
- Up to 12¼ in (31 cm)
- Up to 7½ lb (3.5 kg)
- Solitary
- Near threatened

Sahara Desert (N. Africa)

FROM THE WILD

Cats were first domesticated in ancient Egypt when grain stores attracted rats and mice, which drew wildcats. The wildcats were tolerated because they kept the rodents under control. Today, there are more than 100 million domestic or feral cats in the United States alone. Although wildcats tend to be much fiercer than domestic cats, domestic cats retain the hunting instinct and readily revert to a feral state.

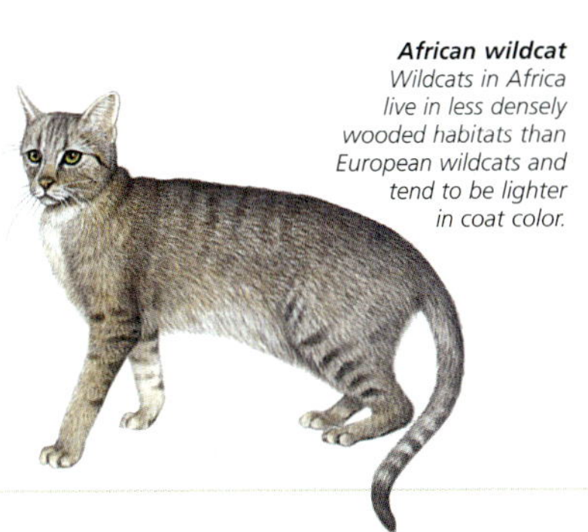

African wildcat
Wildcats in Africa live in less densely wooded habitats than European wildcats and tend to be lighter in coat color.

Marbled cat
Pardofelis marmorata

Blotches on coat merge, giving a marbled appearance

Asiatic golden cat
Catopuma temminckii

Coat can be reddish, golden, or grayish brown

Leopard cat
Prionailurus bengalensis

Rusty-spotted cat
Prionailurus rubiginosus

Iriomote cat
Prionailurus bengalensis iriomotensis

Island form of leopard cat

Bay cat
Catopuma badia

Flat-headed cat
Prionailurus planiceps

Webbed toes on front paws

Fishing cat
Prionailurus viverrinus

Asiatic golden cat This felid mostly preys on small mammals and birds, but will hunt in pairs to bring down larger quarry such as buffalo calves. Females give birth to one or two young. Unusually among cats, the male helps to raise the kittens.

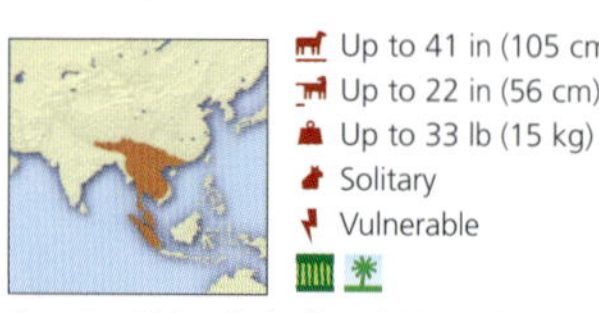

- Up to 41 in (105 cm)
- Up to 22 in (56 cm)
- Up to 33 lb (15 kg)
- Solitary
- Vulnerable

Nepal to China, Indochina, Malaya, Sumatra

Leopard cat The swimming ability of the leopard cat may help to explain its presence on many islands in Asia. A subspecies, the Iriomote cat, is found only on the small Japanese islands of Iriomote and Ryukyu.

- Up to 42 in (107 cm)
- Up to 17½ in (44 cm)
- Up to 15½ lb (7 kg)
- Solitary
- Locally common

Pakistan & India to China, Korea & S.E. Asia

Bay cat This species is so rare that it was not photographed until 1998. It lives in jungles and among limestone outcrops near forests on the island of Borneo. The coat is most commonly chestnut red but can also be gray.

- Up to 26½ in (67 cm)
- Up to 15½ in (39 cm)
- Up to 9 lb (4 kg)
- Solitary
- Endangered

Borneo

Elongated rather than rounded head

Coat can be chestnut or brownish gray

Jaguarundi
Herpailurus yaguarondi

Oncilla
Leopardus tigrinus

Ocelot
Leopardus pardalis

Margay
Leopardus wiedii

Pampas cat
Oncifelis colocolo

Kodkod
Oncifelis guigna

Geoffroy's cats have ocher coats in northern part of range, silvery gray coats in south

Geoffroy's cat
Oncifelis geoffroyi

Long, thick fur provides protection against exposed mountain environment

Andean cat
Oreailurus jacobita

Jaguarundi This species is not closely related to other South American felids and does not look like a typical cat. Sometimes called the weasel cat or otter cat, its long, slender body, short legs, and very long tail do resemble those of a mustelid.

Up to 25½ in (65 cm)
Up to 24 in (61 cm)
Up to 20 lb (9 kg)
Solitary
Uncommon

Arizona & Texas to S. Brazil & N. Argentina

Ocelot The wide variety of prey species taken by this opportunistic hunter has allowed it to live in a variety of habitats, from lush rain forests to semiarid brush. The ocelot's striking coat provides camouflage among thick vegetation.

Up to 18½ in (47 cm)
Up to 16 in (41 cm)
Up to 2½ lb (12 kg)
Solitary
Locally common

S.E. Texas to N. Argentina

Geoffroy's cat This agile cat can walk upside down along a branch and hang from a tree by its feet. Its populations were decimated by the fur trade, with more than 340,000 skins exported from Argentina between 1976 and 1979.

Up to 26½ in (67 cm)
Up to 14½ in (37 cm)
Up to 13 lb (6 kg)
Solitary
Near threatened

S. Bolivia & Paraguay to Argentina & Chile

UNGULATES

CLASS	Mammalia
ORDERS	7
FAMILIES	28
GENERA	139
SPECIES	329

Some 65 million years ago, an order of hoofed mammals known as Condylarthra began to evolve into many diverse orders, of which seven remain. These surviving orders are all now referred to as ungulates, meaning "provided with hoofs." In fact, only two orders possess true hoofs: odd-toed ungulates (Perissodactyla) include horses, tapirs, and rhinoceroses, while even-toed ungulates (Artiodactyla) encompass pigs, hippopotamuses, camels, deer, cattle, sheep, and goats. The other five orders, each with its own specializations, are elephants (Proboscidea), aardvark (Tubulidentata), hyraxes (Hyracoidea), dugong and manatees (Sirenia), and whales and dolphins (Cetacea).

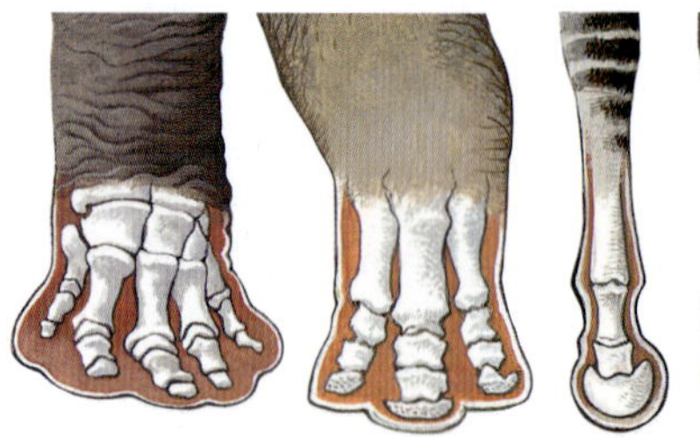

Toes and hoofs Elephants have a broad foot with five toes encased in a fatty matrix. In the "true" ungulates, at least one toe has been lost and the remaining toes form a hoof. The hoofs of odd-toed ungulates have three toes (as in the rhinoceros) or a single toe (as in the horse), while even-toed ungulates have two or four toes, which may be fused to form a cloven hoof (as in deer).

Waterhole A survival strategy employed by many grassland ungulates is the formation of large herds. Living in a group increases the chance that a predator will be detected, and reduces any one individual's chance of being taken. Mixed groups of hoofed grazers often come together to drink and feed, offering security in numbers to all.

HOOFS AND HERDS

Although less closely related to each other than to other members in the ungulate group, odd-toed and even-toed ungulates both stand on the tips of their toes, which are encased in hoofs. Together with their elongated metapodials (the hand and foot bones in humans), this unguligrade stance extends the leg, providing a longer stride and greater speed. They also have mobile ears, sharp, binocular vision, and a keen sense of smell to help them detect danger.

Almost all ungulate species are herbivores, with teeth adapted for grinding. Their specialized digestive systems can break down cellulose, the usually indigestible component of plant cell walls. Food is fermented by microorganisms in the hindgut or in a special stomach chamber.

Elephants, hyraxes, and aardvarks lack true hoofs and do not have an unguligrade stance. While elephants have just the bones of their toes (encased in a fatty matrix) touching the ground, hyraxes and aardvarks walk on the entire foot.

Whales, dolphins, dugongs, and manatees have all evolved for life in the water, with a streamlined body and flippers. Whales and dolphins were only recently classified as ungulates, when genetic studies showed their close relation to hippopotamuses.

Mountain goat Spending most of their time on severely steep, rocky slopes in alpine areas of the northern Rockies, where they are safe from predators, mountain goats nibble small plants that grow among the rocks. Powerful front legs help them climb and descend the steep slopes, and rough pads on the undersides of their split hooves provide traction. The hooves are capable of pinching around a rock edge or spreading out for braking.

Open wide Widespread throughout Africa, the hippopotamus commonly grazes on dry land at night. Hinged far back in the skull, a hippo's jaw can open remarkably wide, achieving a gape of 150 degrees, more than 100 degrees wider than the human gape. A male's large lower canines are used as weapons in battles for mating rights.

Fermenting stomachs Ungulates rely on internal microorganisms to break down plant cellulose. Many even-toed ungulates, such as deer, cattle, and sheep, are ruminants. Their complex stomachs include a rumen, a chamber in which food is fermented by microorganisms before being regurgitated and chewed a second time. A ruminant's digestive process can take up to four days and gains maximum nutrition from food. Odd-toed ungulates, such as horses, tapirs, and rhinoceroses, ferment their food in the hindgut (the cecum and large intestine). The two-day digestive process is less efficient than that of ruminants, requiring odd-toed ungulates to consume greater quantities of food.

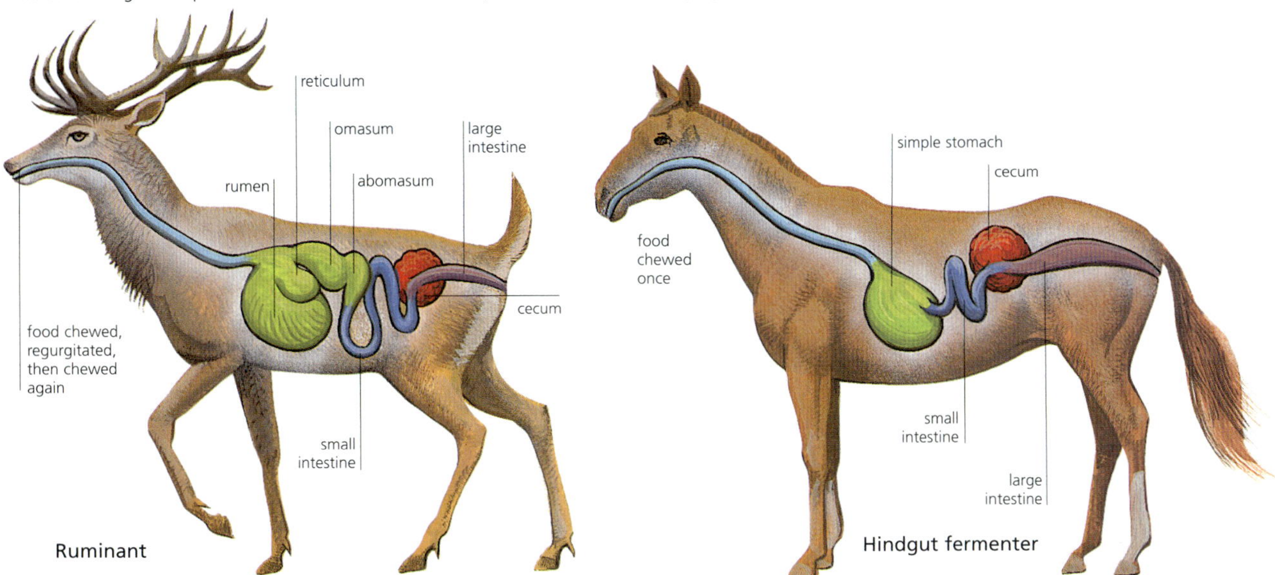

Elephants

CLASS	Mammalia
ORDER	Proboscidea
FAMILY	Elephantidae
GENERA	2
SPECIES	2

Weighing up to 7 tons (6.3 tonnes), elephants are the world's largest land animals. Their massive bulk is supported by four pillar-like legs and broad feet. An enormous head features immense fan-shaped ears and a long, flexible trunk. Full of blood vessels, the ears help to disperse the animal's body heat and will be flapped on hot days. The trunk, a union of the nose and upper lip, has more than 150,000 muscle bands and can both pick up twigs and lift heavy logs. Elephants can live for up to 70 years, longer than any land mammal apart from humans. As well as their longevity and strength, elephants are very social, intelligent, and quick to learn, traits that encouraged their domestication.

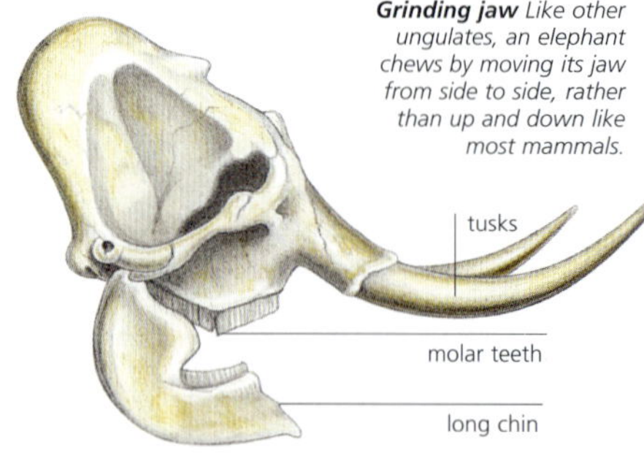

Grinding jaw *Like other ungulates, an elephant chews by moving its jaw from side to side, rather than up and down like most mammals.*

Inside the head The elephant's massive skull contains pockets of air to minimize its weight. The tusks are elongated incisor teeth emerging from deep sockets. The molar teeth are replaced horizontally in a conveyor-belt fashion, with new teeth developing from behind and slowly moving forward to take the place of worn teeth.

MATRIARCHS AND MUSTH

The order Proboscidea appeared about 55 million years ago and once included immense mastodons and woolly mammoths. Its members inhabited all kinds of habitat, from polar regions to rain forest, and at some time lived on all continents apart from Australia and Antarctica. Today's elephants are restricted to the forests, savanna, grasslands, and deserts of Africa and Asia.

To fuel their immense bodies, elephants spend 18–20 hours a day foraging or traveling toward food. An adult may eat up to 330 pounds (150 kg) of vegetation and drink 40 gallons (160 l) of water per day.

The basic social unit of elephants is a family group of related females and offspring, led by a matriarch. Adult males visit these groups only to mate, spending the rest of their time alone or in bachelor bands. A number of family groups may form larger herds. To maintain social ties, elephants rely on a range of communication methods, primarily using touch (greeting one another by intertwining trunks, for example), sound (some vocalizations are much lower than human ears can hear and travel up to 2½ miles, or 4 km), and postures (a raised trunk, for instance, can be used as a warning).

Mature male elephants have periods of musth, a time of high testosterone levels when the musth gland between the eye and ear secretes fluid. The animals become more aggressive and more likely to win fights, and wander greater distances in search of females.

Ivory trade Poaching reduced African elephant populations from about 20 million to 500,000. The 1989 Convention on International Trade in Endangered Species ban on ivory trade reversed the trend, but elephant conservation remains challenging.

Flexible trunk Used for caressing, lifting, feeding, dusting, smelling, as a snorkel, as a weapon, and in sound production, the elephant's trunk is superbly dextrous. Asiatic elephants have one finger-like projection at the tip, while African elephants have two.

Dugong and Manatees

CLASS	Mammalia
ORDER	Sirenia
FAMILIES	2
GENERA	3
SPECIES	5

Credited with inspiring the myth of the mermaid, the aquatic mammals of the order Sirenia are languid, docile creatures that never venture onto land. They are the only mammals to feed primarily on grasses and plants in shallow waters, a unique niche that may explain the order's lack of diversity, since seagrasses are much less varied than terrestrial grasses. The four living sirenian species are all found in the warm waters of tropical and subtropical regions. Because they have few natural predators, dugongs and manatees have developed no defenses other than their large size. This has made them easy targets for human hunters. Only about 130,000 individual sirenians remain, far fewer than in any other order of mammals.

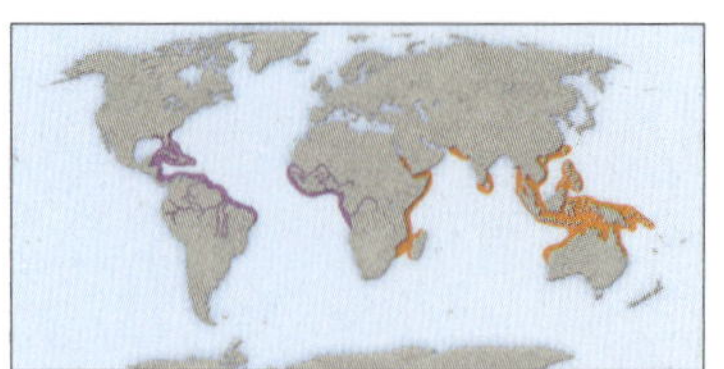

Range Dugong range (orange) is restricted to the sea. Manatee range (purple) is distributed as follows: the Amazonian manatee is found only in the fresh waters of the Amazon River, while Caribbean and African manatees live in freshwater, estuarine, and marine habitats.

Bristled feeder Like all sirenians, the manatee swims slowly through the water, grazing on aquatic plants and seagrasses. It locates food with the aid of the sensitive, bristle-like hairs on its snout, and uses its muscular lips to grasp plants and pass them to the mouth.

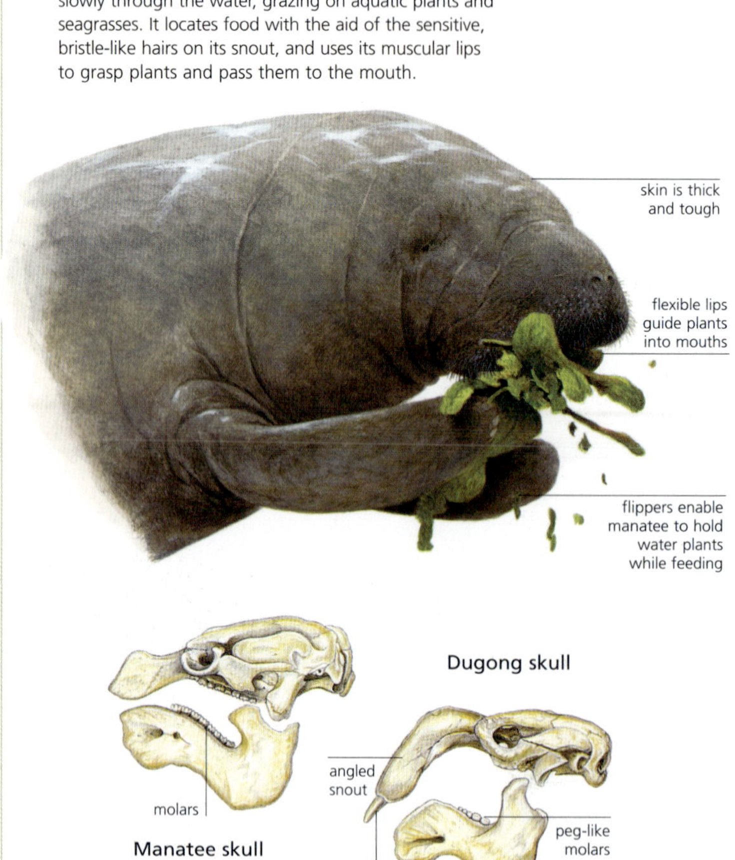

Different skulls Manatees have a row of molar teeth, with new hind teeth constantly moving forward to replace the worn teeth. A dugong skull has an angled snout with only a few peg-like molars, which grow throughout life. In male dugongs, long incisors form tusks.

GENTLE GRAZERS

Like other aquatic mammals, sirenians have a streamlined body, flippers, and flattened tail, and come to the surface to breathe through nostrils on top of the head. Keeping with a grazing lifestyle, the sirenian head is more like that of a pig and is used to root out grass rhizomes (underground stems) from sediment. The angle of the dugong's snout restricts it to feeding from the bottom, but manatees can feed from all levels of the water.

The teeth of sirenians have become specialized in different ways for chewing great quantities of plant matter. The dugong uses rough horny pads in its mouth to crush its food, then grinds it with a few peg-like molars that continue to grow throughout its life. Manatees chew with their front molars, which are replaced by new teeth from behind when they wear out.

Sirenians have simple stomachs with extremely long intestines. Their plant food is broken down by microorganisms in the rear part of the digestive tract, making them hindgut fermenters like horses and other odd-toed ungulates. To counter the buoyancy provided by their gas-producing diet, sirenians have unusually dense, heavy bones.

Because their eyesight is generally poor, sirenians rely heavily on touch to find food. Their hearing is good underwater, with sounds transmitted through the skull and jaw bones. They emit squeaks to communicate, but how these sounds are produced is an enigma as they lack vocal cords.

Some sirenians live a solitary life, but more often, they are found in loosely structured groups of a dozen or so. Occasionally, such groups come together to form large herds of 100 or more. Social bonds are reinforced through playful nuzzling.

African manatee
Trichechus senegalensis

Paddle-like tail

Less bulky than manatee

Fluked tail

Dugong
Dugong dugon

No nails on flippers

Thick, tough skin is often wrinkled

Nostrils closed by valves

Caribbean manatee
Trichechus manatus

Nails on flippers

Amazonian manatee
Trichechus inunguis

Stiff bristles on large, mobile lips

African manatee This poorly studied species is believed to be at least partly nocturnal. Found from coastal seas to rivers, it tends to feed from waters near or at the surface.

Up to 13 ft (4 m)
Up to 1,100 lb (500 kg)
Solitary, family group
Vulnerable

W. African coast, Niger River

Caribbean manatee This species moves freely between freshwater and marine habitats. When a female is ready to breed, a mating herd of up to 20 males will pursue her, competing for her attention for up to a month.

Up to 15 ft (4.5 m)
Up to 1,320 lb (600 kg)
Solitary
Vulnerable

Georgia & Florida to Brazil; Orinoco River

Amazonian manatee Few plants grow under the murky waters of the Amazon River, so this manatee feeds mainly on surface vegetation such as floating grasses and water hyacinths.

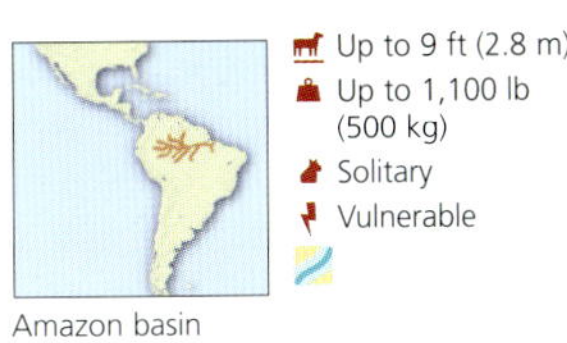
Up to 9 ft (2.8 m)
Up to 1,100 lb (500 kg)
Solitary
Vulnerable

Amazon basin

HORSES, ZEBRAS, & ASSES

CLASS	Mammalia
ORDER	Perissodactyla
FAMILY	Equidae
GENUS	1
SPECIES	9

The horses, zebras, and asses of the family Equidae rely on their large size, swift running style, and herding behavior to evade predation. An equid bears its weight on the tip of a single toe on each foot, a stance that provides a springy gait. The slender legs lock when at rest rather than relying on muscle contraction, a mechanism that minimizes the energy used during the many hours that the animal feeds. The teeth are highly specialized for a diet of grasses and other plants, with incisors to clip the vegetation, and complex, ridged cheek teeth for grinding. Plant cellulose is fermented in the hindgut, allowing equids to live on the abundant low-quality food of arid lands.

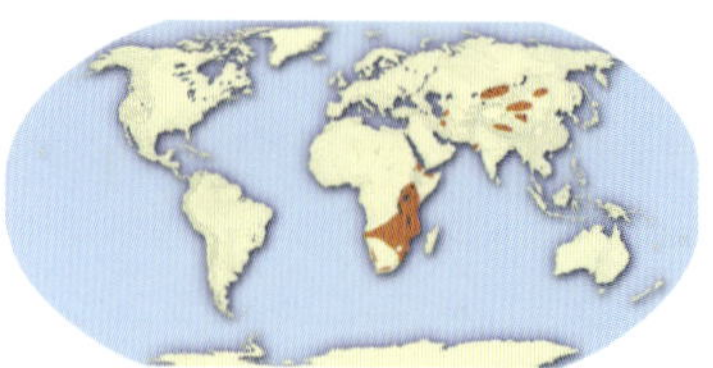

Wild and feral Wild equids live in grasslands, savanna, and deserts in Africa and Asia. They generally congregate in herds. Hunted for meat or hides and persecuted as competitors for grazing land, wild equids have declined and most are at risk of extinction.

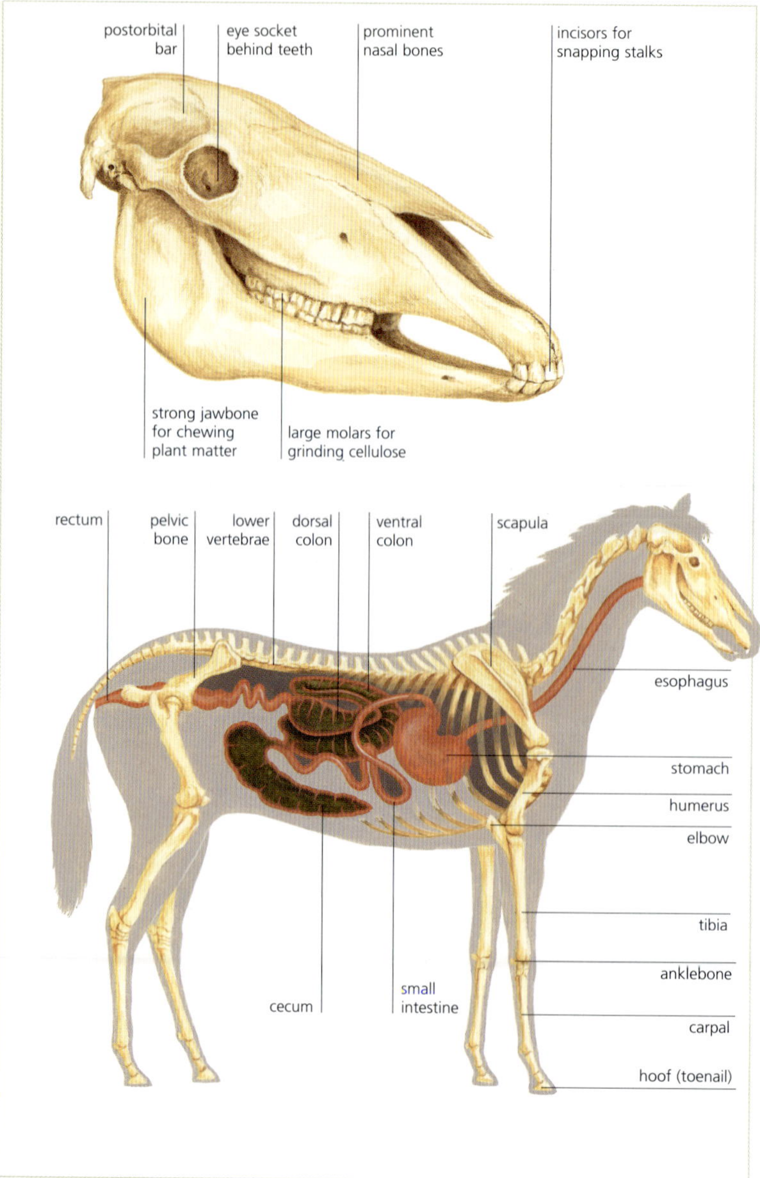

Horse anatomy Horses chew their food once only, and have a single stomach with a long intestinal system and large, long teeth to break down tough, fibrous grasses.

EVOLVING EQUIDS

All equids are covered in thick hair, with a mane along their long neck. Most species have single-colored coats, but zebras can be instantly recognized by their striking black and white stripes. Eyes at the side of the head provide good all-round vision by night and day, while erect, mobile ears with acute hearing listen out for danger. Equids tend to run away from predators, but will kick and bite to defend themselves if necessary. These highly social animals may whinny, bray, nicker, or squeal to communicate. Visual cues involving tail, ear, and mouth positions also play a role in equid communication, as does scent.

The first horse-like animal, a dog-sized mammal that walked on the soft pads of its feet, appeared about 54 million years ago. North America was the center of equid evolution, which resulted in a single-toed horse-like creature by at least 5 million years ago. Equids migrated to Africa and Asia, where modern species of zebras and asses later emerged. By the end of the last ice age, horses had vanished from North America, and were only reintroduced there by Europeans.

Humans first domesticated asses in the Middle East before 3000 BC, but within 500 years the faster and stronger domestic horse had arrived from Central Asia. Domestic horses revolutionized agriculture, transport, hunting, and warfare. Today, virtually all wild horses are feral populations of the domestic horse. Feral herds can be found on every continent except Antarctica.

Kiang
Equus kiang

Upper coat red in summer, becomes browner and longer in winter

Largest wild ass

Onager
Equus onager

Ass (donkey, burro) domestic form
Equus africanus

Ass
wild form
Equus africanus

Coat can be grayish, brownish, or reddish, with white underparts

Some asses have bands on legs

Stallion curls back upper lip to help detect a female's reproductive status from her scent

Mongolian wild ass
Equus hemionus

Large head with short, erect mane and no forelock

Wild form is shorter and stockier than domestic form

Mongolian wild ass
Equus hemionus

Horse
Equus ferus

Kiang Herds of up to 400 mares and their offspring are led by an old female. Stallions usually live alone but several may live together in winter. In the summer mating season, stallions follow the large herds and fight one another for the right to mate.

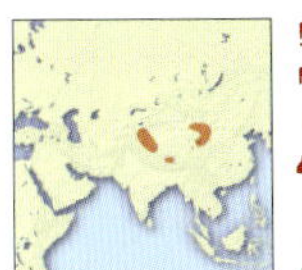

Up to 8 ft (2.5 m)
Up to 4½ ft (1.4 m)
Up to 880 lb (400 kg)
Herd
Declining

Tibetan Plateau

Onager This wild ass is the fastest of the equids, reaching speeds of 40 miles per hour (70 km/h) in short bursts. It lives in arid environments, gathering in large, unstable herds with no lasting bonds between adults.

Up to 8 ft (2.5 m)
Up to 4½ ft (1.4 m)
Up to 575 lb (260 kg)
Herd
Rare

India, Iran; reintrod. in Turkmenistan

Ass In its wild form, the ass is rare, with only 3,000 individuals remaining. It was domesticated 6,000 years ago and is still used as a beast of burden. The ass has great endurance and can survive for a long time in hot conditions with little food or water.

Up to 6½ ft (2 m)
Up to 4¼ ft (1.3 m)
Up to 550 lb (250 kg)
Harem
Critically endangered

Ethiopia, Eritrea, Somalia, Djibouti

Largest of the wild equids

Grevy's zebra
Equus grevyi

Finely divided pattern of black on white

Zebra southern form
Equus burchelli

Shorter ears than those on other zebra species

Zebra northern form
Equus burchelli

Wide stripes on body

Mountain zebra
Equus zebra

Each individual zebra's pattern of stripes is unique

Stripes wider on rump than on body

Foals can walk within an hour of birth

Mountain zebra foal

MOOD DISPLAYS

Like all equids, zebras use a range of visual signals to express their moods. Competing stallions will shake their head, arch their neck, and stamp their feet before resorting to physical contests in which they bite each other's neck and legs. In general, equids tend to run away from predators, but both mares and stallions will bite and kick to try to defend themselves.

Prepared to bite
Stallions will display a bite threat before making physical contact.

Ready to kick
A threatened equid will kick out its back legs in a defense display.

Detecting danger
By curling back the upper lip, an equid directs a scent to the Jacobson's organ in the roof of its mouth.

TAPIRS

CLASS Mammalia
ORDER Perissodactyla
FAMILY Tapiridae
GENERA 1
SPECIES 4

Appearing in the fossil record before either horses or rhinoceroses, tapirs as a group have changed little in the past 35 million years. These reclusive tropical-forest browsers are about the size of a donkey, with a squat, streamlined body for moving through the thick undergrowth, and a sensitive, prehensile trunk that is used for grasping food, detecting danger through scent, and as a snorkel. Tapirs emerge from the shelter of thickets at night to feed on the leaves, buds, twigs, and fruit of low-growing plants. Being strong swimmers and fond of water, they also consume aquatic vegetation. A newborn's patterned coat provides effective camouflage in the dappled light of the forest. After a week, it will accompany the mother, eventually going off on its own by the time it is 2 years of age.

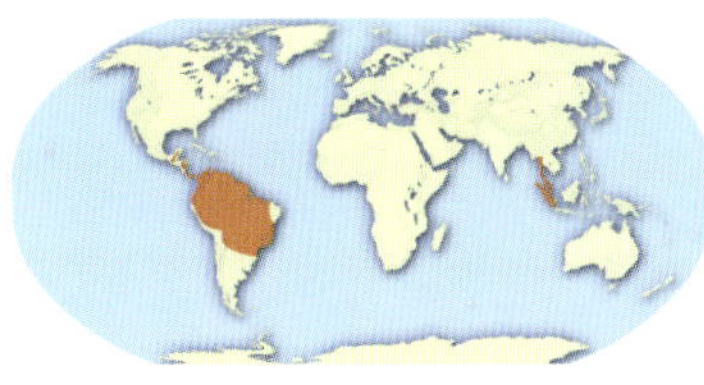

Reduced range Found at various times throughout much of North America, Europe, and Asia, tapirs are now restricted to three species in Central and South America and a single species in Southeast Asia. Usually solitary and sparsely distributed, tapirs communicate through whistles and scentmarking.

Brazilian tapir When this tapir plunges into water to escape a jaguar, it risks being taken by a crocodile. Its chief predator, however, is the human hunter, who follows its clear foraging trails.

- Up to 6½ ft (2 m)
- Up to 3½ ft (1.1 m)
- Up to 550 lb (250 kg)
- Solitary
- Vulnerable

Tropical South America (E. of Andes)

Malayan tapir This species is the only living tapir in Asia. Courtship is marked by whistling pairs walking around in circles as they attempt to sniff each other's genitals.

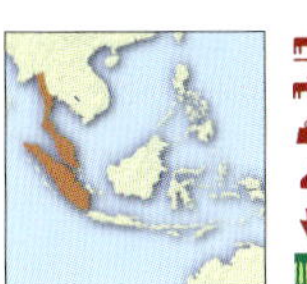

- Up to 8 ft (2.5 m)
- Up to 4 ft (1.2 m)
- Up to 705 lb (320 kg)
- Solitary
- Vulnerable

Myanmar, Thailand, Malaya, Sumatra

WATER REFUGE

Never far from water, tapirs spend much of their time submerged, often with just the trunk appearing above the surface as a snorkel. Water provides cover from predators and relief from the heat. Tapir numbers are in decline as their forest habitat is lost or fragmented, they are hunted for their meat, and livestock compete for similar food soures.

RHINOCEROSES

CLASS	Mammalia
ORDER	Perissodactyla
FAMILY	Rhinocerotidae
GENERA	4
SPECIES	5

The snout of a rhinoceros features its most distinguishing characteristic—a great horn or two made up of fibrous keratin. Rhinoceroses use their remarkable horns to fight rivals, to defend their young against predators, to guide their young, and to push dung into piles as scented signposts. Although keratin is a common substance that is also found in human fingernails, rhino horns are attributed great potency in traditional Asian medicine. Such has been the demand that many thousands of rhinoceroses have been poached, and all five species are now at risk of extinction. Today, there are fewer than 15,000 wild rhinos in Africa, and no more than 3,000 in Asia.

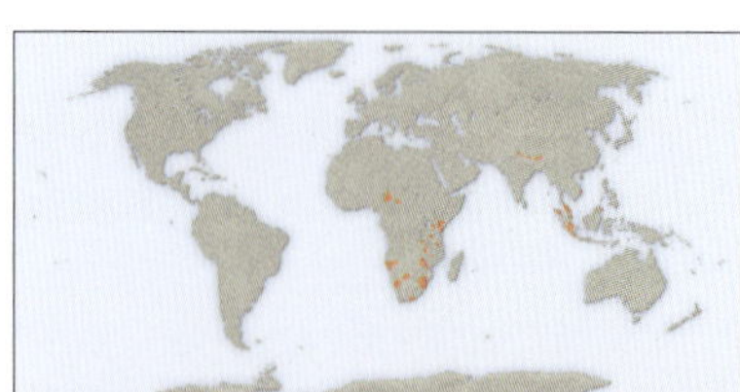

Pockets The family Rhinocerotidae was once abundant, widespread, and diverse. The woolly rhinoceros, for example, roamed Europe until the end of the last ice age 10,000 years ago. Today, only five species remain: two in Africa (the white rhinoceros and the black rhinoceros) and three in Asia (the Indian, the Javan, and the Sumatran).

Defense tactics Rhinoceroses have two forms of defending themselves—their great bulk and long horn. Males use their horn when competing for females, and females use their horn to defend their young from predators.

MASSIVE HERBIVORES

Rhinoceroses have massive bodies supported by four stumpy legs. Three hoofed toes on each foot leave an "ace-of-clubs" print in their tracks. Their thick, wrinkled skin can be gray or brown, but its true color is often masked by dried mud, since rhinos like to wallow in muddy pools. Despite their names, there is no clear distinction in skin color between white and black rhinos: both are grayish, with the precise tinge determined by local soil color.

With a lifespan of up to 50 years, rhinos are slow breeders, a fact that has made them especially vulnerable to habitat loss and overhunting. After a gestation of about 16 months, a female bears a single young and nurses it for more than a year. Their association lasts for 2–4 years, until the next calf is born. Most adult rhinos are solitary, but breeding pairs may stay together for a few months, while females or immature males sometimes form temporary herds. White rhinos are known to form a circle around the young to protect them from predators.

Both African rhino species use their horns to fight rivals, while Asian species use their sharp incisor or canine teeth. Before resorting to physical attack, rhinos will engage in a series of gestures, including pushing horns together, wiping horns on the ground, and spraying urine. Black rhinoceroses are especially aggressive, with half of all males and a third of females dying after fights.

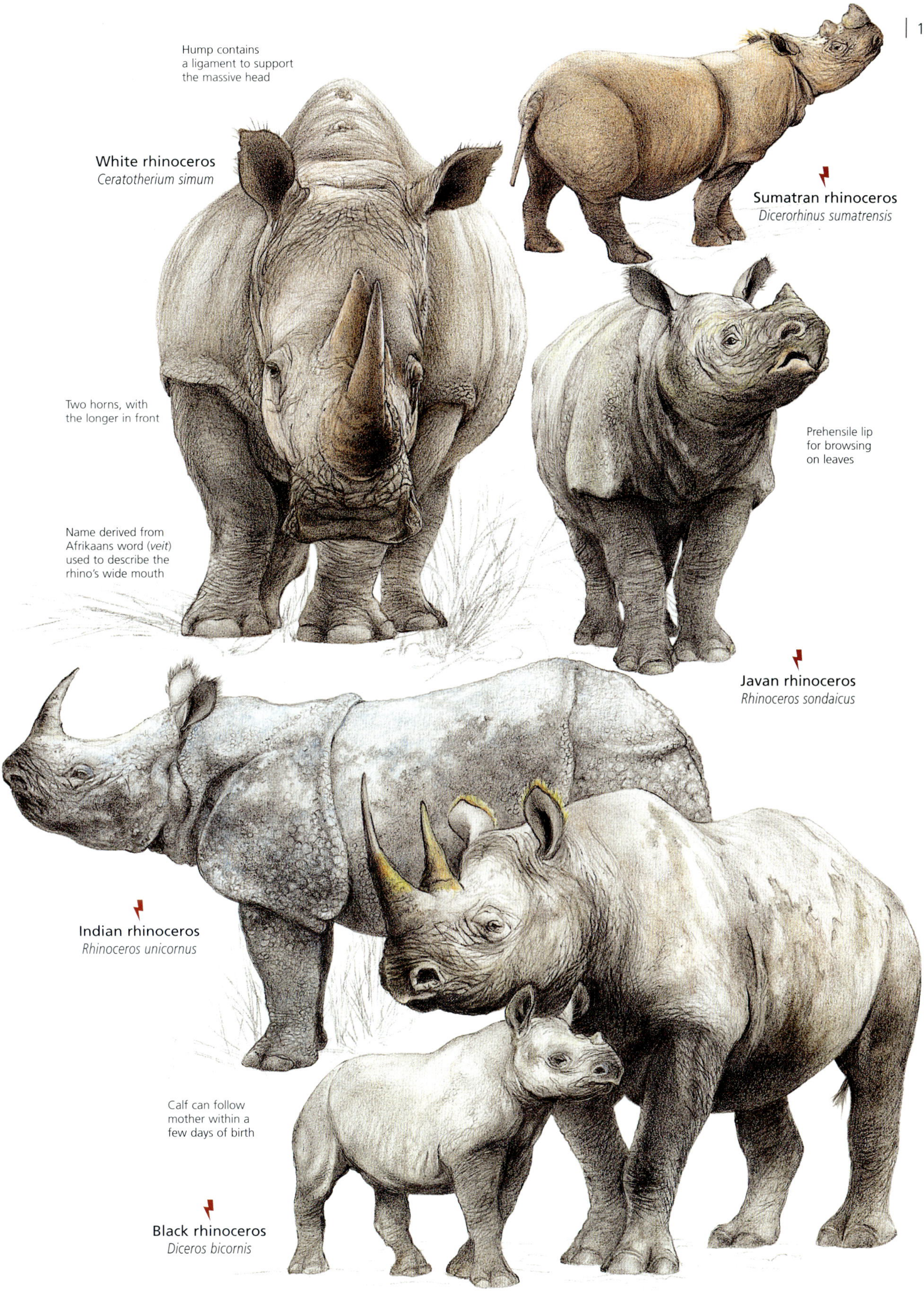

Hump contains a ligament to support the massive head

White rhinoceros
Ceratotherium simum

Two horns, with the longer in front

Name derived from Afrikaans word (*veit*) used to describe the rhino's wide mouth

Sumatran rhinoceros
Dicerorhinus sumatrensis

Prehensile lip for browsing on leaves

Javan rhinoceros
Rhinoceros sondaicus

Indian rhinoceros
Rhinoceros unicornus

Calf can follow mother within a few days of birth

Black rhinoceros
Diceros bicornis

White rhinoceros The largest surviving rhino, this species has a long head and squared upper lip for efficient cropping of short grasses. Despite its size, the white rhino is usually a placid animal.

Up to 14 ft (4.2 m)
Up to 6¼ ft (1.9 m)
Up to 4 tons (3.6 t)
Solitary, family group
Near threatened
Former range

Sub-Saharan Africa

Sumatran rhinoceros The smallest of the rhinos, this is also one of the most threatened—only about 300 individuals survive in the wild. A browser that feeds mainly on saplings, it is the only Asian rhinoceros with two horns.

Up to 10½ ft (3.2 m)
Up to 5 ft (1.5 m)
Up to 2¼ tons (2 t)
Solitary
Critically endangered
Former range

Thailand, Myanmar, Malaya, Sumatra, Borneo

Indian rhinoceros The larger of the one-horned rhinos, this species prefers to graze on tall grasses, but will also browse shrubs, crops, and aquatic plants. It avoids the daytime heat by foraging late or early, or at night.

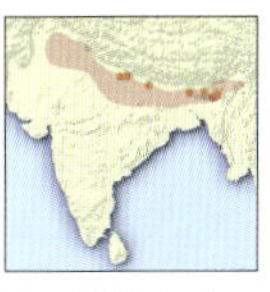

Up to 12½ ft (3.8 m)
Up to 6¼ ft (1.9 m)
Up to 2½ tons (2.2 t)
Solitary
Endangered
Former range

Nepal, N.E. India

HYRAXES

CLASS	Mammalia
ORDER	Hyracoidea
FAMILY	Procaviidae
GENERA	3
SPECIES	7

About the size of a rabbit and looking rather like a large guinea pig, the hyraxes are often mistaken for rodents, but are in fact ungulates, with flattened hoof-like nails on their feet. Millions of years ago, hyraxes, some as large as tapirs, were the dominant grazing mammals of North Africa. They were displaced by larger ungulates such as antelopes and cattle. The surviving hyraxes are robust, agile creatures that scamper and leap along steep rocks and tree branches. The soles of their feet are uniquely equipped to provide traction, with soft pads kept moist by a glandular secretion, and muscles that retract the middle of the sole to form a suction pad. Some hyrax species live in colonies of up to 80 animals.

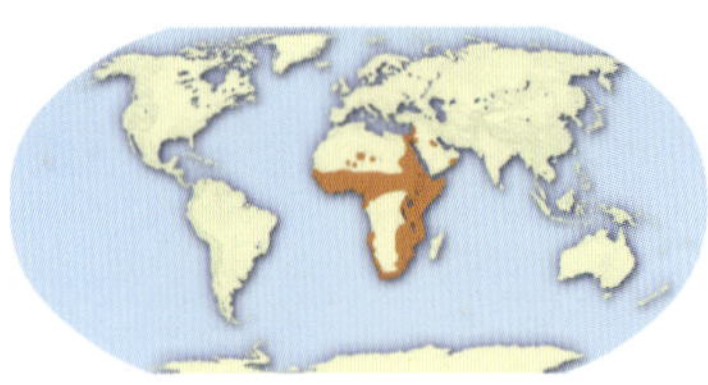

Africa to the Middle East Once more diverse and widespread, there are just seven species of hyrax in three genera. Rock hyraxes (*Procavia*) live in much of Africa and parts of the Middle East. Yellow-spotted hyraxes (*Heterohyrax*) are largely confined to East Africa. Tree hyraxes (*Dendrohyrax*) tend to live in African forests.

Southern tree hyrax
Dendrohyrax arboreus

Yellow-spotted hyrax
Heterohyrax brucei

Rock hyrax
Procavia capensis

WARM HUDDLE

Most small mammals are nocturnal, but hyraxes are diurnal, active by day. Unable to regulate their body temperature well, they conserve heat by huddling together and warm up by basking in the sun. Hyraxes live in family groups of several females and their offspring, headed by a territorial male. Females usually stay with their family for life, but males disperse at about 2 years of age. They may live on the edge of a family group, hoping to take over the territorial male's position.

All hyrax species will feed both in trees and on the ground, and have been known to travel almost a mile (1.3 km) in search of food. Rock hyraxes mainly eat grasses, while yellow-spotted hyraxes and tree hyraxes tend to browse leafy plants. They rely on microorganisms in their multichambered stomach to digest the plant cellulose. Hyraxes are highly vocal creatures that sound like no other animal. The gregarious ground-dwellers chatter, whistle, and scream. At night, tree hyraxes begin a series of loud croaks that end in a scream. Hyrax vocabulary changes and expands throughout its lifetime. Young hyraxes utter long chatters that progressively intensify.

MIXED LIVING

Hyraxes include one of the few examples of two mammal species cohabiting. The yellow-spotted hyrax and the rock hyrax have been found living together on the same kopje (rocky outcrop), sharing holes for shelter by night and then huddling together to warm up by day. They do not interbreed, but females tend to give birth at the same time, and members of both species show great interest in the young. The two species avoid competing with each other by eating different foods.

AARDVARK

CLASS	Mammalia
ORDER	Tubulidentata
FAMILY	Orycteropodidae
GENERA	1
SPECIES	1

A medium-sized pig-like animal with a stocky body, long snout, and large ears, the aardvark is the only living member of the order Tubulidentata, which evolved from an early hoofed mammal. Solitary and nocturnal, this highly specialized creature emerges from its burrow after dark to search for ants and termites, taking up to 50,000 insects a night. A keen sense of smell helps the aardvark to detect prey, while its powerful, clawed feet can excavate a termite mound in a few minutes. The hair-lined nostrils can be contracted and the ears folded back to keep out dirt. A long, sticky tongue snatches the insects, which are usually swallowed without chewing and ground up in the animal's muscular stomach.

Determined digger An aardvark uses the four shovel-shaped claws on each forefoot to dig out food and excavate burrows.

Aardvark Aardvarks are most likely to be found near colonies of termites, their favorite food. Although the aardvark is not considered to be threatened, its highly specialized diet makes it vulnerable to habitat alteration. Nocturnal and shy, they are rarely seen in the wild.

Aardvark
Orycteropus afer

THE CATTLE FAMILY

CLASS	Mammalia
ORDER	Artiodactyla
FAMILY	Bovidae
GENERA	47
SPECIES	135

The family Bovidae includes many millions of domesticated cattle, sheep, goats, and water buffalo. The 135 species of wild bovid are much more diverse, ranging from the dwarf antelope, just 10 inches (25 cm) tall and weighing as little as 5 pounds (2.3 kg), to massive animals such as bison, which can be more than 6 feet (2 m) high at the shoulder and weigh up to 1 ton (1 tonne). Cattle and their kin occur naturally throughout much of Eurasia and North America, and introduced species have formed feral populations in Australasia, but bovids reach their greatest numbers and variety in the grasslands, savanna, and forests of Africa.

Wide spread Bovids are distributed from hot deserts and tropical forests to arctic and alpine regions. There are no native bovids in Australia or South America, but domestic species are found worldwide. There are more than a billion domestic cattle in the world, and all are descended from aurochs, wild cattle that were once widespread but became extinct in 1627.

Perilous journey The most dangerous part of the great wildebeest migration is the crossing of the Mara River, which can be a raging torrent. The migrating herd gathers at the banks until forced by the surge of animals to cross. Many wildebeest drown or are swept away. Enormous crocodiles feast during the river crossing, while lions and hyenas target stragglers.

A DIVERSE FAMILY

The cattle, buffalo, antelopes, sheep, goats, and other members of Bovidae are ruminants, with a four-chambered stomach for fermenting plant matter. This efficient digestive system has allowed bovids to make the most of low-nutrient foods such as grasses and colonize a wide array of habitats. Grazing bovids tend to have a stocky build to house their large stomachs. Antelopes and other slender bovids are often more selective browsers.

All male and many female bovids have horns made up of a bony core covered in a layer of keratin, which is never shed. Always unbranched with pointed tips, bovid horns vary in size and can be straight, curved, or spiral. Adapted for running away from danger, bovids bear their weight on the two middle toes of each foot, which form a cloven hoof. The main bones in the feet are fused to form the cannon bone, which helps to absorb the impact of running.

While some bovids are solitary or live in pairs, most are gregarious. Some species live in harems led by a male, while others have herds made up of females and young, with males being largely solitary or associating in bachelor herds. Group living not only reduces the risk of predation, but also allows members to share information about feeding sites.

Abbott's duiker
Cephalophus spadix

Yellow-backed duiker
Cephalophus silvicultor

Duiker species vary in size but share distinctive body shape

Red-flanked duiker
Cephalophus rufilatus

Common duiker
Sylvicapra grimmia

Large gland beneath each eye produces secretion used in scentmarking

Banded duiker (zebra duiker)
Cephalophus zebra

Short, conical horns

White band breaks up animal's outline

Ader's duiker
Cephalophus adersi

Ogilby's duiker
Cephalophus ogilbyi

Banded duiker This muscular species is easily identified by the striking stripes of its coat. As in zebras, each banded duiker's pattern of stripes is unique. Banded duikers are diurnal creatures that usually live in pairs, with bonding reinforced by grooming.

Up to 35½ in (90 cm)
Up to 19½ in (50 cm)
Up to 44 lb (20 kg)
Solitary, pair
Vulnerable

Liberia

Common duiker This nocturnal antelope lives at higher altitudes than other African hoofed mammals. It relies on remarkable speed and stamina to escape its predators, which include big cats, dogs, baboons, pythons, crocodiles, and eagles.

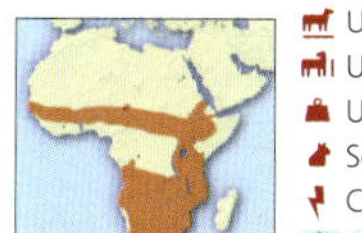

Up to 45½ in (115 cm)
Up to 19½ in (50 cm)
Up to 46½ lb (21 kg)
Solitary
Common

Sub-Saharan Africa except rain-forest zone

Ader's duiker Active by day, this species tends to live in territorial pairs. It feeds mainly on flowers, fruit, and leaves from the forest floor, taking advantage of food scraps discarded by monkeys and birds in the trees above.

Up to 28½ in (72 cm)
Up to 12½ in (32 cm)
Up to 26½ lb (12 kg)
Solitary, pair
Endangered

Zanzibar, coastal S.W. Kenya

Lichtenstein's hartebeest
Sigmoceros lichtensteinii

Named after W.H.C. Lichtenstein, a famous naturalist who explored southern Africa in 1803–06

Hunter's hartebeest (hirola)
Damaliscus hunteri

Males have thick skin on neck for protection during sparring contests

Topi
Damaliscus lunatus korrigum

Tsessebe
Damaliscus lunatus lunatus

Blesbok calves have lighter coats and dark faces

Blesbok (bontebok)
Damaliscus pygargus

Ringed, S-shaped horns

Rump lower than shoulders

Coat color varies from brown to bright red

Common hartebeest
Alcelaphus buselaphus

Lichtenstein's hartebeest Males of this savanna species are territorial, marking their claim in the ground with their horns. Rivals will fight for the right to mate, with the victor leading a harem of 3–10 females and their young.

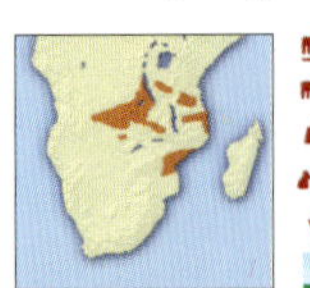

Up to 7 ft (2.1 m)
Up to 4¼ ft (1.3 m)
Up to 375 lb (170 kg)
Small group
Conserv. dependent

S. Africa

Topi and tsessebe These antelopes breed once a year. In a large herd, the calves are likely to be followers, protected within a group of adults. In smaller herds, the calves may be hiders, concealed in dense vegetation while the adults forage.

Up to 8½ ft (2.6 m)
Up to 4 ft (1.2 m)
Up to 310 lb (140 kg)
Herd
Conserv. dependent

Savanna zone of sub-Saharan Africa

Common hartebeest With its sloping back, this large animal can appear awkward, but it is capable of reaching speeds of 50 miles per hour (80 km/h). It lives on open plains, preferring edge habitats near forest.

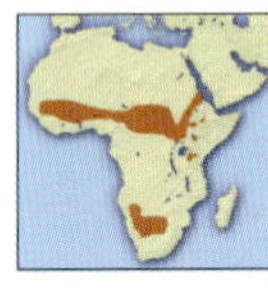

Up to 6¼ ft (1.9 m)
Up to 4¼ ft (1.3 m)
Up to 330 lb (150 kg)
Herd
Conserv. dependent

Sahel, Serengeti, Namibia to Botswana

Horns of a male gemsbok can be up to 5 feet (1.5 m) long

Gemsbok
Oryx gazella

Black-tufted tail

Dark markings on legs, flanks, and face

S-shaped, ridged horns only on male

Impala
Aepyceros melampus

Can leap up to 10 feet (3 m) off the ground

Wildebeest
Connochaetes taurinus

Scent glands beneath patches of black hair on rear feet

Vertical stripes of longer hair on back

White-tailed gnu (black wildebeest)
Connochaetes gnou

Gemsbok This species usually lives in herds of about 40 individuals, but hundreds may congregate during the wet season. During dry periods, the gemsbok can go without drinking for several days, surviving on the moisture content of fruit and roots.

E. & S.W. Africa

Up to 5¼ ft (1.6 m)
Up to 4 ft (1.2 m)
Up to 530 lb (240 kg)
Herd
Conserv. dependent

Impala Mainly grazing on new grasses in the wet season, the impala switches to browsing woody plants in the dry season. When threatened, this animal usually tries to outrun the danger but may also leap in random directions to confuse a predator.

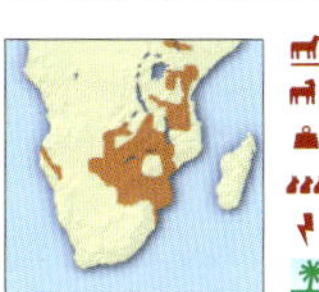

Savanna zone of E. & S. Africa

Up to 5 ft (1.5 m)
Up to 35½ in (90 cm)
Up to 110 lb (50 kg)
Herd
Conserv. dependent

Wildebeest Most wildebeest young are born within a single 3-week season, after a gestation of about 8 months. Within a few minutes of its birth, a calf can stand and nurse. About 40 minutes later, it can run.

E. & S. Africa

Up to 7½ ft (2.3 m)
Up to 5 ft (1.5 m)
Up to 550 lb (250 kg)
Herd
Conserv. dependent

Beira antelope
Dorcatragus megalotis

Gerenuk
Litocranius walleri

Can stand on hindlimbs to browse on leaves that are out of reach of most antelopes

Springbok
Antidorcas marsupialis

Chiru (Tibetan antelope)
Pantholops hodgsonii

Blackbuck
Antilope cervicapra

Large, fleshy nose filters dust from air in summer and warms air in winter

Saiga
Saiga tatarica

Dibatag
Ammodorcas clarkei

PRONKING

Gazelles sometimes leap repeatedly, keeping their back arched and legs stiff and landing on all fours—a habit known as "pronking" or "stotting." This behavior may occur when an animal is excited or alarmed and possibly distracts predators or warns them that they have been detected. During the pronk, a fold of skin with a crest of white fur is exposed on a springbok's rump.

Leaping springbok
Named after its habit of pronking, the springbok leaps up to 13 feet (4 m) in the air.

Saiga Apart from its ridged horns and the long, mobile nose that hangs over its mouth, this antelope resembles a small sheep. Thousands of saigas may form mixed herds to migrate to summer pastures.

Up to 4½ ft (1.4 m)
Up to 31½ in (80 cm)
Up to 152 lb (69 kg)
Herd
Critically endangered

Russia, Kazakhstan

Oribi
Ourebia ourebi

Scent gland beneath eye

Steenbok
Raphicerus campestris

Short, spike-like horns on males

Klipspringer
Oreotragus oreotragus

Guenther's dik-dik
Madoqua guentheri

Waterbuck
Kobus ellipsiprymnus

Peg-like hoofs for traveling over rocks

Bohor reedbuck
Redunca redunca

Reedbucks usually live near water

Mountain reedbuck
Redunca fulvorufula

Steenbok Exclusively a browser, this swift antelope prefers nutritious new growth such as young leaves, flowers, fruit, and shoots. It is the only bovid known to scrape the ground before and after urinating and defecating.

- Up to 33½ in (85 cm)
- Up to 19½ in (50 cm)
- Up to 24½ lb (11 kg)
- Solitary, pair
- Common

E. & S. Africa

Waterbuck Old, weak herd animals are usually easy targets for predators. As a waterbuck ages, however, the secretions released by its sweat glands gradually impart an unpleasant taste to the flesh, discouraging predators.

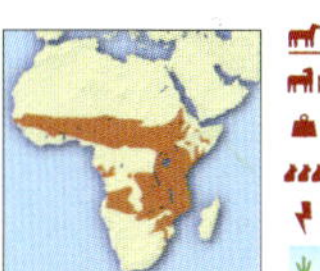

- Up to 8 ft (2.4 m)
- Up to 4½ ft (1.4 m)
- Up to 660 lb (300 kg)
- Herd
- Conserv. dependent

Savanna zone of sub-Saharan Africa

Mountain reedbuck This species can breed at any time, taking advantage of favorable conditions whenever they occur. Like many antelopes and deer, it has a white patch under the tail that is displayed as it runs away from danger.

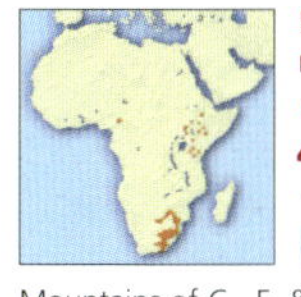

- Up to 4¼ ft (1.3 m)
- Up to 28½ in (72 cm)
- Up to 66 lb (30 kg)
- Harem
- Conserv. dependent

Mountains of C., E. & S. Africa

Thomson's gazelle
Gazella thomsonii

Dama gazelle
Gazella dama

Male horns are thicker and longer than those of female

Grant's gazelle
Gazella granti

Serow
Capricornis sumatraensis

MALE COMPETITION

Male gazelles will defend their territory and harem against challengers during the breeding season. Males mark territory with secretions from the glands beneath each eye, as well as with urine and feces. Threat displays begin with the head raised so the horns lie along the back, and progress to the head tucked in with the horns vertical. Head lowered with horns pointed at the opponent is usually the last display before physical contact.

Thomson's gazelle With 90 percent of its diet consisting of grass, this species is almost exclusively a grazer. Mature males establish territories, which are crossed by small, loosely structured groups of foraging females and offspring.

E. Africa

- Up to 3½ ft (1.1 m)
- Up to 25½ in (65 cm)
- Up to 55 lb (25 kg)
- Herd
- Conserv. dependent

Grant's gazelle This large species has a unique courting ritual in which the male makes sputtering sounds as he follows the female with his head and tail raised. Grant's gazelle can stand on its hindlimbs to reach moisture-rich leaves.

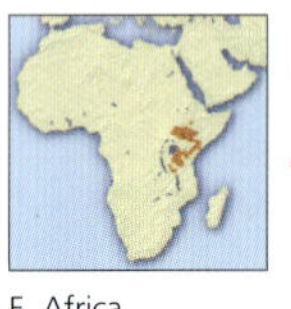

E. Africa

- Up to 5 ft (1.5 m)
- Up to 37½ in (95 cm)
- Up to 175 lb (80 kg)
- Herd
- Conserv. dependent

Spanish ibex
Capra pyrenaica

West Caucasian tur
Capra caucasica

Coat becomes redder in summer

Long outer coat protects warm, dense underfur

Mountain goat
Oreamnos americanus

Flexible hoof pads to grip uneven ground

Chamois
Rupicapra rupicapra

Takin
Budorcas taxicolor

West Caucasian tur This animal lives in stable groups of a few dozen, which may form herds of up to 500 animals. It migrates to higher ground for a diet of grasses in summer, switching to tree and shrub foliage from lower slopes in winter.

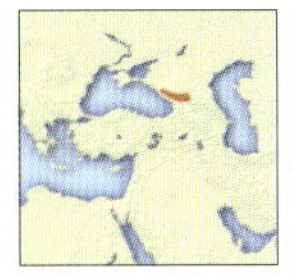

- Up to 5½ ft (1.7 m)
- Up to 3½ ft (1.1 m)
- Up to 220 lb (100 kg)
- Herd
- Endangered

W. Caucasus Mts

DEFENDERS AND GRAZERS

Sheep, goats, musk oxen, and their kin are collectively called goat-antelopes. They belong to the subfamily Caprinae, which first appeared in the tropics, but gradually moved into extreme locations such as deserts and mountains. Today's species range from resource defenders, such as serows, which live in productive habitats, to grazers, such as chamois, which tend to be wide-ranging and adapted to harsh conditions.

Alpine life *Living above the treeline in Europe and Western Asia, chamois announce danger with foot stamping and whistles, then flee in great leaps until reaching an inaccessible spot.*

Male has thick mane of fur around neck and shoulders

Nilgiri tahr
Hemitragus hylocrius

Himalayan tahr
Hemitragus jemlahicus

Arabian tahr
Hemitragus jayakari

Horns can be 5 feet (1.5 m) long

Inside edge of horns is sharp

Markhor
Capra falconeri

Bezoar (wild goat)
Capra aegagrus

Saola (Vu Quang ox)
Pseudoryx nghetinhensis

Nilgiri tahr Vast herds of this species once roamed the grass-covered hills of southern India, but hunting and habitat loss reduced its numbers to only about 100 animals. As a result of conservation efforts, the total population has now risen to about 1,000.

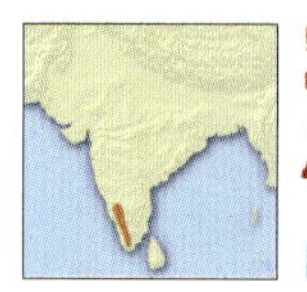

Up to 4½ ft (1.4 m)
Up to 3¼ ft (1 m)
Up to 220 lb (100 kg)
Herd
Endangered

Nilgiri Mts (S. India)

Markhor This animal has suffered from excessive hunting and now survives only in fragmented populations in isolated, rugged terrain above the treeline. Its spiraling horns are prized both by trophy hunters and in Chinese medicine.

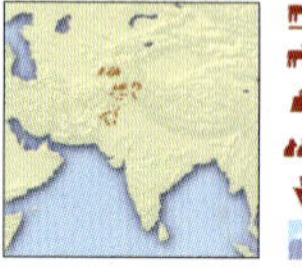

Up to 6 ft (1.8 m)
Up to 3½ ft (1.1 m)
Up to 245 lb (110 kg)
Herd
Endangered

Turkmenistan to Pakistan

Saola In 1992, the saola was discovered in Vietnam. Considered one of the world's rarest mammals, the saola is a nocturnal, forest-dwelling ox found only in remote mountainous regions. It travels in small groups of a few animals.

Up to 6½ ft (2 m)
Up to 35½ in (90 cm)
Up to 220 lb (100 kg)
Solitary, family band
Endangered

Laos, Vietnam

Male's horn size determines rank

Bighorn sheep
Ovis canadensis

Female horns have the same shape as male horns but are smaller

Aoudad (Barbary sheep)
Ammotragus lervia

Ventral mane of long white hair

Siberian bighorn sheep
Ovis nivicola

Dall sheep
Ovis dalli

Long guard hairs can almost reach the ground

Male horns almost meet at top of head in a "boss;" female horns are smaller with no boss

Musk ox
Ovibos moschatus

Bighorn sheep To win the right to mate, male bighorn sheep fight sometimes for more than 24 hours. Their massive horns can weigh up to 30 pounds (14 kg), and these sheep have a reinforced skull that is linked to the spine by a thick tendon.

W. North America

- Up to 6 ft (1.8 m)
- Up to 4 ft (1.2 m)
- Up to 300 lb (135 kg)
- Herd
- Conserv. dependent

ARCTIC SURVIVORS

With a range of arctic Alaska, Canada, and Greenland, musk oxen must survive long winters with subzero temperatures and little light. Their long guard hairs shield a dense underfur that is shed in spring. To reach grasses and sedges under the snow, musk oxen clear circular feeding patches with their horns and feet. A herd will form a protective circle around a young musk ox to safeguard it from a predator.

Mountain anoa
Bubalus quarlesi

Anoa
Bubalus depressicornis

Horns can be held flat along back to avoid becoming entangled in forest undergrowth

Tamaraw
Bubalus mindorensis

Hump of raised muscle

Bison
Bison bison

Hair longer at front than at rear of body

Large lungs and high red blood cell count allow yaks to flourish at high altitudes

Gaur (seladang)
Bos frontalis

Yak
Bos grunniens

Wild males are three times heavier than wild females, and two to three times heavier than domestic males

Kouprey
Bos sauveli

Bison About 60 million bison once lived in North America, but the species now survives in the wild only in two national parks. Bison live in groups of females, offspring, and a few older males. Other mature males may live alone or form bachelor bands.

Up to 11½ ft (3.5 m)
Up to 6½ ft (2 m)
Up to 1 ton (1 t)
Herd
Conserv. dependent

Canada, N.W. USA

Gaur Herds of females and juveniles gaur led by a single male emerge from the forest early in the day to graze on nearby grassy slopes. They return to the forest at night to sleep.

Up to 11 ft (3.3 m)
Up to 7¼ ft (2.2 m)
Up to 1 ton (1 t)
Herd
Vulnerable

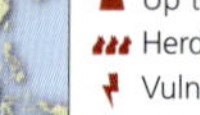

India to Indochina & Malaya

Yak While domesticated yaks are found throughout much of Asia, wild yaks are restricted to a small area of uninhabited alpine tundra and cold steppe regions.

Up to 11 ft (3.3 m)
Up to 6½ ft (2 m)
Up to 1 ton (1 t)
Herd
Vulnerable

Tibet

BOVID HORNS

All bovids have hollow, unbranched horns made up of a layer of keratin surrounding a bony core. Unlike tusks, horns leave the mouth free for grazing. Horns are sometimes used for defense, although bovids generally prefer to flee a predator whenever they can. The bovid species with the most elaborate horns are those that feature territorial or harem males, which must compete with one another in order to mate.

Greatest span
The African buffalo (Syncerus caffer) has a horn span of 50 inches (1.3 m).

Corkscrew horns
The eland of Africa's savanna has tightly twisted horns with sharp points.

Deer

CLASS	Mammalia
ORDER	Artiodactyla
FAMILIES	4
GENERA	21
SPECIES	51

The largest family in the deer group, Cervidae contains deer and their allies, including moose, caribou, and elk. In many ways, deer resemble antelopes, with long bodies and necks, slender legs, short tails, large eyes on the side of the head, and high-set ears. They are distinguished, however, by the often spectacular antlers borne by the males of most species (and also by female caribou). Unlike horns, which are permanent and made of keratin, antlers are made of bone and are shed once a year. Growing antlers are covered in skin known as "velvet," which dies and is rubbed off once the antlers reach full size. Antlers can be small, simple spikes or enormous, branched structures.

Deer distribution Deer never moved into sub-Saharan Africa, but they occur naturally in northwest Africa, Eurasia, and the Americas, and some have been introduced elsewhere. The species in the family Cervidae fall into two groups: Old World deer first evolved in Asia, while the New World group began in the Arctic.

Group grazers Most deer are found in temperate or tropical forest, but some species have adapted to colder conditions. Some species such as white-tailed deer live in groups, which offers a better chance of avoiding predation, as predators are more likely to be detected and tend to target only the weakest members.

LARGE AND SMALL

The family Cervidae ranges from the southern pudu, weighing as little as 17½ pounds (8 kg), to the moose, at 1,760 pounds (800 kg). A moose's antlers can have a span of 6½ feet (2 m). One species, the Chinese water deer, has no antlers at all, but its elongated canine teeth form knife-like tusks. Southeast Asia's muntjaks have only spikes for antlers but also bear tusks.

As prey species, deer have evolved various escape strategies. Some leap and dodge into a hiding spot. Others rely on great speed and stamina to outrun the threat. The moose can easily trot over obstacles that slow down its shorter predators.

All deer are ruminants with a four-chambered stomach, but, unlike bovids, they are not adapted to a diet of coarse grasses and rely on more easily digested food such as shoots, young leaves, new grasses, lichens, and fruit. Even those species that do graze on grasses need large amounts of high-quality browse as well.

The deer family is comprised of Cervidae and three other families of ungulates that superficially resemble them. These include the chevrotains of Tragulidae and the musk deer of Moschidae. North America's pronghorn is in its own family, Antilocapridae.

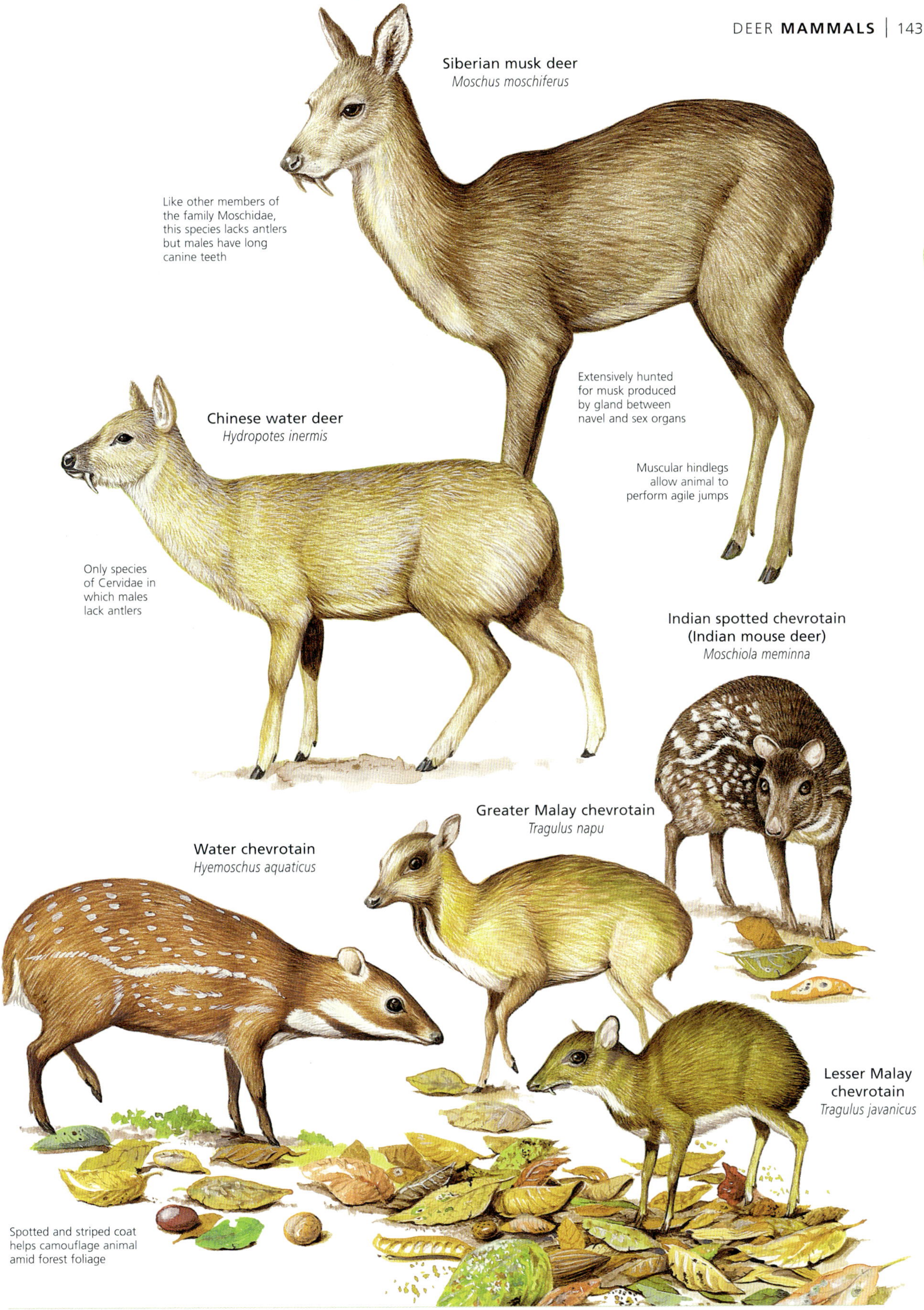

Siberian musk deer
Moschus moschiferus

Like other members of the family Moschidae, this species lacks antlers but males have long canine teeth

Extensively hunted for musk produced by gland between navel and sex organs

Muscular hindlegs allow animal to perform agile jumps

Chinese water deer
Hydropotes inermis

Only species of Cervidae in which males lack antlers

Indian spotted chevrotain (Indian mouse deer)
Moschiola meminna

Greater Malay chevrotain
Tragulus napu

Water chevrotain
Hyemoschus aquaticus

Lesser Malay chevrotain
Tragulus javanicus

Spotted and striped coat helps camouflage animal amid forest foliage

Water chevrotain Entirely nocturnal, this solitary animal hides in the dense undergrowth of the tropical forest by day. It usually lives near water and will enter water to avoid predators, but cannot swim for long periods.

Up to 37½ in (95 cm)
Up to 16 in (40 cm)
Up to 28½ lb (13 kg)
Solitary
Data deficient

Tropical W. Africa

Greater Malay chevrotain Because this species breeds at any time of year and will mate within a couple of hours of giving birth, females spend most of their life pregnant.

Up to 24 in (60 cm)
Up to 14 in (35 cm)
Up to 13 lb (6 kg)
Solitary
Uncommon

Indochina, Thailand to Malaya, Sumatra, Borneo

Lesser Malay chevrotain The smallest of all even-toed ungulates, this animal has legs about as thick as a pencil. It may live alone or in small family groups and feeds on fallen fruit and leaves.

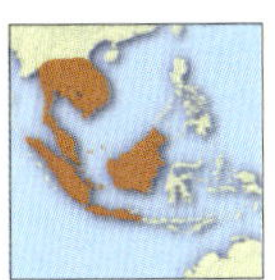

Up to 19 in (48 cm)
Up to 8 in (20 cm)
Up to 4½ lb (2 kg)
Solitary, small group
Uncommon

Indochina, Thailand to Malaya & Indonesia

Antlers can be up to 3 feet (1 m) long

Sambar
Cervus unicolor

Barasingha
Cervus duvaucelii

When tail is raised, the white underside acts as a "follow me" signal

Eld's deer
Cervus eldii

Rusa
Cervus timorensis

Fawn is spotted for camouflage

Roosevelt elk
Cervus elaphus roosevelti

IN A RUT

Known as wapiti or elk in North America and as red deer in Europe, *Cervus elaphus* is the noisiest of all deer species. After beginning courtship with a bugling call, males collect a harem, which they vigorously defend against other males throughout the rutting, or mating, season. A harem male and his challenger will roar in warning for minutes before resorting to physical aggression.

Vocal assault *A harem male will roar for several minutes before engaging in a physical contest.*

Antler lock *After performing a ritual parallel walk, competing males will lock antlers and wrestle until one is pushed back and flees.*

Critically endangered Chinese deer; disappeared from the wild around AD 200, but survived because a captive herd was maintained by the Chinese royal family and breeding pairs were reared in Europe; reintroduced into two Chinese national parks in the 1980s

Mesopotamian fallow deer
Dama mesopotamica

Père David's deer
Elaphurus davidianus

Chital
Axis axis

Palm-shaped antlers have numerous points

Philippine hog deer
Axis calamianensis

Fallow deer
Dama dama

Tufted deer
Elaphodus cephalophus

Tuft of hair hides male's antlers

Giant muntjak
Megamuntiacus vuquangensis

Tusk-like canine teeth on male

Indian muntjak
Muntiacus muntjak

Chital This deer is mainly a grassland grazer, but will enter nearby forests to browse fallen fruit and leaves. Herds of females and young are followed by dominant males in the breeding season.

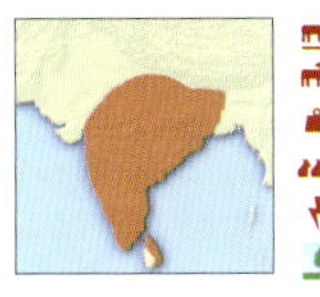

Up to 6 ft (1.8 m)
Up to 3¼ ft (1 m)
Up to 245 lb (110 kg)
Herd
Common

Sri Lanka, India, Nepal

Tufted deer The male tufted deer has simple, spiked antlers and long, tusk-like upper canines. The antlers are often hidden by the hair on its forehead.

Up to 5¼ ft (1.6 m)
Up to 27½ in (70 cm)
Up to 110 lb (50 kg)
Solitary
Data deficient

E. Tibet & N. Myanmar to S.E. China

Indian muntjak Also called the barking deer, this species may bark for more than an hour if it senses a predator. It is omnivorous, using kicks of its forelimbs and bites to subdue small prey.

Up to 3½ ft (1.1 m)
Up to 25½ in (65 cm)
Up to 61½ lb (28 kg)
Solitary
Uncommon

Sri Lanka, India, Nepal to S. China, S.E. Asia

Marsh deer
Blastocerus dichotomus

Moose (elk)
Alces alces

Caribou (reindeer)
Rangifer tarandus

White-tailed deer
Odocoileus virginianus

Pampas deer
Ozotoceros bezoarticus

Mule deer (black-tailed deer)
Odocoileus hemionus

Roe deer
Capreolus capreolus

ANTLER CYCLE

Antlers are used in contests between male deer, but the reason they grow so large appears to be that they advertise the male's healthy genes to females. In species with the largest antlers, much of the courtship routine involves elaborate antler displays.

Spring *In temperate species, new antlers begin growing in late spring. They are covered in sensitive skin known as "velvet."*

Summer *By late summer, antlers are fully grown and have hardened. The velvet begins to dry and loosen.*

Fall *The male rubs the velvet off on shrubs and small trees. The antlers are now ready for the contests and displays of the mating season.*

Winter *Following the mating season, the two antlers are shed within days of each other.*

Red brocket
Mazama americana

Both male and female bear horns, but the male's horns are longer and forked

Black markings only on male

Pronghorn
Antilocapra americana

Little red brocket
Mazama rufina

Peruvian guemal
Hippocamelus antisensis

Appears on Chile's coat of arms

Andean guemal
Hippocamelus bisulcus

Northern pudu
Pudu mephistophiles

Southern pudu
Pudu puda

Smallest of all deer in Cervidae family

Red brocket An elusive creature that most often lives in dense, tropical forest, this small deer can hide in undergrowth or swim to escape predators. It lives alone or in pairs and feeds on fruit, leaves, and fungi.

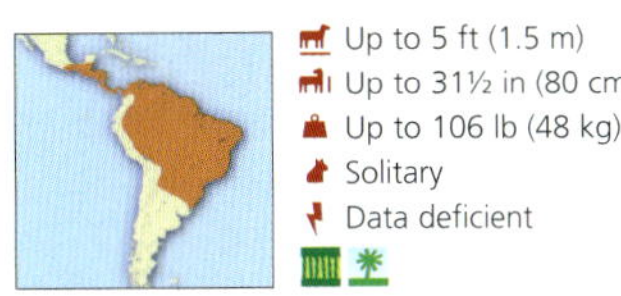

- Up to 5 ft (1.5 m)
- Up to 31½ in (80 cm)
- Up to 106 lb (48 kg)
- Solitary
- Data deficient

S. Mexico to N. Argentina

Southern pudu The smallest of all true deer species, the southern pudu will stand on its hindlegs to reach the leaves of trees or test the wind. Solitary except during the mating season, it follows well-marked trails to its feeding and resting spots.

- Up to 32½ in (83 cm)
- Up to 17 in (43 cm)
- Up to 28½ lb (13 kg)
- Solitary
- Vulnerable

S. Chile, S.W. Argentina

Pronghorn The sole species in the family Antilocapridae, the pronghorn has unusual forked horns, which are made from keratin surrounding a bony core, like the horns of antelope. As with the antlers of deer, however, the keratin is shed annually.

- Up to 5 ft (1.5 m)
- Up to 3¼ ft (1 m)
- Up to 154 lb (70 kg)
- Herd
- Locally common

W. North America

GIRAFFE AND OKAPI

CLASS Mammalia
ORDER Artiodactyla
FAMILY Giraffidae
GENERA 2
SPECIES 2

With its head hovering up to 18 feet (5.5 m) above the ground, the giraffe is the tallest animal in the world. Along with its only close relative, the okapi, it is classified in the family Giraffidae. Both the giraffe and the okapi have a long neck, tail, and legs, with the forelimbs longer than the hindlimbs, creating a sloping back. Their small, constantly growing horns consist of bone covered by furred skin and are unique among mammals. The lips of giraffids are thin and mobile; and the tongue is long, prehensile, and black. Both species are found only in sub-Saharan Africa, where their strikingly patterned coats help them blend into their habitat—the giraffe's blotches mimic the dappled light of savanna woodland, while the okapi's rear stripes break up its outline amid the dense vegetation of the rain forest.

DIFFERENT LIFESTYLES

Beyond their similarities, giraffes and okapi differ significantly. The most obvious contrast is in size and shape, with the okapi appearing rather horse-like, while the extreme elongation of the giraffe makes it instantly identifiable. The giraffe has only seven vertebrae in its neck, the same number as almost all other mammals, but each vertebra is lengthened. A specialized circulatory system powerfully pumps blood all the way up to its brain, but a series of valves adjusts the pressure when the animal leans down to drink.

The giraffe's extraordinary stature has allowed it to fully exploit the resources of its savanna woodland home. Because it is able to reach the leaves of tall acacia trees throughout the dry season, a giraffe can grow to a dramatic height and reproduce year-round. It is most vulnerable to lions and other predators when lying down or drinking. To avoid predation, a giraffe depends on its acute senses of vision, smell, and hearing. It may run away at speeds of more than 30 miles per hour (50 km/h) or deliver sharp kicks to the foe with its forefeet.

Living in dense, dark tropical forest, the okapi has poor vision but sharp hearing and a good sense of smell. It is extremely wary and will disappear into thick cover at the first hint of danger. Mostly solitary, this species marks its territory with urine or by rubbing its neck on trees.

The more open savanna habitat has encouraged giraffes to be social, and most live in small, loose herds of about a dozen animals. Young males may live in bachelor bands but tend to become solitary as they age. Males may fight each other for the right to mate, repeatedly swinging their long neck to deliver powerful head-butts to the rival's underbelly. A reinforced skull usually absorbs the impact of these blows, but occasionally an animal is knocked unconscious.

Necking rivals To establish rank, young male giraffes engage in ritualized necking contests. Much like arm wrestles, these involve two giraffes intertwining necks and pushing each other until one gives way. Competing okapi bulls also neck-wrestle before resorting to more aggressive contact.

Kenyan giraffe
Giraffa camelopardalis tippelskirschi

Southern giraffe
Giraffa camelopardalis giraffa

Horns on both males and females

Short mane along neck

Long, tufted tail used to whisk away flies

Forelimbs longer than hindlimbs

Reticulated giraffe
Giraffa camelopardalis reticulata

Only males have horns

Okapi
Okapia johnstoni

Giraffe These gregarious animals most often live in loose, temporary herds of about a dozen cows and offspring led by an adult male. Giraffes can breed at any time of year. They are able to survive for long periods without drinking.

Sub-Saharan Africa

Up to 18½ ft (5.7 m)
Up to 11½ ft (3.5 m)
Up to 1½ tons (1.4 t)
Variable
Conserv. dependent

IDENTIFYING MARKS

The blotched pattern on a giraffe's coat helps to camouflage the animal in the dappled light of the savanna woodlands. Each individual giraffe has a unique pattern, but broad similarities define the various subspecies. Recent genetic evidence suggests that there may be up to six subspecies, with color and pattern variations.

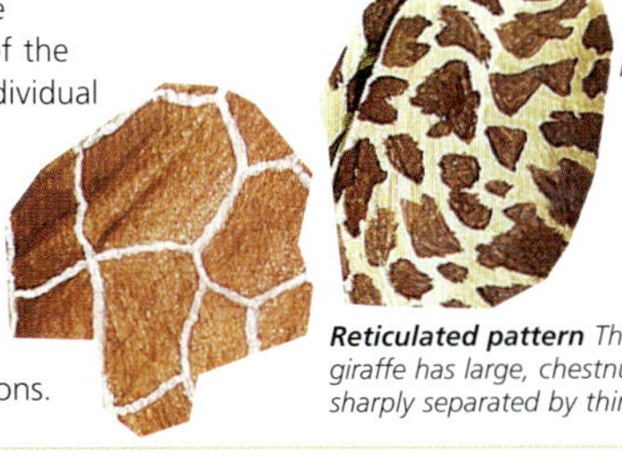

Kenyan contrast *Fuzzier, smaller, darker blotches separated by larger areas of white identify the Kenyan subspecies.*

Reticulated pattern *The reticulated giraffe has large, chestnut patches sharply separated by thin white lines.*

THE CAMEL FAMILY

CLASS	Mammalia
ORDER	Artiodactyla
FAMILY	Camelidae
GENERA	3
SPECIES	6

Famed for their humps and the ability to survive for long periods without drinking, the two species of camel are the single-humped dromedary, now found only in domesticated opulations in northern Africa and the Middle East, and the Bactrian camel, domesticated in northern Asia, but also found in small numbers in the wild. Their relatives in the family Camelidae are the four camelids of South America—the wild guanaco and vicuña, and the domesticated llama and alpaca. Camelids first appeared some 45 million years ago in North America, but disappeared from there about 10,000 years ago at the end of the ice age. By then, they had dispersed to other parts of the world.

Old and new The two Old World camels occur in northern Africa and central Asia. The four South American species range from the foothills to the alpine meadows of the Andes. Domesticated camelids have been introduced in many places, including Australia, where feral dromedaries roam the central desert.

ROBUST CAMELIDS

All camelids are adapted to arid or semiarid regions. A three-chambered ruminating stomach extracts nutrition from their main food of grasses. Their feet are unique among hoofed mammals in that only the front of the hoof touches the ground, and the animal's weight rests instead on a fleshy sole-pad. In camels, the feet are broad, helping it to travel over soft sand without sinking. The four South American species have a narrower foot for walking securely up rocky slopes. A thick double coat insulates against both heat and cold.

While Old World camelids are distinguished from the New World species by their much larger size and prominent humps, the overall anatomy is similar. All species have long, slender legs, a short tail, a long, curved neck, and a relatively small head with a split upper lip. When camelids walk, the front and back legs on the same side of the body move in unison, a distinctive gait known as pacing. Camelids are social animals and tend to live in harems of females and young led by a dominant male. Males without a harem may form bachelor bands.

By herding camelids, which provide meat, milk, wool, fuel, and transportation, humans have been able to make a living in extreme locations, from the hot Sahara to the cool high plains of the Andes. There are more than 20 million camelids in the world, but roughly 95 percent are domestic animals.

Precocious young In all camelid species, a single, well-developed young is born after a long gestation period, which lasts 11 months in South America's guanaco. Grass constitutes the bulk of their diet and guanacos can survive for long periods without drinking.

Bactrian camel
Camelus bactrianus

Fat stored in humps is used when food is scarce, causing humps to shrink

Long winter coat shed in summer

Narrow nostrils can close during dust storms

Long eyelashes keep desert dust out of eyes

Thick, tough lips can handle thorny vegetation

Dromedary
Camelus dromedarius

Guanaco
Lama guanicoe

Vicuña
Vicugna vicugna

Bactrian camel Along with the dromedary, this species is unique among mammals because its blood cells are oval rather than round. This shape may help the cells travel through thick, dehydrated blood.

Up to 11½ ft (3.5 m)
Up to 7½ ft (2.3 m)
Up to 1,540 lb (700 kg)
Herd
Critically endangered

Kazakhstan to Mongolia

Guanaco Male guanacos, like all male South American camelids, have some sharp, hooked teeth that are used as weapons during fights with rival males.

Up to 6½ ft (2 m)
Up to 4 ft (1.2 m)
Up to 265 lb (120 kg)
Family band
Locally common

S. Peru to E. Argentina & Tierra del Fuego

Vicuña Strictly a grazer, this small camelid has sharp, constantly growing incisors for snipping short grasses. They use one territory for day foraging and a higher, therefore safer, territory in their mountain habitat for sleeping.

Up to 6¼ ft (1.9 m)
Up to 3½ ft (1.1 m)
Up to 143 lb (65 kg)
Family band
Conserv. dependent

S. Peru to N.W. Argentina

Pigs

CLASS	Mammalia
ORDER	Artiodactyla
FAMILY	Suidae
GENERA	5
SPECIES	14

Unlike most other ungulates, which are strictly herbivorous, the pigs, hogs, boars, and babirusa in the family Suidae are omnivores with a diet that includes insect larvae, earthworms, and small vertebrates, as well as a wide array of plants. The nostrils on a pig's prominent snout are enclosed in a disk of cartilage. Supported by a unique prenasal bone, this disk helps locate food by shoveling through leaf litter or dirt. The upper and lower canines in both males and females of most species form sharp tusks, which can be used as weapons. Occurring naturally in the forests of Africa and Eurasia, wild pigs have also been introduced in North America, Australia, and New Zealand.

Wild boar
Sus scrofa

Female has smaller tusks than male

Piglets striped for camouflage; stripes fade with age

Weighs 13–20 pounds (6–9 kg), making it the smallest species in Suidae

Pygmy hog
Sus salvanius

Family ties Male wild pigs tend either to live alone or to belong to a bachelor band, while females and their offspring live in close-knit family groups known as sounders. The wild boar is an aggressive omnivore. Adult males sport tusks that are sharpened by grinding and used as weapons.

Warthog
Phacochoerus africanus

Padded knees allow for kneeling during feeding

Giant hog
Hylochoerus meinertzhageni

Bush pig
Potamochoerus larvatus

Upper tusks can grow to 14 inches (35 cm) long

Large folds and wrinkles in skin

Red river hog
Potamochoerus porcus

Mane and ear tassels can be fluffed out to increase the animal's apparent size

Lower canines used in fighting

Babirusa
Babyrousa babyrussa

Warthog This grassland grazer kneels on its forelimbs to feed, using its specialized incisor teeth to pluck new growth. When the grasses shrivel during the dry season, it digs out grass rhizomes (underground stems).

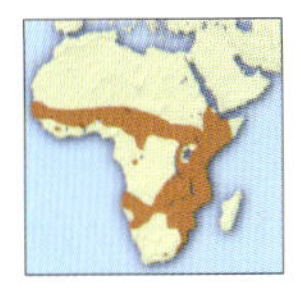

Sub-Saharan Africa

- Up to 5 ft (1.5 m)
- Up to 27½ in (70 cm)
- Up to 230 lb (105 kg)
- Mainly solitary
- Common

Giant hog Around dusk, a mixed group of giant hogs will retreat through dense vegetation to a large sleeping nest. Females share the care of the piglets, suckling and protecting any in the group.

C. & W. Africa

- Up to 7 ft (2.1 m)
- Up to 3¼ ft (1 m)
- Up to 520 lb (235 kg)
- Family band
- Uncommon

Babirusa With a diet of foliage, fruit, and fungi more specialized than that of other pigs, the babirusa also rarely uses its snout to root out food. Fossil studies suggest that it is the most primitive of all pig species.

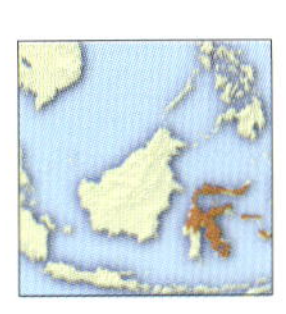

Sulawesi & nearby small islands

- Up to 3½ ft (1.1 m)
- Up to 31½ in (80 cm)
- Up to 220 lb (100 kg)
- Family band
- Vulnerable

PECCARIES

CLASS	Mammalia
ORDER	Artiodactyla
FAMILY	Tayassuidae
GENERA	3
SPECIES	3

Although they resemble the pigs of Suidae in many ways, the three species of peccary in the family Tayassuidae can be distinguished by their long, slender legs, a more complex stomach, and a scent gland on the rump. They are omnivorous like pigs, but prefer fruit, seeds, roots, and vines, with the Chaco peccary depending largely on cacti. These gregarious animals live in herds ranging from 2–10 Chaco peccaries to 50–400 white-lipped peccaries. Social bonds are reinforced by herd members rubbing their cheeks on each other's scent glands. A few white-lipped peccaries will stay behind to fight a predator, allowing the rest of the herd to flee.

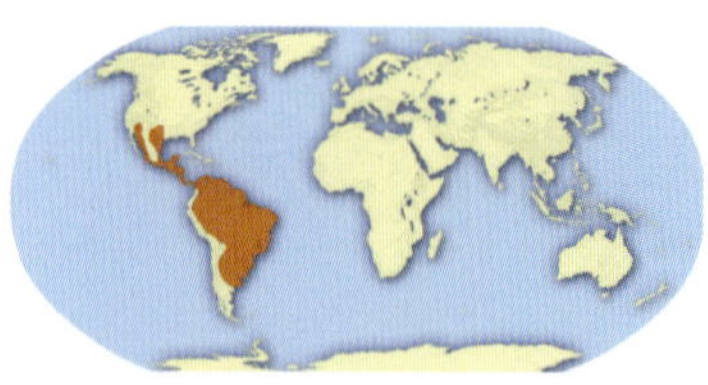

American pigs While pigs occur naturally only in Africa and Eurasia, peccaries are restricted to the Americas, where they range from southwest United States to northern Argentina. The collared peccary and white-lipped peccary are found in tropical forest, wooded savanna, and thorn scrub. The Chaco peccary is found mainly in semiarid thorn forest.

Stinky The Chaco peccary emits a strong odor from a scent gland on its rump when frightened or to mark its territory. It is distinguished from other peccary species by its larger size, long bristles, and shaggy appearance.

Collared peccary (javelina)
Pecari tajacu

White or yellowish collar of hair around shoulders and throat

Incisors form sharp tusks

White-lipped peccary
Tayassu pecari

Disk of cartilage at end of snout

Chaco peccary
Catagonus wagneri

Known only from fossils until its discovery in the wild in 1972

HIPPOPOTAMUSES

CLASS Mammalia
ORDER Artiodactyla
FAMILY Hippopotamidae
GENERA 2
SPECIES 4

Now known to be more closely related to whales than to other ungulates, the two surviving species of hippopotamus lead a semiaquatic life, spending the day resting in water and emerging at night to forage on land. Their thick skin has only a thin outer layer, which rapidly dries out and cracks unless regularly moistened. Both hippopotamus species have large heads, a barrel-shaped body, and surprisingly short legs. There is, however, a huge size disparity, with the grassland-grazing common hippopotamus being seven times heavier than the forest-foraging pygmy hippo.

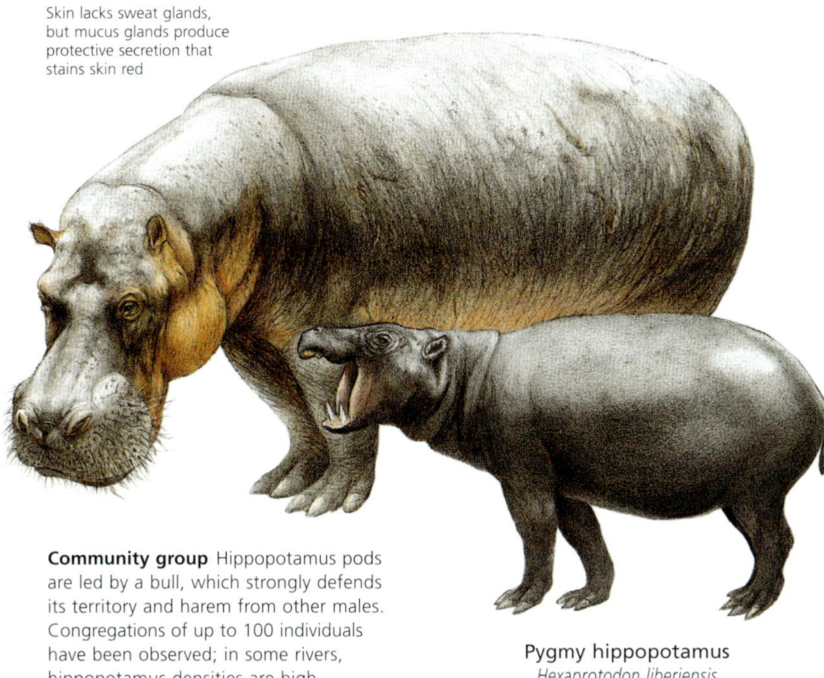

Hippopotamus
Hippopotamus amphibius

Skin lacks sweat glands, but mucus glands produce protective secretion that stains skin red

Pygmy hippopotamus
Hexaprotodon liberiensis

UNDERWATER UNGULATE

Lacking sweat glands, the common hippopotamus relies on water to stay cool. It is a good swimmer and diver, and the density of its body allows it to walk along a river or lake bed and stay submerged for approximately 5 minutes at a time. It can float by filling its lungs with air. A hippo's feet are webbed; the nostrils and ears can close underwater; and the eyes, ears, and nostrils are positioned so that it can see, hear, and breathe with just the top of the head emerging above the surface. Herds of up to 40 hippos may spend the day together in water, devoting most of their time to sleeping or resting. At night, they leave the water to feed on land for about 6 hours. Because their weight is often borne by water, hippos conserve energy and need relatively little food. Young are born and suckled underwater.

Community group Hippopotamus pods are led by a bull, which strongly defends its territory and harem from other males. Congregations of up to 100 individuals have been observed; in some rivers, hippopotamus densities are high.

CETACEANS

CLASS	Mammalia
ORDER	Cetacea
FAMILIES	10
GENERA	41
SPECIES	81

With their entirely aquatic lifestyle, the whales, dolphins, and porpoises of the order Cetacea are perhaps the most specialized of all mammals. They feed, rest, mate, give birth, and raise young in the water, yet they are warm-blooded and breathe air like other mammals. Gregarious and intelligent, cetaceans appear to be descended from the same land mammal that led to hippopotamuses, but their ancestors adapted to a watery life some 50 million years ago. Over time, they became as streamlined as fish, losing their hair and hindlimbs, modifying their arms into flippers, and developing a powerful fluked tail that makes some species the fastest creatures in the sea.

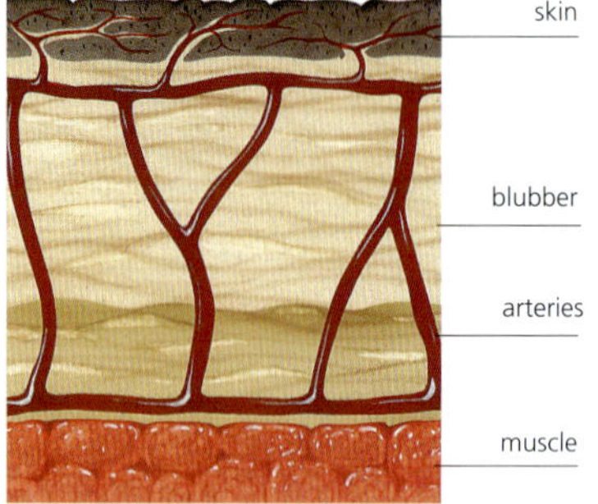

Staying warm and cool Being virtually hairless, a cetacean relies on a layer of blubber beneath the skin for insulation. A network of arteries and veins in the blubber, known as *retia mirabilia,* helps the animal to regulate its temperature.

CETACEAN RECORDS

Cetaceans are found in all of the world's oceans and seas, and in some rivers and lakes. They are split into two living suborders: the toothed whales of Odontoceti, and the baleen whales of Mysticeti. Toothed whales, which include dolphins, porpoises, and sperm whales, have simple, conical teeth that can keep a firm grip on their slippery food of fishes and squids. Baleen whales include blue whales, humpbacks, gray whales, and right whales. They are filter feeders, straining great quantities of tiny plankton, other invertebrates, and small fish through bristled horny plates that hang from the roof of the mouth.

With water supporting their weight, some cetaceans have been able to reach enormous sizes. The blue whale is the largest animal that has ever lived, with a record weight of 209 tons (190 tonnes)—roughly equivalent to the weight of 35 elephants—and a record length of 110 feet (33.5 m).

Another cetacean, the sperm whale, boasts the deepest and longest dives of any mammal. Sperm whales are believed to descend to at least 10,000 feet (3,050 m), and their dives can last for more than 2 hours. When a cetacean dives, its heart rate slows by 50 percent and blood is directed away from the muscles to the vital organs, allowing the animal to survive on very little oxygen until it ascends to breathe.

Most cetaceans are gregarious to some extent. Toothed whales tend to form larger groups than baleen whales and have more complex social structures. Occasionally, thousands of common dolphins

Orca Also called killer whales, orcas are found in all the world's oceans, especially the polar seas. Orcas often skyhop, poking their head above the surface in search of prey that can be captured and consumed.

travel together. Members of a group usually feed at the same time and may hunt cooperatively, herding fish into clusters.

Cetaceans have little or no sense of smell. Their relatively small eyes provide reasonable vision both above and beneath the water's surface. All species lack external ears, but their hearing is highly sensitive, allowing them to pick up distant calls from members of their species. To find prey and avoid obstacles, toothed whales use echolocation, emitting a series of clicks and whistles, and then analyzing the reflected sounds.

Sound is critical in cetacean communication. Blue whales and fin whales emit low-frequency pulses that carry across vast stretches of ocean and can reach 188 decibels—the loudest sound made by an animal. Male humpbacks produce the longest and most complex songs in the animal kingdom.

Because cetaceans spend most of their time underwater, accurate population statistics are difficult to compile. Nevertheless, it is certain that human activities have devastated cetacean numbers. Commercial whaling (now largely banned), driftnet fishing (which inadvertently traps cetaceans), and water pollution have all taken a heavy toll.

Aquatic mammal Although their body shape is highly modified for life in water, dolphins (below) and other cetaceans are warm-blooded and breathe air through lungs. They have a four-chambered heart and a three-chambered stomach.

Humpback song Male humpback whales sing complex songs that can be made up of nine themes and last half an hour. All males in an ocean basin sing the same song, but it may gradually change over time.

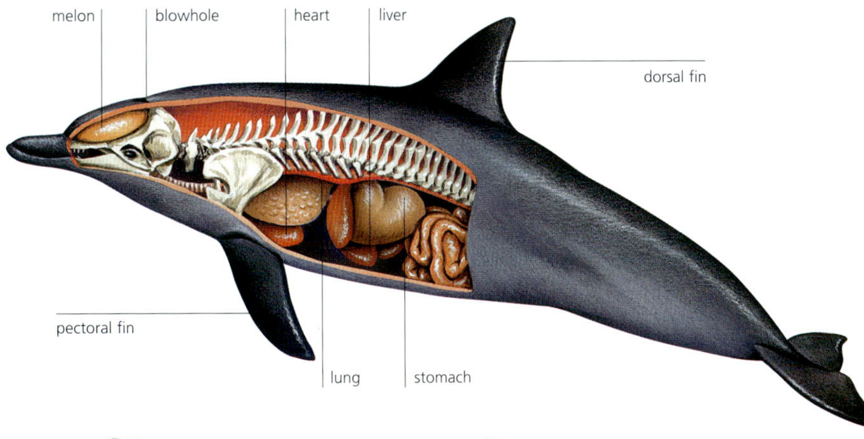

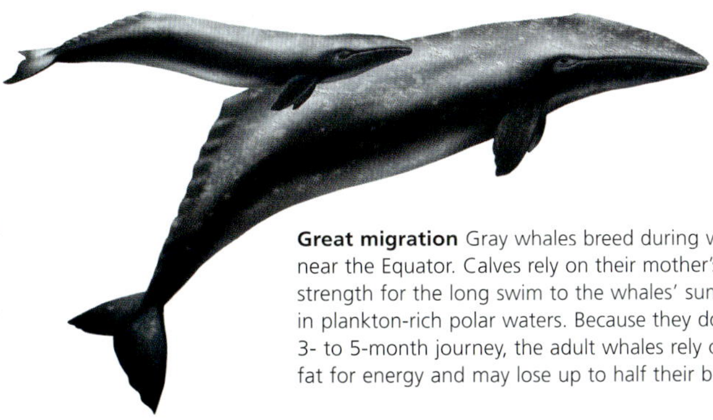

Great migration Gray whales breed during winter in warm waters near the Equator. Calves rely on their mother's rich milk to build up strength for the long swim to the whales' summer feeding grounds in plankton-rich polar waters. Because they do not eat during the 3- to 5-month journey, the adult whales rely on their blubber and fat for energy and may lose up to half their body weight.

TOOTHED WHALES

CLASS	Mammalia
ORDER	Cetacea
FAMILIES	6
GENERA	35
SPECIES	68

About 90 percent of all cetaceans are toothed whales belonging to one of the six families in the suborder Odontoceti. In contrast to the enormous baleen whales, toothed whales tend to be medium-sized, although the largest of them, the sperm whale, is a massive creature. Their brains are relatively large, making them the most intelligent mammals other than primates. While some species are solitary, most are highly social and tend to be very vocal and playful. Members of a group may hunt cooperatively and help care for each other's young. Most toothed whales feed on fishes or squids, but one species, the orca, actively pursues warm-blooded prey such as seals and other whales.

Teeth Toothed whales have sharp, conical teeth. In fish-eating dolphins, such as a bottle-nosed dolphin (below), these are small and numerous. Orcas, which hunt marine mammals, have fewer but larger teeth. Squid-eating beaked whales have just a single tooth per jaw.

SOCIAL CETACEANS

The diverse members of Odontoceti include sperm whales; narwhals and belugas; beaked whales; dolphins, orcas, and pilot whales (grouped together in the family Delphinidae); porpoises; and river dolphins. Most have an elongated, beak-like head with sharp, conical teeth that can firmly seize prey but cannot chew it. Because there is only a single blowhole, the skull is asymmetrical. It supports a fluid-filled organ called the melon, which is thought to focus the clicks used in echolocation and communication. In sperm whales, the melon is greatly enlarged and filled with an oil known as spermaceti. This spermaceti organ may also help to focus sounds.

There is considerable variety in the social organization of toothed whales. Most groups are centered on the females, with males leaving the group at puberty. Orcas and pilot whales, however, never leave their birth group. River dolphins tend to form small groups or even live alone. Coastal dolphins form larger groups because their prey is concentrated in particular areas and they face more predators.

Despite their reputation for being gentle and playful, dolphins do fight. Group living often involves rivalry for food or mates, and this may lead to physical clashes. Many toothed whales bear tooth rake marks as scars of such encounters.

vertebrae
fused neck vertebrae
extended beak-like head
sternum
rib cage
enlarged mandible holds muscles used in echolocation
"fingers" within flipper

Long and narrow The skeleton of a toothed whale has been greatly modified from that of its land mammal ancestor. The hindlimbs have disappeared, while the forelimbs have become flippers, although the bones for five fingers remain. The head is usually long and narrow, forming a beak.

Ganges dolphin Almost completely blind, this freshwater dolphin relies on echolocation to navigate its murky habitat, and uses its snout to probe the muddy bottom for shrimps and fishes. Only a few thousand individuals survive.

- Up to 10 ft (3 m)
- Up to 200 lb (90 kg)
- Solitary
- Endangered

Rivers in India, Bangladesh & Nepal

UNICORNS OF THE SEA

In male narwhals, the left of its two teeth erupts and grows into a very long, tightly spiraled tusk, which can reach 10 feet (3 m) in length.

- Up to 20 ft (6 m) + tusk 10 ft (3 m)
- Up to 3,530 lb (1,600 kg)
- Variable
- Data deficient

Arctic Ocean

Tucuxi (river dolphin)
Sotalia fluviatilis

Found in both marine and freshwater environments

Striped dolphin
Stenella coeruleoalba

Short, stubby beak

Bottle-nosed dolphin
Tursiops truncatus

Largest of the beaked dolphins

Rough-toothed dolphin
Steno bredanensis

Common dolphin
Delphinus delphis

Criss-crossing scars from fights with squid prey or other dolphins

Risso's dolphin
Grampus griseus

Commerson's dolphin
Cephalorhynchus commersonii

Atlantic white-sided dolphin
Lagenorhynchus acutus

Two-tone coloration helps to camouflage animal in its marine environment

Pacific white-sided dolphin
Lagenorhynchus obliquidens

ENTANGLED VICTIMS

The vast nets used in commercial fisheries pose a great risk to dolphins, which follow their prey into the nets and become entangled. Unable to surface to breathe, the dolphins soon drown. Measures to make the nets more visible have helped, but many thousands of dolphins are still accidentally captured each year.

BOTTLE-NOSED DOLPHIN

This is the species made famous by the *Flipper* television series and is most often seen performing in marine parks. In the wild, it is found both inshore and offshore living in groups of about a dozen animals, which sometimes form herds of hundreds. Present in most temperate and tropical waters, the bottle-nosed dolphin ranges widely for food, swimming at an average speed of 12 miles per hour (20 km/h).

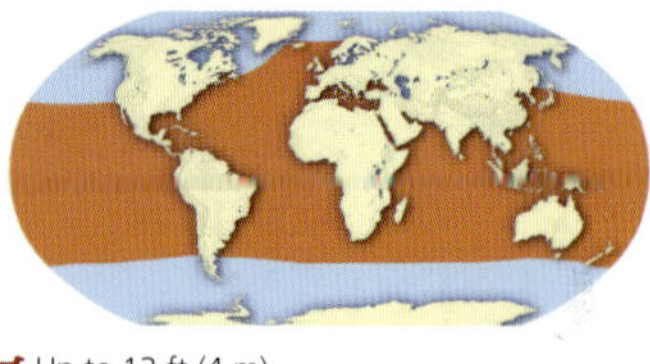

Up to 13 ft (4 m)
Up to 605 lb (275 kg)
Variable
Data deficient

Temperate to tropical oceans & seas

Irrawaddy dolphin This species prefers shallow coastal waters, but can live permanently in freshwater rivers. Like other toothed whales, it emits a series of clicks and then detects the echoes to find its way around and locate prey.

- Up to 9 ft (2.8 m)
- Up to 440 lb (200 kg)
- Family band
- Data deficient

Seas & rivers from Asia to Australia

ORCA

This versatile predator will hunt in packs to trap fishes and squids, beach itself to grab sea lions from the shore, or tip over ice floes to send seals and penguins into the sea.

- Up to 32 ft (9.8 m)
- Up to 10 tons (9 t)
- Variable
- Conservation dependent

Worldwide, but especially polar seas

Spectacled porpoise
Australophocaena dioptrica

Named for circles around the eyes

Dorsal fin set farther back than on any other small cetacean

Burmeister's porpoise
Phocoena spinipinnis

Gulf porpoise (vaquita)
Phocoena sinus

Common porpoise (harbor porpoise)
Phocoena phocoena

Tail sends up spray in the shape of a rooster's tail

Less shy and slow than other porpoises

Dall's porpoise
Phocoenoides dalli

Dorsal ridge rather than fin

Finless porpoise
Neophocaena phocaenoides

Gulf porpoise Found only in the northern part of the Gulf of California, this species has a more restricted range than any other cetacean. It appears to have evolved from Burmeister's porpoise, but was isolated when tropical waters became warmer.

- Up to 5 ft (1.5 m)
- Up to 121 lb (55 kg)
- Not known
- Critically endangered

Estuary of Colorado River; Gulf of California

COMMON PORPOISE

A rounded shape and small flippers, tail, and fin minimize this cetacean's surface area. Along with a layer of blubber, this helps it survive in cooler waters despite its small size.

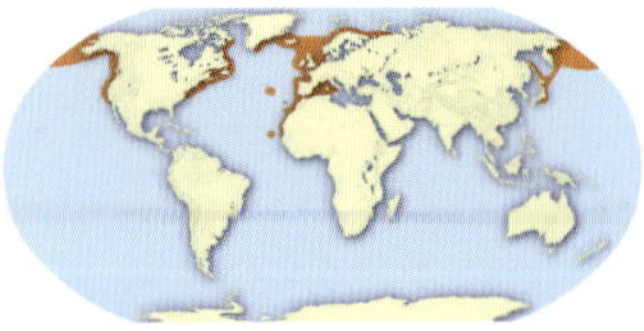

- Up to 6¼ ft (1.9 m)
- Up to 143 lb (65 kg)
- Variable
- Vulnerable

Temperate waters of Northern Hemisphere

CETACEAN STRATEGIES

To reap a rich harvest from the ocean, cetaceans have developed remarkably varied physical characteristics and behavioral strategies. Baleen whales have enormous mouths that can engulf huge quantities of tiny animals, while toothed whales pursue individual prey, relying on echolocation to find it. A dolphin may emit up to 600 clicks a second, analyzing the echoes to build up a picture of its surroundings, including the position of prey. Orcas use echolocation when pursuing fish, but rely more on vision to hunt other cetaceans or seals, which would be alerted by the clicks. Many species practise cooperative hunting, using a range of vocalizations to communicate the next move.

Humpback feeding Humpbacks will synchronize their feeding, lunging at prey shoals together or herding scattered prey into clusters. In bubblenet feeding (below), a humpback spirals toward the surface while exhaling, producing a large "net" of bubbles that traps small prey. The whale lunges through the center of the net to capture its meal.

3. Lunge feeding *Once the bubblenet has trapped the fishes, the humpback swims through the center, lunging to the surface with its mouth open to engulf the prey.*

1. Slow exhale *As a humpback whale spirals to the surface, it slowly lets out its breath, creating columns of bubbles. Small schooling fishes are trapped inside the net of bubbles.*

2. Group effort *Bubblenetting may be carried out by a single whale, or several whales may cooperate to create the net.*

Pygmy sperm whale
Kogia breviceps

Sperm whale
Physeter catodon

Hump and ridges on back rather than dorsal fin

Males weigh three times as much as females

Baird's beaked whale
Berardius bairdii

Both males and females have two pairs of protruding teeth

Northern bottle-nosed whale
Hyperoodon ampullatus

Blainville's beaked whale
Mesoplodon densirostris

May spend 30 minutes plunging to depths of 3,300 feet (1,000 m)

Cuvier's beaked whale
Ziphius cavirostris

Female usually a little larger than male

Stubby beak

Sowerby's beaked whale
Mesoplodon bidens

TUSKED BEAKS

A diet of squids captured through suction has rendered beaked whales virtually toothless. In males, however, one or two pairs of teeth protrude from the mouth to form tusks, which appear to be used as weapons.

Wrap-arounds *In the male strap-toothed whale* (Mesoplodon layardii)*, the tusks are especially long and wrap around the upper jaw. As a result, the mouth can only open an inch (2.5 cm) or so.*

BAIRD'S BEAKED WHALE

Tightly knit groups of 6–30 Baird's beaked whales are led by a dominant male. Rivalry for this position often leads to physical conflict, and most males bear scars on their beak and back.

- Up to 43 ft (13 m)
- Up to 16½ tons (15 t)
- Variable
- Conservation dependent

North Pacific

Baleen whales

CLASS	Mammalia
ORDER	Cetacea
FAMILIES	4
GENERA	6
SPECIES	13

The giants of the ocean, the baleen whales of the suborder Mysticeti feed on tiny prey, filtering aquatic invertebrates and small fishes through their sieve-like baleen plates. Their remarkable size is a great advantage in cooler waters, since, relative to body mass, they have a small surface area from which to lose heat. A thick layer of blubber provides insulation and can act as a food store for the epic annual migrations that many species undertake. Found in all the world's oceans, baleen whales include the gray whale, the right whales, the bowhead whale, and the rorquals—which comprise the blue whale, fin whale, sei whale, Bryde's whale, humpback whale, and minke whales.

Great gulp Blue whales have a pleated throat that expands into a great pouch as they gulp in water and plankton. The throat contracts again as they force the water back out and trap the prey in their bristled baleen plates.

BALEEN AND BLUBBER

Baleen whales can be skimmers or gulpers. Right whales move slowly along the surface, skimming little animals from the water that crosses their long baleen plates. Rorquals lunge at shoals of prey with open mouths, gulp in water, and then force the water back out with their tongues, trapping krill and other creatures in their short plates. Gray whales are bottom feeders, filtering crustaceans and mollusks from the sediment with their heavy baleen.

Most of a baleen whale's prey species are minuscule, so it needs to consume vast quantities to stay alive. During summer, a large blue whale may eat 4½ tons (4 tonnes) of krill per day. It feeds very little during the rest of the year and lives off the fat and blubber reserves laid down in summer.

Baleen and blubber, both crucial to the survival of these giants, also attracted commercial whalers. Since 1985, there has been a worldwide moratorium on all commercial whaling, but this is not observed by a few countries.

Light bones Rather than supporting the animal's body weight, a cetacean's skeleton simply anchors the muscles. The bones are light, spongy, and filled with oil.

Blue whale
Balaenoptera musculus

50–90 throat pleats

Fin whale
Balaenoptera physalus

80 percent of fin whales were killed by 20th-century whalers

Differs from other right whales by having two throat grooves and a dorsal fin

Small dorsal fin positioned far along the back

Pygmy right whale
Caperea marginata

Bowhead whale
Balaena mysticetus

Longest baleen plates of any whale

24 inch (60 cm) layer of blubber keeps bowhead warm during total darkness of Arctic winter

"Necklace" of black spots

Callosities covered in whale lice

Only whale in temperate waters without dorsal fin

Northern right whale
Eubalaena glacialis

Fin whale Outsized only by the blue whale, the fin whale can attain speeds of 23 miles per hour (37 km/h), making it one of the fastest cetaceans. While groups of 300 or more may migrate together, the species is usually seen in pairs or small pods of several animals.

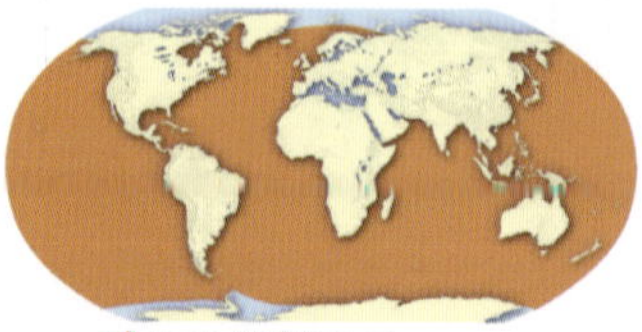

Up to 82 ft (25 m)
Up to 88 tons (80 t)
Variable
Endangered
All oceans except high Arctic

Blue whale The blue whale is the largest animal to have ever lived. A newborn calf measures at least 19½ feet (5.9 m) and guzzles 50 gallons (190 l) of milk per day, adding 8 pounds (3.6 kg) to its weight every hour. Relentlessly hunted in the 20th century, the blue whale population now comprises only about 6,000–14,000 individuals.

Up to 110 ft (33.5 m)
Up to 209 tons (190 t)
Variable
Endangered
All oceans except high Arctic

99 percent of blue whales were killed by 20th-century whalers

Female larger than male

Fin whale's lower jaw has asymmetrical color pattern: white on the right side and dark on the left

Minke whale
Balaenoptera acutorostrata

Sei whale has short throat pleats and fine baleen fringes for skimming food from water as whale swims on its side

Sei whale
Balaenoptera borealis

Mature gray whales are covered in lice and barnacles

Reduced dorsal fin forms small hump on back

Gray whale
Eschrichtius robustus

Humpback whale
Megaptera novaeangliae

97 percent of humpbacks were killed by 20th-century whalers

Minke whale The smallest and most abundant rorqual, this acrobatic whale is found in all the world's oceans. The minke is often seen in pairs, but up to 100 may gather in a rich feeding area. They travel to the Arctic in summer to graze on zooplanton.

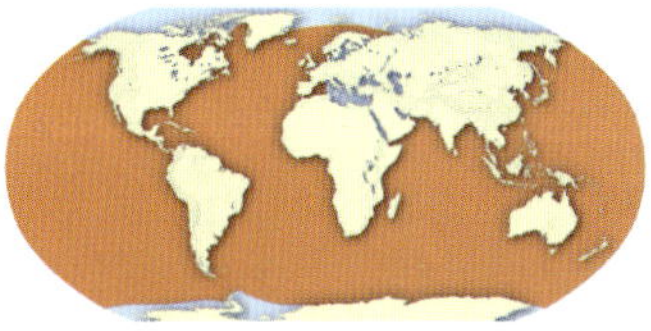

Up to 36 ft (11 m)
Up to 11 tons (10 t)
Variable
Near threatened
Most oceans except coldest

Humpback whale This active animal is famous for spectacular breaching, when it leaps clear of the water. Other surface behaviors include rolling, pec-slapping (lying on its side and loudly slapping its pectoral fins against the water), spyhopping (sticking its head vertically out of the water), and tail splashing.

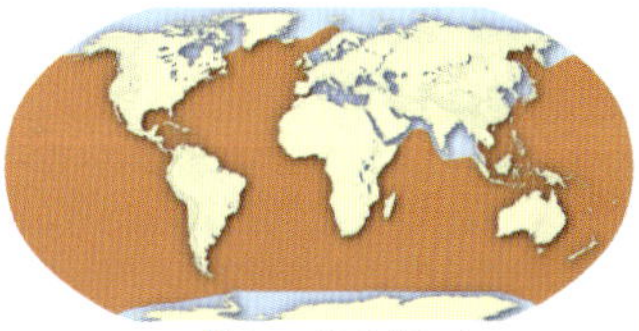

Up to 49 ft (15 m)
Up to 71½ tons (65 t)
Variable
Vulnerable
Feeds near Arctic & Antarctic

RODENTS

CLASS	Mammalia
ORDER	Rodentia
FAMILIES	29
GENERA	442
SPECIES	2,010

With roughly 2,000 species, rodents comprise more than 40 percent of all mammal species and have colonized almost every habitat on Earth. A key to their extraordinary success is the ability to reproduce quickly and abundantly, allowing species to survive harsh conditions and take full advantage of favorable ones. In addition, the small size of most rodents has helped them to exploit many microhabitats. Although rodents are among the earliest of placental mammals, with the oldest rodent fossils dating back some 57 million years, the largest family, Muridae (rats and mice), did not appear until 5 million years ago. It now contains almost two-thirds of all species in the order.

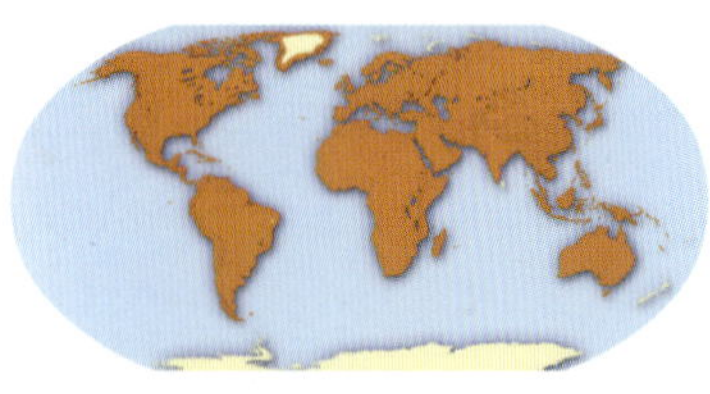

Successful spread Members of the order Rodentia are distributed throughout all the world's continents, except for Antarctica. Their association with humans has even helped them to reach isolated islands. They have adapted to a wide range of habitats, including arctic tundra, tropical forests, deserts, high mountains, and urban areas.

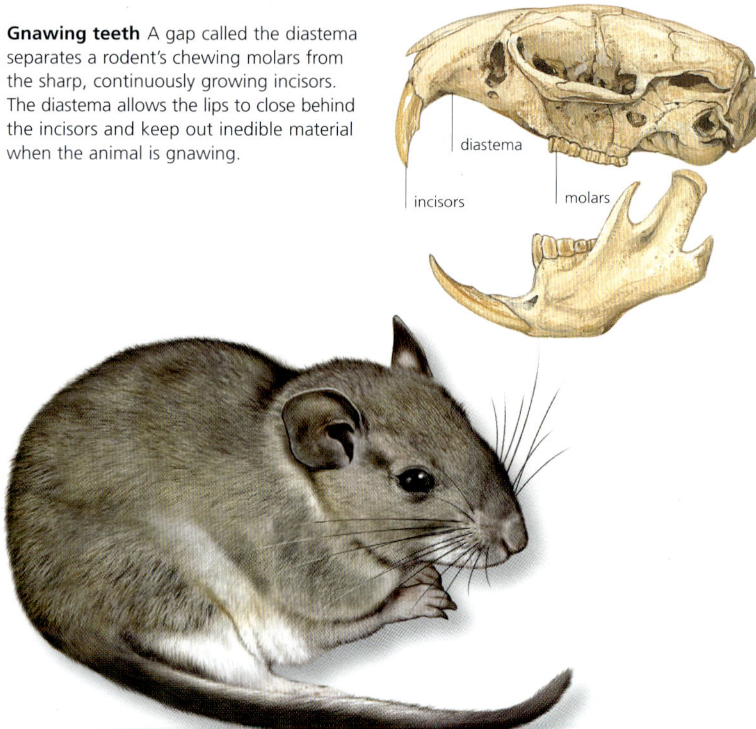

Gnawing teeth A gap called the diastema separates a rodent's chewing molars from the sharp, continuously growing incisors. The diastema allows the lips to close behind the incisors and keep out inedible material when the animal is gnawing.

Body type The white-throated wood rat displays typical rodent anatomy, with a compact body, short legs, clawed feet, long tail, and sensitive whiskers. A keen sense of smell and sharp hearing help rodents find food and avoid predators.

UNIFORM ANATOMY

Rodent size ranges from the tiny pygmy jerboa, less than 2 inches (5 cm) long and weighing less than ¼ ounce (5 g), to the substantial capybara, more than 50 inches (1.3 m) in length with a weight of 140 pounds (64 kg). Typically, however, rodents are small with squat bodies, short legs, and a tail.

The most distinguishing rodent feature is the arrangement of the teeth. All rodents have two pairs of extremely sharp incisors at the front of the mouth that can gnaw through seedpods, nut shells, and other tough matter to get at the nutritious food inside. The incisors grow continuously and are "self-sharpened" against each other. There are no canine teeth behind the incisors. Instead, a gap known as the diastema allows the lips to close during gnawing so inedible material is kept out of the mouth. At the back of the mouth, a series of molars grind plant matter, which is most of a rodent's diet.

While a few rodents are mainly carnivorous, most are opportunistic feeders and eat leaves, fruit, nuts, and seeds, as well as caterpillars, spiders, and other small invertebrates. The indigestible cellulose in plant walls is broken down by bacteria in the large cecum (appendix) of the rodent digestive system. Some species then take this treated food from the anus and eat it again, gaining maximum nutrition from the meal before passing feces as dry pellets, a process known as refection.

Intelligent and resourceful, rodents have put their relatively uniform anatomy to diverse use. Many species are terrestrial, finding their food in forests, grasslands, deserts, or human settlements. Others spend most of their time in trees, scampering over branches and, in some cases, gliding from one tree to another. Some species make their life underground in extensive networks of burrows. A few are excellent swimmers and pursue a semiaquatic lifestyle. While a minority of rodents are solitary, most are highly social, a trait culminating in the townships that contain thousands of prairie dogs.

The order Rodentia was once split into three suborders according to the arrangement of jaw muscles: squirrel-like rodents, mouse-like rodents, and cavy-like rodents. These categories are still used informally for ease, but genetic evidence points to a division into just two suborders. The suborder Sciurognathi includes all the squirrel-like and mouse-like rodents, plus the gundis, a family of cavy-like rodents. The other suborder, Hystricognathi, includes all other cavy-like rodents.

Flexible feeder A resident of temperate forests, the red squirrel eats mainly seeds and nuts, but will also consume flowers, shoots, fungi, and small invertebrates. It builds up a cache of buried seeds and nuts that it can raid during the cold winter.

A rodent's tail The diversity of rodent lifestyles is reflected in the various shapes and purposes of the tail (below). The northern flying squirrel (*Glaucomys sabrinus*) (left) uses its tail for steering and stability as it glides from tree to tree.

Parachuting Widely distributed in northern North America, the northern flying squirrel is found in isolated populations in the southern Appalachians, where it overlaps with the smaller southern flying squirrel. Active only at night, they forage on the ground and in the treetops, where they are hunted by predators, such as owls and hawks. Flying squirrels do not fly, but glide 65–295 feet (20–90 m) from tree to tree. The gliding membrane that stretches between their limbs, called a patagium, acts like a parachute. It keeps the squirrel aloft after it launches itself from on high.

SQUIRREL-LIKE RODENTS

CLASS	Mammalia
ORDER	Rodentia
FAMILIES	8
GENERA	71
SPECIES	383

The squirrels, beavers, and other animals collectively known as squirrel-like rodents share an arrangement of jaw muscles that gives them a strong forward bite. They have simple teeth and have retained one or two premolar teeth in each row, a characteristic not found in other rodents. Aside from the jaw muscles and premolar teeth, the families of squirrel-like rodents share few characteristics and probably diverged from each other early in rodent evolution. They include beavers (family Castoridae), mountain beaver (Aplodontidae), squirrels (Sciuridae), pocket gophers (Geomyidae), pocket mice (Heteromyidae), scaly-tailed squirrels (Anomaluridae), and springhare (Pedetidae).

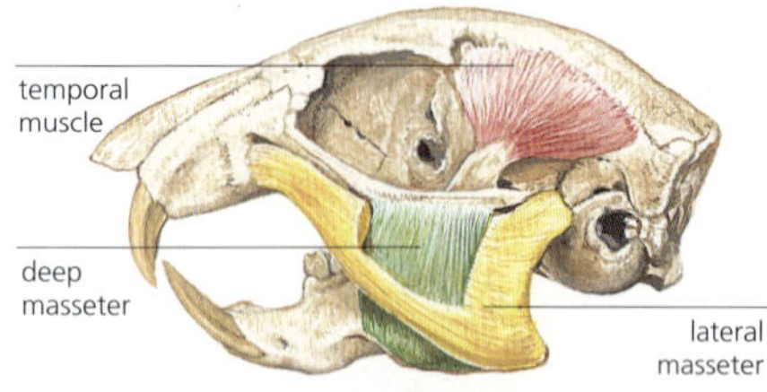

Strong forward bite The chewing muscles are known as masseters. In squirrel-like rodents, the lateral masseter extends to the snout and pulls the jaw forward when biting. The deep masseter is short and direct and simply closes the jaw.

BURROWERS AND LEAPERS

The squirrels in the family Sciuridae make up about three-quarters of all squirrel-like rodents. Active by day, tree squirrels have long, lightweight bodies, sharp claws for clinging to bark, and excellent eyesight for judging distances. They move about by scampering along branches, climbing headfirst down trunks, or leaping from tree to tree. The nocturnal flying squirrels glide through the air, aided by a furred membrane along either side of the body. Arboreal squirrels feed mostly on fruit, nuts, seeds, shoots, and leaves, but may supplement this diet with insects. Ground-dwelling squirrels, which include ground squirrels, prairie dogs, marmots, and chipmunks, tend to favor grasses and herbs. Many of these terrestrial species are gregarious with complex social organization.

Scaly-tailed squirrels are only distantly related to true squirrels. Like tree squirrels, almost all species of scaly-tailed squirrels possess a membrane for gliding—an example of convergent evolution.

Beavers are superbly equipped for a life spent largely in water, with a streamlined body, flattened tail, and webbed feet. Their large incisor teeth allow them to cut down trees and build dams and lodges.

Pocket gophers, pocket mice, mountain beavers, and springhares are all burrowing rodents. The pocket gophers and pocket mice both carry food in cheek pouches on either side of the mouth.

Aquatic adaptations A beaver uses its flat, scaly tail and webbed rear feet to propel its body through the water. Clear eyelids shield the eyes underwater, while valved nostrils and ears keep out water. Thick, oil-coated fur insulates the animal in cold water.

Cheek pouches inside mouth for carrying food

European souslik
Spermophilus citellus

Stands on hindlegs to watch for predators

Spotted souslik
Spermophilus suslicus

Whistles to warn other marmots of danger

Bobak marmot
Marmota bobak

Alpine marmot
Marmota marmota

Hoary marmot
Marmota caligata

Sturdy, slightly curved claws for digging

Thirteen-lined ground squirrel
Spermophilus tridecemlineatus

13 stripes alternating between light and dark fur

Black-tailed prairie dog
Cynomys ludovicianus

Poor eyesight but keen hearing and sense of touch

Mountain beaver
Aplodontia rufa

Black-tailed prairie dog Members of this gregarious species play together, groom one another, and communicate with a range of calls. An alarm bark warns of danger, while another type of bark gives the "all clear" signal.

- Up to 13½ in (34 cm)
- Up to 3½ in (9 cm)
- Up to 3½ lb (1.5 kg)
- Family band, colonial
- Near threatened

Shortgrass prairies of W. North America

MARMOTS

Restricted to the Northern Hemisphere, marmots are found mainly in mountain habitats. They are true hibernators, and spend the harsh winter at rest in their burrows, living off body fat. All marmots, except for the woodchuck, live in family groups. Young female marmots often stay with their parents to help raise their younger siblings.

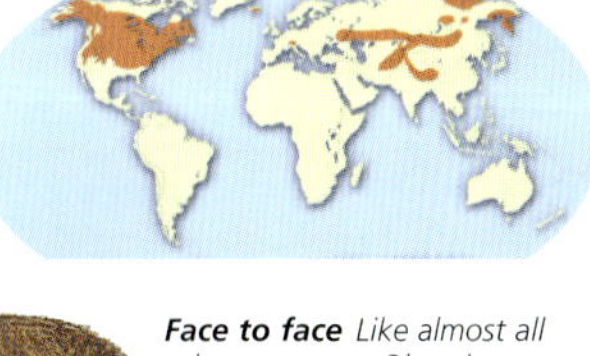

Face to face *Like almost all other marmots, Olympic marmots (Marmota olympus) are highly social. Young remain dependent on the mother for 2 years.*

Variegated squirrel
Sciurus variegatoides

American red squirrel
Tamiasciurus hudsonicus

White band around eye

Eastern gray squirrel
Sciurus carolinensis

Long tufts on ears in winter

Variegated squirrel
Sciurus variegatoides

Relies on ponderosa pines for food and shelter

Tassel-eared squirrel (Abert's squirrel)
Sciurus aberti

Eurasian red squirrel
Sciurus vulgaris

Eurasian red squirrel coat can be red or black

Coat thickens in winter

Guayaquil squirrel
Sciurus stramineus

SQUIRREL LARDER

Like many tree squirrels in climates with harsh winters, the American red squirrel prepares for the cold months by storing food. It collects thousands of pine and spruce cones and caches them in a larder, known as a midden.

Eastern gray squirrel this squirrel builds a nest of twigs and leaves lined with grasses and shredded bark in the branches of a tree, in addition to having a den in a hollow log. The nest is used for resting and feeding, and may also serve as a nursery.

- Up to 11 in (28 cm)
- Up to 9½ in (24 cm)
- Up to 26½ oz (750 g)
- Solitary
- Common

S. Canada to Texas & Florida

Eurasian red squirrel Using its strong incisors, this squirrel can crack a tough nut in a few seconds. It spends much of the day collecting seeds and nuts, as well as fungi, birds' eggs, and tree sap.

- Up to 11 in (28 cm)
- Up to 9½ in (24 cm)
- Up to 10 oz (280 g)
- Solitary
- Near threatened

W. Europe to E. Russia, Korea & N. Japan

Prevost's squirrel
Callosciurus prevostii

Nests high in forest canopy but feeds at lower levels

Horse-tailed squirrel
Sundasciurus hippurus

Southern flying squirrel
Glaucomys volans

Gliding membrane, or patagium, is folded when squirrel is sitting

Siberian flying squirrel
Pteromys volans

Red bush squirrel
Paraxerus palliatus

Handles food with dextrous forepaws

Soil often tints coat

Striped ground squirrel
Xerus erythropus

Gambian sun squirrel
Heliosciurus gambianus

Sits up on hindlegs to eat or look for danger

Eastern chipmunk
Tamias striatus

Smallest squirrel in the world

African pygmy squirrel
Myosciurus pumilio

Five black stripes on back

Southern flying squirrel While this nocturnal glider eats mainly nuts and acorns, it also consumes many insects and young birds. It often lives in pairs, but larger groups may den together during the winter months.

S. Canada to E. USA

- Up to 5½ in (14 cm)
- Up to 4½ in (12 cm)
- Up to 3 oz (85 g)
- Pair, small group
- Locally common

Striped ground squirrel Like prairie dogs, this gregarious animal lives in colonies and is highly vocal, warning of danger with an alarm call. It is common within its range, from western Africa to Kenya.

W. Africa to Kenya

- Up to 16 in (40 cm)
- Up to 12 in (30 cm)
- Up to 2 lb (1 kg)
- Colonial
- Locally common

Eastern chipmunk This usually solitary species shelters in a burrow. When its cheek pouches are stuffed with food, they can expand to be as large as the usual size of its head.

S.E. North America

- Up to 6½ in (17 cm)
- Up to 4½ in (12 cm)
- Up to 5½ oz (150 g)
- Solitary
- Locally common

LUMBERJACK RODENTS

The great engineers of the animal world, beavers deliberately alter their environment by building dams, canals, and lodges. While often causing conflict with farmers and other humans, this construction work has an important ecological function, helping to reduce erosion and flooding and creating new habitats for aquatic species. Beavers live in family groups of a monogamous pair and several offspring. Litters contain an average of two to four kits, which are nursed for 6–8 weeks. Beavers communicate using various calls and postures and will slap their tails against the water to warn of danger. Although similar in appearance and behavior, the two species of beaver—the North American *Castor canadensis* and the Eurasian *C. fiber*—do not interbreed.

Hidden access *Often built before the dam, the lodge needs to be positioned carefully to avoid flooding when dam construction raises the surrounding water.*

Lodges and dams A beaver colony may share a riverbank burrow system or they may build a lodge in the water. A lodge is a dome of sticks and mud with underwater entrances leading to a vegetation-lined living area above the water level. To create a calm pond for their lodge, beavers will construct dam walls that stop the flow of water. They will also dig out canals to link their dam to food sources and construction material.

Chopping chisels *Like all rodents, beavers have self-sharpening incisor teeth that never stop growing. The outer surface is protected by tough enamel, but the inner surface is softer and wears away as the beaver gnaws, creating a sharp, chiseled edge.*

Rapid recovery *Beavers prefer aspens, poplars, alders, and willows. These are all trees that grow rapidly and may even be reinvigorated after being felled by a beaver.*

Stopping the flow *Beavers use mud, stones, sticks, and branches to construct a dam wall. The pond this creates acts as a moat around their lodge and deters most predators.*

Lord Derby's anomalure
Anomalurus derbianus

Flightless scaly-tailed squirrel
Zenkerella insignis

Only scaly-tailed squirrel that does not glide

Pel's anomalure
Anomalurus pelii

Can glide more than 330 feet (100 m) between trees by spreading gliding membrane

Speke's pectinator
Pectinator spekei

Springhare
Pedetes capensis

Long, bushy tail provides balance when hopping

Gundi
Ctenodactylus gundi

Toes of hindfeet bear comb-like bristles

Eurasian beaver
Castor fiber

Large incisors used to chop down trees

Flattened, scaly tail used for propulsion and steering when swimming

Webbed toes

Gundi Once classified as a cavy-like rodent, but now grouped with squirrel-like and mouse-like rodents in the suborder Sciurognathi, the gundi eats a variety of desert plants and never drinks, extracting water from its food instead.

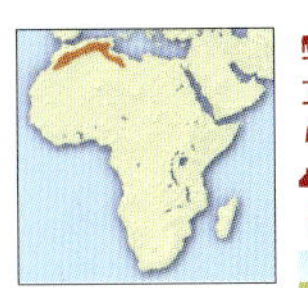

N. Africa

- Up to 8 in (20 cm)
- Up to 1 in (2.5 cm)
- Up to 10 oz (290 g)
- Family group, colonial
- Locally common

Springhare Like a little kangaroo, the springhare has long hindlimbs that it uses for hopping. After sheltering from the heat of the day in its burrow, this rodent emerges at night to forage on grasses and crops.

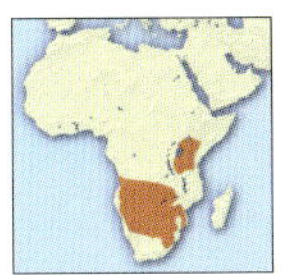

E. & S. Africa

- Up to 17 in (43 cm)
- Up to 18½ in (47 cm)
- Up to 9 lb (4 kg)
- Solitary
- Vulnerable

SCALY TAILS

The scaly-tailed squirrels of the Anomaluridae family are not directly related to the true squirrels of Sciuridae. All but one species moves about by gliding, an adaptation that developed independently in flying squirrels. Scales near the tail's base help scaly-tailed squirrels cling to trees at the end of a glide and then climb back up the trunk.

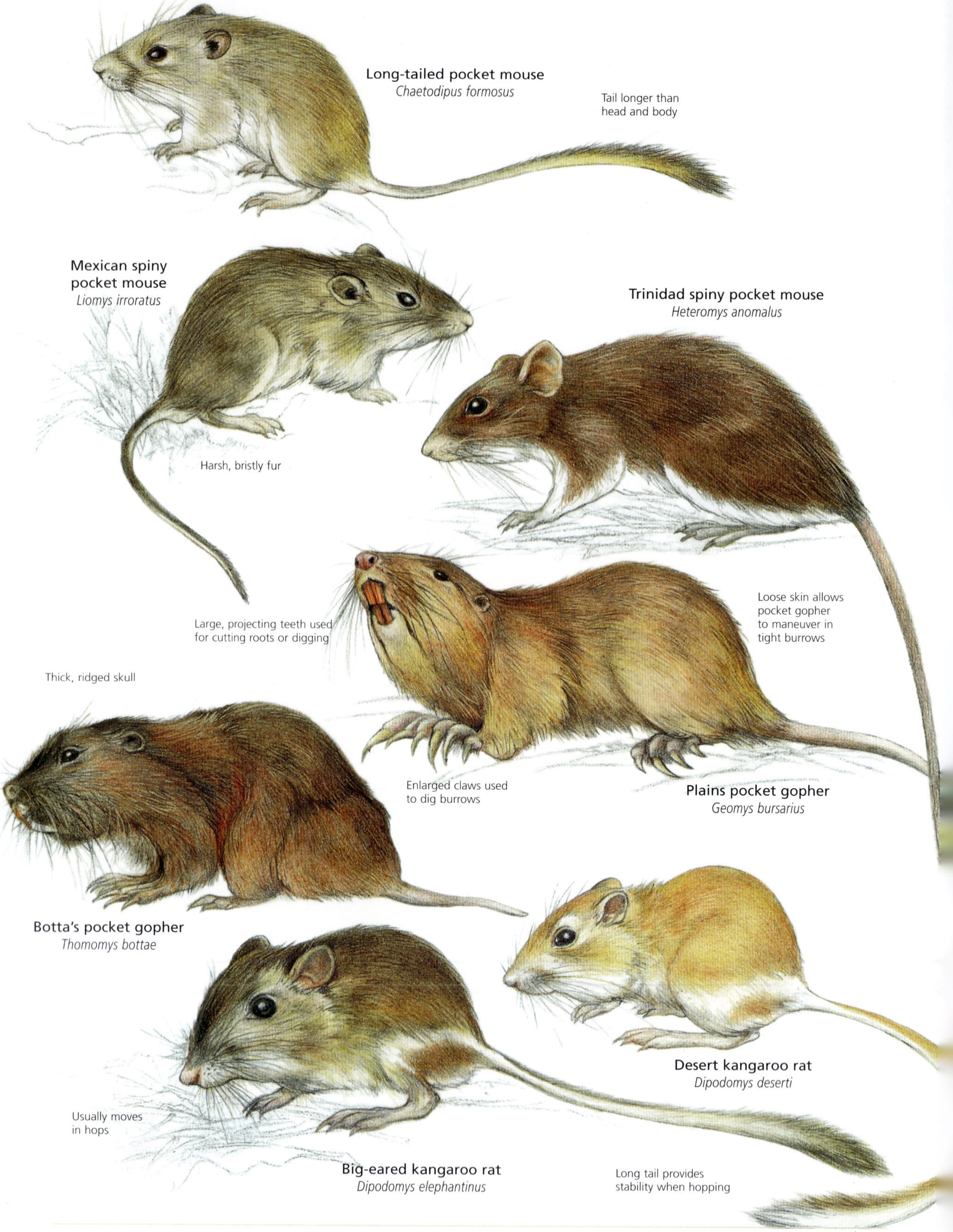

Long-tailed pocket mouse
Chaetodipus formosus

Tail longer than head and body

Mexican spiny pocket mouse
Liomys irroratus

Harsh, bristly fur

Trinidad spiny pocket mouse
Heteromys anomalus

Large, projecting teeth used for cutting roots or digging

Loose skin allows pocket gopher to maneuver in tight burrows

Thick, ridged skull

Enlarged claws used to dig burrows

Plains pocket gopher
Geomys bursarius

Botta's pocket gopher
Thomomys bottae

Desert kangaroo rat
Dipodomys deserti

Usually moves in hops

Big-eared kangaroo rat
Dipodomys elephantinus

Long tail provides stability when hopping

Long-tailed pocket mouse This species moves by using all four limbs, like true mice. It is most often found in gravelly desert areas. During drought, females may avoid producing a litter.

Up to 4 in (10 cm)
Up to 4½ in (12 cm)
Up to 1 oz (25 g)
Solitary
Common

Nevada & Utah to Baja California

Plains pocket gopher This solitary rodent digs an extensive burrow, with tunnels leading to a central chamber. During the mating season, a male may tunnel through to a female's burrow.

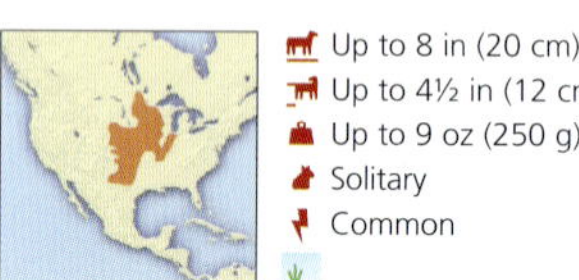

Up to 8 in (20 cm)
Up to 4½ in (12 cm)
Up to 9 oz (250 g)
Solitary
Common

Tallgrass prairies from S. Canada to Texas

Desert kangaroo rat To conserve water in its arid environment, this rodent emerges from its burrow only at night, when humidity is highest. It rarely drinks, obtaining most of its water from food.

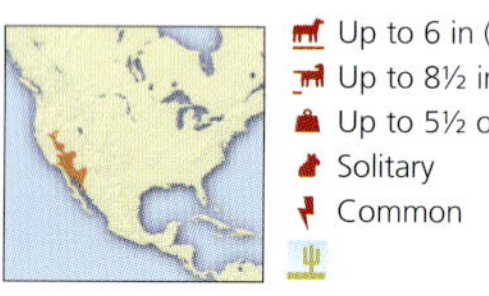

Up to 6 in (15 cm)
Up to 8½ in (21 cm)
Up to 5½ oz (150 g)
Solitary
Common

Nevada to N. Mexico

MOUSE-LIKE RODENTS

CLASS Mammalia
ORDER Rodentia
FAMILIES 3
GENERA 306
SPECIES 1,409

More than a quarter of all mammal species are mouse-like rodents. Once grouped within their own suborder, these rodents share an arrangement of the chewing muscles that provides a versatile gnawing action. All also have a maximum of three cheekteeth in each row. While their lifespans tend to be short, most are early and prolific breeders. The group is dominated by the Muridae family, which has more than 1,000 species, including Old World and New World rats and mice; voles and lemmings; hamsters; and gerbils. The other families of mouse-like rodents are the dormice of Myoxidae and the jumping mice, birchmice, and jerboas of Dipodidae.

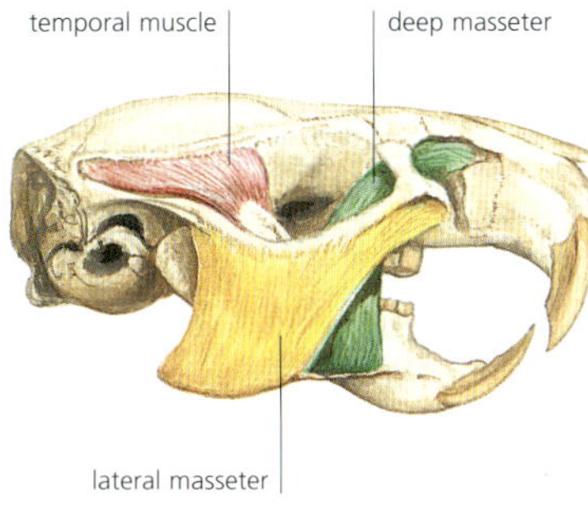

Versatile action The arrangement of their jaw muscles provides mouse-like rodents with a versatile gnawing action. The deep masseter extends onto the upper jaw and works together with the lateral masseter to pull the jaw forward for chewing.

RAPID RADIATION

The majority of members of the Muridae family are small, nocturnal, seed-eating ground-dwellers with a pointed face and long whiskers, but some spend much of their time in water or trees and others live underground.

There are more than 500 species of Old World rats and mice. These include the house mouse and brown rat, both of which are urban pests. New World rats and mice range from climbing rats to fish-eating rats.

While rats and mice account for 80 percent of species in the family Muridae, voles and lemmings, hamsters, and gerbils form distinct subfamilies. Voles and lemmings are found throughout the Northern Hemisphere. Many spend winter living in tunnels beneath the snow. Although popular as pets, the hamsters of Eurasia are decidedly solitary in the wild and will react very aggressively to intruders. Gerbils are found mainly in arid parts of Africa and Asia.

The Myoxidae and Dipodidae families are smaller and more specialized than Muridae. Dormice tend to live in trees and hibernate through cold winters. Jumping mice, birchmice, and jerboas all have long back feet and long tails that enable them to move by hopping. Jerboas have evolved to survive in some of the world's harshest deserts.

Eating habits Dormice and most other mouse-like rodents are mainly herbivorous, existing on a diet of seeds, fruit, and buds supplemented by the occasional insect. Voles and lemmings, like the Norwegian lemming (left), have specialized to feed on grasses. A few species are more carnivorous. Water rats will sometimes add a turtle or bat to their diet of aquatic invertebrates, while brown rats may even attack poultry or rabbits.

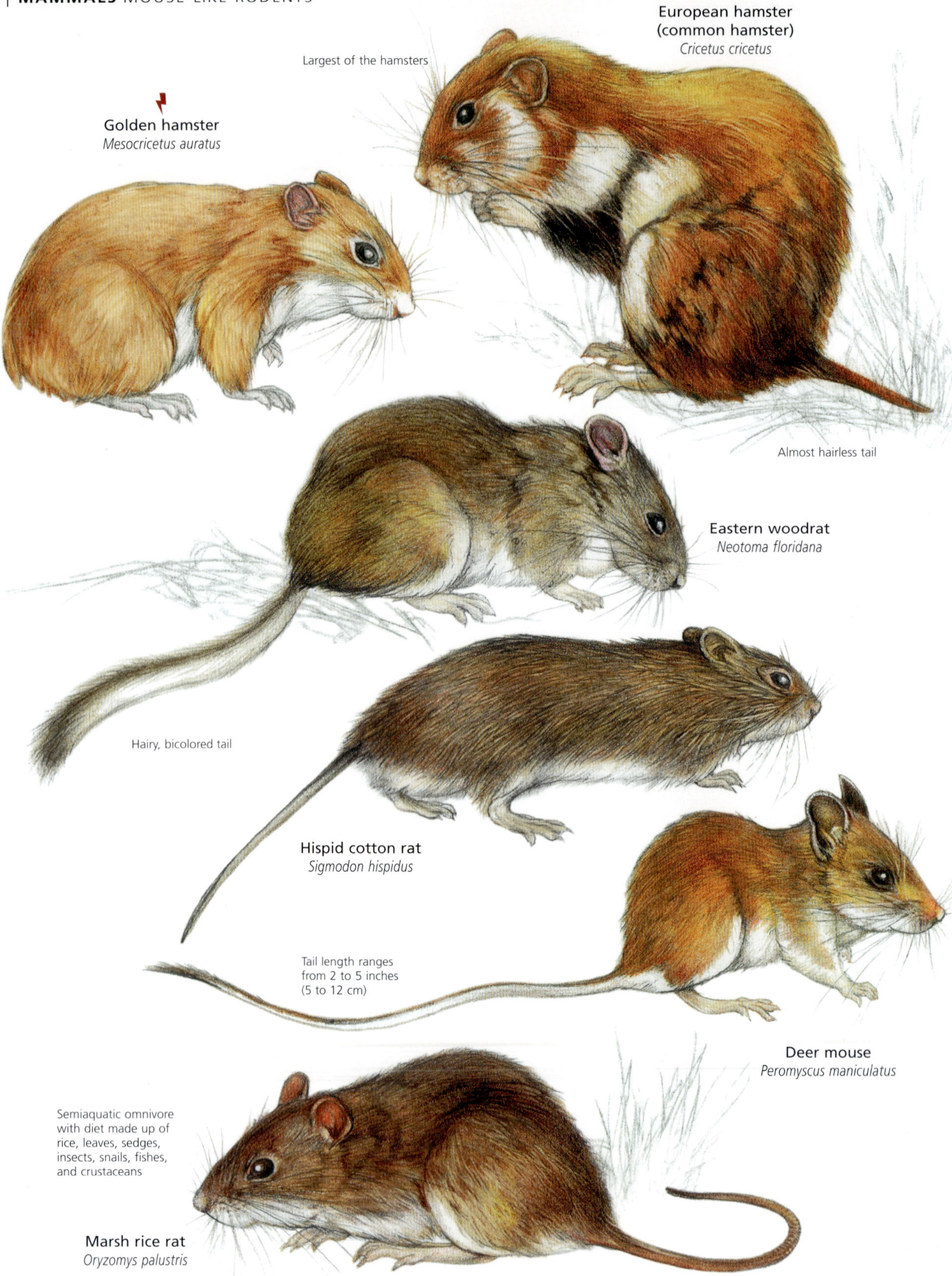

European hamster (common hamster)
Cricetus cricetus

Largest of the hamsters

Golden hamster
Mesocricetus auratus

Almost hairless tail

Eastern woodrat
Neotoma floridana

Hairy, bicolored tail

Hispid cotton rat
Sigmodon hispidus

Tail length ranges from 2 to 5 inches (5 to 12 cm)

Deer mouse
Peromyscus maniculatus

Semiaquatic omnivore with diet made up of rice, leaves, sedges, insects, snails, fishes, and crustaceans

Marsh rice rat
Oryzomys palustris

Golden hamster Now a popular pet and the best known of the hamsters, this species is endangered in the wild. It was introduced to the United States and England in the 1930s and has since proliferated in captivity.

- Up to 7 in (18 cm)
- Up to ¾ in (2 cm)
- Up to 5½ oz (150 g)
- Solitary
- Endangered

Middle East, S.E. Europe, S.W. Asia

European hamster This solitary burrower hibernates through winter, waking once a week or so to feed on its massive hoard of seeds and roots. During warmer months, it stocks up this winter food supply, carrying plant matter in its cheek pouches.

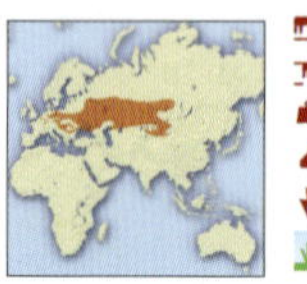

- Up to 12½ in (32 cm)
- Up to 2½ in (6 cm)
- Up to 13½ oz (385 g)
- Solitary
- Common

Belgium to Altai Mts of C. Asia

Hispid cotton rat After a gestation of just 27 days, females of this species give birth to several fully furred young. The female is ready to mate again almost immediately, and the young are sexually mature within 40 days.

- Up to 8 in (20 cm)
- Up to 6½ in (16 cm)
- Up to 8 oz (225 g)
- Solitary
- Common

S.E. USA to N. Venezuela & N. Peru

Long-clawed mole-vole
Prometheomys schaposchnikowi

Long claws for digging burrows

European water vole
Arvicola terrestris

Siberian collared lemming (Arctic lemming)
Dicrostonyx torquatus

White winter coat

Lives farther north than any other rodent

Brown summer coat

Bank vole
Clethrionomys glareolus

Great gerbil
Rhombomys opimus

Wood lemming
Myopus schisticolor

Norway lemming
Lemmus lemmus

Libyan jird
Meriones lybicus

Tail flattened vertically for use as rudder

Small webs between toes

Largest of the voles

Muskrat
Ondatra zibethicus

Muskrat This semiaquatic rodent swims with its large, webbed back feet and uses its naked tail as a rudder. Like the beaver, it lives in a group in a riverbank burrow or a lodge of twigs and mud.

Up to 13 in (33 cm)
Up to 12 in (30 cm)
Up to 4 lb (1.8 kg)
Small to large group
Common

USA & Canada except tundra; introd. Eurasia

FLUCTUATING LEMMINGS

Contrary to popular myth, Norway lemmings are not deliberately suicidal, but every 3 or 4 years, their number rises. Intolerant of one another at the best of times, the lemmings become highly aggressive. Such conflicts may trigger mass movements from the crowded alpine tundra to lower forests. When the lemmings meet obstacles such as rivers, panic can cause them to take flight, with some ending up in the sea.

Fighting techniques *Norway lemmings may wrestle, box, or adopt dominating postures.*

Striped grass mouse This savanna resident shelters in a burrow or abandoned termite nest. It builds a roundish nest for its litters, which are born during the rainy season after a gestation of 28 days.

Up to 5½ in (14 cm)
Up to 6 in (15 cm)
Up to 2½ oz (68 g)
Solitary
Common

Sub-Saharan Africa

House mouse Through its association with humans, this species has been able to spread throughout the world. This widespread pest nests in buildings or nearby fields and will eat almost any human food as well as items such as glue and soap.

Up to 4 in (10 cm)
Up to 4 in (10 cm)
Up to 1 oz (30 g)
Variable
Abundant, often regarded as pest

Worldwide except tundra & polar regions

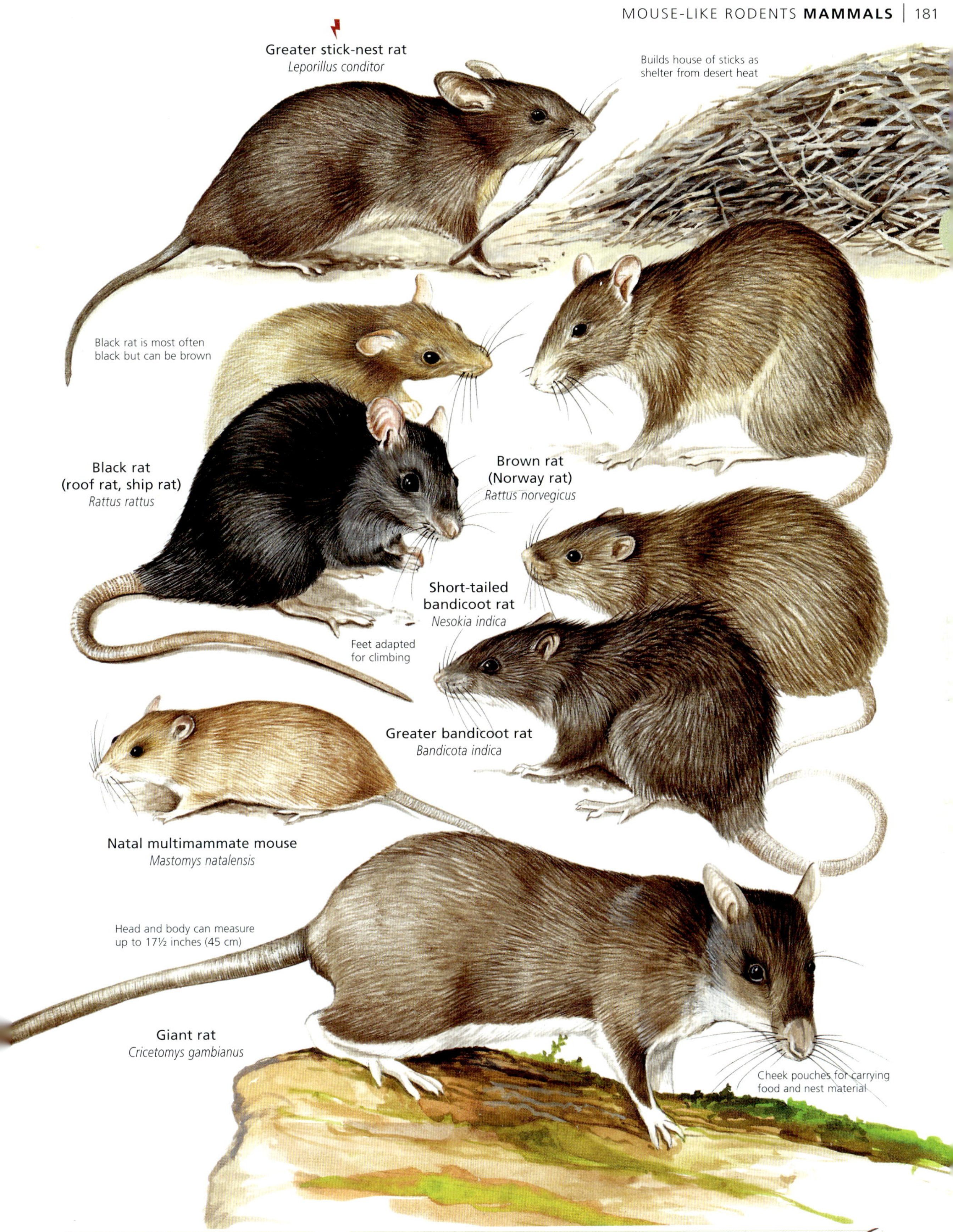

Greater stick-nest rat
Leporillus conditor

Builds house of sticks as shelter from desert heat

Black rat is most often black but can be brown

Black rat
(roof rat, ship rat)
Rattus rattus

Brown rat
(Norway rat)
Rattus norvegicus

Short-tailed bandicoot rat
Nesokia indica

Feet adapted for climbing

Greater bandicoot rat
Bandicota indica

Natal multimammate mouse
Mastomys natalensis

Head and body can measure up to 17½ inches (45 cm)

Giant rat
Cricetomys gambianus

Cheek pouches for carrying food and nest material

Greater stick-nest rat Once found throughout the shrublands of southern Australia, this small rat became extinct on the mainland when rabbits and sheep overgrazed their habitat. Various reintroduction campaigns are underway.

Up to 10 in (26 cm)
Up to 7 in (18 cm)
Up to 16 oz (450 g)
Family group
Endangered
Former range Franklin I. (Australia); reintrod. in several sites

RATS AS PESTS

Along with the house mouse, the black rat and the brown rat have had a close relationship with humans, living off crops and stored food. They have caused untold damage and spread serious disease, including the bubonic plague, which killed a third of Europe's population in the Middle Ages. Both rats live in large packs and will vigorously defend their feeding area. Brown rats will eat almost anything and may attack rabbits or human babies.

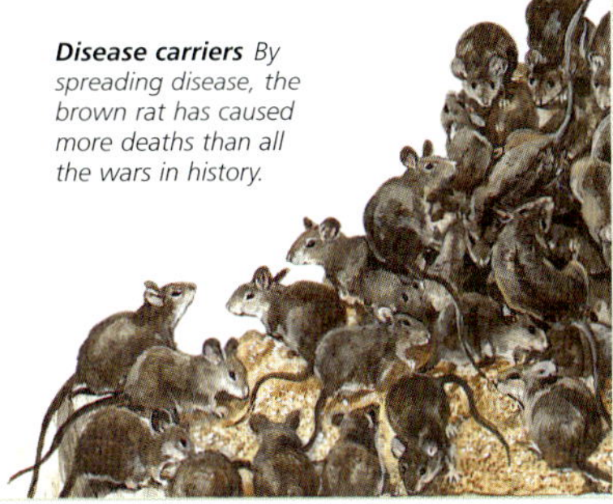

Disease carriers *By spreading disease, the brown rat has caused more deaths than all the wars in history.*

Gray climbing mouse
Dendromus melanotis

Cloud rat
Phloeomys cumingi

Long tail is semiprehensile

Luzon bushy-tailed cloud rat
Crateromys schadenbergi

Long muzzle and small eyes resemble those of a shrew

Mount Data shrew rat
Rhynchomys soricoides

Vlei rat
Otomys irroratus

Australian water rat
Hydromys chrysogaster

Thick tail with white tip

Fawn hopping mouse
Notomys cervinus

Golden-backed tree rat
Mesembriomys macrurus

WATER RAT

The Australian water rat lives in burrows that run along the banks of rivers or lakes. Able to withstand pollution, it is often found in urban areas. It depends on fresh water for the bulk of its diet, which includes crustaceans, mollusks, and fish. Broad, webbed feet act as paddles in the water. Its coat is not waterproof, but a pad of fat helps keep the heart warm.

Aquatic predator *The water rat often takes its catch to a preferred feeding platform.*

Gray climbing mouse With its long, prehensile tail, this little mouse can easily climb the grasses and shrubs of its savanna home. It may dig a simple burrow to shelter from seasonal fires, but more often occupies a globe-shaped grass nest above the ground.

Sub-Saharan Africa

- Up to 2¾ in (7 cm)
- Up to 3 in (8 cm)
- Up to 1¼ oz (8 g)
- Solitary
- Common

Fat dormouse
Myoxus glis

Squirrel-like body and tail

Highly arboreal with short, curved claws for climbing

Garden dormouse
Eliomys quercinus

Forest dormouse
Dryomys nitedula

Flattened, bushy tail

Northern three-toed jerboa
Dipus sagitta

Woodland dormouse
Graphiurus murinus

Lesser Egyptian jerboa
Jaculus jaculus

Meadow jumping mouse
Zapus hudsonius

Long tail provides stability when hopping

Garden dormouse This noisy rodent lives in large colonies, making globe-shaped nests of leaves and grass in tree hollows, among shrubs, or in rock crevices. As well as eating acorns, nuts, and fruit, it hunts insects and small rodents and birds.

Up to 6½ in (17 cm)
Up to 5 in (13 cm)
Up to 4 oz (120 g)
Colonial
Vulnerable

Europe

Lesser Egyptian jerboa With back legs four times longer than its front legs, this solitary desert animal will hop its way out of danger. It spends the day in a burrow, plugging the entrance with dirt in summer to make a cooler, humid microclimate.

Up to 4 in (10 cm)
Up to 5 in (13 cm)
Up to 2 oz (55 g)
Solitary
Not known

Morocco & Senegal to S.W. Iran & Somalia

Meadow jumping mouse Although it usually moves in short hops, this mouse can leap up to 3 feet (1 m) in the air when startled. It breeds very soon after emerging from its winter hibernation.

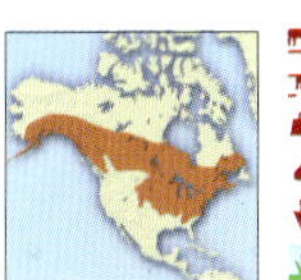

Up to 4 in (10 cm)
Up to 5 in (13 cm)
Up to 1 oz (30 g)
Solitary
Locally common

N. & E. North America

Cavy-like rodents

CLASS	Mammalia
ORDER	Rodentia
FAMILIES	18
GENERA	65
SPECIES	218

With their large heads, plump bodies, small legs, and short tails, the cavies of the Caviidae family—more commonly known as guinea pigs—are typical of cavy-like rodents. There are exceptions to this general body plan: some cavy-like rodents, such as the spiny rats of the Echimyidae family, look more like common mice and rats. All cavy-like rodents share a distinctive arrangement of the jaw muscles that gives them a strong forward bite. Unlike most other rodents, they also tend to have small litters of well-developed young. There are cavy-like rodents in both the Old World and the New World, but their relationship continues to be debated.

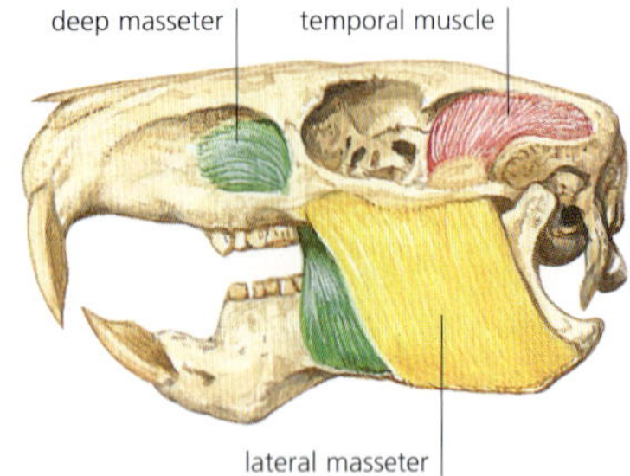

Powerful bite Like squirrel-like rodents, cavy-like rodents have a strong forward bite, but it is produced by a different arrangement of the jaw muscles. The lateral masseter closes the jaw, while the deep masseter extends past the eye and pulls the jaw forward for biting.

Spiky protection Porcupines give birth to well-developed young and have low infant mortality rates. Newborns have open eyes and can walk almost straight away. With their spines forming protective armor, porcupines have few natural predators. They are, however, killed by humans for food, sport, and pest control.

CAVIES AND THEIR KIN

Although controversy persists as to whether the South American cavy-like rodents came from North America or rafted over from Africa, most of today's cavy-like rodents live in Central and South America. The cavies of Caviomorpha include not only guinea pigs and similar species, but also the deer-like mara, a long-legged grazer. The semiaquatic capybara is more than 3 feet (1 m) long, making it the largest of all rodents. Chinchillas and viscachas live mainly at high elevations, kept warm by a thick, soft coat. Agoutis have long, slender limbs that allow them to run swiftly from danger. While most of South America's cavy-like rodents are terrestrial, tuco-tucos dig complex burrows. Other South American cavy-like rodents include degus, hutias, pacas, coypu, and pacarana.

New World porcupines, found in both North and South America, are arboreal and climb trees easily, aided in some species by a prehensile tail. As well as porcupines, the Old World's cavy-like rodents include African mole-rats, cane rats, and dassie rat. The gundis of North Africa (family Ctenodactylidae) are in the suborder Sciurognathi. All other cavy-like rodents are grouped in the suborder Hystricognathi.

Bristle-like fur

Thumps ground with hindfeet and whistles when alarmed

Greater cane rat
Thryonomys swinderianus

Large bamboo rat
Rhizomys sumatrensis

Uses claws and teeth to dig burrow network

Dassie rat
Petromus typicus

Bristles on back feet used for grooming

Lesser mole-rat
Nannospalax leucodon

Cape dune mole-rat
Bathyergus suillus

Sensory whiskers on face and tail

Hair between toes acts as broom

Naked mole-rat
Heterocephalus glaber

Greater cane rat This robust nocturnal rodent eats mainly grasses and cane, and lives in small family groups. When threatened, it grunts and stamps its hindfeet, or runs into water. It is a good swimmer and diver.

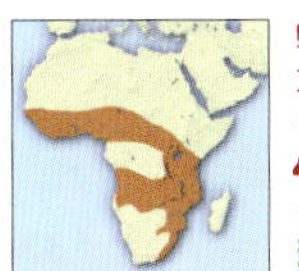

C. & S. Africa

- Up to 24 in (60 cm)
- Up to 7½ in (19 cm)
- Up to 10 lb (4.5 kg)
- Small groups
- Locally common

COOPERATIVE COLONY

Naked mole-rats survive in arid regions of Ethiopia, Somalia, and Kenya by living underground in highly structured colonies of about 70 animals. As with social insects such as ants and bees, a single female mole-rat known as the queen and a few males do all the breeding. The others are either workers or soldiers, and do not produce young.

Diggers *Workers in a naked mole-rat colony form a digging chain to create tunnels and foraging burrows.*

Crested porcupine
Hystrix cristata

Dark quills along neck can be raised into a crest

Indian porcupine
Hystrix indica

Sumatran porcupine
Hystrix sumatrae

Tail quills rattle when shaken

Malayan porcupine
Hystrix brachyura

Quills cannot bristle or rattle

Scaly tail with bristled tip

African brush-tailed porcupine
Atherurus africanus

Long-tailed porcupine
Trichys fasciculata

Partially webbed feet

Crested porcupine This porcupine has been known to kill lions, hyenas, and humans. When threatened, it raises its quills to make itself seem larger, but if this display fails to deter, it will then back itself into the predator so that the quills become embedded.

- Up to 27½ in (70 cm)
- Up to 4½ in (12 cm)
- Up to 33 lb (15 kg)
- Family group
- Near threatened

Italy, Balkans, Africa

African brush-tailed porcupine This species lives in a family group made up of a breeding pair and several offspring. They spend the day together hiding in caves, crevices, or logs, emerging at night to forage alone for roots, leaves, fruit, and bulbs.

- Up to 22½ in (57 cm)
- Up to 9 in (23 cm)
- Up to 9 lb (4 kg)
- Family group
- Common

Equatorial Africa

Long-tailed porcupine Unlike other Old World porcupines, this forest-dweller cannot bristle or rattle its quills, but the long tail can break off from the body if snatched by a predator. It climbs trees to reach fruit and other food.

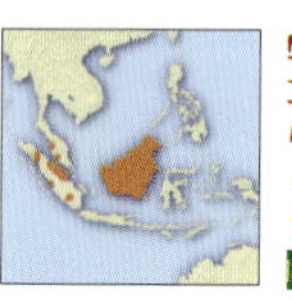

- Up to 19 in (48 cm)
- Up to 9 in (23 cm)
- Up to 5 lb (2.3 kg)
- Not known
- Not known

Malaya, Sumatra, Borneo

Brazilian porcupine
Coendou prehensilis

Prehensile tail lacks spines

North American porcupine
Erethizon dorsatum

Up to 30,000 sharp, barbed quills

Bicolor prehensile-tailed porcupine
Coendou bicolor

Naked pad on underside of tail for grip

Stump-tailed porcupine
Echinoprocta rufescens

Can swim with just eyes, nose, and ears protruding, or stay fully submerged for up to 5 minutes

Male has scent gland on snout

Capybara
Hydrochaeris hydrochaeris

Webbed feet for swimming

North American porcupine This species dens in a cave, rock crevice, or fallen log, emerging at night to browse on the bark of trees and shrubs.

Up to 3½ ft (1.1 m)
Up to 10 in (25 cm)
Up to 39½ lb (18 kg)
Solitary
Locally common

N. & W. North America

THE BIGGEST RODENT

The largest of all rodents, capybaras are barrel-shaped grazers that feed mainly on grasses growing in or near water. Capybaras are extremely agile in the water, using their partially webbed toes like tiny paddles. These social animals usually live in family groups of one male, several females, and their young, but in dry times may form larger temporary herds of up to a hundred individuals.

Cooling off *Capybaras enter the water to seek refuge from the midday heat, to escape predators, and to mate.*

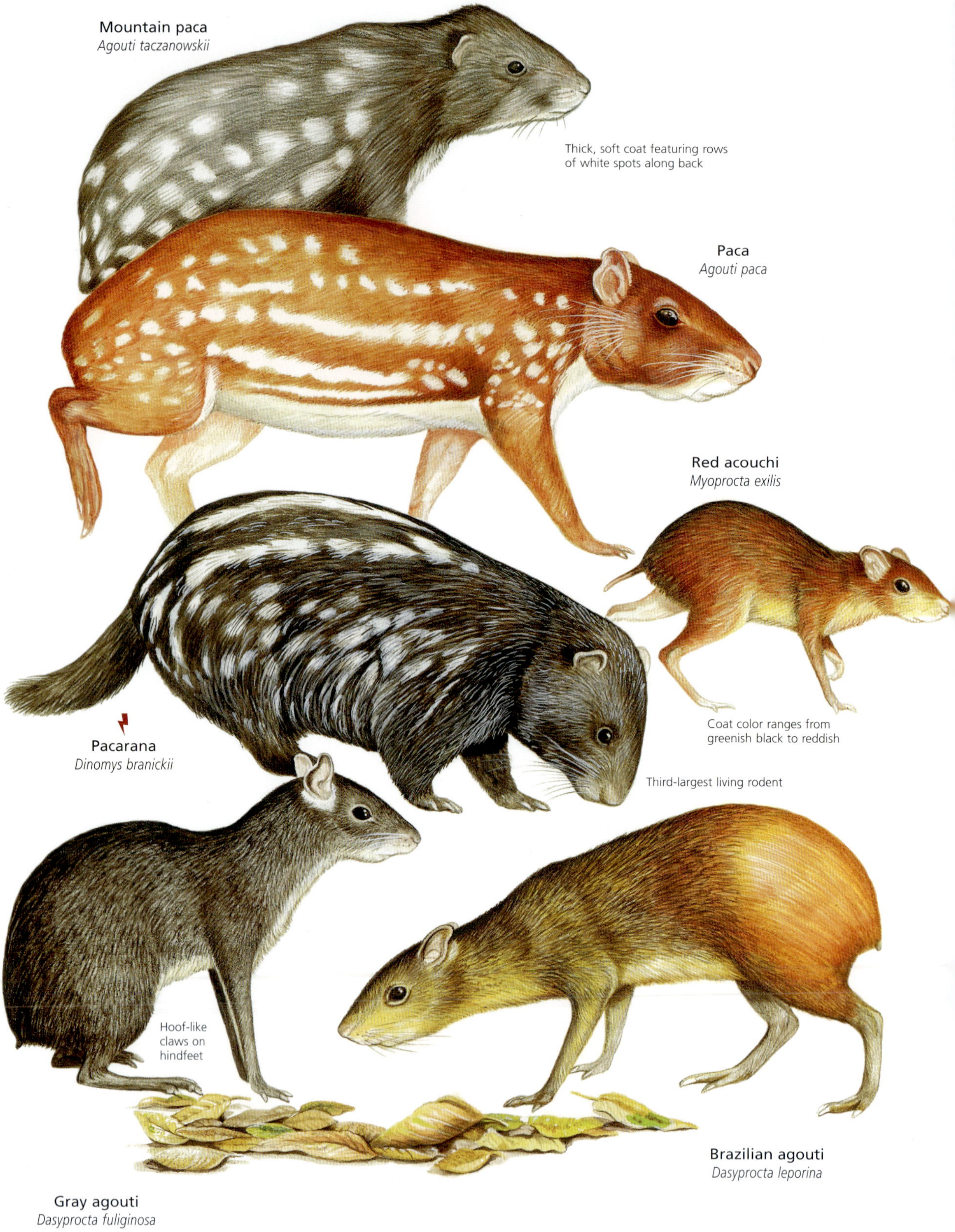

Mountain paca
Agouti taczanowskii

Thick, soft coat featuring rows of white spots along back

Paca
Agouti paca

Red acouchi
Myoprocta exilis

Coat color ranges from greenish black to reddish

Pacarana
Dinomys branickii

Third-largest living rodent

Hoof-like claws on hindfeet

Brazilian agouti
Dasyprocta leporina

Gray agouti
Dasyprocta fuliginosa

Red acouchi Active by day, the red acouchi will bury some food during times of abundance. This ensures the animal a food supply in leaner times, but also helps to disperse seeds throughout the forest.

- Up to 15½ in (39 cm)
- Up to 3 in (8 cm)
- Up to 3½ lb (1.5 kg)
- Solitary
- Data deficient

S. Colombia to Guianas, Amazon basin

Gray agouti This species can leap more than 6 feet (2 m) in the air. It walks on its digits and will gallop when in a hurry. Courtship involves the male spraying the female with urine, causing her to make frenzied movements.

- Up to 30 in (76 cm)
- Up to 1½ in (4 cm)
- Up to 13 lb (6 kg)
- Solitary, pair
- Common

Upper Amazon basin

Brazilian agouti As in other agoutis, the front part of this animal is slender, while the rear is bulkier, an adaptation to foraging through the undergrowth for its preferred food of fallen fruit.

- Up to 25 in (64 cm)
- Up to 1 in (3 cm)
- Up to 13 lb (6 kg)
- Pairs
- Common

E. Venezuela & Guianas to S.E. Brazil

Demarest's hutia
Capromys pilorides

Hispaniolan hutia
Plagiodontia aedium

Scaly tail

Hairy tail

Prehensile-tailed hutia
Mysateles prehensilis

Brown's hutia
Geocapromys brownii

Coypu (nutria)
Myocastor coypus

Fringes of bristles on hindfeet used for grooming

White-bellied tuco-tuco
Ctenomys colburni

Demarest's hutia While this species has strong claws and easily climbs trees, it spends more time on the ground than most other hutias. It shares with other hutias a stomach divided into three compartments, the most complex stomach of any rodent.

Up to 24 in (60 cm)
Up to 12 in (30 cm)
Up to 18½ lb (8.5 kg)
Pair
Common, declining

Cuba & adjacent offshore islands

Brown's hutia This terrestrial animal usually lives alone, but can be found in family groups of up to 10 animals. It hides in rock crevices by day, but waddles out at night to search the forest for leaves, roots, bark, and fruit.

Up to 17½ in (45 cm)
Up to 2½ in (6 cm)
Up to 4½ lb (2 kg)
Not known
Vulnerable

Jamaica

White-bellied tuco-tuco This robust burrower digs tunnels with the strong claws on its forefeet, cutting through any roots with its prominent incisors. Considered a pest, it has been hunted in great numbers.

Up to 6½ in (17 cm)
Up to 3 in (8 cm)
Not known
Not known
Not known

Known from only two localities in Argentina

Coruro
Spalacopus cyanus

Prominent incisors and strong front legs used for digging tunnels

Degu
Octodon degus

Skin of tail is shed if seized by predator

Chinchilla-rat
Abrocoma cinerea

Chilean rock rat
Aconaemys fuscus

Large ears listen for predators

Chinchilla
Chinchilla lanigera

Peruvian mountain viscacha
Lagidium peruanum

Striped face unusual among rodents

Plains viscacha
Lagostomus maximus

DEGUS AND DOGS

Degus are the ecological equivalent of North America's prairie dogs (right). Both are diurnal rodents that live in large colonies in extensive burrow systems and communicate through a range of vocalizations. Degus and prairie dogs are only distantly related. Their similarities are the result of convergent evolution, in which both groups adapted in the same way to their semiarid habitats.

Alarmed *With many prairie dogs active during the day at the same time, some act as sentinels, watching for predators.*

Chinchilla Most commonly found in barren, rocky mountain areas, this nocturnal species lives in large colonies of up to 100 animals. Popular as pets, chinchillas are now rare in the wild.

Andes of N. Chile

- Up to 9 in (23 cm)
- Up to 6 in (15 cm)
- Up to 1 lb (500 g)
- Colonial
- Vulnerable

LAGOMORPHS

CLASS	Mammalia
ORDER	Lagomorpha
FAMILIES	2
GENERA	15
SPECIES	82

The rabbits, hares, and pikas of the order Lagomorpha were once considered a suborder of Rodentia, and they do bear some resemblance to large rodents. They are gnawing herbivores, with large, constantly growing incisors, no canine teeth, and a gap between the incisors and molars. Like rodents, they can close their lips in this gap and gnaw on material without taking it into the mouth. Unlike rodents, they have a second, smaller pair of upper incisors, known as peg teeth, behind the first pair. All lagomorphs are terrestrial. They are found almost worldwide in a diverse range of habitats, from snowy arctic tundra to steamy tropical forest and scorching desert.

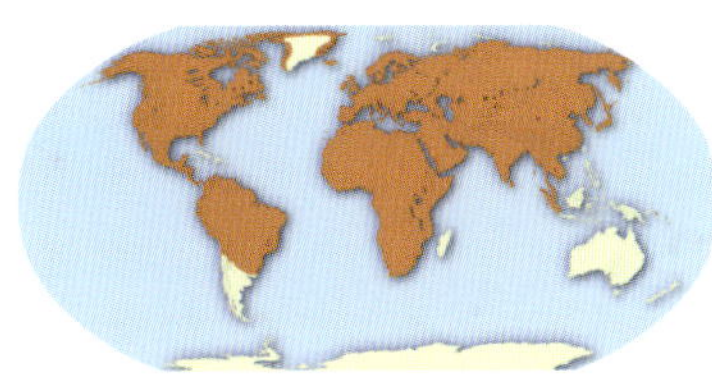

In human company While their close association with humans has allowed lagomorphs to spread almost worldwide, they are absent from southern South America and many islands. Where they have been introduced, as in Australia and New Zealand, lagomorphs have often had a devastating effect, outcompeting both native animals and livestock for food.

Cool ears A resident of North America's deserts, the black-tailed jackrabbit relies on its long ears to keep cool. Hundreds of tiny blood vessels radiate heat to the surface of the ears, allowing the blood to cool before it returns to the heart. A jackrabbit spends the hottest part of the day resting in the shade of a shrub or long grass.

RABBITS, HARES, AND PIKAS

Lagomorphs are divided into two families: the rabbits and hares of Leporidae, and the pikas of Ochotonidae. As key prey species for many birds and carnivores, all lagomorphs have eyes on the side of the head that provide a wide field of vision, and relatively large ears that contribute to their sharp hearing. In pikas, the ears are short and round, while rabbits and hares have very long ears.

Many lagomorphs are social species, and all use their scent glands in communication. In pikas, these are supplemented by a variety of vocalizations, which inspired their nickname of "calling hares."

To escape predators, both rabbits and hares have long hindlegs that allow them to run at speed—rabbits tend to flee to cover, while hares try to outrun the danger over open ground. Pikas have shorter legs, but usually live in rocky terrain where they can quickly slip into a crevice.

In spite of their anti-predator tactics, all lagomorph species are an important food source for other animals and suffer a high mortality rate. To cope with this, they tend to be prolific breeders. Gestation periods are short, usually lasting just 30–40 days, and litters are often large. Many species reach sexual maturity quickly—female European rabbits can breed at only 3 months. The female's eggs are released in response to copulation, and she is able to become pregnant almost immediately after giving birth. In some species, the female can conceive a second litter while still pregnant with the first. These reproductive strategies have allowed some species, such as the European rabbit, to become so successful that they are considered pests.

Large-eared pika
Ochotona macrotis

Nostrils can be completely closed

Northern pika
Ochotona alpina

Turkestan red pika
Ochotona rutila

Calls include a one- or two-note squeak that expresses fear

Royle's pika
Ochotona roylei

Daurian pika
Ochotona daurica

Two main vocalizations are an alarm call and a mating song

Steppe pika
Ochotona pusilla

American pika
Ochotona princeps

Heavily furred feet

Northern pika Active through long, cold winters, this species stays on the surface until the snow cover is about 12 inches (30 cm) deep, and then retreats into tunnels under the snow.

Up to 8 in (20 cm)
None
Up to 7 oz (200 g)
Solitary
Not known

Mongolia, Siberia

Royle's pika Although this pika normally nests in natural rock piles, it has been known to live in the rock walls of huts built by humans. Because it forages throughout the year, it does not create a haystack as other pikas do.

Up to 8 in (20 cm)
None
Up to 7 oz (200 g)
Solitary
Not known

Himalayas of Pakistan, India, Nepal & Tibet

Daurian pika This steppe-dweller is a highly social burrower, living in large colonies made up of family groups. Family members communicate through a repertoire of calls, groom each other, rub noses, and play together.

Up to 8 in (20 cm)
None
Up to 7 oz (200 g)
Colonial
Common

Steppes of Mongolia & S. Siberia

Arctic hare
Lepus timidus

Brown summer coat blends in with tundra vegetation

Black ear tips

White winter coat provides camouflage against snow cover

Grows white coat and eats barks and buds in winter

Snowshoe hare
Lepus americanus

Grows brown coat and eats green plants and berries in summer

Antelope jackrabbit
Lepus alleni

Asiatic brown hare
Lepus tolai

Black-tailed jackrabbit
Lepus californicus

White-tailed jackrabbit
Lepus townsendii

Coat lightens in summer

Brown hare
Lepus europaeus

Can flee predators at speeds of 35 miles per hour (56 km/h)

Arctic hare To survive the harsh Arctic winter, this usually solitary hare may gather in flocks of several hundred individuals, and smaller groups may cooperate to build a protective wall of snow. The Arctic hare changes color with the season.

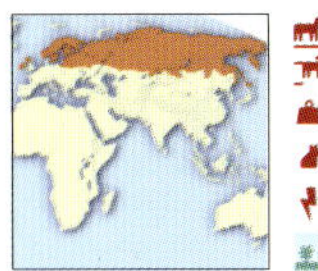

- Up to 24 in (60 cm)
- Up to 3 in (8 cm)
- Up to 13 lb (6 kg)
- Solitary
- Common

Iceland, Ireland, Scotland, N. Eurasia

Antelope jackrabbit Like an antelope, this large hare is able to make great leaps. A desert species, it is nocturnal and can survive without drinking water, obtaining all its water from plants.

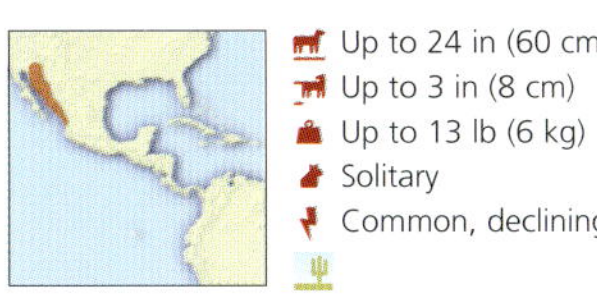

- Up to 24 in (60 cm)
- Up to 3 in (8 cm)
- Up to 13 lb (6 kg)
- Solitary
- Common, declining

S. Arizona to N. Mexico

Brown hare Female brown hares can produce four litters a year. For the first month, the young are left in a form, a shallow depression in long grass, and visited once a day for feeding.

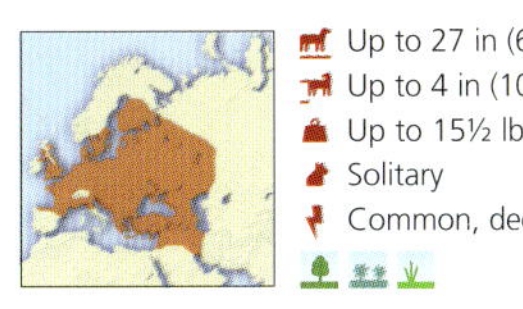

- Up to 27 in (68 cm)
- Up to 4 in (10 cm)
- Up to 15½ lb (7 kg)
- Solitary
- Common, declining

Europe to Middle East; introd. widely

Forest rabbit
Sylvilagus brasiliensis

Eastern cottontail
Sylvilagus floridanus

Pygmy rabbit
Brachylagus idahoensis

White tail displayed when running

Brush rabbit
Sylvilagus bachmani

European rabbit
Oryctolagus cuniculus

Volcano rabbit
Romerolagus diazi

No visible tail

Thumps hindleg on ground to warn others of danger

Coarse, bristly outer fur, with softer underfur

Hispid hare
Caprolagus hispidus

Sumatran rabbit
Nesolagus netscheri

The rarest lagomorph, with just one record from 1972 and a photograph taken by remote camera in 1998

Central African hare
Poelagus marjorita

RABBIT WARREN

The European rabbit is one of the few rabbit or hare species to dig its own burrow, and is the only one to live in stable, territorial breeding groups. Large numbers of young are raised in the shelter of an underground warren. Rabbit warren entrances are small to keep them hidden, but the tunnel width expands below the surface.

Nursery *Litters of kits, baby rabbits, are kept warm and safe in grass-lined chambers.*

Eastern cottontail This rabbit spends the daytime resting in a hollow beneath a log or brush. Although blind and naked at birth, young cottontails can leave the nest at 2 weeks of age, disperse at 7 weeks, and are sexually mature at only 3 months.

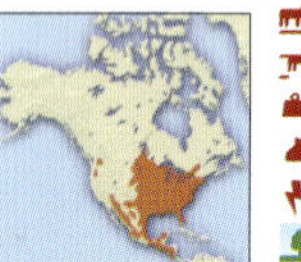

- Up to 19½ in (50 cm)
- Up to 2½ in (6 cm)
- Up to 3½ lb (1.5 kg)
- Solitary
- Common

E. & S. North America

ELEPHANT SHREWS

CLASS Mammalia
ORDER Macroscelidea
FAMILY Macroscelididae
GENERA 4
SPECIES 15

In some Asian tropical forests, small, squirrel-like mammals known as tree shrews scurry along the ground and up trees, foraging for insects, worms, small vertebrates, and fruit. Their sharp claws and splayed toes keep a firm grip on bark and rock alike, while a long tail helps with balance. When eating, they hold food in their hands and may sit on their haunches, alert for predators such as birds, snakes, and mongooses. An average of three young are born in a nest of leaves, which is often made by the father in a tree hollow. Maternal care tends to be limited, with some females visiting only once every 2 days. Considered a primitive form of placental mammal, tree shrews have no direct relationship to true shrews.

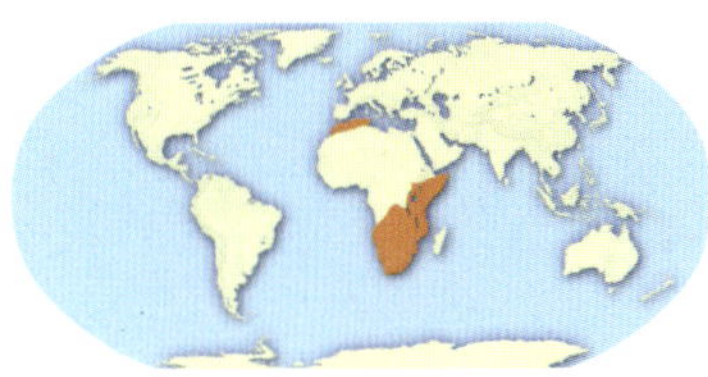

African habitats Elephant shrews are found throughout much of Africa, but are missing from West Africa and the Sahara. They live in diverse habitats, including rocky outcrops, desert, savanna, grasslands, and tropical forest. Although terrestrial and often active by day, these secretive, swift mammals are not common and are rarely seen.

Elephant shrews live in monogamous pairs, usually mating for life. The pairs rarely meet, but share and defend the same territory. They scentmark to communicate, and maintain a network of trails that allows fast escape from predators. Trespassing male elephant shrews are driven off by the male of the pair, while the female deters female intruders.

Checkered elephant shrew
Rhynchocyon cirnei

Highly sensitive, flexible snout

Back legs longer than front legs

Large eyes and ears

Rufous elephant shrew (spectacled elephant shrew)
Elephantulus rufescens

Four-toed elephant shrew
Petrodromus tetradactylus

Rat-like, bristled tail

BIRDS

PHYLUM	Chordata
CLASS	Aves
ORDERS	29
FAMILIES	194
GENERA	2,161
SPECIES	9,721

Descended from reptiles that developed the ability to fly, birds are among the most mobile of all animals alive today. Most numerous in marshes, woodlands, and rain forests, they have also adapted to live in large cities, inhospitable deserts, and even the North and South poles. Some never stray from home; others cross entire oceans or continents, sometimes in a single flight. In size, they range from the tiny bee hummingbird to the imposing ostrich. Among the more than 9,700 species are birds of almost every color and a dazzling array of patterns. While some hover on the verge of extinction, others, such as domestic chickens, outnumber humans.

On the wing The wingbeat of a European robin (left) is typical of flying birds: a smooth alternation between the wings moving upward (upstroke or recovery stroke) and downward (downstroke or powerstroke). Penguins (right) are in the minority of flightless birds.

ON THE WING

Birds' flying and singing abilities have awed humans through the millennia. These distinctive animals have been the inspiration not only for the invention of mechanical flying machines, but also for myths, musical works, and other art forms.

Although some species of birds have lost the ability to fly, they all have feathers. In general, no other types of animals can travel so far or so fast under their own power.

Birds can perform an astonishing range of aerial maneuvers, from swooping and soaring to hovering or even, in some cases, flying backwards. They may fly to search for food and more favorable climates, or else to ward off predators or communicate with each other.

Some birds seek out territories for temporary breeding, roosting, or feeding purposes, while others spend their entire lives in one area. Many species of birds migrate over great distances to escape harsh winters before returning to their starting places when the weather begins to improve again. They are guided by instinct and experience.

Not all types of birds sing, but more than half of the species are considered songbirds. Some of these have very large repertoires, which are used primarily by males to court females and deter other males from their territories.

Some birds are solitary; others live in large flocks and may even help raise other parents' young. More than 30 species of birds became extinct during the 20th century, mostly due to human activities. Today, one out of every nine species of birds is threatened, and many more species are in decline. Past species extinctions were mostly due to overhunting, as in the case of the passenger pigeon.

Today, habitat loss and destruction are by far the most serious factors in the decline of birdlife worldwide. Pesticides and other pollutants also take their toll, as do introduced species such as cats and rats, especially on populations of island birds that never evolved defenses against such animals.

Although many types of birds appear on the threatened list, parrots feature very prominently, because their bright feathers attract poachers and because they tend to live in tropical rain forests, which are disappearing at a fast rate.

ENGINEERING MARVELS

Birds come in all shapes and sizes. Yet, although external features vary markedly among species, all birds share a remarkably similar anatomy. All birds are descended from flying ancestors and have front limbs in the form of wings. A flying bird's body is designed to enable it to move efficiently through the air. For example, most of a bird's weight is in the center of its body, which is light

FEATHER TYPES

There are three kinds of feathers. Closest to the body are the fluffy down feathers that protect a bird from the cold. Over these are the contour feathers. These are short and round, and give the bird its streamlined shape. The longer feathers on a bird's wing and tail enable the bird to take off, fly, maneuver, and land.

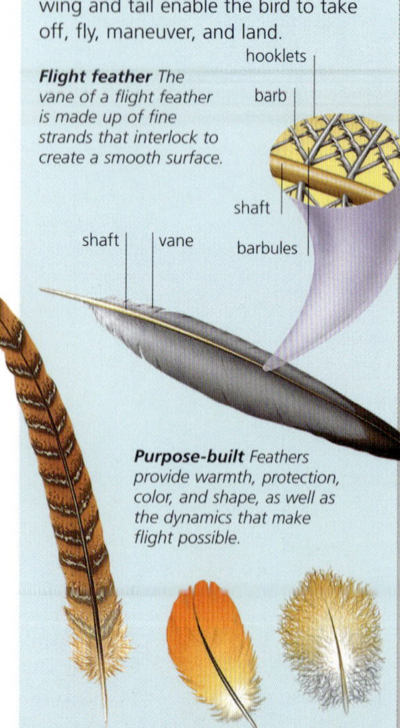

Flight feather *The vane of a flight feather is made up of fine strands that interlock to create a smooth surface.*

Purpose-built *Feathers provide warmth, protection, color, and shape, as well as the dynamics that make flight possible.*

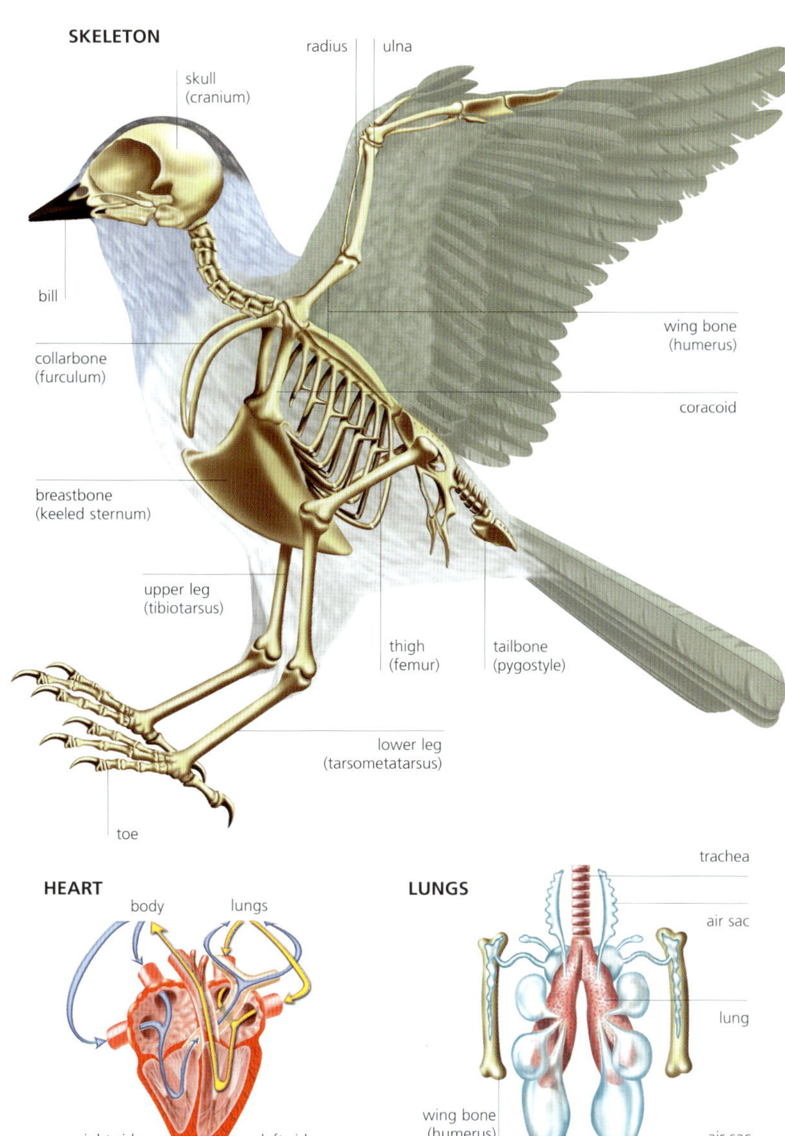

Adapted to flight A bird's skeleton (right) is typical of vertebrates, with modifications to support powered flight. Parts of the backbone are fused into a sturdy frame. The collarbone is fused into a furculum (commonly known as a wishbone). As the bird flies, this bone acts like a spring. The breastbone, or sternum, is broad and curved to provide a secure anchor for the strong flight muscles. Lungs continuously transfer oxygen to the blood, which the powerful heart pumps to the muscles.

and strong, with fewer bones than the bodies of reptiles or mammals. Flight is a very demanding activity that requires a high metabolism, so the respiratory system and heart of birds are very efficient.

A bird's digestive tract is fairly typical of vertebrates, although some birds have a crop for the temporary storage of food (for subsequent grinding in the gizzard and for absorption through the intestine walls, or for regurgitation to feed the young).

Some avian features—such as the scales on their legs and feet and the laying of eggs—are reminiscent of their reptilian ancestors. However, because birds are warm-blooded, they must spend a large part of their waking time searching for food. Their diets range from seeds and fruits to insects, small mammals, and even other birds.

Bills serve similar functions for birds as hands do for humans; their primary function is food gathering. There is enormous variation in the structure and shape of birds' bills and feet, reflecting the ecological niches species have adapted to fill.

The forms of a bird's wings and tail determine the amount of lift, thrust, and maneuverability it will have in flight. An improvement in one of these aspects can generally be achieved only with a sacrifice of performance in the other two. Small birds that need to fly off quickly to evade predators tend to have short wings; large birds such as albatrosses that regularly fly very long distances have long, thin wings for gliding.

Eyesight is birds' most well-developed sense, with hearing ranking second, despite the lack of any external ear structure.

Birds' feathers come in many colors (which may vary with the seasons) and are frequently used for social displays as well as for flying. Males tend to be more colorful than females.

To keep feathers in fine shape, birds preen regularly, running their beaks along the length of their feathers to smooth them out and to remove foreign particles. But even the most well-tended bird's feathers eventually become old and tattered, so at least once a year birds shed their feathers and grow new ones in a process known as molting. Birds also clean themselves by bathing in water or dust.

Most birds build nests to house their young. They lay from one to a dozen or more eggs, either once or several times a year.

Some species are born well-developed enough to fend for themselves right away; others are born blind and helpless. Until they are strong enough to fly and be independent, they remain in the nest, relying on one or both of their parents for food and for protection from predators. This nestling period can range from about a week to more than 5 months.

Keeping warm Birds such as the northern cardinal (*Cardinalis cardinalis*) (left) that don't live in or migrate to warmer climates struggle to find enough food and maintain a constant body temperature during the winter months. Fluffing up their feathers traps air and helps insulate their bodies.

RATITES AND TINAMOUS

CLASS Aves
ORDERS 5
FAMILIES 6
GENERA 15
SPECIES 59

Ratites are flightless birds with flat sternums, or breastbones, that lack the prominent, keel-like sternums of flying birds. Like other flightless birds, they are believed to have lost the ability to fly because they either lacked predators or could evade them by less energy-intensive means. This category includes two groups of large extinct birds, the moa of New Zealand and the elephant bird of Madagascar. Most ratites are believed to have evolved along similar lines rather than from one distinct ancestor. Tinamous are a related order of flying birds with keeled sternums that share some unusual anatomical characteristics—such as a distinctive palate structure—with ratites.

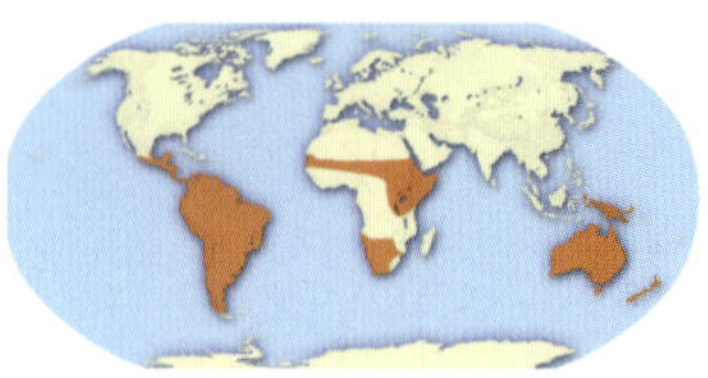

Southern exposure Ratites and tinamous are found only on southern Gondwanan continents. The various species are adapted to grasslands, jungles, woodlands, or high mountain ranges. The emu is the most versatile, roaming in diverse habitats.

GIANTS AND RUNNERS

Adult ratites and tinamous—among them the largest of all birds, the ostrich, which may exceed 9 feet (2.8 m) in height—resemble overgrown chicks, with underdeveloped, stubby wings and soft flight feathers.

Powerful runners, ratites have as few as two toes per foot, rather than the standard four toes of other birds. Ostriches can run faster than a racehorse and sometimes defend themselves by kicking. Rheas, sometimes called South American ostriches, occasionally raise one of their wings while running, but in this case the front limb serves the function of a sail rather than a wing.

The kiwis of New Zealand have vestigial wings. These three species of nocturnal birds—the country's national symbol—live in burrows and find their invertebrate prey by scent, using their long, sensitive bills. They are one of the few birds with a good sense of smell.

Tinamous are an ancient order of diverse, abundant smaller birds with plump bodies, short, rounded wings, and loose plumage. Some of the 47 species roost in trees, others on the ground. Although tinamous can fly, they usually escape predators by stealing through cover or by freezing.

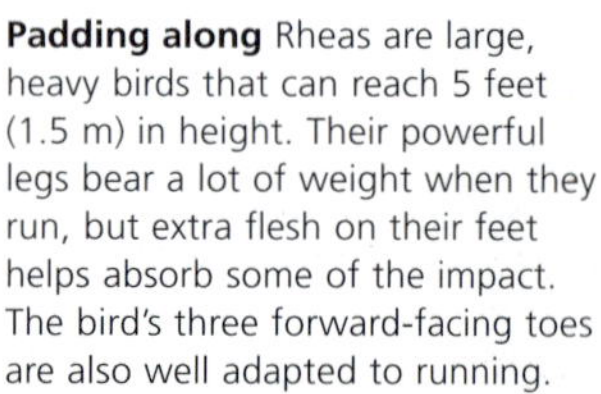

Padding along Rheas are large, heavy birds that can reach 5 feet (1.5 m) in height. Their powerful legs bear a lot of weight when they run, but extra flesh on their feet helps absorb some of the impact. The bird's three forward-facing toes are also well adapted to running.

Ostrich
Struthio camelus

Contrastingly colored tail is raised and fanned in display

Broad, flat bill for feeding on seeds and fruit

The only ratites in Africa, ostriches have only two toes

♂ ♀

Lesser rhea
Pterocnemia pennata

Greater rhea
Rhea americana

Emus, like cassowaries, have twinned hairy feathers

Long neck for reaching food

Emu
Dromaius novaehollandiae

Northern cassowary
Casuarius unappendiculatus

Southern cassowary
Casuarius casuarius

Powerful short legs are well adapted for running through rain forest

Brown kiwi
Apteryx australis

Little spotted kiwi
Apteryx owenii

Kiwis, found only in New Zealand, have sensory nostrils at the tip of the bill

Ostrich This large bird once ranged to the Middle East and Asia, but is now limited mainly to national parks in eastern and southern Africa. Males keep a temporary harem when breeding.

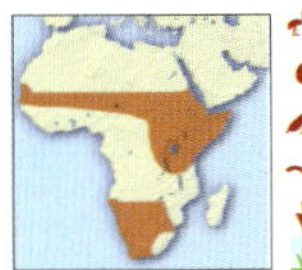

- Up to 9½ ft (2.9 m)
- 5–11 (major hen)
- Sexes differ
- Nomadic
- Locally common

C., E. & S. Africa

Emu This species is the sole surviving member of its family. Adaptable birds, emus live in a variety of habitats. They form temporary pairs to mate, and walk vast distances in search of food.

- Up to 6½ ft (2 m)
- 7–11
- Sexes alike
- Nomadic
- Locally common

Australia, Tasmania (extinct)

Greater rhea The larger of two species of rhea, the greater rhea has a wide, flat beak that is well adapted for grazing on the plains. It is a very agile runner, able to twist and turn abruptly to elude its pursuers.

- Up to 5¼ ft (1.6 m)
- 13–30
- Sexes alike
- Sedentary
- Near threatened

E., S.E. & C.W. South America

Gamebirds

CLASS	Aves
ORDERS	1
FAMILIES	5
GENERA	80
SPECIES	290

These familiar birds include chickens (domesticated forms of central Asian jungle fowl) and turkeys. Humans also hunt and eat pheasants, partridges, grouse, and quail. Peafowl are sometimes kept in captivity because of their beautiful plumage and ornate displays. These birds vary widely in size, but they all have stocky frames, relatively small heads, and short, broad wings. They typically fly low and fast. They are a favorite prey animal of many wild predators. To elude them, gamebirds rely on the camouflage effect of their dull-colored plumage, or else they fly or scurry away quickly. They have large clutches of up to 20 eggs, but species' populations tend to fluctuate greatly.

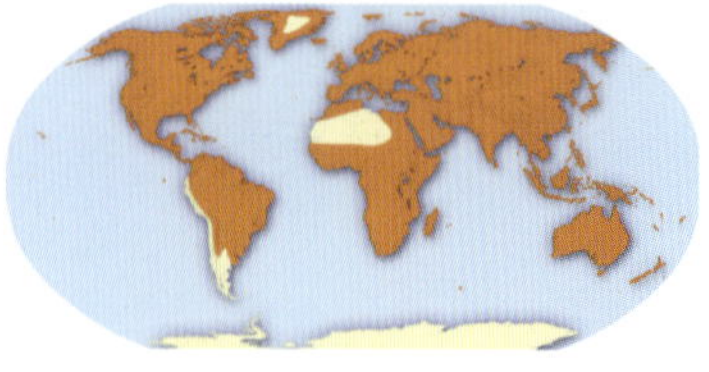

Far and wide Gamebirds live in a variety of climatic zones and, depending on the species, prefer either forests, scrub, open habitats, or grasslands. Some types of gamebirds are much more widespread than others: for example, the quails and partriges are found on several continents across the world, whereas turkeys occur only in North America.

California quail *Callipepla californica*

California qualis are popular cage-birds worldwide

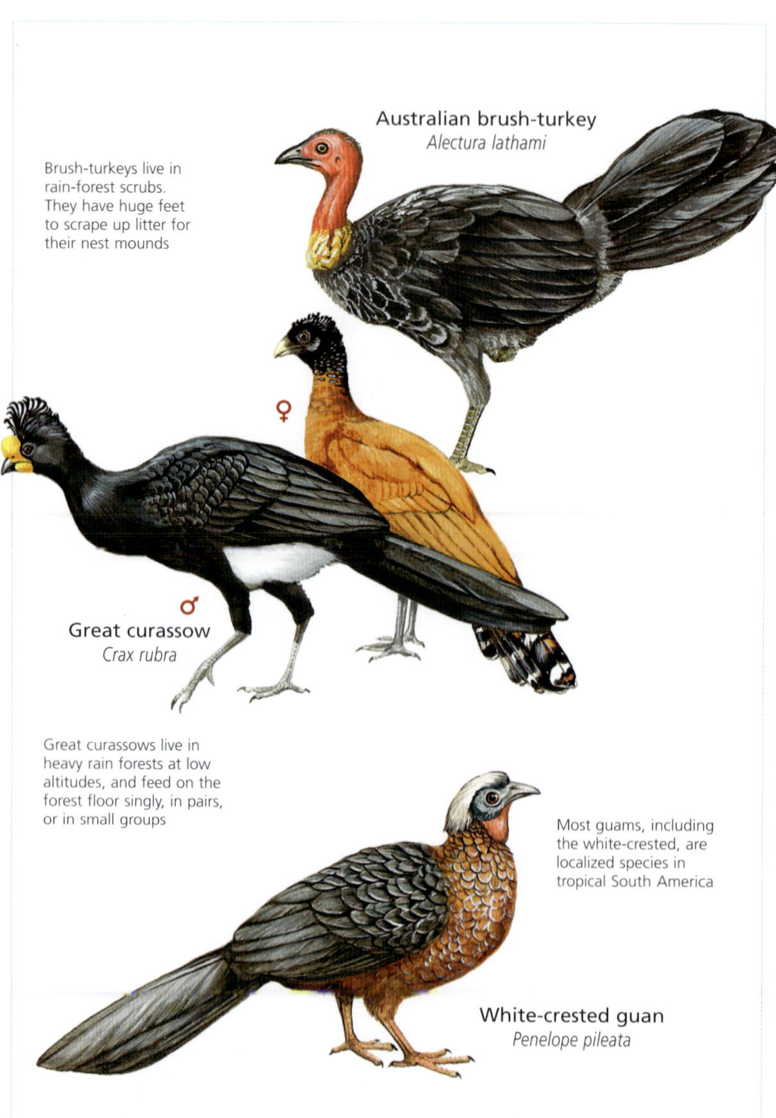

Australian brush-turkey *Alectura lathami*

Brush-turkeys live in rain-forest scrubs. They have huge feet to scrape up litter for their nest mounds

Great curassow *Crax rubra*

Great curassows live in heavy rain forests at low altitudes, and feed on the forest floor singly, in pairs, or in small groups

White-crested guan *Penelope pileata*

Most guams, including the white-crested, are localized species in tropical South America

On show Many gamebirds stage elaborate courtship displays. Black grouse cocks fan their lyre-shaped tails and make a bubbling, spitting call while competing for hens. While some species dwell mainly in trees, most gamebirds spend much of the time on the ground and can run well on their strong legs.

CONSERVATION WATCH

Under threat Gamebirds and waterfowl are the most widely and heavily hunted birds. Because they rear large broods, they can usually withstand such pressure; but when habitat destruction is added, serious declines follow. The most threatened gamebirds are those with very limited distributions.

Harlequin quail
Coturnix delegorguei

Painted bush-quail
Perdicula erythrorhyncha

Most bush-quails have striped faces

Red spurfowl
Galloperdix spadicea

Males have two elongated spurs on the back legs

Red-legged partridge
Alectoris rufa

Partridges are Eurasian; all have barred plumage on the sides of the body

Red-necked francolin
Francolinus afer

Chestnut-naped francolin
Francolinus castaneicollis

This species is found only in Ethiopia and Somalia, in northeast Africa

Double-spurred francolin
Francolinus bicalcaratus

This boldly striped francolin inhabits underbrush in tropical west Africa

Francolins are large, stout, quail-like birds that forage in underbrush for seeds, bulbs, and insects. The gray-striped francolin is confined to west Angola, in Africa

Gray-striped francolin
Francolinus griseostriatus

Black francolin
Francolinus francolinus

California quail Both sexes of this species have a head plume. Although normally herbivorous, this bird may eat various invertebrates. The young develop feathers quickly and can fly within weeks.

Up to 11 in (28 cm)
13–17
Sexes differ
Sedentary
Locally common

W. USA, N.W. Mexico, S.W. Canada

Red-legged partridge The female can lay two clutches in two different nests, leaving one clutch for the male to incubate. This species has been introduced to Britain as a gamebird.

Up to 15 in (38 cm)
11–13
Sexes alike
Sedentary
Locally common

Iberia & France to N. Italy

Black francolin The six subspecies of this bird differ slightly in size and plumage hue; these differences are most apparent in the females. Black francolins have been introduced in many countries.

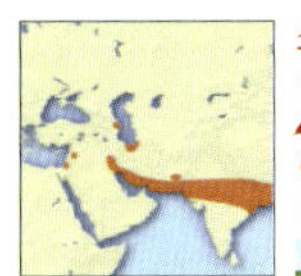

Up to 14 in (36 cm)
6–12
Sexes differ
Sedentary
Locally common

Middle East & trans-Caucasia to N. India

Blue eared-pheasant
Crossoptilon auritum

Blue and white eared-pheasants belong to a small group of round-tailed pheasants with projecting white ear tufts. They live only in China

White eared-pheasant
Crossoptilon crossoptilon

The only member of its group with whitish plumage; considered vulnerable

Female red junglefowl carry out all the brooding of eggs and chicks

♀ ♂

Red junglefowl
Gallus gallus

Males crow to advertise their territory

Koklass pheasant
Pucrasia macrolopha

♂ ♀

Eurasian black grouse
Lyrurus tetrix

Rock ptarmigan
Lagopus muta

Summer breeding plumage

Winter plumage for camouflage

Ocellated turkey
Meleagris ocellata

SPURRED ON

In the junglefowl (and some of its close relatives within this order), the male bird has spurs on the back of its legs, just above its toes. These are used as weapons when fighting other males. The sport of cockfighting, using domesticated birds, has been practiced for millennia. It is illegal in most countries but takes place in others.

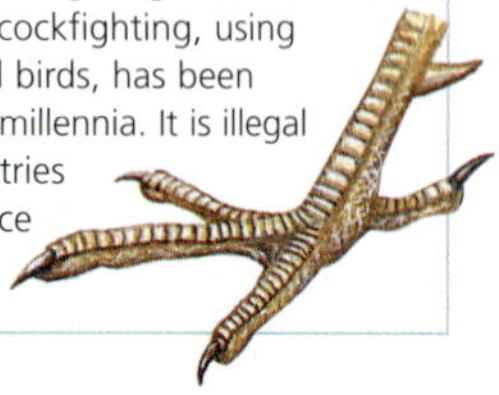

Black grouse This bird lives in a variety of habitats, from wooded areas to marshy ground. The black grouse nests in a hollow that it digs in the ground, where the female incubates her eggs.

- Up to 23 in (60 cm)
- 6–11
- Sexes differ
- Sedentary
- Locally common

Britain & N. Eurasia to E. Siberia & N. Koreav

Red junglefowl This colorful bird was domesticated as the chicken at least 5,000 years ago in the Indus valley. The meat and eggs of chickens are now eaten by people the world over.

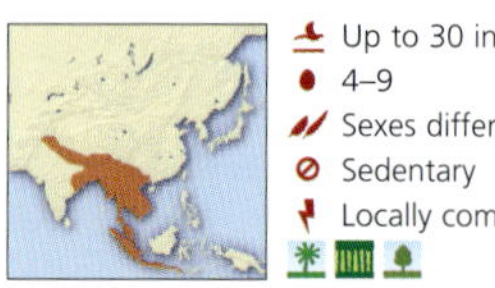

- Up to 30 in (75 cm)
- 4–9
- Sexes differ
- Sedentary
- Locally common

N. India & S.E. Asia to W. Lesser Sundas

Vulturine guineafowl
Acryllium vulturinum

Helmeted guineafowl
Numida meleagris

All guineafowl have spotted plumage and bare, colored skin on the head

Mikado pheasants live only on Taiwan; they are considered vulnerable

This is a black male. Note the red facial skin

Mikado pheasant
Syrmaticus mikado

This species, endemic to northeast China, has perhaps the longest tail of all pheasants

Reeves' pheasant
Syrmaticus reevesii

Pheasants have powerful feet for scratching in litter on the ground to find their food of seeds and invertebrates

Indian peafowl
Pavo cristatus

In display, the male raises his huge tail of occelated feathers in a giant fan

Feeds on dropped fruits on the forest floor

♂ Great argus ♀
Argusianus argus

Reeve's pheasant This beautiful bird lives only in the hill forests of central eastern Asia. For many centuries, the Chinese used its impressively long tail feathers and other plumage in decorative, ceremonial, or religious motifs.

- Up to 7 ft (2.1 m)
- 6–9
- Sexes differ
- Sedentary
- Vulnerable

N. & C. China & Mongolia

Helmeted guineafowl This ancestor of the domesticated guineafowl nests on the ground amid vegetation. After they are reared by the female, nestlings sometimes remain together in groups.

- Up to 25 in (63 cm)
- 6–12
- Sexes alike
- Sedentary
- Locally common

Sub-Saharan Africa

Great argus During this pheasant's courtship display, the male clears a hilltop in a forest. Dancing around the female, he opens his wings to reveal "eye" patterns that appear three-dimensional.

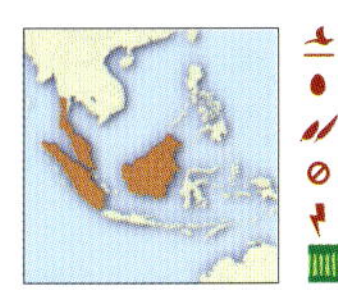

- Up to 6½ feet (2 m)
- 2
- Sexes differ
- Sedentary
- Near threatened

Malay Peninsula, Sumatra, Borneo

WATERFOWL

CLASS	Aves
ORDERS	1
FAMILIES	3
GENERA	52
SPECIES	162

Among the first birds to be domesticated, ducks and geese were raised for food more than 4,500 years ago. Swans have also been kept in captivity, because of their beauty and grace. Several species are flightless, but the rest are powerful fliers; many northern species migrate in family groups over great distances. In the air, they flap their wings continuously and can attain maximum speeds greater than 68 miles per hour (122 km/h). Some have been observed flying at an altitude of 28,000 feet (8,485 m), near the summit of Mount Everest. Some doze on water; others come ashore to rest. Their calls range from quacks to barks, hisses, whistles and even trumpeting sounds.

Citizens of the world Waterfowl predominate in the Northern Hemisphere (with the largest number of species being in North America), but they can be found worldwide, except in Antarctica. They occur in almost every type of wetland, from city ponds to Arctic sea inlets. Some species spend a great deal of time at sea.

AT HOME ON THE WATER

All species are remarkably similar in form, with short legs, webbed feet, relatively long necks, and flattened, broad bills. Most are excellent swimmers, although several species have adapted to life on land and have less webbing on their feet.

To protect them from the cold water, these birds rely on their dense, waterproof feathers and a thick coat of insulating down. Their plumage is often brightly colored and patterned.

Domestic waterfowl breeds are descended from mallards, muscovies, graylags and swan geese. Mallards, native to the Northern Hemisphere, have been widely introduced in other parts of the world, which has led to unwanted hybridization and genetic dilution of local species.

Many waterfowl consume grass, seeds, grain, and other vegetation, but some species eat fishes, insects, mollusks, and crustaceans.

The South American screamers bear little superficial resemblance to other waterfowl, but they all share certain anatomical similarities.

Canada geese are among the waterfowl that migrate flying in wedge formation. The birds take turns flying in the lead, and do so as they fly by day and night to reach summer or winter quarters.

Southern screamer
Chauna torquata

The three species of screamers all live in tropical and subtropical South America. They feed on succulent vegetation on stream sides

Vulturine bill is adapted for grasping and pulling

Magpie goose
Anseranas semipalmata

Magpie geese are found only in the swamps of northern Australia and New Guinea. They breed in huge colonies on floating vegetation

Although its feet are only part-webbed, this bird is a good swimmer

Snow goose
Anser caerulescens

Snow geese are migratory, breeding in the north American tundra and wintering on the east and west coasts of the United States

White-faced whistling-duck
Dendrocygna viduata

This species has an unusual distribution, with widespread populations in both Africa and South America

Red-breasted goose
Branta ruficollis

Canada goose
Branta canadensis

The only goose with a black head and neck ornamented with a white bib

Bean goose
Anser fabalis

Bean geese breed in the high arctic zone of Eurasia and fly south to winter in China and the Mediterranean

Southern screamer This bird lives in marshlands and along watercourses. It typically lives in groups and is a fine swimmer, despite the lack of webbing on its feet. Some indigenous people raise domesticated southern screamers.

- Up to 37 in (95 cm)
- 3–5
- Sexes alike
- Sedentary
- Locally common

C. & E. South America

Snow goose This bird has two color forms, one all white with black wing feathers, the other white-headed with a blue-gray body and wings. It breeds in the Arctic tundra and can be heard cackling as it flies.

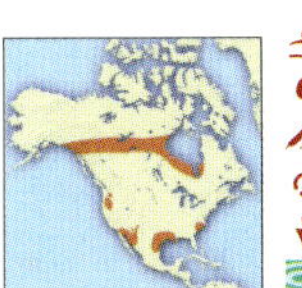

- Up to 32½ in (80 cm)
- 4–5 usually
- Sexes alike
- Migrant
- Common

Arctic & S. North America

Canada goose There are about a dozen races of Canada goose, which vary in size, color, and geographic distribution. Originally native to North America, where it is a common sight in city parks.

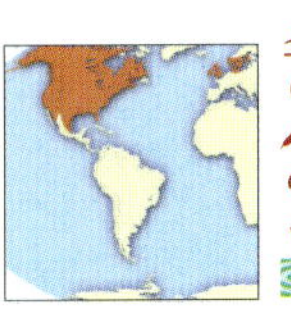

- Up to 3¾ ft (1.15 m)
- 4–7 usually
- Sexes alike
- Migrant
- Common

North America, N. Europe, N.E. Asia

Torrent duck
Merganetta armata

Lives only in the Andes of South America, and feeds by diving in swift-flowing mountain streams

Northern shoveler
Anas clypeata

Muscovy duck
Cairina moschata

Freckled duck
Stictonetta naevosa

A rare duck of inland Australia that feeds on microplankton in shallow, freshly flooded swamps

Orinoco goose
Neochen jubata

Lives along the forested sides of rivers in tropical South America

Common shelduck
Tadorna tadorna

Male has an orange knob above the base of the bill; the female's knob is white

Comb duck
Sarkidiornis melanotos

Males are twice as big as females

Magellanic steamer duck
Tachyeres pteneres

Mallard
Anas platyrhynchos

Wood duck
Aix sponsa

SPATULATE BEAK

The northern shoveler gets its name from its distinctive, heavy, shovel-like bill, which is longer than its head. It feeds in shallow water, extending its neck and dabbling its bill just below the water surface. Comb-like serrations along the sides of the bill help it strain food items from the water.

Common shelduck This pugnacious bird looks heavy in flight and beats its wings slowly. It often breeds in old rabbit burrows or in gaps under old buildings. The male shelduck whistles and the female growls.

- Up to 25 in (65 cm)
- 8–10
- Sexes differ
- Migrant
- Common

W. Europe, C. Asia, N.W. Africa

Wood duck This handsome duck nests in tree cavities. Nestboxes have helped to reestablish this threatened species. After nesting, the male morphs into drab, female-like plumage, but retains a red bill.

- Up to 20 in (51 cm)
- 9–15
- Sexes differ
- Partial migrant
- Locally common

C. & S. North America & W. Cuba

Red-crested pochard
Netta rufina

Oldsquaw
(long-tailed duck)
Clangula hyemalis

Despite their distinctively long tails, oldsquaws dive almost exclusively for their food of crustaceans and mollusks in the Arctic

Common pochard
Aythya ferina

Pochards feed mostly by diving and dabbling. They paddle to the floors of lakes and swamps for mainly vegetable matter and seeds

Common mergansers dive for their food of fishes and aquatic invertebrates

Common merganser
Mergus merganser

Musk duck
Biziura lobata

This southern Australian species has a large lobe of skin under the bill

Common eider
Somateria mollissima

Eiders are marine, living in often rough seas around rocky coasts, and diving for mollusks and crustaceans

♀

♂

White-headed duck
Oxyura leucocephala

Males have an eclipse plumage, but are never as dull as the brown females

Male in full breeding plumage

Red-crested pochard This diving duck feeds mainly at night. To fly, it has to run over the surface of the water before it can lift itself into the air. During courtship, the male may bring food or twigs to the female.

Up to 23 in (58 cm)
6–14
Sexes differ
Partial migrant
Locally common

C. & S. Europe to S. Asia, N. Africa

White-headed duck This rare duck breeds in shallow, brackish wetlands with abundant vegetation and reeds. It has declined due to hunting, habitat loss, and hybridization. The male's bill becomes smaller in winter.

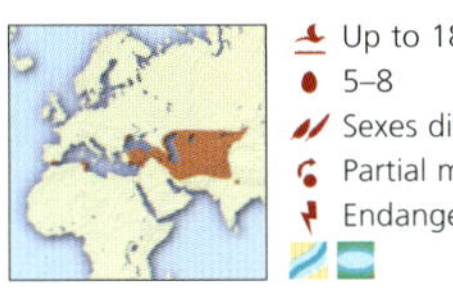

Up to 18 in (46 cm)
5–8
Sexes differ
Partial migrant
Endangered

S. Europe, Middle East, C. Asia to N. India

WEBBED PROPELLERS

Most waterfowl are excellent swimmers, due to their webbed front toes, which act as paddles to propel them through the water. They also help the birds walk on mud. But because the legs are set so far back on the body for propulsion in water, waterfowl can do little more than waddle awkwardly on land.

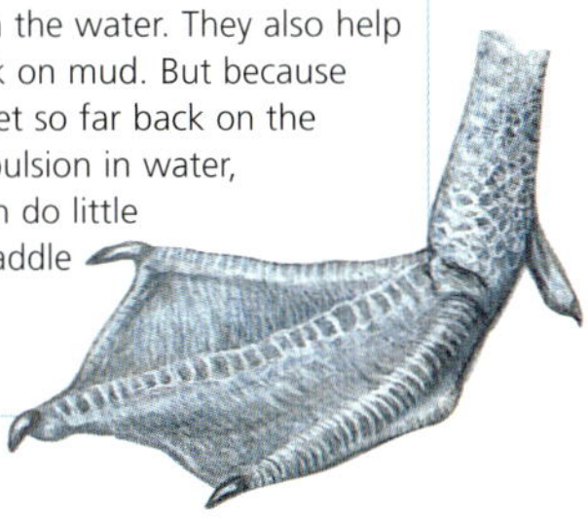

PENGUINS

CLASS	Aves
ORDERS	1
FAMILIES	1
GENERA	6
SPECIES	17

The most aquatic of all birds, penguins have remained unchanged in form for at least 45 million years. Although they evolved from flying birds, none of the 17 species of penguins can fly. Very highly specialized, social seabirds, they take advantage of their streamlined bodies (which minimize drag) and small, flipper-like wings to travel underwater at speeds of up to 15 miles per hour (24 km/h). They spend up to three-quarters of their life in the sea, staying underwater for as long as 20 minutes or more at a time, and coming ashore only to breed and molt. Penguins eat fish, krill, and other invertebrates.

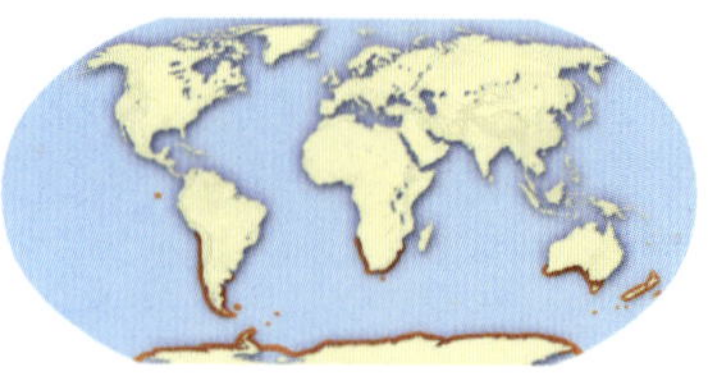

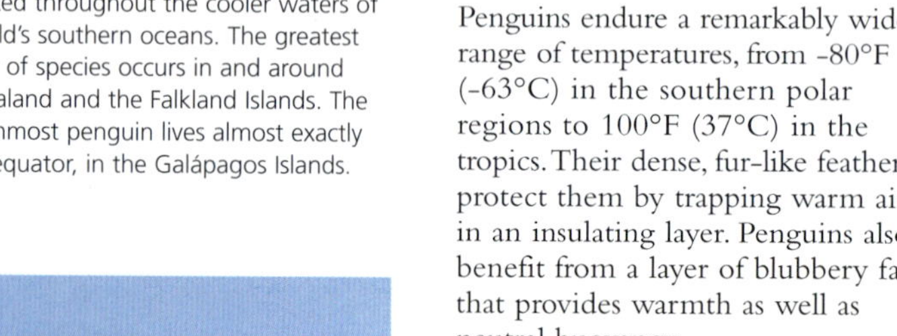

South seas dwellers Penguins are widely distributed throughout the cooler waters of the world's southern oceans. The greatest diversity of species occurs in and around New Zealand and the Falkland Islands. The northernmost penguin lives almost exactly on the equator, in the Galápagos Islands.

GOING TO EXTREMES

Penguins endure a remarkably wide range of temperatures, from -80°F (-63°C) in the southern polar regions to 100°F (37°C) in the tropics. Their dense, fur-like feathers protect them by trapping warm air in an insulating layer. Penguins also benefit from a layer of blubbery fat that provides warmth as well as neutral buoyancy.

The young are raised in crowded colonies. The parents recognize their own chicks by voice. Penguins are born with insulating down, but cannot enter the water until they have grown their first adult layer of waterproof feathers. As adults, they then lose this layer when they molt and must remain ashore for 3 to 6 weeks to regrow it; during this period, they may lose a third or more of their body weight.

Penguins are superb swimmers and divers, propelling themselves with their wings, which have evolved into stiff, flat, paddle-like flippers. They twist their flippers in such a way as to provide thrust on both the upstroke and downstroke. Their bodies also have numerous special physiological features that allow them to regulate their body temperatures and oxygen levels in cold waters.

Warm spot Emperor penguin chicks depend on their parents for warmth and protection. Some penguins nest on the ground or in burrows, but emperor penguins are among those that incubate the eggs on their feet. After 6 months, the chicks molt and mature enough to go to sea.

Emperor penguin The largest penguin, the emperor is unique in that it breeds in the middle of winter. These penguins usually do so on annual fast ice, making them the only birds to not normally set foot on solid ground.

- Up to 4 ft (1.2 m)
- 1
- Sexes alike
- Sedentary
- Locally common

Seas & coasts of Antarctica

GENTOO PENGUIN

These penguins, whose main colony is on the Falkland Islands, gather in colonies of a few hundred breeding pairs. They nest on rocky shores and inland grasslands, building skimpy nests using pebbles and molted feathers (in Antarctica) or vegetation (on sub-Antarctic islands). Gentoo penguins can be very aggressive and will often fight over these materials as they walk around gathering them.

Divers and grebes

CLASS	Aves
ORDERS	2
FAMILIES	2
GENERA	7
SPECIES	27

Although both are aquatic birds that use their webbed feet for propulsion underwater, divers (known as loons in North America) and grebes are only distantly related to each other. Their similarities are believed to be the result of convergent evolutionary paths, over the course of which they both developed features suited to diving for fishes and aquatic invertebrates. Physically, loons resemble stout ducks and cormorants that ride low in the water and have pointed bills. They are thought to be descended from wing-propelled swimming ancestors and therefore may be related to penguins and petrels. Some grebes are slim and elegant; smaller ones resemble ducklings.

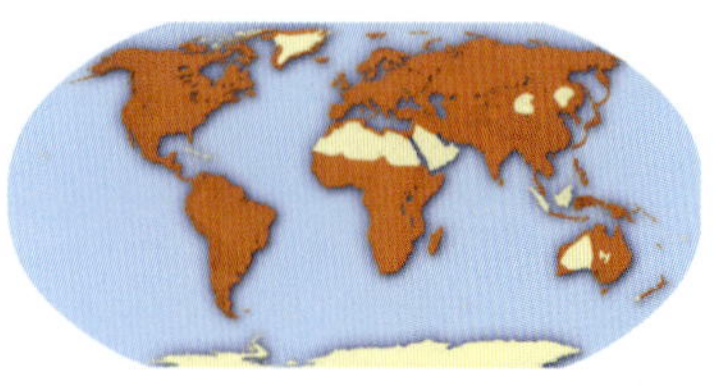

Watery world Loons and grebes spend most of their lives on water, coming ashore only to nest (grebes even nest on the water, on floating platforms of vegetation). Loons live on freshwater lakes; grebes inhabit wetlands in general.

SKILLED SWIMMERS

Loons are shy birds that cannot walk properly. They nest on the edges of lakes, to make it easier to slip into the water, where they can swim rapidly and dive to depths of more than 200 feet (61 m) chasing fishes. They spend their winters at sea.

Grebes can also be elusive. They appear to lack tails; their bodies end in a powder-puff effect of loose feathers. Females often mount males during the breeding season.

Loon and grebe chicks can swim and dive as soon as they are born, but because they are sensitive to cold water, they prefer to hitch rides on their parents' backs or shelter under their parents' wings.

These birds do not fly much on a daily basis and usually avoid danger by diving underwater.

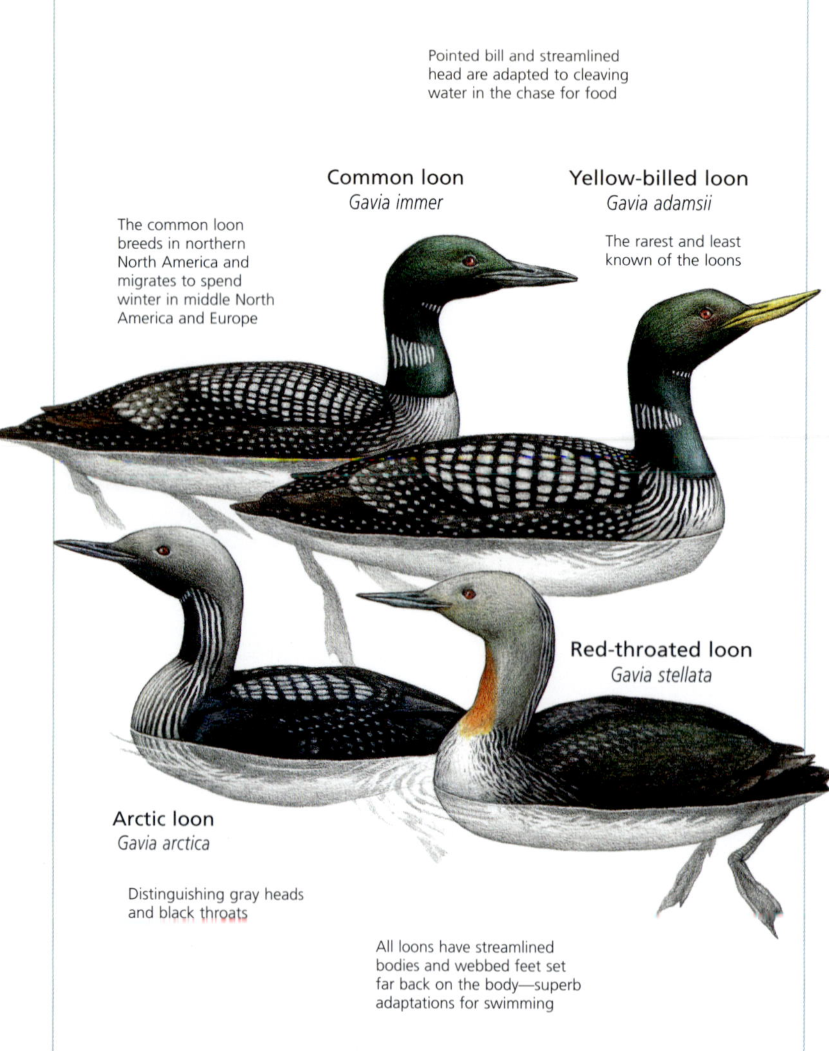

Pointed bill and streamlined head are adapted to cleaving water in the chase for food

Common loon *Gavia immer*

The common loon breeds in northern North America and migrates to spend winter in middle North America and Europe

Yellow-billed loon *Gavia adamsii*

The rarest and least known of the loons

Red-throated loon *Gavia stellata*

Arctic loon *Gavia arctica*

Distinguishing gray heads and black throats

All loons have streamlined bodies and webbed feet set far back on the body—superb adaptations for swimming

Loon call A loon gives an aggressive territorial call. The loud, yodelling calls—audible for a mile (1.6 km) or more—of returning migratory loons are a familiar sign of summer in boreal North American and Eurasia. These shy birds are most easily seen during migrations.

White-tufted grebe
Rollandia rolland

New Zealand grebe
Poliocephalus rufopectus

One of two large species of grebes in North America

Western grebe
Aechmophorus occidentalis

White-tufted grebes feed mainly on the surface of the water on fishes and aquatic invertebrates

Great grebe
Podiceps major

Named for its mostly black head

Hooded grebe
Podiceps gallardoi

Great crested grebe
Podiceps cristatus

Face plumes, called tippets, are flared in sexual display. Like other grebes, this species builds a rough nest of vegetable matter floating in water

Eared grebe
Podiceps nigricollis

Horned grebe
Podiceps auritus

Lobed toes on feet placed far back on the body are much better adapted for swimming than walking

Little grebe
Tachybaptus ruficollis

If forced to leave the nest, this grebe first covers its eggs with wisps of vegetation to hide them

Western grebe This bird is famous for its spectacular courtship displays. It nests in colonies on lakes and migrates to the coast. A nocturnal feeder, it belongs to the only group of grebes that spear fish with the bill.

- Up to 30 in (76 cm)
- 3–4
- Sexes alike
- Migrant
- Locally common

W. & C. North America

Great crested grebe This elegant waterbird, the best-known grebe, was almost hunted to extinction in some countries for its ornate head feathers.

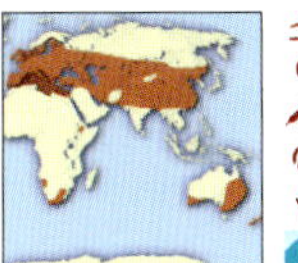

- Up to 25 in (64 cm)
- 3–5
- Sexes alike
- Partial migrant
- Locally common

Eurasia, S. Africa, S. Australia,

Little grebe This small, dumpy bird appears to have a fluffy rear end. If disturbed, it quickly submerges, reappearing some distance away. It makes loud, distinctive whinnying trills.

- Up to 11 in (28 cm)
- 3–7
- Sexes alike
- Partial migrant
- Common

Sub-Saharan Africa, W. & S. Eurasia to N. Melanesia

Albatrosses and petrels

CLASS	Aves
ORDERS	1
FAMILIES	4
GENERA	26
SPECIES	112

These seabirds, collectively known as tubenoses, are highly adapted to life at sea. They can glide for hours without beating their wings, and it is not uncommon for them to fly hundreds of miles (km) in search of squids, fishes, and the large zooplankton that comprise their diets. They seldom come within sight of land, except to breed. The biggest tubenose is the wandering albatross, which has the largest wingspan—up to 11 feet (3.3 m)—of any bird. Giant petrels are the size of some albatrosses, but the smallest storm petrels have an average wingspan of only about 12 inches (30 cm). Diving petrels have small, rigid wings as much suited for underwater propulsion as for flying.

Ocean-going birds Albatrosses are most frequently associated with the windswept expanses of southern oceans.

Royal albatross
Diomedea epomophora

Yellow-nosed albatross
Thalassarche chlororhynchos

Huge long wings enable albatrosses to wander the oceans by effortlessly gliding on wind currents

Northern fulmar
Fulmarus glacialis

Cape petrel
Daption capense

EXPRESSING THEMSELVES

Tubenoses are found throughout the world's oceans. They have large, external, tubular nostrils and a relatively well-developed sense of smell that they may use to locate food, breeding sites, and each other. All species have a musty body odor that persists for decades in museum exhibits.

Most tubenoses store large quantities of oil in their stomachs, which they regurgitate for their young or eject to deter predators.

The male and female perform spectacular courtship displays on the ground, facing each other with their wings spread and their tails fanned, throwing their heads back as they gurgle and bray. They renew their bonds with their life partners with complex head movements and rituals that can last for days.

Tubenoses are long-lived birds, but do not begin breeding until they are at least 10 years old. Some only breed every other year.

THE LONG WAY HOME

Several shearwaters are examples of birds that migrate in a loop pattern. They skirt the coastline of one continent in the spring and a different one in the fall, in a large figure-eight pattern. Loop migrations may be undertaken due to food availability, ocean currents, prevailing winds, and temperature.

A shearwater's long wings help it migrate over very large distances across open oceans.

Gray petrel
Procellaria cinerea

Fiji petrel
Pseudobulweria macgillivrayi

A rare species known only from around Fiji

Jouanin's petrel
Bulweria fallax

Bulwer's petrel
Bulweria bulwerii

Manx shearwater
Puffinus puffinus

Longer, more pointed wings, forked tails, and short feet distinguish northern storm-petrels from those in Southern Hemisphere waters

Band-rumped storm-petrel
Oceanodroma castro

Diving-petrels have small, chunky, bullet-shaped bodies adapted to plunge-diving for food

Wilson's storm-petrel
Oceanites oceanicus

Common diving-petrel
Pelecanoides urinatrix

This transequatorial migrant is the most widespread storm-petrel in the world

Wedge-tailed shearwater
Puffinus pacificus

Wilson's storm-petrel
Like other storm-petrels, this bird is colonial and nests in rock crevices or underground, mostly on isolated islands. It makes long migrations, traveling between the Antarctic and the subarctic.

Antarctica, all oceans to north of equator

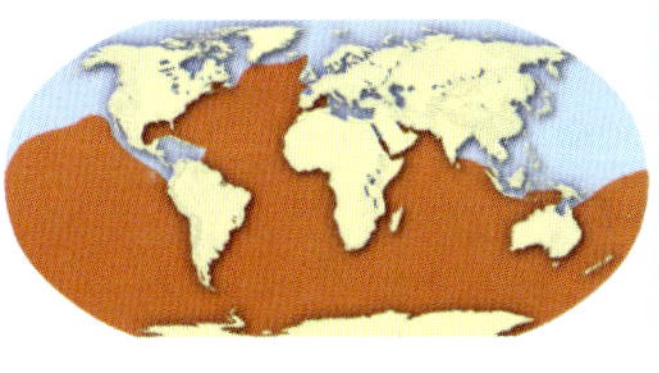

- Up to 7½ in (19 cm)
- 1
- Sexes alike
- Migrant
- Common

Common diving petrel
This bird dives for its food, traveling through water and air with equal ease. Its tubular nostrils open upward rather than forward to prevent water entering.

Southern Ocean of S.E. Australia & New Zealand, S.W. Africa & S.W. South America

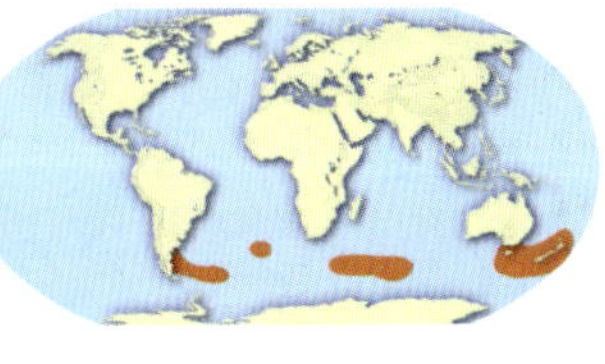

- Up to 10 in (25 cm)
- 1
- Sexes alike
- Sedentary to locally nomadic
- Common

MIGRATIONS

Nearly half of the world's birds divide their time between two main localities. Most migrate because of seasonal fluctuations in the availability of food. Many species travel alone, but others prefer to migrate in flocks comprising one or more species. They travel either by day or night, and either overland or across oceans. Most break the journey into short hops of a couple of hundred miles (km). But terrestrial birds that cross oceans and cannot land on water must complete the entire journey in one epic flight. One such bird is the American golden plover, which flies between Alaska and Hawaii, a journey it undertakes twice a year. To prepare for such energy-intensive journeys, birds load up on food almost equal to their regular weight before they leave.

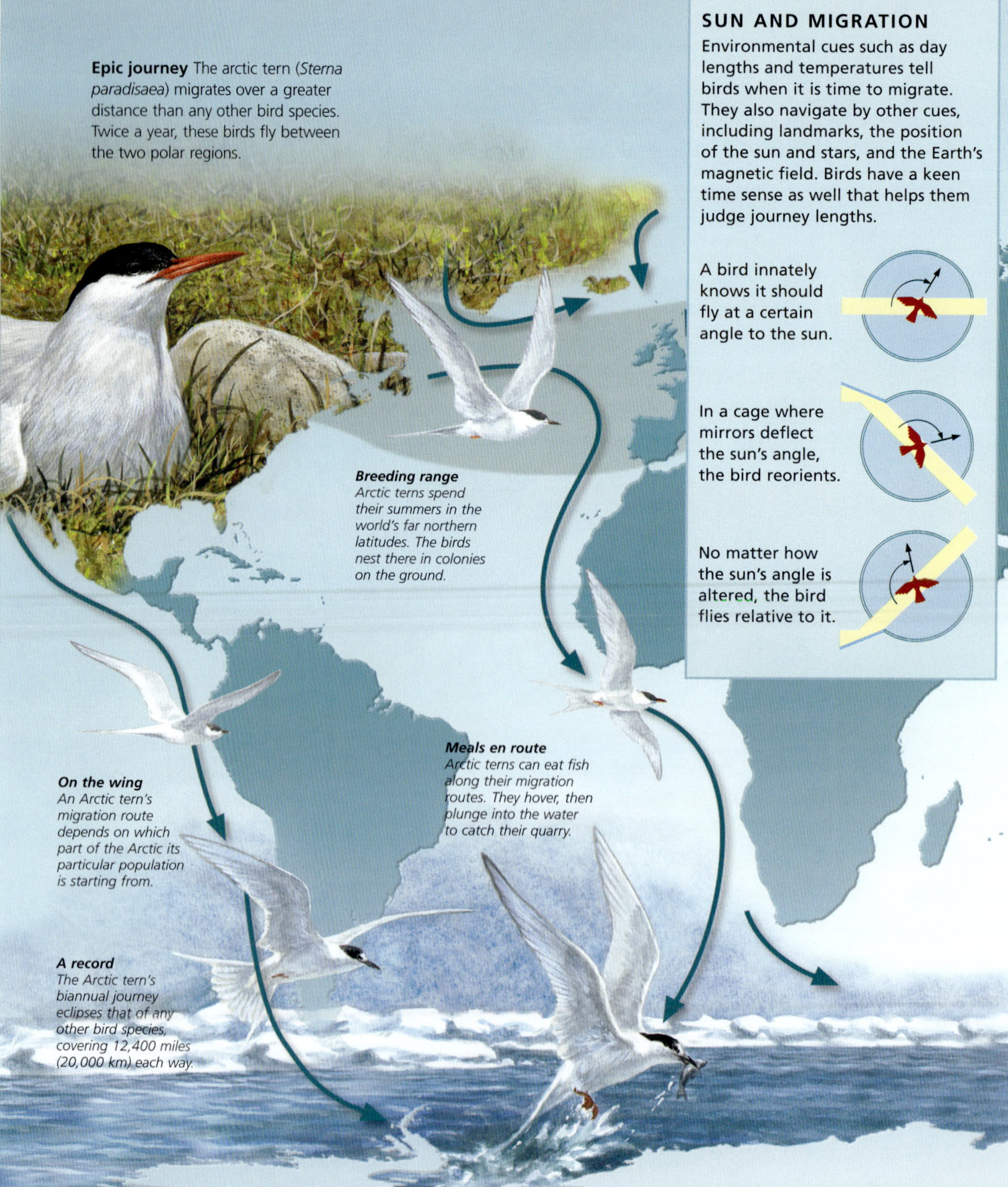

Epic journey The arctic tern (*Sterna paradisaea*) migrates over a greater distance than any other bird species. Twice a year, these birds fly between the two polar regions.

Breeding range *Arctic terns spend their summers in the world's far northern latitudes. The birds nest there in colonies on the ground.*

On the wing *An Arctic tern's migration route depends on which part of the Arctic its particular population is starting from.*

Meals en route *Arctic terns can eat fish along their migration routes. They hover, then plunge into the water to catch their quarry.*

A record *The Arctic tern's biannual journey eclipses that of any other bird species, covering 12,400 miles (20,000 km) each way.*

SUN AND MIGRATION

Environmental cues such as day lengths and temperatures tell birds when it is time to migrate. They also navigate by other cues, including landmarks, the position of the sun and stars, and the Earth's magnetic field. Birds have a keen time sense as well that helps them judge journey lengths.

A bird innately knows it should fly at a certain angle to the sun.

In a cage where mirrors deflect the sun's angle, the bird reorients.

No matter how the sun's angle is altered, the bird flies relative to it.

FLAMINGOS

CLASS	Aves
ORDERS	1
FAMILIES	1
GENERA	3
SPECIES	5

These distinctively beautiful birds are easily recognized by their bright pink or red and white plumage, their long legs and neck (proportionately longer than on any other bird), and their oddly depressed bills. There are five species of flamingo; the largest is the greater flamingo, which reaches a height of nearly 5 feet (1.5 m). Vast flamingo flocks that congregate on the lakes of the Great Rift Valley are one of the famous natural spectacles of Africa. The reddish color of flamingo plumage comes from carotenoid proteins in the birds' diet of plant and animal microplankton. Liver enzymes break down these proteins into usable pigments that are deposited in both skin and feathers.

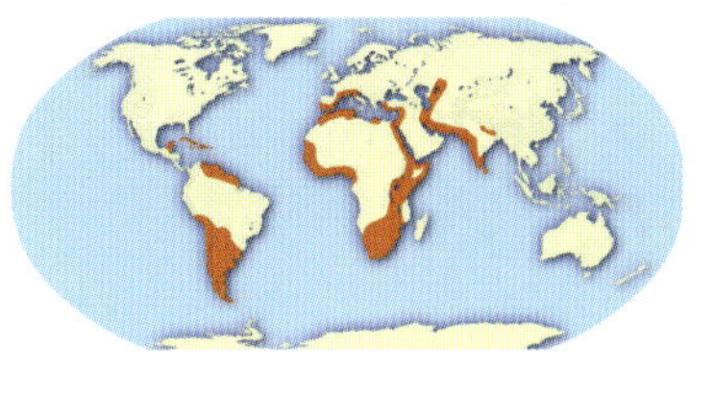

Old salts Flamingos were once found on every continent, but have disappeared from Australasia. These mainly tropical birds live in shallow lakes and coastal regions, preferring salty or brackish water. They inhabit some isolated islands and can also be seen at high altitudes in the Andes.

SPECIALIZED FEEDERS

The evolutionary lineage of flamingos is still something of a mystery, though they may represent the link between herons and their allies, and waterfowl.

The flamingo's strongly hooked bill is well designed for filter feeding on small shellfish, insects, single-celled animals, and algae. To feed, the bird bends forward, turns its head upside down (looking backward between its legs) and drags its opened bill through the water. After closing its bill, it uses its lower jaw and tongue (which has tooth-like projections) to pump water and mud out through the slits lining the upper jaw. The bird then swallows the food that remains.

The shallow waters where flamingos live sometimes drain away, forcing them to travel great distances to more abundant feeding grounds. The flocks travel at night, honking as they fly.

Flamingos nest on lakes and in coastal regions, laying one or two eggs per breeding season. The young can run and swim well at an early age, leaving the nest to follow their parents around by the time they are about 4 days old, and flying by the age of 70–80 days.

Adults sometimes swim rather than walk while they feed in deep waters, often at night.

Unusual angle
The long neck and downturned bill are adapted to feeding by immersing the head deep under water for long periods.

HERONS AND ALLIES

CLASS	Aves
ORDERS	3
FAMILIES	5
GENERA	41
SPECIES	118

This group of long-legged wading birds includes herons, storks, ibises, and spoonbills. They can step through water without getting their plumage wet as they search for fishes, insects, and amphibians to eat. Some types of herons are known as egrets; they were thus named for their special, filamentous breeding plumage, which was highly sought after by 19th-century hat-makers. Many species within this group are gregarious, and large groups comprising several species can sometimes be seen feeding, roosting, or nesting together. The white stork—a migratory bird that often nests in pairs on chimneys—has long been associated with human births in folk tales.

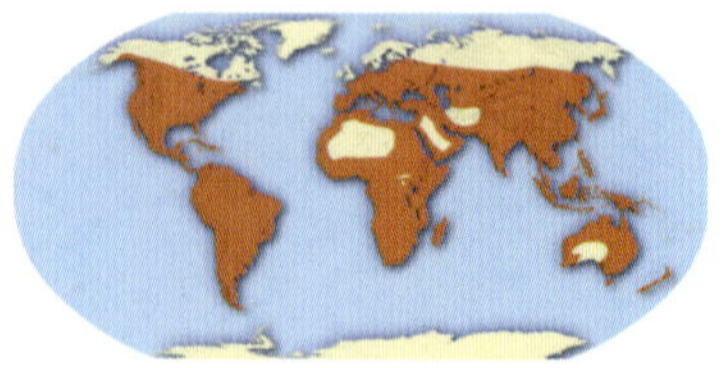

Freshwater dwellers Herons and their allies live worldwide, except near the poles. They are typically found in or near various types of freshwater habitats, including swamps, marshes, rivers, streams, lakes, and ponds. However, several species have adapted to drier environments.

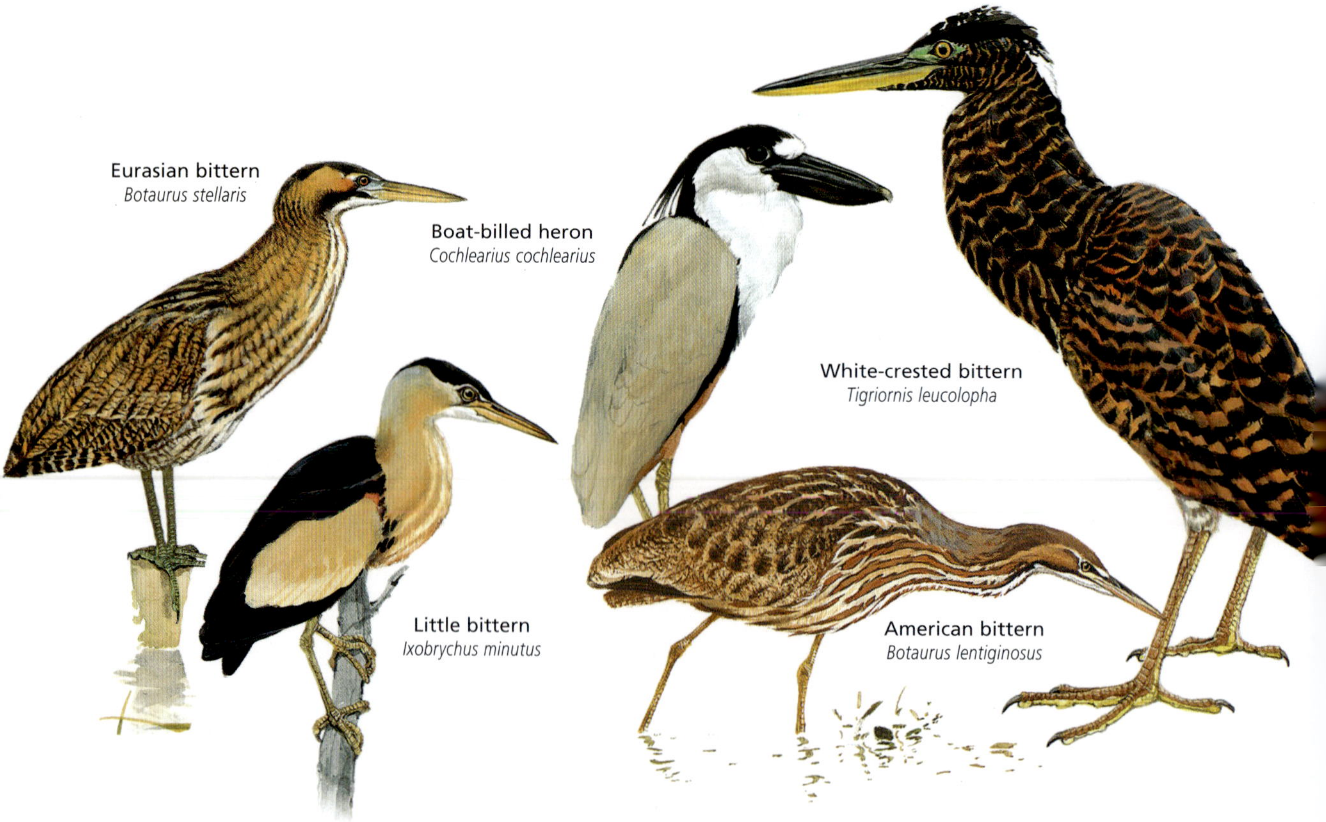

Eurasian bittern *Botaurus stellaris*

Boat-billed heron *Cochlearius cochlearius*

White-crested bittern *Tigriornis leucolopha*

Little bittern *Ixobrychus minutus*

American bittern *Botaurus lentiginosus*

WADING IN

All birds in this group have short tails and long beaks and necks, as well as long legs. Nonetheless, there is considerable variation in size, color, plumage pattern, and feeding behavior among species.

Some species have specialized to fill very specific niches. The cattle egret, for example, follows grazing animals such as buffalo, eating the insects disturbed. Some herons camouflage themselves in marshes; if approached, they try to blend in with the reeds by pointing their heads skyward, compressing their bodies, and swaying with the vegetation. Unlike their close relatives, herons fly with their heads folded back on their shoulders, making them easy to recognize.

Like some other types of birds, herons have specialized patches of feathers called powder-down. These are never molted, but grow continually. As the tips fray, they turn into a fine powder, which the bird picks up in its bill and uses to remove slime and oil from its feathers when grooming. Some storks and ibises have bare necks, perhaps to prevent fouling plumage while eating carrion. Most species in this group migrate long distances, maybe because their large bodies would require a great deal of energy to warm in winter.

Black-headed heron
Ardea melanocephala

Long, articulated neck for stabbing strikes at prey

Gray heron
Ardea cinerea

Long, bare legs and feet for wading

Whistling heron
Syrigma sibilatrix

Dagger-shaped bill for grasping prey from strikes

Great blue heron
Ardea herodias

Goliath heron
Ardea goliath

Purple heron
Ardea purpurea

Cattle egret
Bubulcus ibis

Capped heron
Pilherodius pileatus

Chinese egret
Egretta eulophotes

Cattle egret populations erupted around the world in the 20th century. This species associates in small flocks with domestic and wild stock, often perching on the backs of cattle

Gray heron This colonial bird generally nests high up in trees. Both parents look after the young, which remain in the nest for almost 2 months. The gray heron's diet sometimes includes small birds.

- Up to 3¼ feet (1 m)
- 3–5
- Sexes alike
- Partial migrant
- Common

Sub-Saharan Africa, C. & S. Eurasia to Indonesia

Great blue heron This bird has both white and gray phases in different regions. The most familiar large wading bird in North America, it is often seen stalking the edges of lakes or marshes.

- Up to 4½ ft (1.4 m)
- 3–7
- Sexes alike
- Partial migrant
- Common

Mid-North to Central America, Galápagos Is.

Purple heron The purple heron nests alone or in colonies among reeds. A common bird, it sometimes eats small birds and small mammals, in addition to its staple diet of amphibians, fishes, and invertebrates.

- Up to 35 in (90 cm)
- 2–5
- Sexes alike
- Partial migrant
- Locally common

Europe, Middle East, Sub-Saharan Africa, S. & E. Asia

Crested ibis
Nipponia nippon

Eurasian spoonbill
Platalea leucorodia

Madagascan crested ibis
Lophotibis cristata

Color pattern of rusty brown body with white wings is unique to this species

Shoebill
Balaeniceps rex

Hamerkop
Scopus umbretta

Roseate spoonbill
Platalea ajaja

Sacred ibis
Threskiornis aethiopicus

Black stork
Ciconia nigra

Wood stork
Mycteria americana

Long, bare legs and feet are adapted for wading in shallow waters

These South American birds use their open bills to detect fish in muddy water

Shoebill This rather odd-looking bird is thus named because it appears to be wearing a clog on its face. Its large bill is well adapted for catching the slippery lungfish found in its wetland habitats.

- Up to 4 ft (1.2 m)
- 1–3
- Sexes alike
- Sedentary
- Near threatened

C. Africa

Sacred ibis This distinctive bird was very prominent in ancient Egyptian mythology, symbolizing the god Thoth. The Egyptians mummified these birds. Sacred ibis became extinct in Egypt, but still thrive elsewhere.

- Up to 35 in (90 cm)
- 2–3
- Sexes alike
- Partial migrant
- Common

Sub-Saharan Africa & W. Madagascar

Madagascan crested ibis This large terrestrial bird feeds on moist ground in forests and scrubs. When disturbed, it prefers to run away rather than fly, dodging through trees along the way.

- Up to 20 in (50 cm)
- Sexes alike
- 2–3
- Sedentary
- Near threatened

E. & W. Madagascar

Pelicans and Allies

CLASS	Aves
ORDERS	1
FAMILIES	6
GENERA	8
SPECIES	63

The distinctive pelicans (which have been in existence since mid-tertiary times, up to 30 million years ago) are related to four other families of water birds: tropicbirds; gannets and boobies; cormorants and anhingas; and frigatebirds. They all have webbed feet that allow them to move easily through water, and the webbing extends uniquely across all four toes. Many have large, naked throat sacs that are used to hold fishes or as a sexual attractant in courtship displays. Unusually extensive air-sac systems in their chests and lower necks make those areas well-cushioned (hence protective in diving) and buoyant. They eat primarily fishes, as well as squid and other invertebrates.

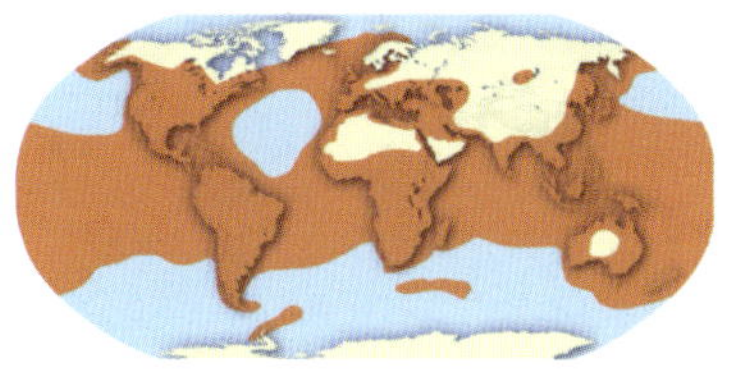

A varied range Pelicans and their close relatives are found in all types of water environments, from open oceans and sea coasts to lakes, swamps, and rivers. Most species live in tropical or temperate areas.

EXPERT FISHERS

These birds are much better suited to water than land. Tropicbirds cannot even walk because their legs are situated too far back on their bodies, and so they have to push themselves forward on their bellies. One species of cormorant in the Galápagos Islands cannot fly. But pelicans are surprisingly graceful in the air, despite being among the largest of flying birds in terms of body weight. Frigatebirds, on the other hand, are very lightweight and can remain in the air for days.

Pelicans and their allies are skilled at catching fish. Some species dive from great heights. For centuries, Chinese fishermen have attached roped collars to cormorants, letting them out to catch (but not swallow) fishes, before pulling them back to the boats and taking their prey.

Cormorants and darters lack waterproofing in their wings. This allows them to dive deeper and in general to move through water more efficiently. But their plumage eventually becomes waterlogged, forcing them to spend considerable time ashore waiting for it to dry.

All of the group breed in large colonies with other species, but may vigorously defend their individual patches. All stages of the breeding cycle may be synchronized within a colony. Many of these birds reuse the same nest sites year after year.

Lesser frigatebird
Fregata ariel

Darters and anhingas may look like cormorants, but instead of diving underwater to chase their prey, they submerge like a submarine

Darter
Anhinga melanogaster

Anhinga
Anhinga anhinga

Pelican party Most pelicans feed as they sit on the surface of the water, dipping down to capture fishes in their pouches. Groups sometimes herd fishes into shallow water where they can be more easily caught. Pelicans can often be seen scavenging near fishing boats and piers.

Great white pelican
Pelecanus onocrotalus

Red-tailed tropicbird
Phaeton rubricauda

Dalmatian pelican
Pelecanus crispus

Northern gannet
Morus bassanus

Broad wings allow Dalmatian pelican to glide long distances, conserving energy during migrations

Brown pelican
Pelecanus occidentalis

Unlike most pelicans, which fish by swimming, this species plunge-dives out of the air into the water for its food

Red-billed tropicbird
Phaeton aethereus

Blue-footed booby
Sula nebouxii

Distinctive black bill

Nests of the blue-footed booby are nothing more than a circle of excreta on the ground or among vegetation there

Peruvian booby
Sula variegata

Great white pelican Because of its size (it is among the heaviest flying birds in the world), the great white pelican relies as much as possible on thermals. For reasons of balance, it cannot fly with a full pouch.

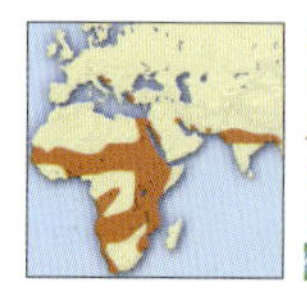

- Up to 5¾ feet (1.75 m)
- 1–3
- Sexes alike
- Partial migrant
- Locally common

S.E. Europe, Africa, S. & S.C. Asia

Northern gannet This bird is distinctive in flight, with its pointed tail and beak, and long, narrow wings. It nests in colonies on rocky coasts, and is renowned for its spectacular high dives in search of fish.

- Up to 36 in (92 cm)
- 1
- Sexes alike
- Partial migrant
- Locally common

N. Atlantic, Mediterranean Sea

Blue-footed booby The name booby is thought to be from the Spanish word for "clown." This booby dives for fish at such a high speed that it passes its prey and must snatch it from below as it resurfaces.

- Up to 33 in (84 cm)
- 1–3
- Sexes alike
- Partial migrant
- Locally common

N.W. Mexico to N. Peru, Galápagos Is.

Imperial shag Imperial shags remain close to the coasts in the cold southern waters where they live. Island-based populations have remained isolated from each other and therefore exhibit slight variations.

- Up to 30 in (76 cm)
- Sexes alike
- 2–3
- Sedentary
- Locally common

Coasts of S. South America, Falkland Is.

GANNET DIVE

Gannets regularly plunge from a height of about 100 feet (30 m). Their three-dimensional vision (a result of having both eyes positioned toward the front of the head) helps them pinpoint prey.

Missiles away
Gannets plunge-dive head-first into a school of fishes. After catching their prey, they surface to eat it or fly off with it.

Birds of prey

CLASS	Aves
ORDERS	1
FAMILIES	3
GENERA	83
SPECIES	304

These skilled hunters are collectively known as raptors, from the Latin word meaning "one who seizes and carries away." They comprise one of the avian world's largest orders, with members ranging from the world's fastest birds to its ugliest scavengers, and varying in standing height from 6 inches (15 cm) to more than 4 feet (1.2 m). The group includes eagles, kites, falcons, buzzards, vultures, and hawks. The raptors' hunting prowess has awed humans throughout history, making them common features on military insignia, national crests, and business logos. They all have large eyes and powerful beaks and claws, but there is considerable variety in behavior among the groups.

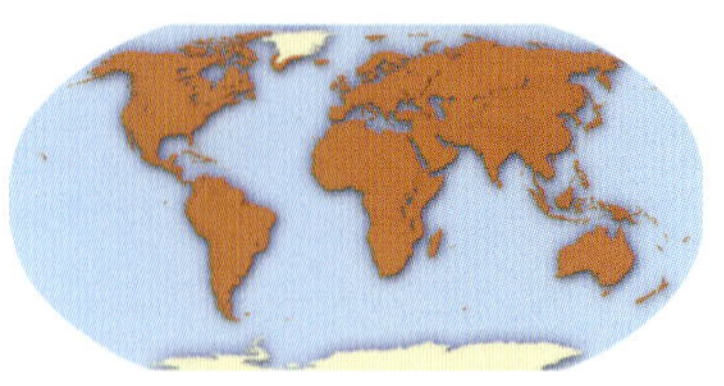

Widespread predators Raptors are found in most habitats, from Arctic tundra to tropical rain forests, deserts, marshes, fields, and cities. Because they need space for their hunting, their presence is determined by physical environment rather than by type of vegetation.

SUPREME HUNTERS

Raptors have sharp, hooked bills adapted for tearing flesh; powerful feet and talons for grabbing prey; and large eyes for spotting their quarry in daylight. Their diets vary from species to species and include insects, birds, mammals, fishes, and reptiles. Anatomical features vary accordingly. Long toes help falcons grab airborne prey. Their sturdy legs allow them to hit birds hard enough to incapacitate them. Large legs and talons help forest eagles capture monkeys, sloths, and other large, tree-dwelling mammals.

Raptors are renowned for their keen eyesight and aerial prowess. A rabbit that a human can barely see at a distance of 1,640 feet (500 m), a wedge-tailed eagle can see from a mile (1.6 km) away. Eagles, hawks, and ospreys often swoop suddenly when they spot a likely target. Female raptors are typically larger than their male counterparts

Many scavenging raptors have bare skin on their heads and necks, perhaps to prevent messing up their feathers when they stick their heads into carcasses, or else to assist with regulating body temperature.

WINGING IT

Some raptors have long, broad wings suited for soaring while looking for carrion or live prey on the ground far below. Others have more pointed wings that allow them to fly quickly and change direction rapidly.

Andean condor *– 9½ feet (2.9 m)*

Bearded vulture *– 8 feet (2.5 m)*

Secretarybird *– 6¾ feet (2.1 m)*

White-bellied sea eagle *– 6½ feet (2 m)*

Rough-legged buzzard *– 5½ feet (1.5 m)*

Peregrine falcon *– 2¼ feet (0.7 m)*

Lesser kestrel *– 2¼ feet (0.7 m)*

Little sparrowhawk *– 1¼ feet (0.4 m)*

A meaty treat Most raptors breed in trees, although some nest among vegetation on the ground or in small hollows scraped on cliff edges. There is usually a clear division of labor, with mothers being the ones who offer torn-up prey to nestlings.

Andean condor
Vultur gryphus

World's largest bird of prey

California condor
Gymnogyps californianus

Black vulture
Coragyps atratus

Like most vultures, these birds are scavengers and have weak talons

Turkey vulture
Cathartes aura

Osprey
Pandion haliaetus

King vulture
Sarcoramphus papa

The ultimate fish-catching raptor, the osprey ranges around the coasts of the world

Black baza
Aviceda leuphotes

African cuckoo-hawk
Aviceda cuculoides

Cuckoo-hawks are inoffensive raptors that hunt low over tree tops in search of large insect food, or dive on small reptiles from set perches

Hook-billed kite
Chondrohierax uncinatus

California condor Condors are carcass feeders. This scavenger has the longest nestling period—5 months—of any bird. During this time, the young are totally dependent on the parents.

S.W. USA

- Up to 4⅓ ft (1.3 m)
- Up to 9 ft (2.7 m)
- 1
- Sedentary
- Critically endangered

POWERFUL TOOLS

Ospreys drop feet first into the water to catch fishes. Their reversible outer toe, long curved talons, and rough, spiny toes (with thorny growths called "spicules") help them to grab and carry prey. Like all raptors, they have strong, curved beaks to tear flesh.

Mississippi kites hunt on the wing, stooping to feed mainly on insects flushed by grazing animals or fires

Mississippi kite
Ictinia mississippiensis

Scissor-tailed kite
Chelictinia riocourii

Distinguished by its deeply forked tail and elegant gray form

European honey-buzzard
Pernis apivorus

Honey-buzzards feed on wasps and bees, attacking their nests and combs

Congregate to roost in flocks of up to 1,000

Distinctive red feet and facial skin

Snail kite
Rostrhamus sociabilis

Bald eagles are sea eagles, and have huge talons and feet bare of feathers that could otherwise become waterlogged

Black kite
Milvus migrans

Bald eagle
Haliaeetus leucocephalus

Short-toed eagle
Circaetus gallicus

Specializes in preying on small reptiles, by plunging from the sky and swallowing them whole

White-tailed eagle
Haliaeetus albicilla

Snail kite This raptor eats only water snails, which it collects in freshwater lowland marshes. Snail kites frequent open places, where they sometimes gather in large flocks. Nests are built among grasses and aquatic bushes.

- Up to 17 in (43 cm)
- Up to 3½ ft (1.1 m)
- 2–3
- Partial migrant
- Locally common

S.E. USA, C. America, N.E. South America

Bald eagle This bird became the national symbol of the United States due to its fierce look. It hunts mainly fishes, though it may also feed on ducks or carrion. It has also been known to steal food from other birds.

- Up to 38 in (96 cm)
- Up to 6½feet (2 m)
- 1–3
- Migrant
- Locally common

North America, S. to N. Mexico

Short-toed eagle This eagle prefers to nest in evergreen trees. Its habitat includes terrain with scrubby vegetation and woods with large clearings. It eats mainly reptiles, even venomous snakes.

- Up to 26½ in (67 cm)
- Up to 73 in (1.85 m)
- 1
- Partial migrant
- Locally common

N.W. Africa, W. to C. Eurasia, W. China & India

Lappet-faced vulture
Torgos tracheliotus

These huge vultures feed mainly on the large bones of animal carcasses, which they break open on rocks

Lammergeier
Gypaetus barbatus

Cinereous vulture
Aegypius monachus

Eurasian griffon
Gyps fulvus

Egyptian vulture
Neophron percnopterus

Hooded vulture
Necrosyrtes monachus

African white-backed vulture
Gyps africanus

Palm-nut vulture
Gypohierax angolensis

Eurasian griffon The largest type of vulture in Europe, this bird prefers to nest, roost, and soar in mountainous terrain, but then moves to open plains to feed. It eats carrion, mainly from large mammals such as sheep.

- Up to 3½ ft (1.1 m)
- Up to 9 ft (2.8 m)
- 1
- Sedentary
- Locally common

N. & S. Africa, S. Europe to Middle East & Caucasia

Egyptian vulture This scavenger builds crude nests using small branches and rubbish, which it typically places in holes and nooks in rocks. Its diet includes rotten fruit, rubbish, carrion, and dung.

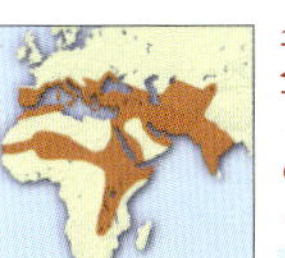

- Up to 27 in (69 cm)
- Up to 5½ ft (1.7 m)
- 2
- Partial migrant
- Locally common

S. Europe, N. & E. Africa, S.W. Asia to India

Hooded vulture This vulture cannot compete with larger scavengers at carcass sites, so it circles around the periphery, picking at scraps. It is the only member of this species group that can live in areas of high rainfall.

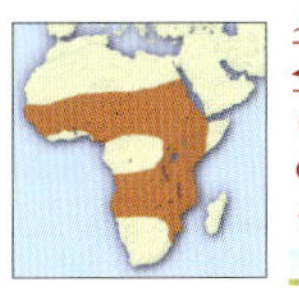

- Up to 27 in (69 cm)
- Up to 6 ft (1.8 m)
- 1
- Sedentary
- Common

Sub-Saharan Africa excl. Congo Basin

FLEDGLINGS

Most raptors breed in trees, but northern harriers nest on the ground among tall vegetation. The female feeds the young with food brought by the male until they are ready to fly and hunt for themselves, about a month after hatching.

Gabar goshawk This bird has two very different color forms, one gray and one mostly black. It hunts birds, mammals, lizards, and insects by swiftly flying out from trees. It prefers lower rainfall areas.

- Up to 14 in (36 cm)
- Up to 24 in (60 cm)
- 2–4
- Partial migrant
- Common

Sub-Saharan Africa excl. Congo, S. Yemen

Eurasian sparrowhawk This raptor lives in forested areas interspersed with open spaces. It hunts its prey by flying low along the edge of the woods, catching mostly birds, as well as small mammals and insects.

- Up to 15 in (38 cm)
- Up to 29 in (74 cm)
- 3–6
- Partial migrant
- Uncommon

Europe, far N. Africa, N. to S. Asia

FIGHTING BIRDS

Raptors sometimes squabble over prey. Eurasian buzzards attack each other with their powerful feet and sharp talons. Beaks, no matter how dangerous looking, are never used.

Foot fighting
Two competing hawks try to get the upper hand the only way they know.

Secretarybird The only living representative of its family, the secretarybird has no close relatives. It is a semi-terrestrial raptor that strides over grassland plains, looking for prey it can subdue with a kick from its long legs, short, stout toes and nail-like claws.

- Up to 5 ft (1.5 m)
- Up to 7 ft (2.1 m)
- 1–3
- Sedentary
- Locally common

Sub-Saharan Africa excl. Congo Basin

Golden eagle This large bird reuses its crude nests, typically building several on rocks or in trees, and using them in turn. It hunts mammals and birds, and also eats carrion. Its defended breeding area can cover 25,000 acres (10,120 ha).

- Up to 3¼ ft (1 m)
- Up to 7 ft (2.2 m)
- 1–3
- Partial migrant
- Locally common

North America, Eurasia, N.W. Africa

HUNTING METHODS

Raptors hunt and kill their prey in a variety of ways, although their powerful feet and sharp talons are their main weapons. Some pursue airborne prey; others capture terrestrial reptiles and mammals. Hawks kill with their strong grips, squeezing their victims to death. Some vultures drop tortoises until they break, then swoop down to eat the flesh inside. Sea eagles and ospreys snatch fishes out of the water. The unusual secretarybirds subdue their prey by kicking it. And the African harrier-hawk has extraordinarily flexible legs which it can bend at extreme angles to grope inside tree hollows for nestling birds and other small animals.

Slippery prey The osprey snatches fish from shallow depths. Before hitting the water, it thrusts its talons forward, pushes out its breast, and holds its wings back, ready to clutch its catch headfirst.

Stages in prey capture Falcons strike and take their prey in mid air. Here we see the sequence of attack by a peregrine falcon: circling to scout prey (top left); the attacking dive (left); the hit, a Eurasian oystercatcher (below).

ALL IN THE FAMILY

The family Falconidae includes true falcons and caracaras. Unlike true falcons, caracaras walk about on the ground, and they eat insects, fruit, and seeds, or scavenge for flesh.

Contrasting ways
Falcons typically hunt prey in the air; but their closest relatives, the South American caracaras, scavenge on the ground.

Yellow-headed caracara Outside the breeding season, this bird roosts in large colonies. In the morning, the individuals head to their respective hunting grounds. The caracara has harsh, cat-like cries. It is often attacked by other birds.

- Up to 17 in (43 cm)
- Up to 29 in (74 cm)
- 1–2
- Sedentary
- Common

Central to N. & E. South America

Cranes and allies

CLASS	Aves
ORDERS	1
FAMILY	11
GENERA	61
SPECIES	212

This ancient bird order, whose members are sometimes known as the misfits of the avian world, comprises a variety of predominantly ground-living birds that prefer walking and swimming to flying. Some species have, in fact, altogether lost the ability to fly. Descended from a ground-dwelling shorebird, cranes and their relatives have filled a variety of ecological niches around the world. They typically nest on the ground or on platform nests in shallow water. Most have loud calls, and in some cases the male and female perform duets. In parts of Asia, cranes are symbols of good luck and long life; one captive crane is known to have lived to the age of 83.

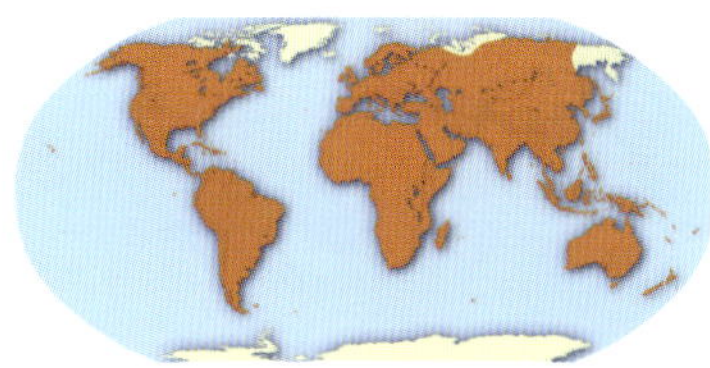

Widespread At least one species of this group of birds can be found on every continent except Antarctica and on many islands. They live in wetlands, deserts, grasslands, and forests. Trumpeters and limpkins occur only in the New World; the majority of bustards inhabit Africa. Some unusual gruids have very narrow ranges.

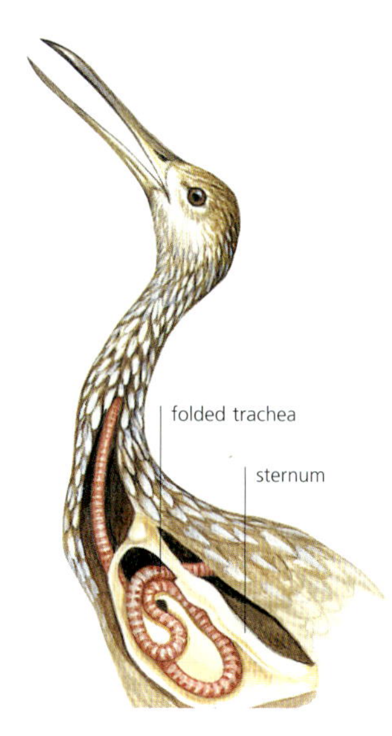

Crane's trachea Cranes can make a wide range of calls, from purrs to screams. Their very long tracheas are coiled and fused with their sternums. In that region, the bony rings of the trachea are like thin plates that vibrate, amplifying the sounds produced in the voicebox and allowing the voice to carry for more than 1 mile (1.6 km) in some circumstances. Crane species with better developed tracheas make higher-pitched calls.

Small buttonquail
Turnix sylvatica

Barred buttonquail
Turnix suscitator
♀

Barred buttonquail
Turnix suscitator
♂

After attracting a mate, a female buttonquail leaves him to incubate and rear the young, and goes off to find another

Hoatzin
Opisthocomus hoatzin

The relationships of the strange, fowl-sized hoatzin are unclear; some research even associates them with cuckoos

White-breasted mesite
Mesitornis variegatus

Subdesert mesite
Monias benschi

Gray-winged trumpeter
Psophia crepitans

Limpkin
Aramus guarauna

Barred buttonquail Although this bird generally lives at lower elevations, it can be found in mountains up to 7,500 feet (2,273 m) in the Himalayas. It often forages for food in sugarcane, tea, and coffee plantations.

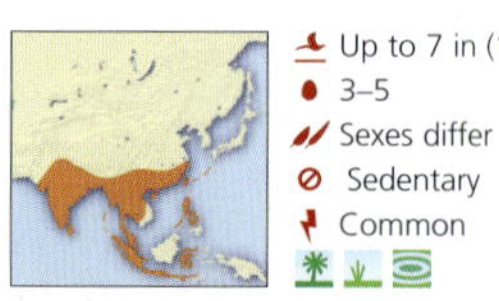

Up to 7 in (17 cm)
3–5
Sexes differ
Sedentary
Common

S., S.E. & E. Asia to Philippines & Sulawesi

Hoatzin This prehistoric-looking bird has an extraordinarily large crop. Just a few days after hatching, chicks can climb trees using their feet, bills, and specially adapted wings; the wings have "claws" that later disappear.

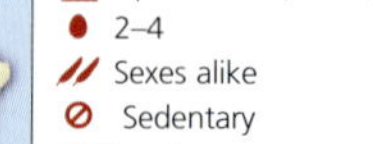

Up to 27 in (70 cm)
2–4
Sexes alike
Sedentary
Locally common

N. South America

Limpkin The sole species within its family, the limpkin, which is closely related to cranes, uses its long, curved bill to extract snails from their shells. It builds large, bulky nests of rushes and sticks. Before it was protected, hunting had reduced its numbers.

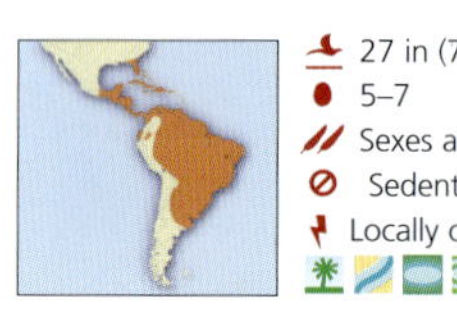

27 in (70 cm)
5–7
Sexes alike
Sedentary
Locally common

C. America to N.E. South America, West Indies

Swamphens flick their tails to flash the white underside as they walk about feeding in wetlands and on verges; this is a social signal to others in their group

Inaccessible rail
Atlantisia rogersi

Coots are the most aquatic of all rails

Horned coot
Fulica cornuta

Purple swamphen
Porphyrio porphyrio

Corn crake
Crex crex

White-breasted waterhen
Amaurornis phoenicurus

Little crake
Porzana parva

Takahes live on high moors in the mountains of southeast New Zealand, digging out and eating the fleshy bases of tussocks of grass and sedges

Gray-necked wood-rail
Aramides cajanea

Takahe
Porphyrio hochstetteri

Wood-rails have very slender bodies and long legs, enabling them to run very quickly through dense undergrowth

Corn crake Active at twilight, this bird spends most of its time in tall grasses. It lays up to two clutches in a nest on the ground, amid vegetation. It eats invertebrates, plants, seeds, and grain.

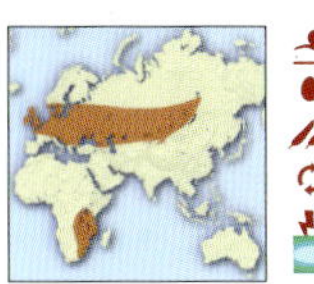

Up to 12 in (30 cm)
8–12
Sexes alike
Migrant
Vulnerable

W. & C. Eurasia, S.E. Africa

Takahe The largest living member of the rail family, the flightless takahe uses its wings only in displays of courtship or aggression. Usually only one chick from each clutch will survive its first winter.

Up to 25 in (63 cm)
1–3
Sexes alike
Non-migrant
Endangered

S.W. South Island (New Zealand)

BACK FROM THE BRINK

Many island-based rail species cannot fly and are vulnerable to predation by introduced animals such as rats. The Lord Howe rail had fallen to 10 breeding pairs, but has been bred in captivity and released successfully.

Kagu
Rhynochetos jubatus

Red-legged seriema
Cariama cristata

Sunbittern
Eurypyga helias

Long legs adapted for walking and feeding on the ground

Great bustard
Otis tarda

Colorful neck plumes are flared in display

Colorful tail is raised in display. Male bustards are much larger and more colorfully plumaged than females

Denham's bustard
Neotis denhami

Houbara bustard
Chlamydotis undulata

One of 20 of the 26 species of bustards that occur in Africa; 18 are only found here

Lesser florican
Sypheotides indica

Bustards have long legs adapted for walking through open habitats to feed; when the birds rise to fly, the feet are tucked up under the body

African finfoot
Podica senegalensis

Sunbittern The sunbittern forages along forest rivers and streams. Its nuptial display involves perching on a conspicuous branch and fanning its colored wings and tail.

Up to 19 in (48 cm)
1–2
Sexes alike
Sedentary
Vulnerable

C. & N. South America

Great bustard Primarily a herbivore, this bird nests on the ground, laying a small clutch of two to three eggs that the female incubates. The young can fly within 5 weeks and are fully independent in 12–14 weeks.

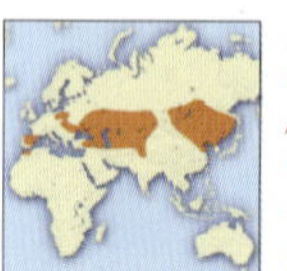

Up to 3¾ ft (1.05 m)
2–3
Sexes differ
Sedentary
Vulnerable

Europe, C. & E. Asia

Denham's bustard When alarmed, this and other bustards often crouch and are difficult to see. Its elaborate ornamental plumes are used in nuptial displays. One male has many mates.

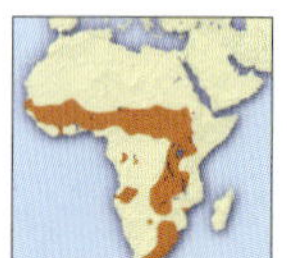

Up to 3¼ ft (1 m) (male)
1–2
Sexes alike
Partial migrant
Locally common

C. to C.S. Africa

WADERS AND SHOREBIRDS

CLASS	Aves
ORDERS	1
FAMILIES	16
GENERA	86
SPECIES	351

The world's shallow waters and shorelines teem with marine and terrestrial creatures that this order of generally sociable birds has evolved to hunt. The groups within the order exhibit significant diversity, allowing them to exploit different resources within aquatic habitats. Typical waders such as sandpipers and plovers patrol shallow waters and shorelines. Gulls scavenge along shorelines, too, but they are also adapted to swim out to feed on the surface of deeper water. Terns venture even further from shore and dive for food. Auks swim underwater after their prey, much like penguins. The eyes of many waders and shorebirds are set on the sides of the head to scan for predators.

Watery world Some species in this group live next to oceans, along estuaries or seashores; others can live far inland in arid climates, in mid deserts.

A WORLD OF DIFFERENCES

Birds within this order exhibit many anatomical differences, as befits their respective niches. Those that scavenge along the mud flats and in the shallows, probing for food, tend to have long, thin legs and long necks and bills. Those that patrol the surf are usually shorter and can scurry out of the way of incoming water. Birds that swim out and scoop up food on the water's surface tend to be stouter and to have webbed feet.

Species that search for marine prey further out over the water are more capable fliers: they have shorter legs and smaller feet, but long, narrow wings. Terns are very agile, and have long, forked tails for quick maneuvering in the air.

Auks have webbed feet set well back on their compact bodies, and use their wings as flippers underwater.

The diets of these birds range from insects and worms to fishes and crustaceans. Some are scavengers.

Shoreline Sandpipers probe rapidly for food, often in tidal flats. Their slender bills are able to sense micro-organisms in the sand and mud.

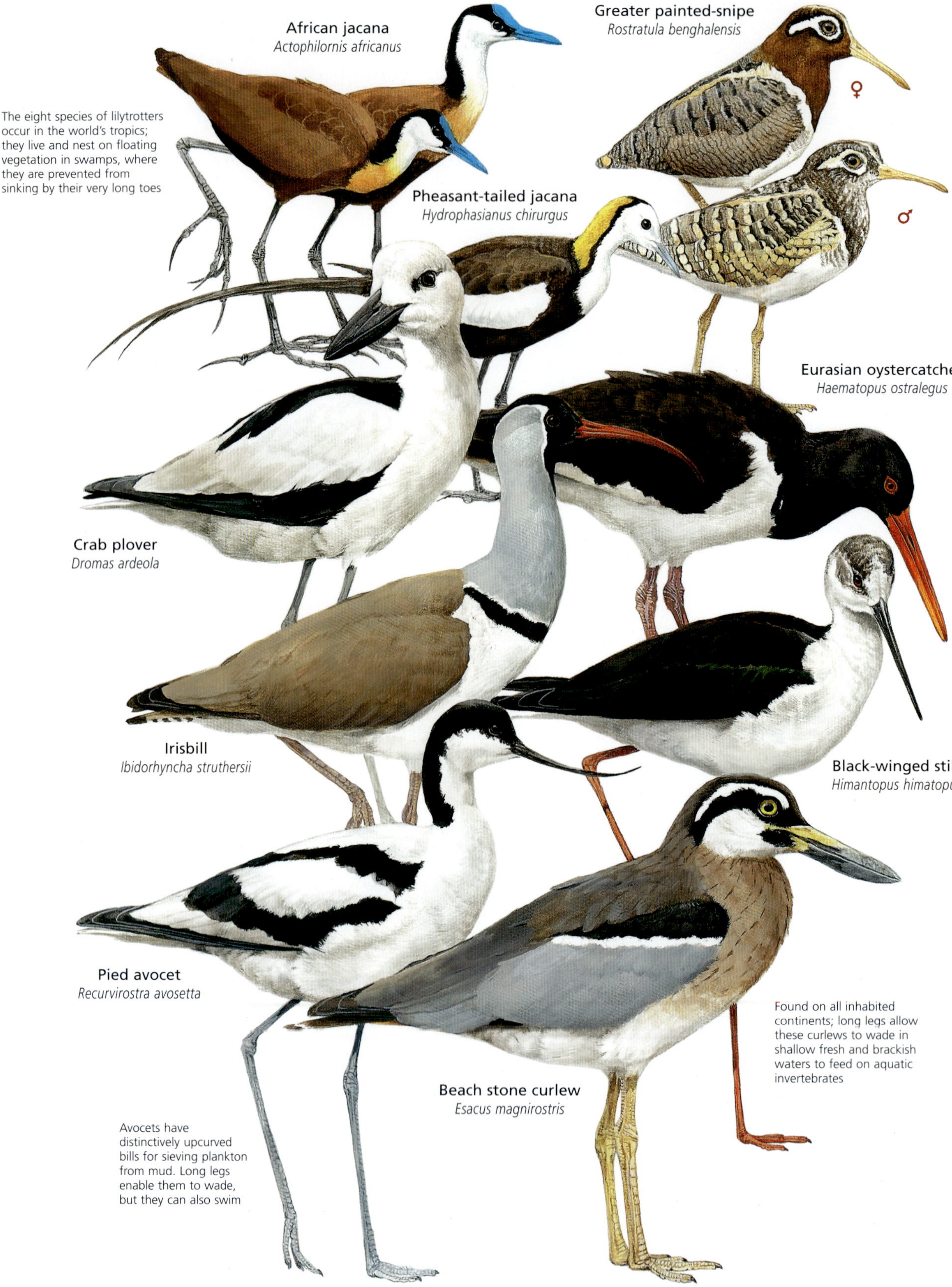

African jacana
Actophilornis africanus

The eight species of lilytrotters occur in the world's tropics; they live and nest on floating vegetation in swamps, where they are prevented from sinking by their very long toes

Greater painted-snipe
Rostratula benghalensis

♀

♂

Pheasant-tailed jacana
Hydrophasianus chirurgus

Eurasian oystercatcher
Haematopus ostralegus

Crab plover
Dromas ardeola

Irisbill
Ibidorhyncha struthersii

Black-winged stilt
Himantopus himatopus

Pied avocet
Recurvirostra avosetta

Found on all inhabited continents; long legs allow these curlews to wade in shallow fresh and brackish waters to feed on aquatic invertebrates

Beach stone curlew
Esacus magnirostris

Avocets have distinctively upcurved bills for sieving plankton from mud. Long legs enable them to wade, but they can also swim

African jacana Also known as a lilytrotter, this bird uses a high-stepping gait to walk on floating vegetation without stumbling on its long toes. It probes under the plants for insects, snails, and other organisms.

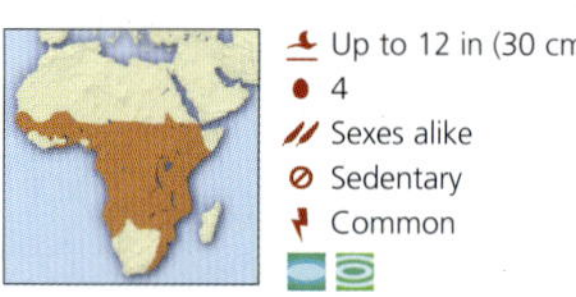

Up to 12 in (30 cm)
4
Sexes alike
Sedentary
Common

Sub-Saharan Africa

Eurasian oystercatcher This elegant bird feeds on mollusks and other invertebrates that it catches along seashores. It gathers in large flocks. A strong flier, it can also swim and dive.

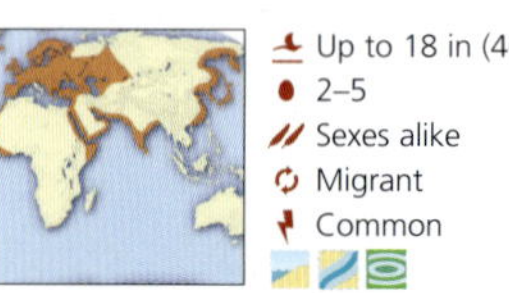

Up to 18 in (46 cm)
2–5
Sexes alike
Migrant
Common

Europe; W., S.W. & E. Asia; N.W., N. & E. Africa

Beach stone curlew This bird's extremely large yellow eyes help it find crabs and other shellfish on the reefs, muddy seashores, and sandy beaches it patrols at night. Its call is harsh and eerie.

Up to 22 in (56 cm)
1
Sexes alike
Sedentary
Vulnerable

Malay Peninsula to Philippines, New Guinea & N. Australia

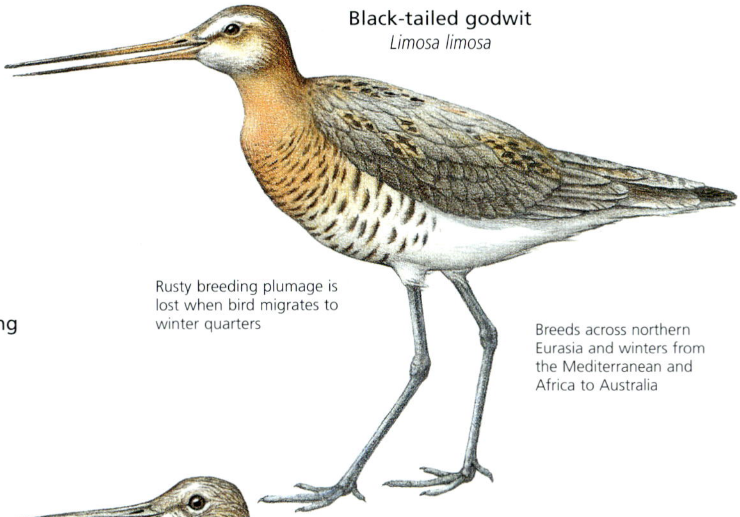

Black-tailed godwit
Limosa limosa

Rusty breeding plumage is lost when bird migrates to winter quarters

Breeds across northern Eurasia and winters from the Mediterranean and Africa to Australia

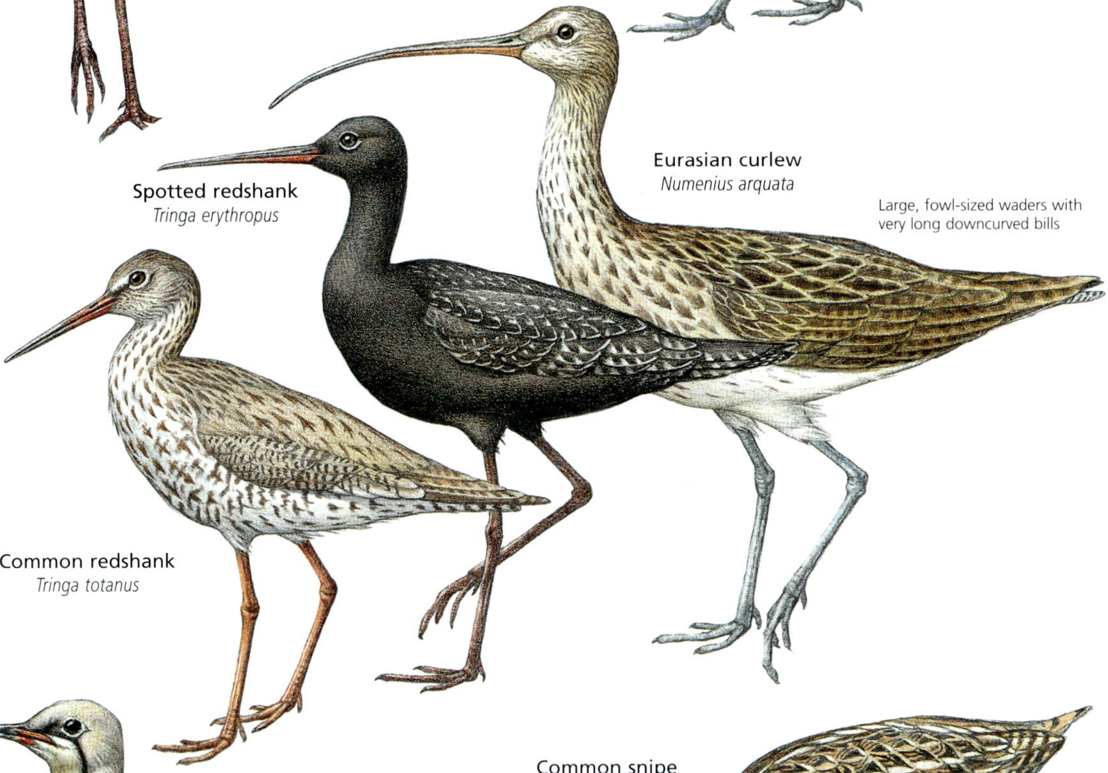

Southern lapwing
Vanellus chilensis

Only gray lapwing with a crest

Eurasian curlew
Numenius arquata

Large, fowl-sized waders with very long downcurved bills

Spotted redshank
Tringa erythropus

Common redshank
Tringa totanus

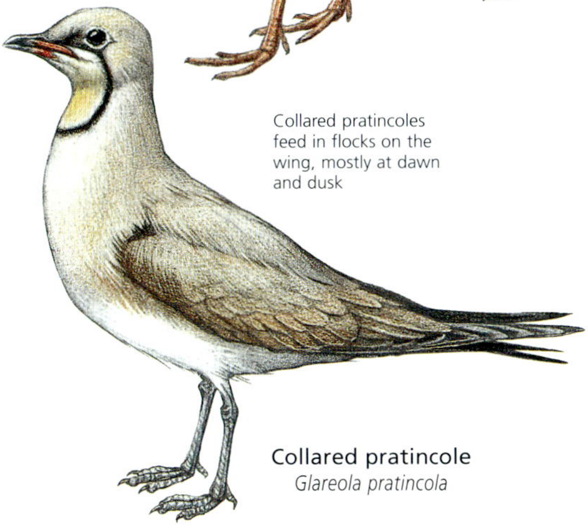

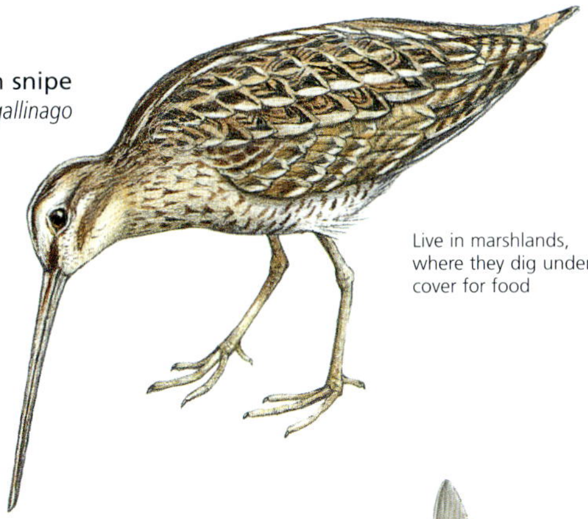

Common snipe
Gallinago gallinago

Collared pratincoles feed in flocks on the wing, mostly at dawn and dusk

Live in marshlands, where they dig under cover for food

Collared pratincole
Glareola pratincola

Southern lapwing This bird frequents damp meadows and agricultural fields, where it catches insects and other small prey. Its nests are scrapes in the ground that it lines with a thin layer of grass.

- Up to 15 in (38 cm)
- 3–4
- Sexes alike
- Sedentary
- Common

N., E. & S. South America

Spotted redshank A shy, solitary bird, generally found only in the company of other waders. It lives near fresh, brackish, or salt water on heathland, marshes, and tundra when breeding.

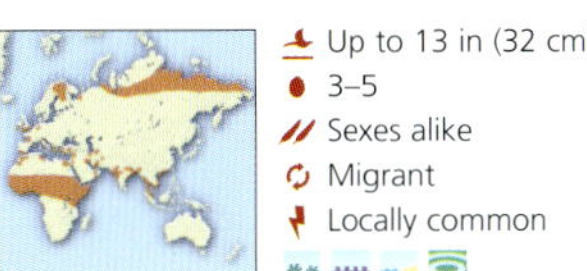

- Up to 13 in (32 cm)
- 3–5
- Sexes alike
- Migrant
- Locally common

Sub-arctic Eurasia to S. Asia & C. Africa

FLIGHT OF THE SNIPE

Common snipes have a conspicuous display involving "drumming" dives. From a considerable height, the male dives steeply with wings beating and tail fanned. The outer tail feathers are held apart from the rest. They vibrate and produce the drumming sound as the bird falls through the air. The snipe then repeats the display.

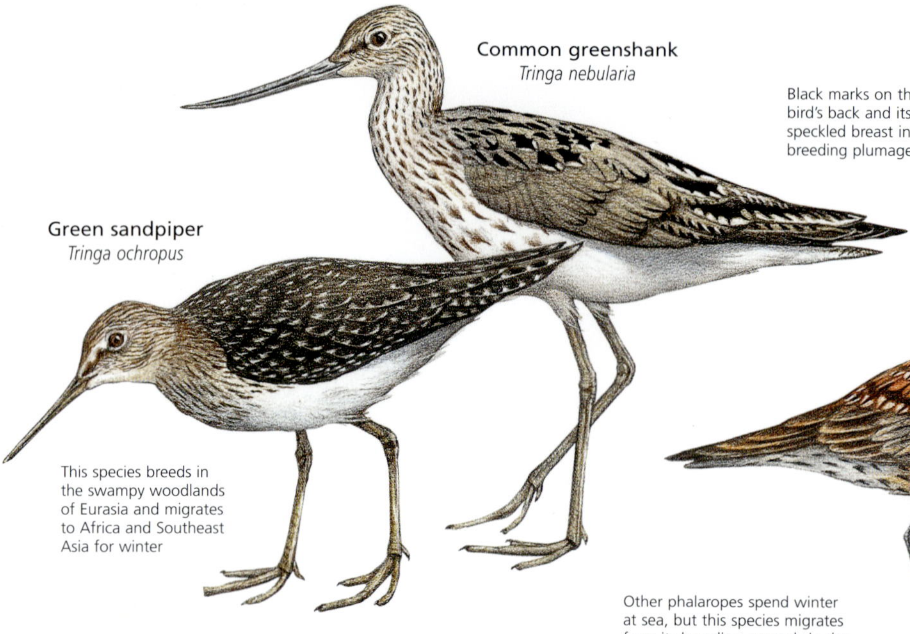

Common greenshank
Tringa nebularia

Black marks on this bird's back and its speckled breast indicate breeding plumage

Green sandpiper
Tringa ochropus

This species breeds in the swampy woodlands of Eurasia and migrates to Africa and Southeast Asia for winter

Curlew sandpiper
Calidris ferruginea

Other phalaropes spend winter at sea, but this species migrates from its breeding grounds in the North American midwest to the lakes and mudflats of Andean South America

Little stint
Calidris minuta

Wilson's phalarope
Phalaropus tricolor

Ruff in full breeding regalia. Ruffs congregate in leks at the beginning of breeding to attract and mate with females

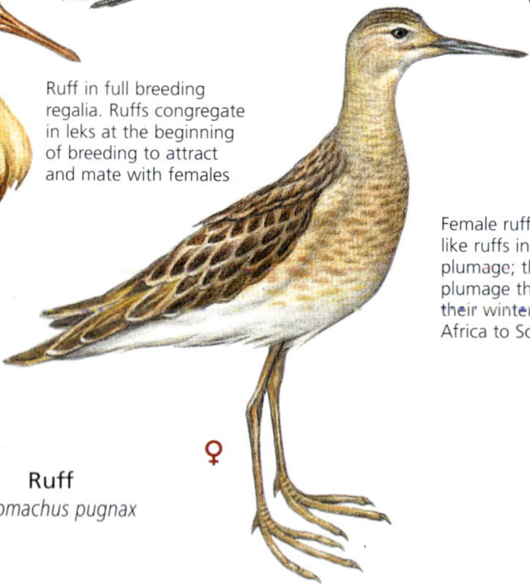

Red-necked phalarope
Phalaropus lobatus

Female ruffs (reeves) look like ruffs in non-breeding plumage; this is the plumage they wear at their winter quarters from Africa to Southeast Asia

♂ **Ruff** ♀
Philomachus pugnax

RUFFING IT

During the spring breeding season, the male ruff has colorful ear tufts and a raised collar of feathers. Males assemble on selected hillocks each morning to display to the females.

Green sandpiper This bird's name derives from the fact that its underparts have a greenish sheen in summer. It makes piping sounds in flight. It feeds by probing deeply into mud along inland waterways.

- Up to 9 in (24 cm)
- 3–4
- Sexes alike
- Migrant
- Locally common

W.C. to E.C Eurasia, C. Africa, S. Asia

Ruff In spring, the ruff can be seen in flocks of hundreds. The female, or reeve, builds a nest out of fine grasses, either among thick grass or in clumps of sedge or rushes.

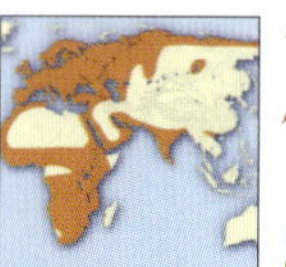

- Up to 13 in (32 cm)
- 3–4
- Sexes differ
- Migrant
- Locally common

Arctic to W. & S. Eurasia, Africa

Gray-breasted seedsnipe
Thinocorus orbignyianus

Seedsnipes browse on the tips of low marshy vegetation in the South American highlands

Black-faced sheathbill
Chionis minor

A fatty layer keeps the bird warm and gives it a chubby look

A rudimentary spur on the wing is used as a weapon

White-eyed gull
Larus leucophthalmus

Jaegers are parasitic; they live in cold seas and chase other seabirds for scraps

Breeding birds have black heads; out of breeding, the head molts to white

Long-tailed jaeger
Stercorarius longicaudus

Bonaparte's gull
Larus philadelphia

Plains-wanderer
Pedionomus torquatus

Great black-backed gulls are omnivorous, and usually breed in small colonies

Great black-backed gull
Larus marinus

Once thought to be a buttonquail, this Australian bird is now know to be a wader related to the South American seedsnipes

Herring gull
Larus argentatus

In Europe and North America, herring gulls can be distinguished from other large gulls by their combination of a pale gray back with black-tipped wings, pink legs, and pale eyes

Long-tailed jaeger Like other skuas, the long-tailed jaeger has acrobatic courtship flights. Its nest is an unlined shallow hollow on the ground. This bird eats lemmings at nesting grounds.

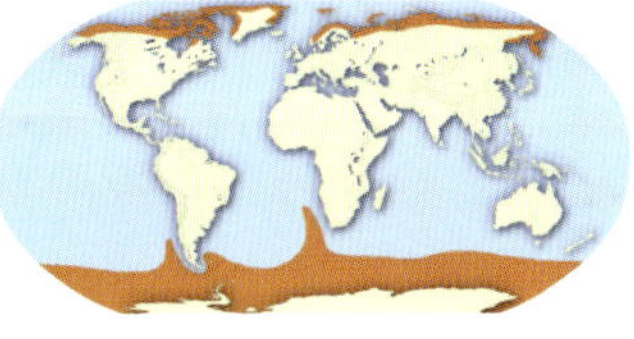

- Up to 21 in (53 cm)
- 2
- Sexes alike
- Migrant
- Common

Arctic, Antarctic, N. & S. Pacific & Atlantic oceans

Herring gull This large, familiar gull can be seen at harbors and on beaches, as well as at garbage dumps and in plowed fields. It flies inland to bathe in fresh water, and roosts communally.

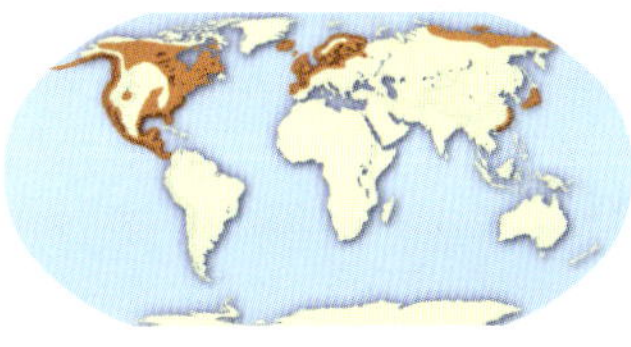

- Up to 26 in (66 cm)
- 2–3
- Sexes alike
- Partial migrant
- Common

North & Central America, W., N. & E. Eurasia

Fairy tern (white tern)
Sterna nereis

Fairy terns have a more deeply and elegantly forked tail than larger terns

Little tern
Sterna albifrons

This tern has faster wingbeats than other terns. It migrates in western and northern parts of its range

Common tern
Sterna hirundo

Kerguelen tern
Sterna virgata

Sandwich tern
Sterna sandvicensis

Sooty tern
Sterna fuscata

Large-billed terns live along the major rivers of South America, where they plunge-dive for fish

Large-billed tern
Phaetusa simplex

Inca tern
Larosterna inca

Distinctive white mustache plumes

Whiskered tern
Chlidonias hybridus

Little tern This is one of the smallest of the terns, a group of birds whose species look very similar to each other. It lives on sea beaches, bays, and large rivers.

- Up to 11 in (28 cm)
- 2–3
- Sexes alike
- Migrant
- Locally common

Europe, Africa, Asia to Australasia, Indian & W. Pacific oceans

Common tern This small bird inhabits lakes, oceans, bays, and beaches in the Northern Hemisphere and migrates to the Southern Hemisphere. It nests in colonies on sandy beaches and small islands.

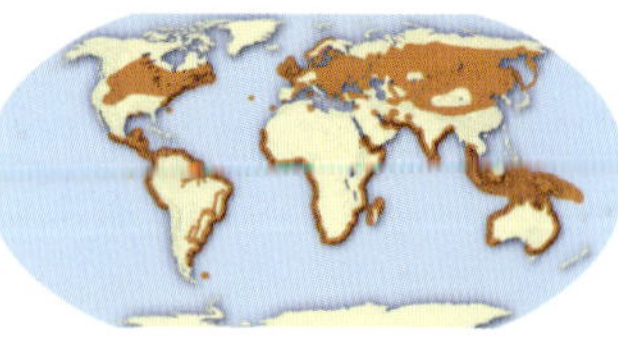

- Up to 15 in (38 cm)
- 2–4
- Sexes alike
- Migrant
- Locally common

Worldwide

Common murre
Uria aalge

Murres live in arctic and subarctic seas, sit on the water in large flocks, and dive for a minute or more to chase and catch fish; their feet are webbed

Ancient murrelet
Synthliboramphus antiquus

Nests colonially like other auks around the north Pacific, but it nests in burrows

Parakeet auklet
Cyclorrhynchus psittacula

Rhinoceros auklet
Cerorhinca monocerata

Crested auklet
Aethia cristatella

Tufted puffin
Fratercula cirrhata

Atlantic puffin
Fratercula arctica

Black skimmer
Rhynchops niger

In skimmers, the lower bill is longer than the upper. They feed by flying just above rather still water with the lower bill immersed, ready to snap up any fishes that come in contact

Atlantic puffin In summer, the Atlantic puffin's beak is red, blue, and yellow. In winter, part of the beak is shed; the remaining part is gray-brown with a yellowish tip. Puffins nest in colonies, sometimes using burrows dug by rabbits or shearwaters.

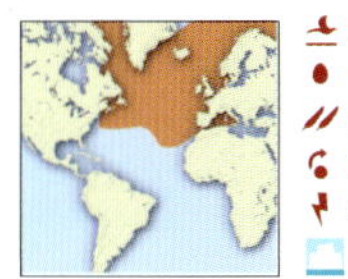

Up to 14 in (36 cm)
1
Sexes alike
Partial migrant
Common

Arctic, N. Atlantic Ocean

A LONG WAY DOWN

Murres typically lay just one egg at a time on precarious-looking bare rock ledges on coasts and cliffs. Unlike most eggs, which are oval, murre eggs are pear-shaped. Therefore, if budged, they will roll around in a small circle.

Egg designs
Murre eggs show a wide variety of colors and patterns.

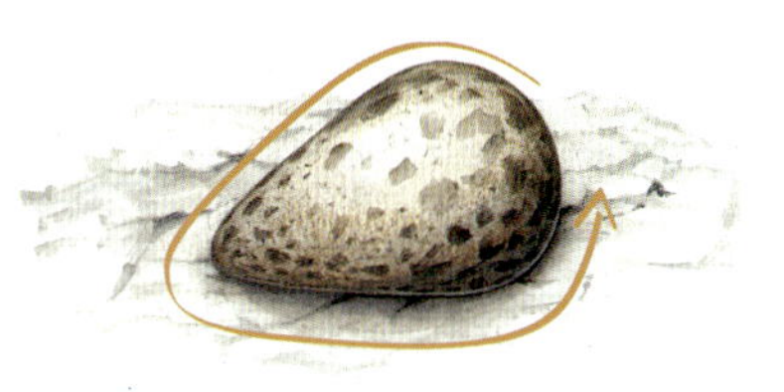

PIGEONS AND SANDGROUSE

CLASS Aves
ORDERS 1
FAMILIES 3
GENERA 46
SPECIES 327

Pigeons and sandgrouse are quite dissimilar and may not even be related. Pigeons and doves are commonplace, tree-dwelling birds that eat fruits and seeds. They have a close association with humans, who have used pigeons for carrying messages. They vary in color from the drab bluish-gray of the familiar street pigeon to the riot of hues that characterize the fruit doves of the Indo-Pacific region. Pigeons feed their young with a milky substance produced in their crops. They also have specialized bills that enable them to suck up water when they drink. By contrast, sandgrouse are dull-colored, fast-flying desert dwellers that are well adapted to arid climates.

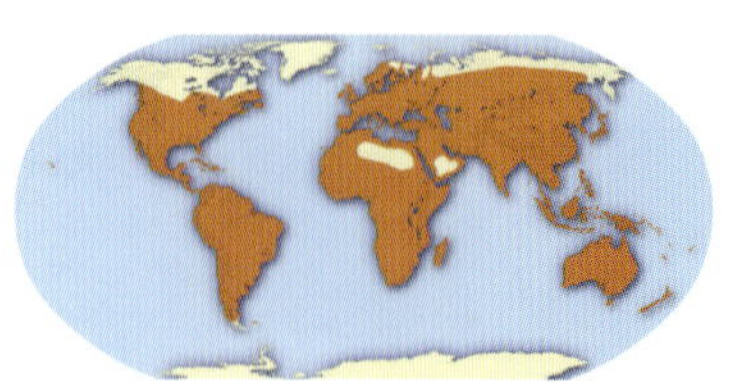

Abundant Pigeons and doves are found worldwide, except in polar regions. Sandgrouse only inhabit Africa and Eurasia.

Emerald doves feed on fallen fruits on the floor of rain forests

Rock dove Also known as a feral pigeon, this bird is familiar to urban dwellers across the world. Originally from Eurasia and North Africa, where it nests on cliffs, it has readily adapted to living on building ledges in cities.

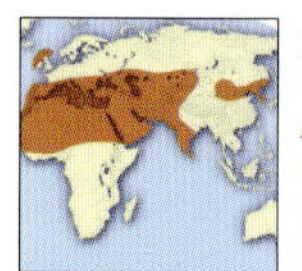

- Up to 13 in (33 cm)
- 2
- Sexes alike
- Sedentary
- Common

S. Europe, Middle East, S.W. & C.E. Asia, N. Africa

Pallas's sandgrouse This rare vagrant bird breeds on open steppes in Central Asia. Each year large numbers fly for long distances outside their main range, to grasslands and beaches, to avoid winter snows that cover feeding areas.

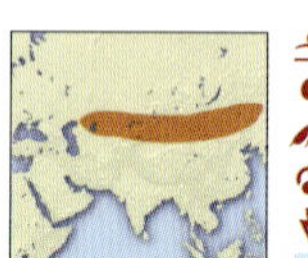

- Up to 16 in (40 cm)
- 2–3
- Sexes differ
- Partial migrant
- Locally common

S. Urals & Transcaspia to Mongolia

Emerald dove This bird is commonly found in rain forests or nearby dense vegetation, where it forages on the forest floor, alone or in pairs. It can also be seen flying swiftly over open ground between foraging areas.

- Up to 11 in (27 cm)
- 2
- Sexes differ
- Sedentary
- Common

India & S.E. Asia to E. Australia & Melanesia

Madagascar green-pigeon
Treron australis

The Indo-African green-pigeons are arboreal fruit-eaters that have a gizzard modified for grinding seeds as well as fruit

Fruit-doves have thin-walled gizzards to allow the ingestion of fruit and the passing of seeds

Black-backed fruit-dove
Ptilinopus cinctus

Seychelles blue pigeon
Alectroenas pulcherrima

Bruce's green pigeon
Treron waalia

The turkey-sized crowned pigeons of New Guinea have a uniquely fan-like crest

Victoria crowned pigeon
Goura victoria

Mourning dove
Zenaida macroura

Gray-headed dove
Leptotila plumbeiceps

Zebra dove
Geopelia striata

Madagascar green-pigeon This handsome, fruit-eating bird is one of several species of green-pigeon. It spends most of its time in trees and bushes, climbing in the manner of a parrot. It has complex, soft calls that are not typical of pigeons and doves.

Up to 13 in (32 cm)
2
Sexes alike
Sedentary
Locally common

Madagascar & Comoro Is.

Mourning dove This dove's name derives from its low, mournful, cooing call. It is a strong, fast flier that may fly considerable distances to find feeding areas or the nearest water source, often at dawn or dusk.

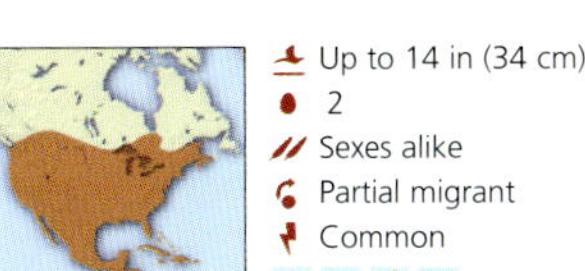

Up to 14 in (34 cm)
2
Sexes alike
Partial migrant
Common

North & Central America, West Indies

Victoria crowned pigeon The only crowned pigeon with a white-tipped crest, the Victoria crowned pigeon lives in lowland rain forest. Hunted for food, it remains common only in remote areas of northern New Guinea.

Up to 29 in (76 cm)
1
Sexes alike
Sedentary
Vulnerable

Lowland N. New Guinea

Parrots

CLASS	Aves
ORDERS	1
FAMILIES	1
GENERA	85
SPECIES	364

Parrots and cockatoos form an ancient and highly distinct order of birds. They are easily recognized by their short, blunt bills, which have downcurved upper mandibles, and also by their feet, which have two toes pointing forward and two pointing backward. Although some species are dull, most have brilliantly colored plumage, predominantly in shades of green accented by splashes of red, yellow, and blue. Their visual appeal is one reason for their popularity as pets over the centuries. Another is their antics: they can perform acrobatics, hanging on to perches with their feet or bills. They can also mimic human voices.

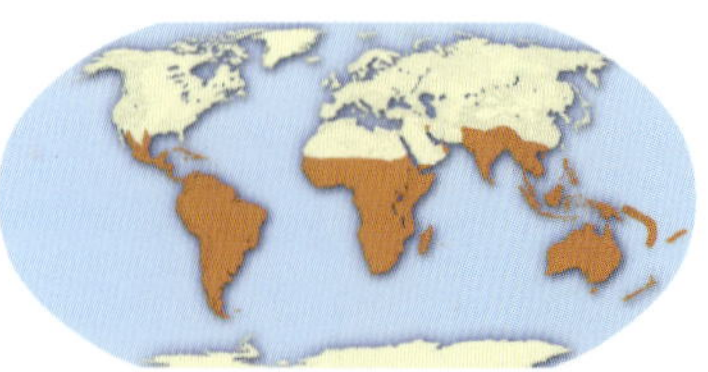

Southerly distibution Parrots live primarily in the Southern Hemisphere. They are most common in tropical rain forests, especially lowland tropical rain forests, but some species prefer open, arid regions. The highest concentrations of species occur in Australasia and South America. The most southerly parrot inhabits Tierra del Fuego.

A SOCIABLE GROUP

Most parrots eat seeds and nuts (which they crack open with their heavy bills) as well as fruit. They forage among the treetops or on the ground. Lorikeets, on the other hand, are strictly arboreal; they eat soft fruit, and harvest pollen and nectar from blossoms.

Although parrots' basic features differ little among species, there is considerable variation in size and shape. Wings can be narrow and pointed, or broad and rounded. Similarly, tails may be long and pointed or short and squarish. Some have ornate feathers. Cockatoos, a separate family from "true parrots," have prominent, erectile head crests. The sexes are usually alike.

Parrots are very social birds. They squawk loudly and frequently, and are heard more readily than observed in the wild. They can be difficult to glimpse as they fly through the forest canopy, camouflaged by their green plumage. That has not stopped many of them from falling prey to poachers, who sell them to the worldwide pet market. That and habitat destruction have made parrots one of the most common groups of birds on lists of threatened species.

Aloft The brilliantly colored red-and-green macaws are among the largest parrots. They display a rainbow of hues when they fly. Each individual bird has its own unique red facial markings.

Kea This is one of the world's few dull-colored parrots. Its coloring, compact body, and beak shape make it resemble a bird of prey. Keas do sometimes eat carrion, and sheep farmers accuse the species of killing their livestock.

Up to 19 in (48 cm)
2–4
Sexes alike
Sedentary
Uncommon

Montane South I. (New Zealand)

DOMESTICATION OF PARROTS

The familiar budgerigar (*Melopsittacus undulatus*) is one of the world's most popular pet species, on par with goldfish. They began to be kept in captivity in the middle of the 19th century. Other popular domesticated parrot species include cockatiels and lovebirds. Although pet parrots can mimic human speech, they do not understand a word they are saying.

Tamed *In the wild in Australia, budgerigars are mainly light green and pale yellow. In captivity, they have been bred in a range of colors.*

Blue-crowned hanging parrot
Loriculus galgulus

Alexandrine parakeet
Psittacula eupatria

Hanging parrots are among the few groups of parrots with short, blunt tails; they may rest and sleep hanging upside-down in foliage

Swift parrot
Lathamus discolor

A common feature of the Afro-Asian parakeets is a ringed neck

Vernal hanging-parrot
Loriculus vernalis

Plum-headed parakeet
Psittacula cyanocephala

Senegal parrot
Poicephalus senegalus

Yellow-collared lovebird
Agapornis personatus

Fischer's lovebird
Agapornis fischeri

Ground parrot
Pezoporus wallicus

BLUE VARIATIONS

Alexandrine parakeets are among those parrot species that also come in a mutant blue color variety due to the suppression of yellow pigmentation. They are popular pets because of their excellent mimicking skills and affectionate behavior.

Alexandrine parakeet This bird normally nests in holes it gnaws in trees, or in naturally occurring hollows. It lives in small groups that join together to form larger groups for the night. Only the male has a prominent black collar around its neck.

S. and S.E. Asia

- Up to 24 in (62 cm)
- 3
- Sexes differ
- Sedentary
- Locally common

Blue-crowned hanging parrot This small parrot is quite common in the forested lowlands of Southeast Asia. It gets its name from its habit of sleeping upside down and is part of a group known as the bat parrots. It wanders in small groups.

Malay Peninsula, Borneo, Sumatra & nearby islands

- Up to 5 in (12 cm)
- 3–4
- Sexes differ
- Sedentary
- Common

Thick-billed parrot
Rhynchopsitta pachyrhyncha

Blue-and-yellow macaw
Ara ararauna

Hyacinth macaw
Anodorhynchus hyacinthinus

Macaws use their massive bills to open hard-shelled nuts and fruits

White-eared parakeet
Pyrrhura leucotis

Scarlet macaw
Ara macao

Military macaw
Ara militaris

Nanday parakeets feed on seeds and fruit, mainly on the ground

Nanday parakeet
Nandayus nenday

Burrowing parakeet
Cyanoliseus patagonus

Thick-billed parrot The bill of this species is pale in young birds and darkens as they mature. The parrot eats pine seeds, juniper berries, and acorns. It gathers in nomadic flocks of about half a dozen up to several hundred birds.

Up to 17 in (43 cm)
1–4
Sexes alike
Nomadic
Endangered

W. Mexico

Hyacinth macaw The parrot family's largest species, the hyacinth macaw is far less common than it once was, due to trafficking in the pet-bird trade and hunting for food and feathers. Males are slightly larger than females.

Up to 3¼ ft (1 m)
2–3
Sexes alike
Sedentary
Endangered

N.E. & C. South America

Scarlet macaw This brightly colored bird sometimes supplements its fruit and nut diet with nectar and flowers. It sometimes ingests clay to aid in the digestion of the harsh chemicals found in the unripe fruit that it eats before other birds can get to it.

Up to 35 in (89 cm)
1–4
Sexes alike
Sedentary
Common

C. & N. South America

Red-fan parrot
Deroptyus accipitrinus

White-crowned parrot
Pionus senilis

Monk parakeet
Myiopsitta monachus

Canary-winged parakeet
Brotogeris versicolurus

Orange-winged parrot
Amazona amazonica

Cuban parrot
Amazona leucocephala

Blue-fronted parrot
Amazona aestiva

Kakapo
Strigops habroptilus

Flightless; has cryptic plumage to help it hide in its scrubby habitat, but this has not stopped introduced stoats, rats, and feral cats from exterminating it in its natural range

Like other parrots, the kakapo's feet have two toes forward and two toes back

Kakapo The world's heaviest parrot is nocturnal and cannot fly because it lacks a sternal keel. It chews leaves or stems to extract their juices. Males court females by booming with their inflated gular air-sacs.

Up to 25 in (64 cm)
1–3
Sexes alike
Sedentary
Extinct in natural range

S. W. South Island, New Zealand; introd. to Little Barrier, Maud, Codfish & Pearl Is.

Blue-fronted parrot This arboreal, climbing parrot lives in forests and builds its nests in hollows in trees. It lays a clutch of two eggs, which are incubated mainly by the female over the course of about 25 days.

Up to 15 in (37 cm)
2–4
Sexes alike
Sedentary
Common

N.E. & C. South America

Monk parakeet This highly adaptable South American species is found in city parks, on farms and in gardens as well as in savannas, forests, and palm groves. Unlike other species of parrots, it builds its own rather complex nests of twigs.

Up to 12 in (29 cm)
1–11
Sexes alike
Sedentary
Common

C. & S.E. South America

Cuckoos and Turacos

CLASS	Aves
ORDERS	1
FAMILIES	3
GENERA	42
SPECIES	162

These two old families of birds are related by virtue of some molecular similarities, but otherwise differ markedly in development, anatomy, and other characteristics. Cuckoos are notorious as parasitical birds that trick other species into raising their chicks. However, less than half of the 140 or so species of cuckoos actually engage in such behavior. All cuckoos have feet with two toes pointing forward and two pointing backward, but otherwise they differ considerably among species. Turacos are more homogeneous: with one exception, they are rather slender-necked birds with long tails, short, rounded wings, and erectile, laterally compressed crests.

Here, there, and everywhere Turacos inhabit savannas and forests. They live only in Africa, south of the Sahara. Cuckoos thrive in a variety of habitats—from open moorland to tropical rain forest—and are virtually cosmopolitan, although they predominate in the tropics and subtropics.

ROBBING THEM BLIND

Some cuckoo species have evolved remarkably devious strategies to get away with their brood parasitism, such as laying eggs that closely resemble those of their host.

Most "true cuckoos" are drab, although the bronze-cuckoos of the Old World tropics are brightly colored. Some cuckoo species are solitary; others breed communally and maintain group territories. Some, such as the roadrunners and ground cuckoos, rarely fly and prefer to run.

Turacos are gregarious, noisy birds whose harsh, barking alarm calls can be heard from afar. They gather in small groups and glide or flap from tree to tree. They are more agile when running among the branches than when flying.

The savanna-based turaco species have dull plumage, but those that favor forests tend to be brightly colored, with special blue-green pigments that are water soluble. Turacos are primarily herbivorous, but they also eat insects.

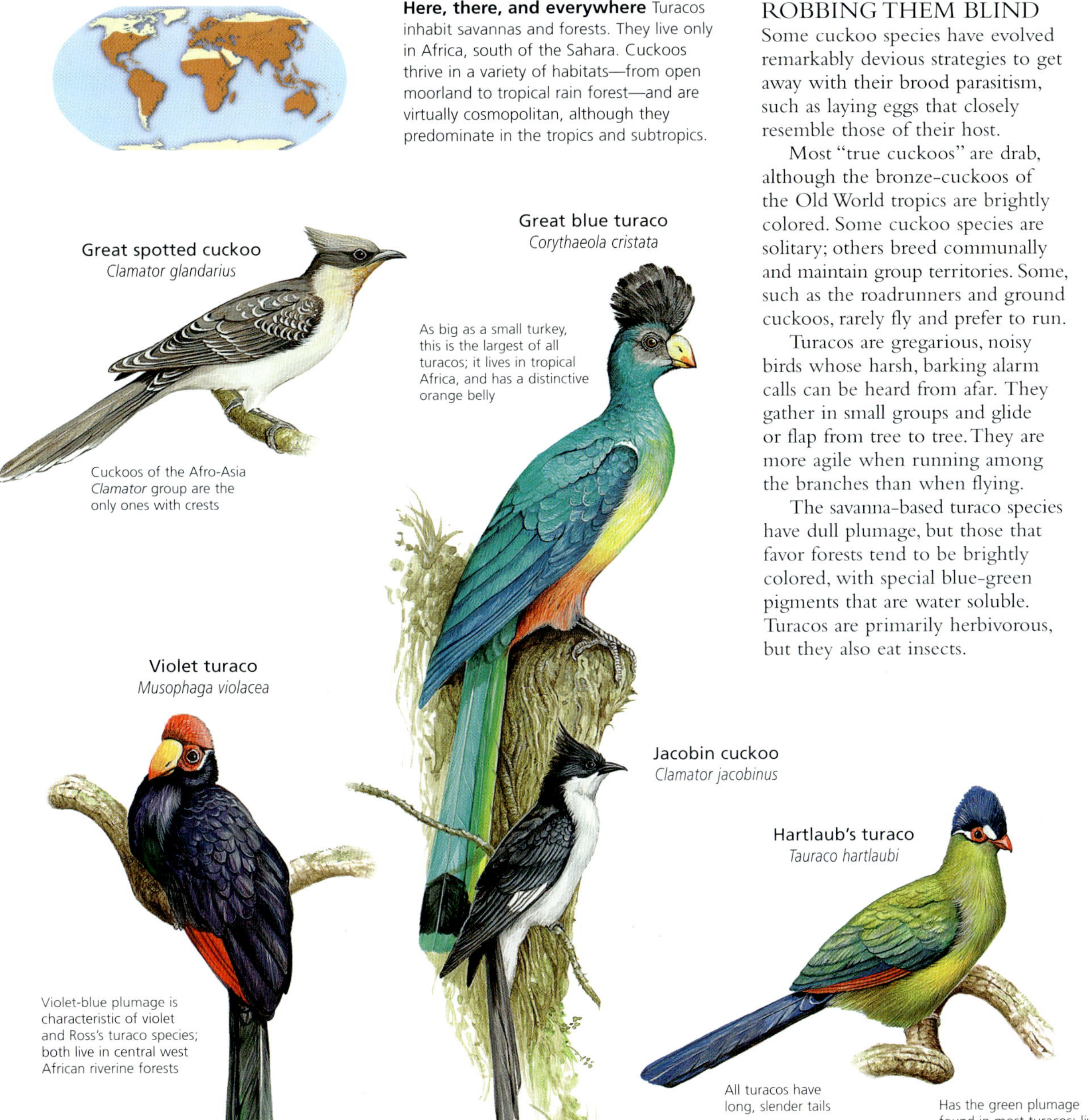

Great spotted cuckoo
Clamator glandarius

Cuckoos of the Afro-Asia *Clamator* group are the only ones with crests

Great blue turaco
Corythaeola cristata

As big as a small turkey, this is the largest of all turacos; it lives in tropical Africa, and has a distinctive orange belly

Violet turaco
Musophaga violacea

Violet-blue plumage is characteristic of violet and Ross's turaco species; both live in central west African riverine forests

Jacobin cuckoo
Clamator jacobinus

Hartlaub's turaco
Tauraco hartlaubi

All turacos have long, slender tails

Has the green plumage found in most turacos; lives in equatorial east Africa

Common koel
Eudynamys scolopaceces

One of the bronze cuckoos, this species has the characteristic brilliantly glossed green back of the group

Dideric cuckoo
Chrysococcyx caprius

Greater coucal
Centropus sinensis

Common cuckoo
Cuculus canorus

A female common cuckoo about to place her egg in the nest of another bird—in this case, a reed warbler

Yellow-billed cuckoo
Coccyzus americanus

Feeds on insects on the floor of rain forests, but roosts and nests in trees

Bornean ground-cuckoo
Carpococcyx radiceus

Rufous-vented ground-cuckoo
Neomorphus geoffroyi

High-ridged bill for parting vegetation as bird forages for foliage

Smooth-billed ani
Crotophaga ani

Greater roadrunner
Geococcyx californianus

Omnivorous, and an efficient predator on small reptiles; also eats insects, mice, and the eggs of other small birds

Common cuckoo This well-known bird inhabits woodland clearings and cultivated fields, where it feeds on insects. It has a familiar call from which its name arose. Its females are polyandrous and parasitic.

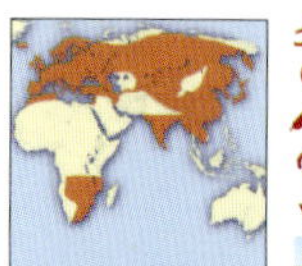

Up to 13 in (33 cm)
Up to 20, in host nests
Sexes differ
Migrant
Common

Eurasia except S.W., N.W. & S. Africa

Greater coucal This non-parasitic bird has terrestrial habits. It is often mistaken for a game bird due to its size and voluminous tail. It has a distinctive dull, booming call, as well as a variety of croaks and chuckles.

Up to 20 in (52 cm)
2–4
Sexes alike
Sedentary
Common

S. & S.E. Asia, Greater Sundas, islands off Philippines

Greater roadrunner This ground-loving bird is the basis for the famous cartoon character. It inhabits arid regions and deserts with shrubs and cacti. It uses its long tail as a rudder to change direction when it runs quickly.

Up to 22 in (56 cm)
2–6
Sexes alike
Sedentary
Common

S.W. USA to C. Mexico

Owls

CLASS	Aves
ORDERS	1
FAMILIES	2
GENERA	29
SPECIES	195

Solitary creatures of the night, owls are instantly recognizable by their forward-facing eyes, face masks, and stout silhouettes. There are two families of owls: barn owls, with heart-shaped faces and elongate bills, and "true owls," with rounded heads and hawk-like bills. They usually roost in out-of-the-way spots. Even when out in the open, they tend to perch stoically, camouflaged by their mottled, earth-toned plumage, so they are more often heard than seen. Their mysterious nocturnal habits, coupled with their far-reaching hoots and other sometimes eerie-sounding calls, have given rise to many folk superstitions. In lifestyle, however, they are no more than nocturnal birds of prey.

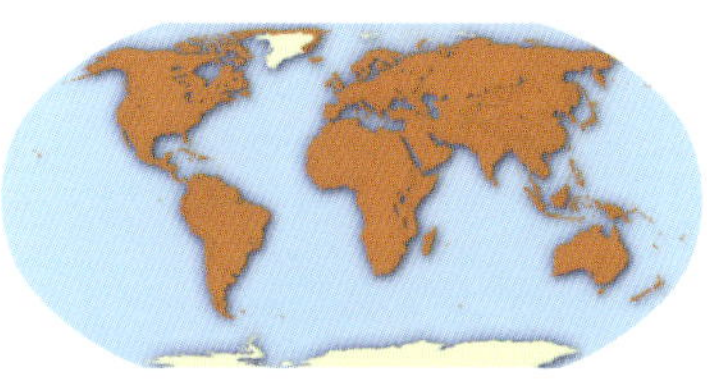

Cosmopolitan birds Both families of owls are widespread; some species such as the barn owl are among the most widely distributed of all birds. Most species inhabit woodlands and forest edges, but some prefer treeless habitats instead.

True owl The large ear tufts are are a distinguishing feature of the great horned owl. This large owl is at home in the woods and deserts of the Americas. Its prey includes the striped skunk and it is not deterred by the foul-smelling spray emitted from the skunk's anal glands when it is threatened.

STEREO HEARING

Many owls have ears that are asymmetrical in both size and shape. This structure enables them to more precisely locate their prey, by noting the subtle difference between the sound signals reaching each ear.

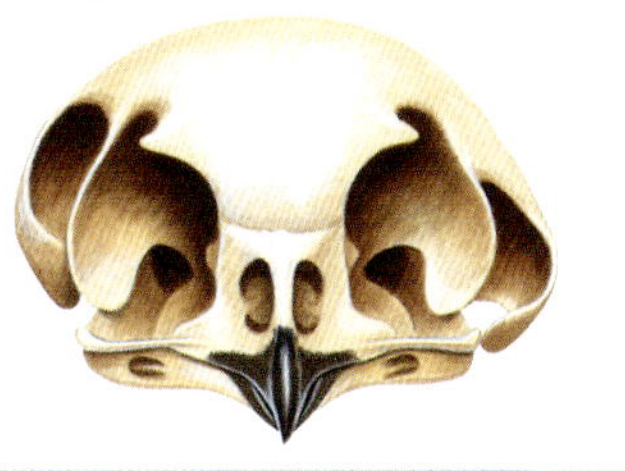

NIGHT HUNTERS

Owls, which typically become active at twilight, are well adapted to a nocturnal predatory lifestyle. Their forward-facing eyes give them binocular vision that helps them judge distances well. They can also rotate their heads to see behind them. They can see well in poor light, thanks to their tubular eyes and abundance of light-sensitive rods in their eyes. They also rely on their acute hearing: some species even have facial masks to help them capture sound waves more effectively.

Owls also have sharp, hooked bills and strong legs with talon-tipped feet. They remain on the alert for prey as they perch on branches or ledges, then swoop down to catch the unwary mammal or insect on the ground. They may also snatch insects or mammals from trees, or catch insects in the air.

When they capture sizable prey with their feet, owls sometimes kill it by reaching down to bite it. Small prey is often lifted to the beak with one foot, in the manner of parrots, and swallowed whole. Larger prey is held with the feet and torn apart with the bill before being eaten.

Owls flatten their plumage to become less conspicuous when they are disturbed. Several species have erectile feathers that resemble "ears," which they raise in emotion.

In most species, females are larger than the males, sometimes weighing twice as much as their mates. The young often leave the nest well before they can fly; their parents continue to feed them until they can fend for themselves.

All owls call, especially as the breeding season approaches, and species are often named for the typical sounds they make. A few owl species almost sing.

Ural owl
Strix uralensis

Ural owls range across the boreal forests of Eurasia; their pale gray plumage is distinctive

Black-banded owl
Ciccaba huhula

Great gray owl
Strix nebulosa

Facial disk assists hearing by conducting sound to the ears, like a satellite dish

Eurasian pygmy-owl
Glaucidium passerinum

Barred owl
Strix varia

Northern hawk owl
Surnia ulula

This predator of forest tundra and taiga preys on mammals and birds, specializing in voles; it nests mostly in hollows and has clutches of 6–10 eggs

Elf owl
Micrathene whitneyi

Snowy owl
Nyctea scandiaca

Great gray owl This large bird sometimes uses the abandoned nests of other birds. It may not lay any eggs in years when food is particularly scarce, whereas in good years, it may lay up to nine eggs.

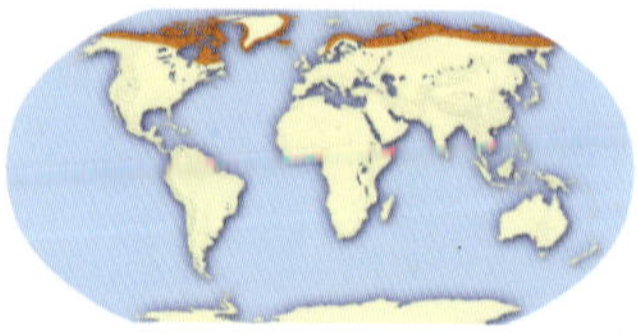

Up to 27 in (70 cm)
2–9
Sexes alike
Sedentary
Uncommon

N. North America, N. Eurasia

Snowy owl This bird migrates during the winter, when it may be found on lake shores, in marshland, or by the sea. It perches on rocks or in trees, and kills its prey—invertebrates, small mammals, or birds—by pouncing in flight.

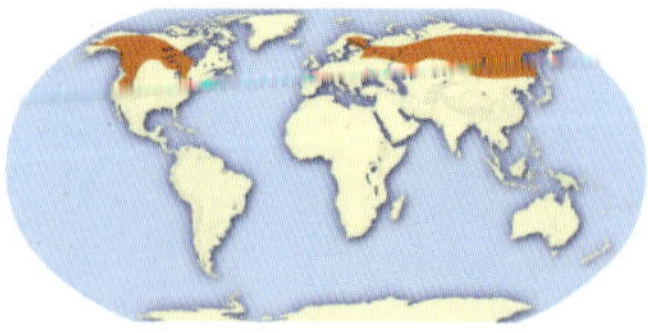

Up to 27 in (70 cm)
3–11
Sexes differ
Partial migrant
Uncommon

N. Eurasia, N. Canada, Arctic

Great horned owl
Bubo virginianus

Verreaux's eagle-owl
Bubo lacteus

The size of a buzzard, and one of the largest owls; confined to sub-Saharan Africa, it preys on medium-sized mammals and rather large birds

Tropical screech-owl
Otus choliba

Spectacled owl
Pulsatrix perscipillata

Barn owl
Tyto alba

Comb-like central toenail is used for preeening

Has the shortest "ears"—twin crests above the eyes—of its group; lives in open country and hunts mostly small mammals

Short-eared owl
Asio flammeus

Northern long-eared owl
Asio otus

Boreal owl
Aegolius funereus

The palest member of a group of small, mostly New World owls with incomplete facial disks

Burrowing owl
Athene cunicularia

Northern saw-whet owl
Aegolius acadicus

Barn owl One of the most widespread birds in the world, the barn owl can often be seen hunting rodents in the grassy shoulders and medians of highways. Barn owls have exceptional hearing, and can catch prey in total darkness, guided by sound alone.

- Up to 17 in (44 cm)
- 4–7
- Sexes alike
- Sedentary
- Common

S. North & South America, sub-Saharan Africa, W. Eurasia to Australia

Northern long-eared owl This bird sometimes nests in squirrels' nests. The young are fed by the female, although it is the male that does most of the hunting and delivers their diet of small mammals, birds, and invertebrates. Females are generally darker and larger than the males.

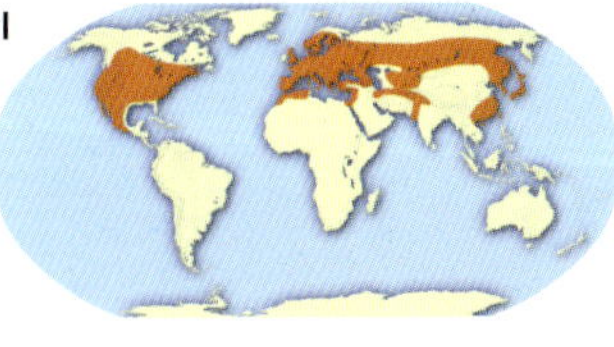

- Up to 16 in (40 cm)
- 5–7
- Sexes alike
- Partial migrant
- Locally common

C. & S. North America, W. to E. temperate Eurasia

NIGHTJARS AND ALLIES

CLASS	Aves
ORDERS	1
FAMILIES	5
GENERA	22
SPECIES	118

Similar to owls in nocturnal lifestyle, this order of birds—which comprises oilbirds, frogmouths, potoos, owlet-nightjars, and nightjars—is thought to be distantly related to them. Like owls, nightjars and their allies are active at twilight and at night. They have soft plumage in patterns and shades that make them difficult to distinguish from the trees or ground. They tend to have rather large heads and large eyes, though the latter are less forward-facing in this group than in owls. Tubular in shape, their eyes are adapted to see well in low light, and they have a keen sense of hearing.

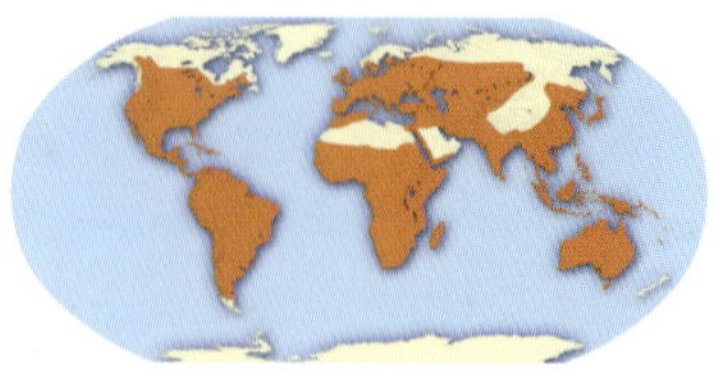

Diverse habitats The oilbird is found only in tropical South American caves. Potoos inhabit open woodlands in Central and South America. Frogmouths and owlet-nighjars live in or near arboreal areas in Australasia. Nightjars live in a variety of warm-climate habitats across the world.

MASTERS OF DISGUISE

These rather unusual-looking night-birds are experts at blending in and often strike curious poses that make them resemble broken-off tree limbs.

They have broad, flattened bills for catching insects. The bristles that surround the bills help funnel food (insects, fruit, or other animals) into their large mouths.

The oilbird is the only member of its family. It has a fan-like tail and long, broad wings. It is a cave dweller that can navigate in total darkness by relying on echolocation, in a similar way to bats.

Frogmouths are odd-looking birds whose bodies taper from their very large, flat, shaggy heads. Their stout bills, hardened like bones, snap shut to trap their prey.

Potoos resemble frogmouths but have thinner bills and fewer bristles. They eat insects while flying.

Nightjars comprise about half the species in this order. They fly swiftly but have weak legs and feet, which they rarely use. They open and display their brightly colored mouths when they feel threatened.

Owlet-nightjars look like a cross between nightjars and owls. Their broad, flat bills are almost hidden by bristles. They have longer legs and run more than the other birds in this group, and hunt by perch-pouncing.

The only regular fruit-eater among nightjar-like birds; feeds mainly on palms, and regurgitates hard seeds

Oilbird
Steatornis caripensis

Perch-and-pounce hunters; they perch motionless on a vantage perch, then, sighting prey, dive down onto it

Tawny frogmouth
Podargus strigoides

Common potoo
Nyctibius griseus

Camouflages itself at roost by imitating a dead branch; plumage pattern is cryptic

Spotted nightjar
Eurostopodus argus

Pennant-winged nightjar
Macrodipteryx vexillarius

Long-tailed nightjar
Caprimulgus climacurus

Males of the *Macrodipteryx* group are polygamous; they grow a pair of prolonged inner flight feathers prior to breeding, which they flaunt in group displays to attract females

Standard-winged nightjar
Macrodipteryx longipennis

Common pauraque
Nyctidromus albicollis

European nightjar (Eurasian nightjar)
Caprimulgus europaeus

Common poorwill
Phalaenoptilus nuttallii

Abyssinian nightjar
Caprimulgus poliocephalus

Australian owlet-nightjar
Aegotheles cristatus

Watches from perch, waiting to pounce on insect prey

Oilbird This bird lives in caves, mostly in mountains but also sometimes along rocky coasts. Its cup-like nests are made of regurgitated fruit. It hovers while plucking fruits and seeds, which it locates by sight and by scent.

- Up to 20 in (50 cm)
- 1–4
- Sexes alike
- Sedentary
- Locally common

N. & N.W. South America

Tawny frogmouth This sociable bird can be seen perching in pairs or in family groups. It searches for food—invertebrates, small mammals, and rodents—on the ground, flying into the trees at any hint of danger.

- Up to 21 in (53 cm)
- 1–3
- Sexes alike
- Sedentary
- Common

Mainland Australia & Tasmania

European nightjar This bird's eggs hatch asynchronously in a nest made in a shallow, unlined scrape in the ground. Eggs are color-camouflaged. The adult uses a special display to distract and drive away any predators that approach the nest.

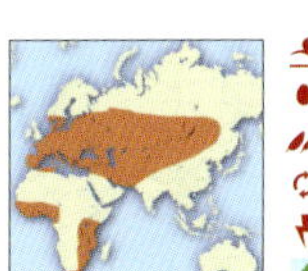

- Up to 11 in (28 cm)
- 1–2
- Sexes differ
- Migrant
- Common

W. & C. Eurasia, W. & S.E. Africa

HUMMINGBIRDS AND SWIFTS

CLASS	Aves
ORDERS	1
FAMILIES	3
GENERA	124
SPECIES	429

Any common ancestry between these dissimilar birds would have to be very old indeed. Nevertheless, hummingbirds and swifts do have significant anatomical similarities, such as the relative length of their wing bones. This is related to two distinguishing features for both types of birds: their very rapid wing beats and flight behavior. Hummingbirds are also known for their tiny size, bright iridescent colors, and hovering flight. The average weight of these birds is less than one-third of an ounce (8.3 g); the bee hummingbird is the smallest known species of bird, at a mere one-tenth of an ounce (2.5 g). Swifts are relatively bigger birds that are the fastest fliers of the avian world.

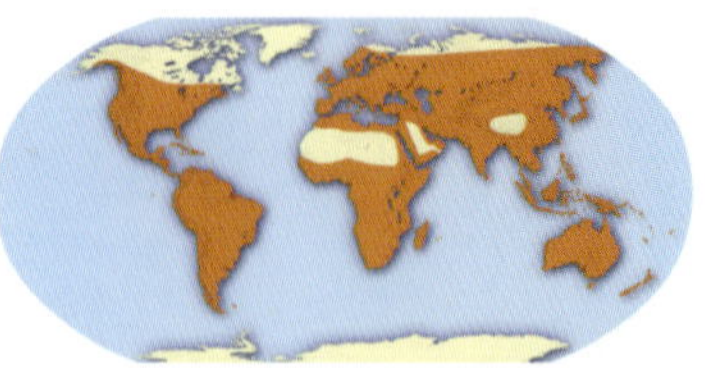

Prevalent throughout Swifts are widely distributed, but are most numerous in the tropics. Hummingbirds are confined to the New World; most live in its tropical belt.

FREQUENT FLIERS

Swifts spend most of their time in the air, taking advantage of their narrow, swept-back wings to maneuver after insects, especially swarming species such as mayflies and termites. They also eat bees and wasps. Several species make long migrations over land and stretches of ocean to reach their Southern Hemisphere wintering grounds.

Predominantly dark colored, swifts have short legs with strong claws. Some cave swiftlets nest and roost in total darkness, deep in caves. They are among the few bird species capable of navigating by echolocation, using clicking calls.

Hummingbirds have long, very narrow bills that they poke into flowers for nectar, hovering in front of the blossoms while they collect the sweet substance with their brush-tipped tongues. They also supplement their diet with insects that they eat for protein.

Swifts and hummingbirds can become torpid to conserve energy, lowering their metabolism and body temperatures while resting.

Eating on the fly Hummingbirds can hover while they feed (left) thanks to the anatomy of their wings. They can turn the wings completely over on the backstroke as well as on the forestroke, thereby counteracting the tendency to move in any one direction. Their broad tails help them make precise maneuvers.

Whiskered tree swift
Hemiprocne comata

Tree swifts build half-saucer nests of feathers and bark agglutined with saliva on the open branches of a tall forest tree; the single egg hatches into a cyptically patterned chick

Square tail and broad white rump bar distinguish this species from other related swifts; it feeds at great heights and lives in Africa and Southwest Asia

Little swift
Apus affinis

Alpine swift
Tachymarptis melba

Breeds at high altitudes across southern Eurasia and migrates to Africa and India; can sleep on the wing

Gray-rumped treeswift
Hemiprocne longipennis

Tree swifts fly out from exposed set perches in tree tops to hunt insects; they all have either short crests or white mustache stripes on the face

Common swift
Apus apus

The common swift is the most familiar and widespread swift in Eurasia. It nests mostly in cavities in buildings. It migrates to winter exclusively in sub-equatorial Africa.

Asian palm-swift
Cypsiurus balasiensis

African palm-swift
Cypsiurus parvus

Chimney swift
Chaetura pelagica

With long forked tail and tapered wings, palm-swifts are the most streamlined of all swifts; they roost and nest mostly in palms

Common swift This bird spends most of its time in flight, in urban and rural areas. It is highly sociable and can often be seen flying in chirruping groups. Nestlings are reared by both parents and can fly within 2 months.

Up to 7 in (17 cm)
1–4
Sexes alike
Migrant
Common

W. & C. Eurasia, S. Africa

Chimney swift This bird can often be heard making high-pitched twittering calls as it flies high overhead. It can often be seen in the company of swallows, a species it resembles but to which it is not closely related.

Up to 5 in (13 cm)
2–7
Sexes alike
Migrant
Common

E. North America, N.W. South America

African palm-swift This bird's nest is a small, flat pad of seed down that it glues to a leaf. The female then uses saliva to glue one or two eggs to the pad. The parents incubate the eggs in a vertical position.

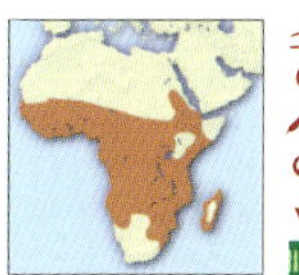

Up to 6 in (16 cm)
2
Sexes alike
Sedentary
Locally common

Sub-Saharan Africa, Madagascar

Crimson topaz
Topaza pella

Great sapphirewing
Pterophanes cyanopterus

Giant hummingbird
Patagona gigas

Collared inca
Coeligena torquata

Lives in the cloud forests of the northern South American Andes, and can be identified by its combination of blue head, white breast, and white sides to the tail

Sword-billed hummingbird
Ensifera ensifera

Green-backed firecrown
Sephanoides sephanoides

Buff-winged starfrontlet
Coeligena lutetiae

Hummingbirds have been given the most imaginative names in the world of birds; a buff wing patch identifies this north Andean species

White-tipped sicklebill
Eutoxeres aquila

Polygamous; males group to display, and females build pendent, often cone-shaped nests attached on the undersides of leaves or twigs

Black-throated mango
Anthracothorax nigricollis

Crimson topaz This visually stunning bird is the second-largest hummingbird. It lives in the Amazon region, where it is often found in the forest canopy, making it difficult to observe. The male's long tail feathers cross halfway down its tail.

- Up to 9 in (22 cm)
- 2
- Sexes differ
- Sedentary
- Locally common

N. South America

Giant hummingbird The largest bird in this order, the giant hummingbird is the size of a large swift. Its plumage is mostly brown, and quite dull compared to that of most hummingbirds. It has a long, slender bill and a forked tongue.

- Up to 9 in (23 cm)
- 1–2
- Sexes differ
- Sedentary
- Locally common

Andean W. South America

Sword-billed hummingbird This bird's bill is longer than its body, which makes it the longest bill relative to body length of any bird. The sword-billed hummingbird has co-evolved with the climbing passion flower.

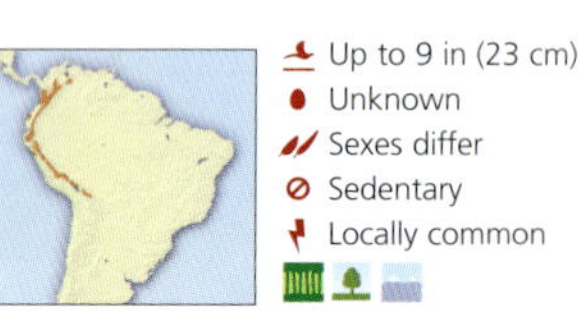

- Up to 9 in (23 cm)
- Unknown
- Sexes differ
- Sedentary
- Locally common

Andean N.W. South America

Uniquely upturned tip on the bill for accessing nectar; curled, bifurcate tongue snakes out into the nectar in the base of a flower and takes it up by capillary action

Fiery-tailed awlbill
Anthracothorax recurvirostris

Purple-throated carib
Eulampis jugularis

Males hold a flower-centered territory year-round, but females only do so when not breeding; the sexes are alike in plumage

Festive coquette
Lophornis chalybeus

Fork-tailed woodnymph
Thalurania furcata

One of a group of fork-tailed hummingbirds, this species is widespread in rain forests across northern South America

Ruby topaz
Chrysolampis mosquitus

Frilled coquette
Lophornis magnificus

Identifiable by its glistening golden gorget. Iridescent plumage in hummingbirds comes from the physical structure of platelets in the feather barbules

Racket-tailed coquette
Discosura longicauda

Racket-tipped tail feathers are present in several hummingbirds; tail feathers play an important role in flight maneuvers

Green-throated carib
Eulampis holosericeus

Purple-throated carib This bird lives in the Lesser Antilles, in high altitude forests and various other habitats. The female's bill is slightly longer and more curved than the male's. Its cup-shaped nest is camouflaged with lichens.

Lesser Antilles

- Up to 5 in (12 cm)
- 2
- Sexes alike
- Sedentary
- Locally common

Festive coquette This hummingbird inhabits humid forests and scrub-covered areas. It can be found in the lowlands, at altitudes up to 3,300 feet (1,000 m), east of the Andes from Colombia down to Argentina.

N.W. South America

- Up to 3¼ in (8.5 cm)
- 2
- Sexes differ
- Sedentary
- Indeterminate

Green-throated carib This primarily green hummingbird has a thin, curved black bill. It lives mainly in dry low-altitude areas, though it can also be found in higher regions, as well as in mangroves. It sometimes builds its nests quite high up in the branches.

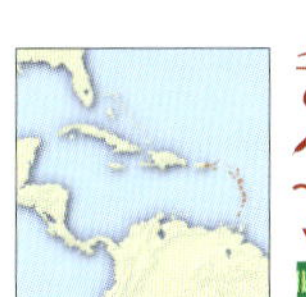
E. Puerto Rico, Lesser Antilles, Grenada

- Up to 5 in (13 cm)
- 2
- Sexes alike
- Locally nomadic
- Locally common

HUMMINGBIRD FLIGHT

Most birds can only fly forward, but hummingbirds can also fly backward, sideways, and straight up or down. They can even switch direction without turning their bodies. They can accomplish this thanks to their specialized physiology and anatomy, with wings that can be turned through 180 degrees. To maintain hovering, they can beat their wings up to 90 times per second. They can also stop and accelerate very quickly. Their flight feathers take up almost the entire wing, and their breastbones are very strong.

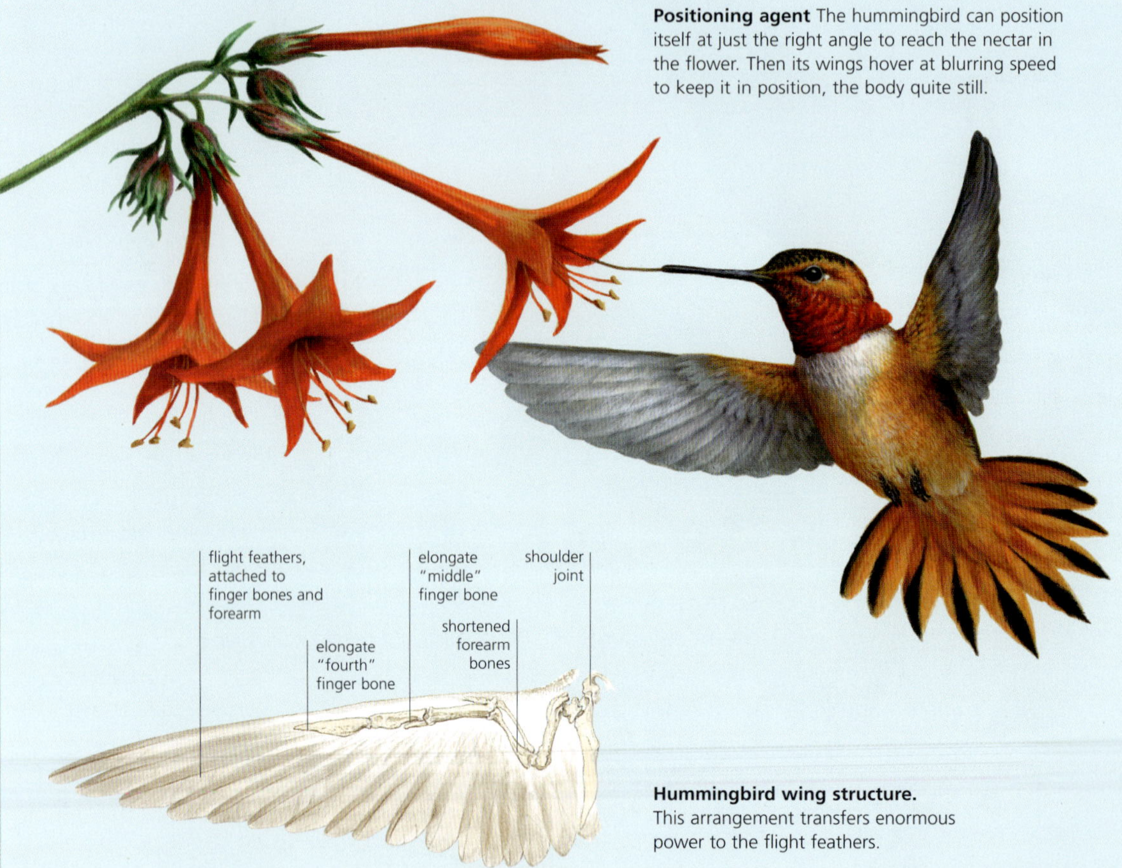

Positioning agent The hummingbird can position itself at just the right angle to reach the nectar in the flower. Then its wings hover at blurring speed to keep it in position, the body quite still.

Hummingbird wing structure. This arrangement transfers enormous power to the flight feathers.

HUMMINGBIRD FLIGHT

Flying forward Hummingbirds flap their wings up and down to move forward.

Hovering Hummingbirds move their wings in a rapid figure-eight motion to hover in place.

Flying upward By altering the wing angle of the movements, many directional changes are possible. Here, a more up-and-down angle allows a vertical rise.

Flying backward Hummingbirds flap their wings above and behind their heads to move backward.

Mousebirds

CLASS	Aves
ORDERS	1
FAMILIES	1
GENERA	2
SPECIES	6

The mousebird gets its name from its odd habit of creeping and crawling among bushes and clinging upside down, with its long tail held high. Herbivores, their diets include wild or cultivated fruit and even seedlings. They are therefore considered pests by gardeners and farmers. They build nests in bushes and sometimes cannibalize their young if they stray from the nest. They dislike rain and cold, huddling together and sometimes becoming torpid. Their feathers are replaced randomly and have long aftershafts.

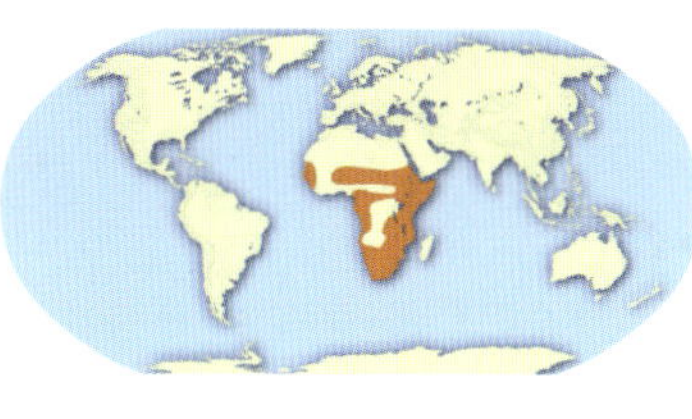

Only in Africa Mousebirds live in a wide range of African habitats, from almost-dry bushland to the edge of forests, south of the Sahara.

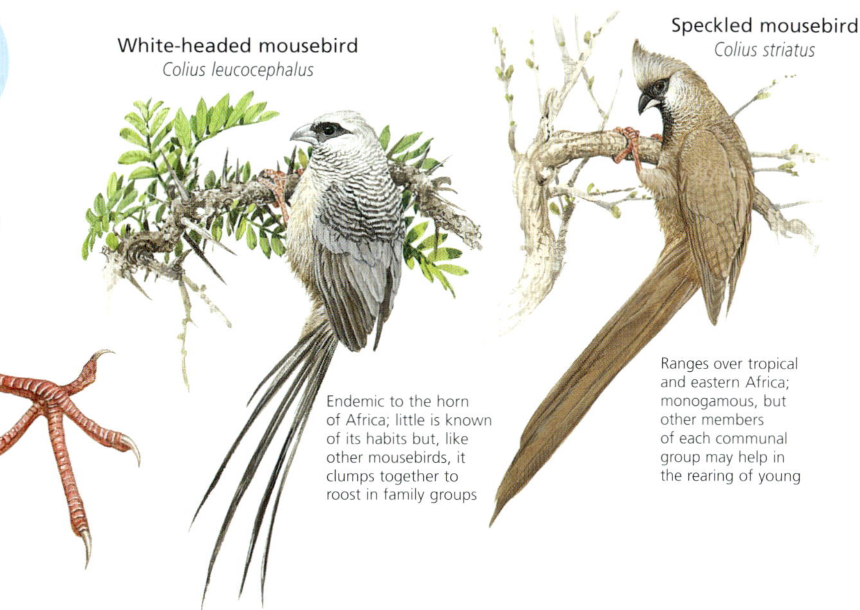

This way or that Mousebirds have unique feet with two reversible outer toes that can point either forward or backward. This helps when scrambling and clinging to vegetation.

White-headed mousebird
Colius leucocephalus
Endemic to the horn of Africa; little is known of its habits but, like other mousebirds, it clumps together to roost in family groups

Speckled mousebird
Colius striatus
Ranges over tropical and eastern Africa; monogamous, but other members of each communal group may help in the rearing of young

Trogons

CLASS	Aves
ORDERS	1
FAMILIES	1
GENERA	6
SPECIES	39

The best known of these brilliantly colored birds is the resplendent quetzal, the national bird of Guatemala, which was considered divine by the Aztecs. Female trogons are generally duller than males, and the Asian species are not as colorful. Trogons are secretive and territorial, perching while they scan for food. They eat insects and small lizards; some species also eat fruit. Males sometimes stage displays that include chases through the trees.

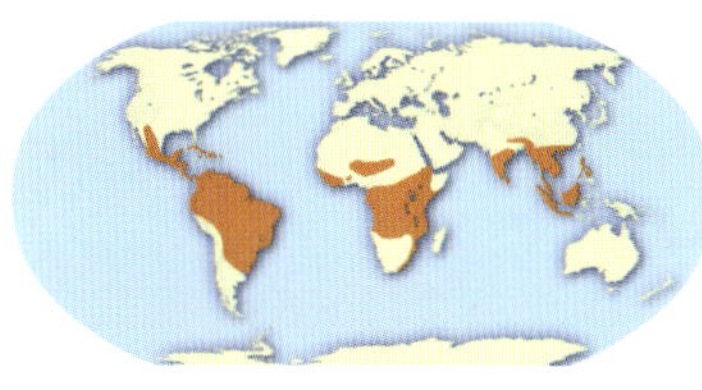

Pan-tropical beauties Trogons are forest dwellers. They can be found in the tropical regions of several continents. Arboreal birds, they live in rain forests and monsoon scrubs.

Resplendent quetzal
Pharomachrus mocinno
The size of a small chicken; lives in canopy of undisturbed montane rain forest in Central America; female is much duller and shorter-tailed

White-tailed trogon
Trogon viridis

Narina's trogon
Apaloderma narina

Red-headed trogon
Harpactes erythrocephalus

KINGFISHERS AND ALLIES

CLASS	Aves
ORDERS	1
FAMILIES	11
GENERA	51
SPECIES	209

Kingfishers are linked to todies, motmots, bee-eaters, rollers, hornbills, the hoopoe, and wood-hoopoes due to certain anatomical and behavioral similarities. All of these birds have small feet with three fused forward toes; they also have affinities in their ear bones and egg proteins. Many species have brilliantly colored plumage, and all nest in holes that they dig in soil or in rotten trees with their bills. Kingfishers are recognized by their short legs and their large, robust, long, and straight bills, which typically have sharply pointed or slightly hooked tips. The former are best suited for striking at and grasping fishes and other prey, and the latter for holding and crushing prey.

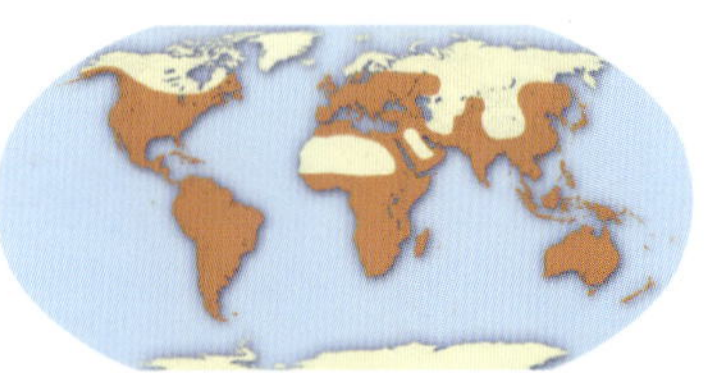

Waters and woods Kingfishers and their relatives can be found in a variety of aquatic and wooded environments around the world. The regions with the most species are Africa and Southeast Asia. Ground rollers are found only in Madagascar. "Non-fishing" kingfishers favor tropical rain forest and woodland.

Fishing equipment A typical kingfisher bill is proportionally large, generally long and straight, with a sharply pointed or slightly hooked tip. Despite their name, most species of kingfishers eat a variety of foods, including insects.

DIFFERENT HABITS

Most kingfisher species are generalized predators that eat a variety of terrestrial and aquatic invertebrates and vertebrates. Their typical foraging technique involves sitting quietly on a perch from which they can quietly survey their surroundings. Spotting prey, they then swoop or dive down and seize it in their bills. After bringing it back to the perch, they immobilize it by striking it repeatedly against a branch before eating it.

Some kingfishers, however, have markedly different diets and exhibit a variety of foraging techniques, such as shoveling into moist soil in search of earthworms.

Todies—very small, stocky birds with flattened bills—capture insect prey from the undersides of leaves; they also catch insects in flight.

Bee-eaters eat stinging insects, squeezing out and wiping off the sting before swallowing their prey.

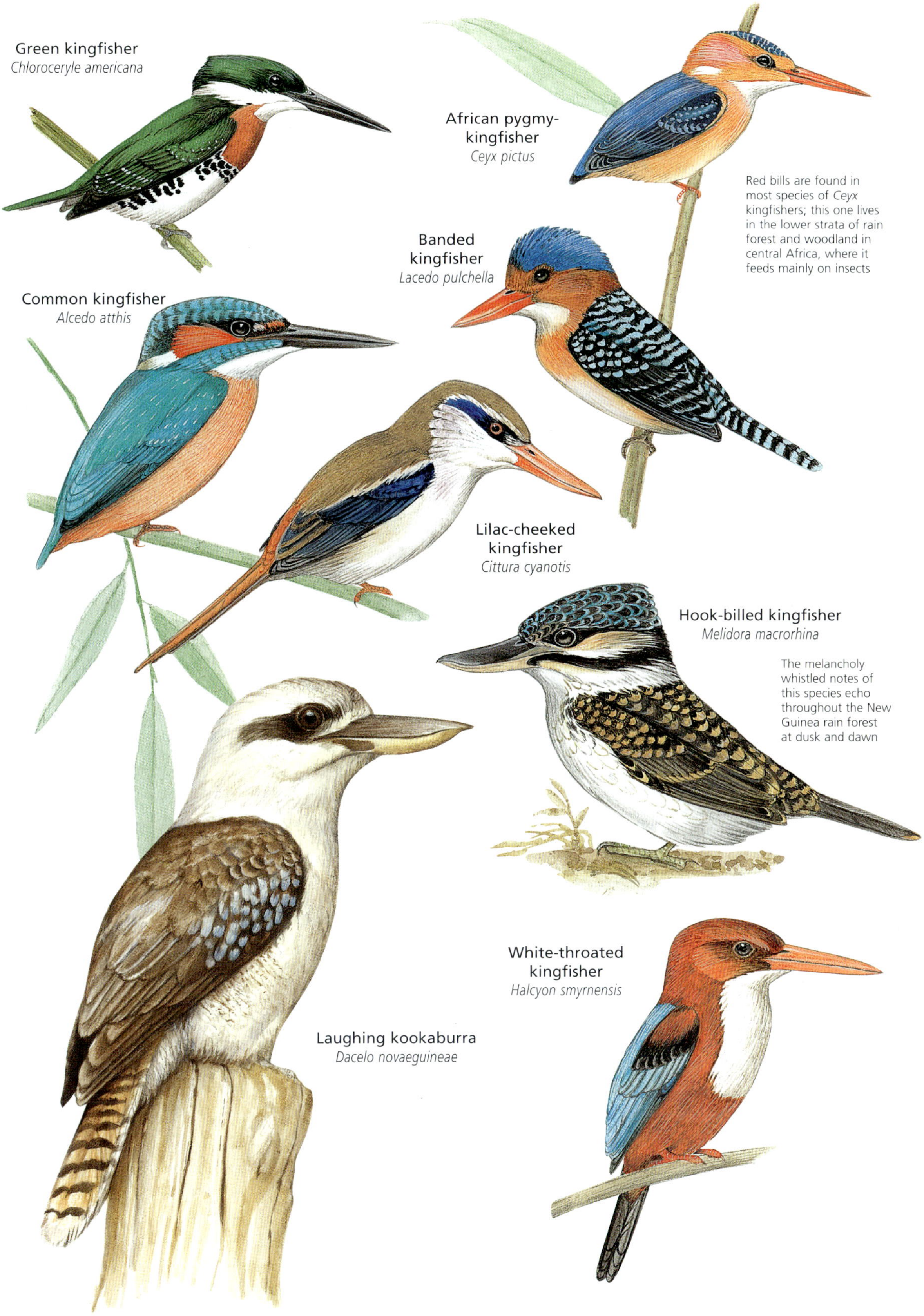

Green kingfisher
Chloroceryle americana

African pygmy-kingfisher
Ceyx pictus

Red bills are found in most species of *Ceyx* kingfishers; this one lives in the lower strata of rain forest and woodland in central Africa, where it feeds mainly on insects

Banded kingfisher
Lacedo pulchella

Common kingfisher
Alcedo atthis

Lilac-cheeked kingfisher
Cittura cyanotis

Hook-billed kingfisher
Melidora macrorhina

The melancholy whistled notes of this species echo throughout the New Guinea rain forest at dusk and dawn

White-throated kingfisher
Halcyon smyrnensis

Laughing kookaburra
Dacelo novaeguineae

Common kingfisher This bird's upper parts can appear either brilliant blue or emerald green, depending on the refraction of the light. It flies low over the water, and nests inside tunnels on river banks.

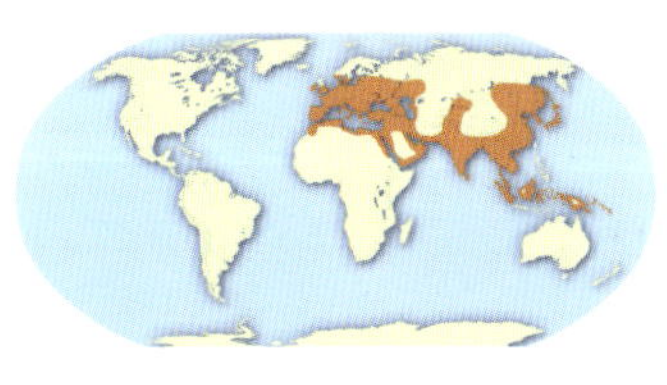

- Up to 6 in (16 cm)
- 4–10
- Sexes alike
- Partial migrant
- Common

Europe, N. Africa, W. Eurasia to N. Melanesia

Laughing kookaburra This bird's unmistakable loud cry sounds like laughter. It nests in hollow trunks or holes in trees, or sometimes in emptied-out termite nests. The young remain in the nest for over a month.

- Up to 17 in (43 cm)
- 1–4
- Sexes alike
- Sedentary
- Common

E. Australia; introd. S.W. Australia, Tasmania

Cuban tody
Todus multicolor

All todies have short, straight tails and the same plumage pattern. Only the Cuban has a pale blue wash on the cheeks

Red-bearded bee-eater
Nyctyornis amictus

♂

Lives in the upper stratum of tall Southeast Asian rain forests; sits on bare vantage perches and nests in burrows in earth banks

Blue-crowned motmot
Momotus momota

All motmots have racket-tipped tails, but only this species has both a blue crown and rusty breast

Blue-bellied roller
Coracias cyanogaster

European bee-eater
Merops apiaster

Carmine bee-eater
Merops nubicus

Hawk for aerial insect prey from high bare perches, in gracefully elaborate maneuvers; East Asian and Australian populations are migratory, but others in the tropical zones are sedentary

Dollarbird
Eurystomus orientalis

European roller
Coracias garrulus

Like most bee-eaters, the carmine grows prolonged central tail feathers when breeding. It nests colonially in burrows in earth banks in drier parts of sub-Saharan Africa. Breeding is tied to the rainy season

Cuban tody All five species of tody are confined to the West Indies. This bird typically catches insects from the underside of a leaf in the course of an arc-like flight. Its nest is a burrow.

Up to 4½ in (11 cm)
3–4
Sexes alike
Non-migrant
Locally common

Cuba, Isla de la Juventud

Dollarbird This bird gets its name from circular patches of silvery white on its wings. When the bird flies, these patches resemble American silver dollar coins. It eats mainly insects caught on the wing.

Up to 13 in (32 cm)
3–5
Sexes alike
Partial migrant
Common

S. & S.E. Asia to E. Australia & N. Melanesia

Carmine bee-eater The two regional populations of this species undergo three-stage migrations each year. They live in large colonies of up to 1,000 pairs. They ride on the backs of large mammals, waiting to catch insects.

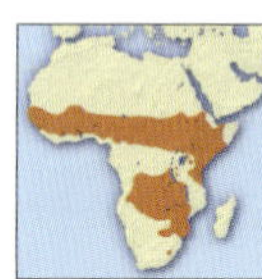

Up to 11 in (27 cm)
2–5
Sexes similar
Migrant
Common

Sub-Saharan & S.C. Africa

Long-tailed ground-roller
Uratelornis chimaera

Rufous-headed ground-roller
Atelornis crossleyi

Common hoopoe
Upupa epops

Cuckoo-roller
Leptosomus discolor

♂

Cuckoo-rollers give loud whistling and cackling cries as they glide over Madagascan forests; the calls are among the most characteristic for the island

Green wood-hoopoe
Phoeniculus purpureus

Red-billed hornbill
Tockus erythrorhynchus

♂

Northern ground-hornbill
Bucorvus abyssinicus

Feeds on the ground, walking about in pairs or small groups, picking up a range of vertebrate and invertebrate prey

♂

Great hornbill
Buceros bicornis

Monogamous, territorial birds that nest in big hollows in tall rain-forest trees; the male seals the female in to incubate by blocking up the entrance hole with mud, then feeds his mate and chicks through a small hole in the entrance for the period of incubation and rearing, up to about 100 days

Great hornbill This bird has a casque on the upper half of its bill. Males and females have different coloration in their bills, their eyes, and in the rings around their eyes. They eat fruit, invertebrates, and small vertebrates.

- Up to 3½ ft (1.1 m)
- 1–4
- Sexes alike
- Sedentary
- Rare

W. India & Himalayas to S.E. Asia, Sumatra

CAVITY NESTS

The hoopoe nests in cavities. Some sites are reused. The young are initially cared for by the female and later by both parents. In most species of the related hornbills, which also nest in natural cavities, the female closes the entrance hole and remains with the chicks for months. The male feeds the entire family through the hole.

WOODPECKERS AND ALLIES

CLASS	Aves
ORDERS	1
FAMILIES	5
GENERA	68
SPECIES	398

These six families of birds—woodpeckers, honeyguides, jacamars, puffbirds, barbets, and toucans—look different but share certain anatomical features, such as zygodactyl feet, with two toes in front and two at the back. They also generally lack down feathers and lay white eggs. Most species are colorful; some are even gaudy. They are primarily tropical birds that nest in cavities in trees, in termite mounds, or in the ground. Woodpeckers and barbets typically excavate their own cavities; after they are abandoned, the holes are often used by other species. Woodpeckers sometimes live within sight of large cities and are frequent visitors to bird-feeding stations.

Wide ranges Woodpeckers live in forests; toucans, jacamars, and puffbirds in the New World tropics. Honeyguides are mainly African. Barbets are more widely dispersed.

Spotted puffbird *Bucco tamatia*

Black nunbird *Monasa atra*

Nunbirds differ from other puffbirds in their black plumage, red bills, and more slender shape

Red-and-yellow barbet *Trachyphonus erythrocephalus*

Lives in open wooded and riverine habitats in central east Africa; forages in groups over ground and through trees for fruit and insects

Rufous-tailed jacamar *Galbula ruficauda*

♂

A white-throated male; females have rufous-buff throats

A glossed dark green-blue breast and back identifies the Paradise jacamar.

Paradise jacamar *Galbula dea*

Blue-throated barbet *Megalaima asiatica*

Most Southeast Asian barbets are green, matching the forest foliage in which they live

SPECIALIZED FEATURES

Woodpeckers are often heard before they are seen. They use their strong, tapering, often chisel-tipped bills to peck at the bark of trees, probing for insects. Bony, muscle-cushioned skulls help protect them from the impact. They use their strong, long-clawed toes to grip the bark and their straightened tail feathers to prop themselves against the trees. Most of them hop rather than walk.

Despite their large, gaudy-colored bills, which can be up to a third or more of their body length, toucans are often surprisingly inconspicuous as they perch on branches amid foliage. They feed mainly on fruit, as do the brightly colored and patterned barbets, which have relatively stout, sometimes notched bills.

Jacamars and puffbirds eat insects, flying forth from perches to catch them. They nest in tunnels that they excavate; some species cover the entrance with a pile of sticks. The typically inconspicuous and inactive puffbirds are named for their unusually loose, fluffy plumage.

Honeyguides also eat insects, as well as wax from beehives. Their name comes from the fact that one species leads humans to beehives.

Some species within these families are solitary; others, such as the nunbirds and puffbirds, are social and even breed cooperatively. Some African barbets nest in pair groups of 100 or more birds, all residing in one dead tree. The greater honeyguide is a brood parasite, laying one egg in the nest of each host bird.

Barbets tend to be highly vocal; some are called "brain-fever birds" because of their unceasing, monotonous calls. Uniquely among birds, woodpeckers have developed instrumental drumming on wood. The tone and pattern are species-specific.

Yellow-rumped honeyguide
Indicator xanthonotus

This Himalayan honeyguide is one of only two species of the group to occur outside Africa

Greater honeyguide
Indicator indicator

Lesser honeyguide
Indicator minor

Honeyguides are rather small gray-green birds with distinctive white flashes at the sides of the tail; wax from insects or their nests is their staple diet

Black-necked aracari
Pteroglossus aracari

Curl-crested aracari
Pteroglossus beauharnaesii

Emerald toucanet
Aulacorhynchus prasinus

Toco toucan
Ramphastos toco

Distinctively striped bill; this species lives in montane rain forest in the northern Andes

Gray-breasted mountain-toucan
Andigena hypoglauca

Zygodactyl toes with two forward and two back, an arrangement better fitted for clasping perches than for running

Channel-billed toucan
Ramphastos vitellinus

Identified by its broad crimson breast band and dark bill. It lives in the lowland rain forests of tropical South America. It eats palm nuts, fruits, and insects and drinks from bromeliads or by holding the bill open in rain

Moderately alternated tail that is often cocked temporarily at perches or during vocalizations

Greater honeyguide This bird is known for leading humans to the hives of wild honeybees. The humans harvest the comb and leave the remains for the bird, which is particularly fond of the wax.

- Up to 8 in (20 cm)
- Sets of up to 5
- Sexes alike
- Sedentary
- Common

Sub-Saharan Africa excl. Congo basin & S.W.

Lesser honeyguide Like other honeyguides, the lesser honeyguide is a brood parasite. Newly hatched young use a hook on their bills to kill the host's chicks and push them out of the nest, or they break the eggs before hatching.

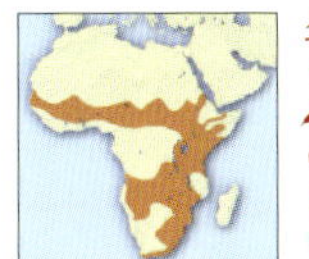

- Up to 6½ in (16 cm)
- Sets of 2–4
- Sexes alike
- Sedentary
- Locally common

Sub-Saharan Africa excl. Congo basin & S.W.

Emerald toucanet The male and female of this small green toucan are alike in color but differ in size, the male being slightly larger. The emerald toucanet sometimes appropriates the previously used nesting holes of other species, or excavates its own.

- Up to 14½ in (37 cm)
- 1–5
- Sexes alike
- Sedentary
- Locally common

Mexico, S. Central & N.W. South America

Northern wryneck
Jynx torquilla

Soft, cryptic plumage; one of only three migratory woodpeckers

Feeds mostly on ants, hopping around mainly on the ground; curved bill is used to open anthills

Acorn woodpecker
Melanerpes formicivorus

Acorn woodpeckers are well known for their habit of storing nuts in holes in trees for winter

A white rump helps to identify this species

Gray woodpecker
Dendropicos goertae

Feeds by pecking on the middle branches of trees on ants, termites, and other insects

Yellow-bellied sapsucker
Sphyrapicus varius

Red-headed woodpecker
Melanerpes erythrocephalus

Feeds on both insects and seeds in forests, and stores nuts and seeds in cracks and crevices for winter

Golden-tailed woodpecker
Campethera abingoni

Brown-capped woodpeckers live in woodlands and forage on trees for ants and other insects

Brown-capped woodpecker
Dendrocopos moluccensis

Ground woodpecker
Geocolaptes olivaceus

Acorn woodpecker This unmistakable bird is common in oak woods. It is a social bird that can be heard making chattering and rhythmic laughing calls. The birds eat insects (pried from tree trunks or caught in flight) and acorns.

Up to 9 in (23 cm)
4–6
Sexes alike
Sedentary
Common

W. North to N.W. South America

Yellow-bellied sapsucker The male and female of this species have different colored throat patches. At nesting grounds, the bird makes distinctive drumming sounds, with alternating rapid and slow thumps.

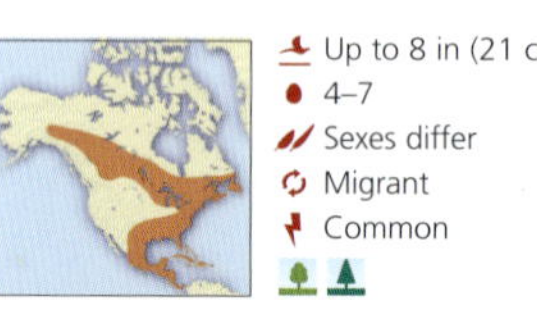

Up to 8 in (21 cm)
4–7
Sexes differ
Migrant
Common

C.N. to S.E. North America, Central America

Ground woodpecker This species feeds almost entirely on the ground. It is specialized to eat ants by probing crannies and digging its bill into ants' nests, head down. It also nests on the ground, in 3¼ ft (1 m) long burrows, dug mostly by the male.

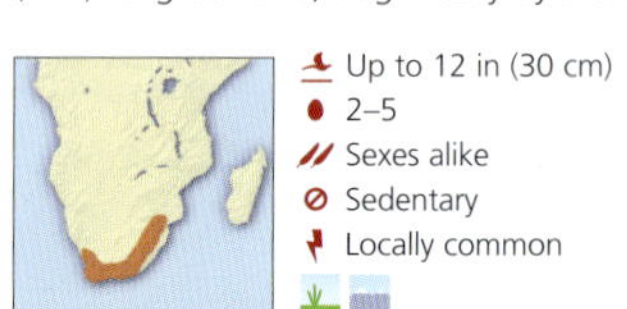

Up to 12 in (30 cm)
2–5
Sexes alike
Sedentary
Locally common

Cape Province to Transvaal & Natal (Africa)

Rufous woodpecker
Celeus brachyurus

Blond-crested woodpecker
Celeus flavescens

Yellow head, red cheek splash, and gold mottled back identify this species; it lives in the forests and savannas of central east South America

Black woodpecker
Dryocopus martius

Pileated woodpecker
Dryocopus pileatus

This Indian species lives in dry to moist woodlands; it forages at all levels in trees on ants and other insects

Black-rumped woodpecker
Dinopium benghalense

This widespread Eurasian woodpecker has similar habits to the European green woodpecker

A stiffened, tapered tail helps to prop woodpeckers as they clasp upright trunks and branches in their feeding

Green woodpecker
Picus viridis

Greater yellow-naped woodpecker
Picus flavinucha

Gray-headed woodpecker
Picus canus

Rufous woodpecker This bird appears frequently in gardens. It has the unusual habit of building its nest by digging a burrow in the middle of a tree ant nest, thus giving it a ready food supply at its doorstep.

- Up to 10 in (25 cm)
- 2–3
- Sexes alike
- Sedentary
- Common

S. & S.E. Asia to Borneo & Sumatra

Pileated woodpecker This adaptable, widespread large bird can even be seen within sight of skyscrapers. It eats primarily ants, which it extracts from rotting trees and fallen logs, using its frilled, sticky tongue.

- Up to 18 in (46 cm)
- 2–6
- Sexes differ
- Sedentary
- Rare

C.W. & E. North America

GREEN WOODPECKER

This bird has a distinctive flight pattern, alternating three or four wingbeats with undulations with wings closed. It sometimes lays up to 11 eggs.

Grounded
The European green woodpecker feeds mainly on the ground.

Passerines

CLASS	Aves
ORDER	Passeriformes
FAMILIES	96
GENERA	1,218
SPECIES	5,754

Passeriformes is the largest order of birds by far, encompassing more than 5,700 species. At the same time, this order includes a disproportionately small percentage of bird families, meaning that each passerine family has a relatively high number of species. These two facts together attest to the great evolutionary success of the passerines. Relative newcomers—they are believed to have evolved about 75 million years ago on the southern super-continent Gondwana—they have proven themselves remarkably adaptable, having spread to every continent except Antarctica. Sometimes known as perching birds or songbirds, they are distinguished by their distinctive palates, voice box, and feet.

Robin redbreast Escaping harsher winters in Eurasia and western Siberia, the European robin (*Erithacus rubecula*) migrates to Britain and northern Europe.

SKILLED VOCALIZERS

The passerines are typically small or medium-sized birds, which nevertheless vary greatly in shape, size, and plumage colors (the last ranges from very drab to brilliant).

Their feet, unlike those of other bird orders, have four toes, each of which joins the leg at the same level; three of the toes are directed forward, whereas the innermost one is directed backward. This foot structure and the flexibility of the toes are ideally suited to grasping perches, whether they be trees, shrubs, or even blades of grass.

Most passerines have wings that taper to a point at the outer tip. This shape makes them well suited for rapid take-offs and good in-air maneuverability, which is useful for catching prey and also for avoiding predators. They are less capable of sustaining high flight speeds.

Most passerines are songbirds, or oscines. Unlike mammals, which produce sound in their larynx, birds make sound in a voice box known as a syrinx. This unique structure at the base of the windpipe, or trachea, consists of two chambers and relies on about half a dozen pairs of muscles to control the tension of its three elastic, vibrating membranes. The bird's trachea, interclavicular air sac and mouth modify the sound that is produced. The two chambers allow the bird effectively to sing a duet by itself.

One large group of passerines includes birds with poorly muscled voice boxes and simple, stereotyped songs, among them the pittas and broadbills of the Old World tropics, and the ovenbirds, antbirds, tyrant flycatchers, and cotingas of Central and South America.

In the past, many families in this order were grouped together due to morphological similarities. These were in many cases later attributed to convergent evolution rather than very similar genetics; for example, the wrens of Australia are not close relatives of the wrens that live in the Northern Hemisphere.

Most passerines eat plants or insects or both. During each breeding season, adults and young alike may consume more protein.

Most of the birds in this order are monogamous, and both parents help rear the young, although the female tends to do more brooding.

Feeding time Many passerines migrate to warmer climates to breed and rear their young (below) in regions where the food supply is more plentiful.

Migrating seasonally The Asian paradise flycatcher (*Terpsiphone paradisi*) (left) spends the winter season in tropical Asia. This male guards his nest fiercely.

Bright songbird The yellow warbler (*Dendroica petechia*) (above) sings from a branch. These warblers are members of the suborder Oscines, whose muscled voice boxes make them the best of singers.

Stunning plumage Amongst Australia's best-known native birds, the superb lyrebird (*Menura novaehollandiae*) (below) is known for its plumage. Awkward in flight, the lyrebird generally scratches for food at ground level, in the scrub.

Black-bellied gnateater
Conopophaga melanogaster,
family Conopohagidae

Barred antshrike
Thamnophilus doliatus,
family Thamnophilidae

Red-billed scythebill
Campylorhamphus trochilirostris,
family Dendrocolaptidae

Slaty spinetail
Synallaxis brachyura,
family Furnariidae

Slaty breasts and rufous wings characterize the species of *Synallaxis*

Gleans for insects on tree branches

Black-throated huet-huet
Pteroptochos tarnii,
family Rhinocryptidae

Ruddy treerunner
Margarornis rubiginosus,
family Furnariidae

Rufous-winged antwren
Herpsilochmus rufimarginatus,
family Thamnophilidae

Rufous hornero
Furnarius rufus,
family Furnariidae

Horneros build ovenlike nests

Streak-chested antpitta
Hylopezus perspicillatus,
family Formicariidae

Family Formicariidae Antbirds get their name from some species' habit of clinging to vertical stems above ant swarms. Some prefer to walk as they forage; others fly around in search of prey. Their songs are usually whistles.

Genera 7
Species 62

C. Mexico to subtropical South America

Family Furnariidae The rufous hornero and some of its relatives build mud nests that resemble traditional clay ovens, hence the name ovenbirds. Such nests protect the young from sun, cold, and predators. Ovenbirds can move in narrow places.

Genera 55
Species 236

S. Mexico to South America

Family Dendrocolaptidae Woodcreepers are sometimes confused with woodpeckers due to their common habit of using their tails to brace themselves against tree trunks. Woodcreepers have passerine feet and are all brown with often pale-streaked plumage.

Genera 13
Species 50

Mexico to N. temperate South America (C. Argentina)

Guianan cock-of-the-rock
Rupicola rupicola,
family Cotingidae

Throat ornamentation characterizes the Central American bellbirds

Bearded bellbird
Procnias averano,
family Cotingidae

Turquoise cotinga
Cotinga ridgwayi,
family Cotingidae

Male cock-of-the-rocks display in groups called leks

Crimson fruitcrow
Haematoderus militaris,
family Cotingidae

Ranges from Mexico to Brazil

Capuchinbird
Perissocephalus tricolor,
family Cotingidae

Bearded manakin
Manacus manacus,
family Pipridae

Most manakins have contrastingly colored crowns or heads

Blue-crowned manakin
Lepidothrix coronata,
family Pipridae

Three-wattled bellbird
Procnias tricarunculatus,
family Cotingidae

Golden-headed manakin
Pipra erythrocephala,
family Pipridae

Family Cotingidae These fruit-eating birds are typically found in the upper portions of the forest canopies in Central and South America. They also congregate in low densities and are therefore poorly known.

Genera 33
Species 96

Mexico to South America

FAMILY PIPRIDAE

Manakins are small, brightly colored fruit-eaters with short tails and short, wide bills. They live in lowland forests of Central and South America. Their complex social behavior includes elaborate courtship rituals in which males congregate at leks, where they prepare dance arenas or perches.

Genera 13
Species 48

S. Mexico to subtropical South America

Ritual brushing
During courtship, the male wire-tailed manakin rapidly brushes the female's throat with the wire-like tips of his tail feathers.

Scissor-tailed flycatcher
Tyrannus forficatus,
family Tyrannidae

Great kiskadee
Pitangus sulphuratus,
family Tyrannidae

Scarlet crown plumage spread in display

Royal flycatcher
Onychorhynchus coronatus,
family Tyrannidae

Flattened, slightly hooked bill for catching insects in flight

Eastern kingbird
Tyrannus tyrannus,
family Tyrannidae

Olive-sided flycatcher
Contopus cooperi,
family Tyrannidae

Sulfur-bellied flycatcher
Myiodynastes luteiventris,
family Tyrannidae

Erectile crest is raised in excitement

Tufted flycatcher
Mitrephanes phaeocercus,
family Tyrannidae

Sallies out to catch insects on the wing

Black phoebe
Sayornis nigricans,
family Tyrannidae

Long tail for maneuverability in flight

DIET SPECIALIZATION

Different species within the Tyrannidae family exploit very particular niches within their environments, allowing dozens of species to coexist. In the majority of cases this can be attributed to slightly different combinations of prey size, habitat, vegetation type, foraging position, and capture technique.

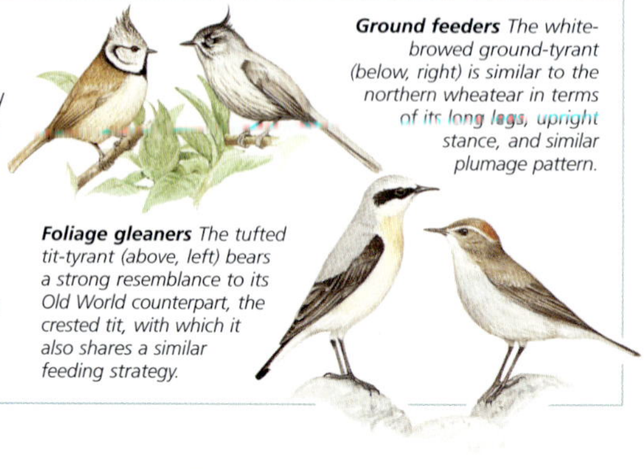

Ground feeders *The white-browed ground-tyrant (below, right) is similar to the northern wheatear in terms of its long legs, upright stance, and similar plumage pattern.*

Foliage gleaners *The tufted tit-tyrant (above, left) bears a strong resemblance to its Old World counterpart, the crested tit, with which it also shares a similar feeding strategy.*

Family Tyrannidae This largest of the passerine families includes the tyrant flycatchers as well as phoebes, flatbills, elaenias, kingbirds, and wood-peewees. The sexes usually look alike, with green, brown, yellow, and white plumage.

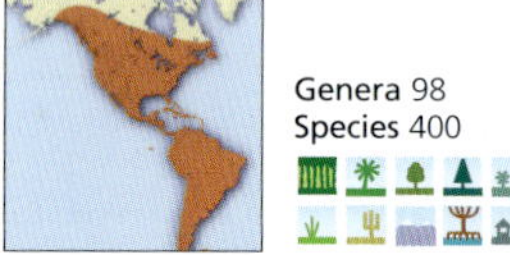

South America, West Indies, North America

COOPERATIVE BREEDING

About 3 percent of all bird species are classified as cooperative breeders, meaning that additional birds—either juveniles or adults—help the parents raise their young. Recent studies have shown that this phenomenon is more prevalent among passerines than was previously thought. There are two forms of cooperative breeding. One involves non-breeding birds helping parents protect and rear their nestlings. The other, called communal breeding, involves some shared parentage (whether shared maternity, paternity, or both) of the offspring that a group of adults are raising together.

Staying behind *Like many other Australian birds, juvenile superb fairy-wrens (left and right) from an earlier brood frequently remain with their parents to help raise their younger siblings. This helps increase the chances of similar genes surviving.*

Tidying up *An immature (or eclipse) male superb fairy-wren (right) removes a fecal sac from the nest. Helpers carry out various chores, including tidying the nest, procuring food, and protecting the nestlings from predators.*

A lighter load *The mother (right) brings food to her nestlings. Her workload is considerably eased by the presence of her helpers. Studies have shown that chicks raised in communal settings are fed more insects per hour than those raised in a nuclear family setting.*

A group effort The superb fairy-wren (*Malurus cyaneus*) (right) is one of the many communally breeding species of passerines. It is common in southeastern Australia, including in suburban gardens. Why cooperative breeding evolved in some species of birds and not in others is still something of a mystery, especially in light of the patchy distribution of the phenomenon among genetically related types of birds. Ornithologists believe it is probably related to a combination of factors, including environmental constraints (with respect to the availability of unoccupied breeding territories or sexually mature mates) and life-history traits, such as mortality rates and dispersal tendencies. Helpers may include birds whose own attempts at nesting have failed that season.

Standing guard *The father (above) of the brood perches near the nest. Among superb fairy- wrens, fertilization is more likely to occur with a partner outside the group than within the pair itself. Studies have shown that per capita reproductive success does not increase with group size.*

Green broadbill
Calyptomena viridis,
family Eurylaimidae

Cream head-spots shine in the forest, possibly for social signalling

Sharpbill
Oxyruncus cristatus,
family Cotingidae

Long-tailed broadbill
Psarisomus dalhousiae,
family Eurylaimidae

Red-bellied pitta
Pitta erythrogaster,
family Pittidae

Red-bellied pittas range from the Philippines to Australia

Rufous-tailed plantcutter
Phytotoma rara,
family Cotingidae

Rifleman
Acanthisitta chloris,
family Acanthisittidae

Velvet asity
Philepitta castanea,
family Philepittidae

Noisy scrub-bird
Atrichornis clamosus,
family Atrichornithidae

A lyrebird singing in display posture

Superb lyrebird
Menura novaehollandiae,
family Menuridae

FAMILY PITTIDAE

Pittas are thrush-sized, insect-eating birds with short tails. They are brilliantly colored forest-dwellers that prefer to hop or run rather than fly. Their large nests, built on the ground or in low vegetation, resemble piles of plant debris. All give loud, melodious double whistles.

Genera 1
Species 30

C. Africa, S.E. Asia to N. & E. Australia, Melanesia

Stealthy bird
The banded pitta, though colorful, is surprisingly inconspicuous on the forest floor. It feeds quietly and seldom calls.

Family Eurylaimidae Known as broadbills, the members of this family are stout birds with broad heads, broad, flattened bills, and short legs. They have striking plumage. Most species eat insects, as well as small lizards, frogs, and fruit.

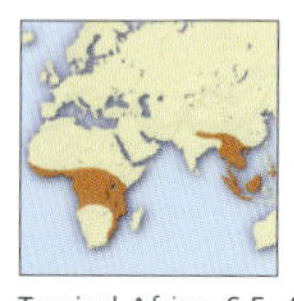

Genera 9
Species 14

Tropical Africa, S.E. Asia, Gr. Sundas, S. Philippines

Long, curved bill adapted for nectar feeding

Scarlet-chested sunbird
Chalcomitra senegalensis,
family Nectariniidae

Cape sugarbird
Promerops cafer,
family Promeropidae

Brown-throated sunbird
Anthreptes malacensis,
family Nectariniidae

Sugarbirds have long, plumed tails

Stripe-headed creeper
Rhabdornis mystacalis,
family Rhabdornithidae

Red-browed treecreeper
Climacteris erythrops,
family Climacteridae

Orange-bellied flowerpecker
Dicaeum trigonostigma,
family Dicaeidae

Treecreeper's toes are fused at base to work like calipers

Crimson-breasted flowerpecker
Prionochilus percussus,
family Dicaeidae

Fine, toothed bill for skinning fruit

Spotted pardalote
Pardalotus punctatus,
family Pardalotidae

Mistletoebird
Dicaeum hirundinaceum,
family Dicaeidae

Family Dicaeidae The flowerpeckers are dumpy birds with pointed wings, stubby tails, and short, conical bills. Their cleft tongues are adapted for feeding on nectar as well as fruit.

Genera 2
Species 44
India, S.E. Asia & its archipelagos to Melanesia & Australia

Family Climacteridae This small family of bark-climbing birds known as tree-creepers is found only in Australia and New Guinea. They have slender bills, short, square tails, and strong legs.

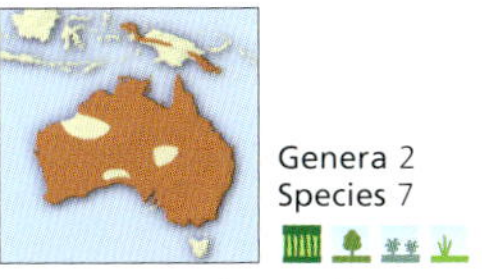

Genera 2
Species 7
Mainland Australia excl. deserts, New Guinea

Family Pardalotidae These common, multicolored birds closely resemble flowerpeckers in size and proportions. They feed almost entirely on larvae, and on an exudate produced by the larvae, of leaf-eating insects in foliage.

Genera 1
Species 4
Australia, Tasmania

Brown honeyeater
Lichmera indistincta,
family Meliphagidae

This species calls like a reed-warbler

Red-headed honeyeater
Myzomela erythrocephala,
family Meliphagidae

Pied honeyeater
Certhionyx variegatus,
family Meliphagidae

A nomadic honeyeater found in Australian deserts

Most Myzomela honeyeaters have glistening red plumage on the belly

Lewin's honeyeater
Meliphaga lewinii,
family Meliphagidae

Black-chinned honeyeater
Melithreptus gularis,
family Meliphagidae

Silver-eye
Zosterops lateralis,
family Zosteropidae

White eye-rings are found in most white-eyes

Oriental white-eye
Zosterops palpebrosus,
family Zosteropidae

Blue-faced honeyeater
Entomyzon cyanotis,
family Meliphagidae

FAMILY MELIPHAGIDAE

Honeyeaters, friarbirds, wattlebirds, miners, and spinebills have unique tongues that are deeply cleft and fringed at the tip to help the birds collect nectar. Honeyeaters often specialize in certain types of plants, especially eucalypts.

Genera 44
Species 174

Australia & Melanesia to L. Sundas, Sulawesi & Oceania

Family Zosteropidae White-eyes are found in almost every kind of wooded habitat throughout their range, and in suburban parks and gardens. Many species are similar in appearance, with citrine plumage, and white feathers around their eyes.

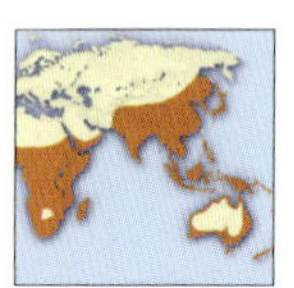

Genera 14
Species 95

Sub-Saharan Africa, S. & E. Asia to Australasia

Asian paradise-flycatcher
Terpsiphone paradisi,
family Monarchidae

Chestnut-breasted whiteface
Aphelocephala pectoralis,
family Acanthizidae

Crimson chat
Epthianura tricolor,
family Meliphagidae

Penduline tit
Remiz pendulinus,
family Remizidae

Black-throated tit
Aegithalos concinnus,
family Aegithalidae

Penduline tits are found in Eurasia, Africa, and southern North America

Yellow-bellied fantail
Rhipidura hypoxantha,
family Rhipiduridae

Crested tit
Parus cristatus,
family Paridae

This species gleans over ground and trees in the Australian bush

Gray shrike-thrush
Colluricincla harmonica,
family Colluricinclidae

FAMILY PARIDAE

Tits, chickadees, and titmice tend to have black caps, some with crests, and white cheeks. Most are sedentary, forest-dwelling birds that use their short, straight bills to feed on insects and seeds in foliage.

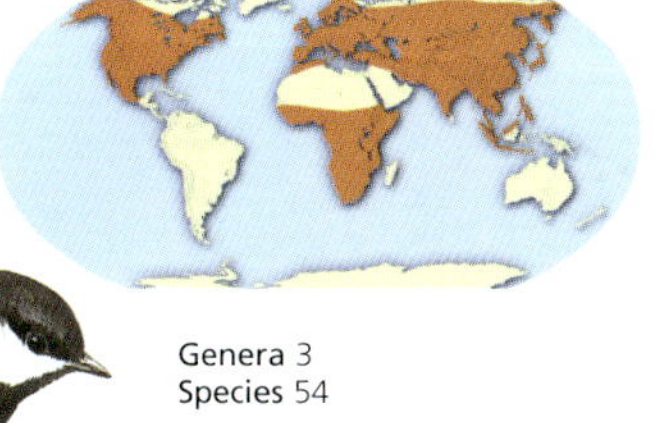

Standing out
The colorful great tit inhabits woods, parks, and gardens. It nests in trees and holes, including artificial nesting boxes.

Genera 3
Species 54

North America, Africa & Eurasia to Japan & L. Sundas

Family Rhipiduridae Fantails get their common name from their habit of fanning or waving their long tails. They also pirouette through the air and appear hyperactive. Their broad bills are surrounded by bristles that aid in catching insect prey.

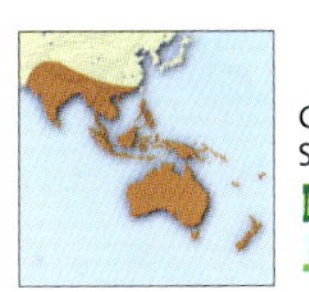

Genera 1
Species 43

Australasia & Melanesia, Oceania, S.E. Asia & India

Wallcreeper
Tichodroma muraria,
family Sittidae

Wallcreepers nest in rock crevices in cliff faces

Chinspot batis
Batis molitor,
family Platysteiridae

Males have black breast band

Varied sittella
Neositta chrysoptera,
family Neosittidae

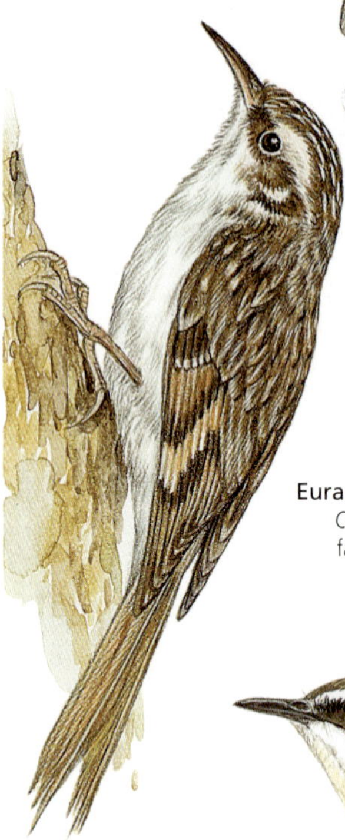

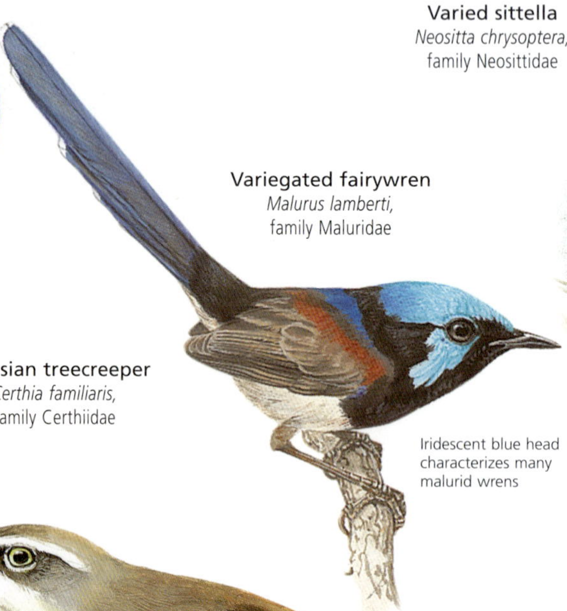

Variegated fairywren
Malurus lamberti,
family Maluridae

Eurasian treecreeper
Certhia familiaris,
family Certhiidae

Iridescent blue head characterizes many malurid wrens

White-breasted nuthatch
Sitta carolinensis,
family Sittidae

Fairy gerygone
Gerygone palpebrosa,
family Acanthizidae

White-browed scrubwren
Sericornis frontalis,
family Acanthizidae

Yellow-rumped thornbill
Acanthiza chrysorrhoa,
family Acanthizidae

Gerygones, scrub-wrens, and thornbills live in Australia and New Guinea

FAMILY SITTIDAE

Nuthatches are found throughout the Northern Hemisphere. These bark-climbing birds are superficially similar to the Australasian sittellas in appearance, but the two families are not closely related, representing an example of evolutionary convergence, a response to similar environmental factors.

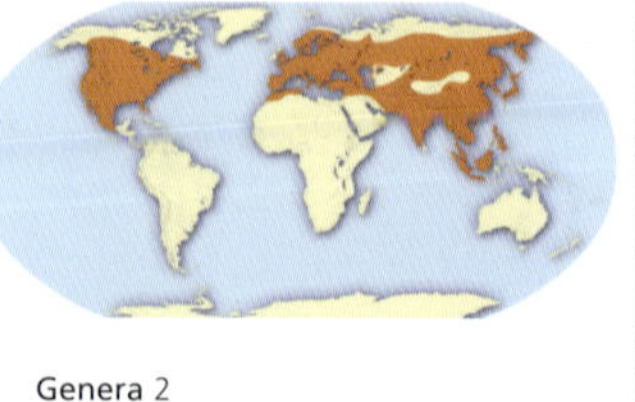

Genera 2
Species 25

North America, Eurasia to Japan, Philippines & Gr. Sundas

Family Maluridae Like the northern wrens to which they are not closely related, the fairy-wrens, grass-wrens, and emu-wrens that make up this family are small birds that cock their tails. They are insectivores; most forage on the ground and in shrubbery.

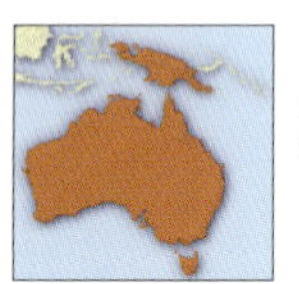

Genera 5
Species 28

Australia, Tasmania, New Guinea, Aru Is.

Brubru
Nilaus afer, family Malaconotidae

White-browed woodswallow
Artamus superciliosus, family Artamidae

Red-backed shrike
Lanius collurio, family Laniidae

Greater racket-tailed drongo
Dicrurus paradiseus, family Dicruridae

Eurasian golden oriole
Oriolus oriolus, family Oriolidae

Magpie-lark
Grallina cyanoleuca, family Monarchidae

Saddleback
Philesturnus carunculatus, family Callaeatidae

Tail streamers

This species carols in duet with others of its flock

Strong feet adapted for ground feeding

Australian magpie
Gymnorhina tibicen, family Cracticidae

FAMILY ORIOLIDAE

The orioles in this family are found in Old World forests and woodlands. Sturdy birds with often yellow-plumaged males, they tend to spend most of their lives high in the treetops.

An unusual bird *The figbird differs from other orioles in several ways. It is gregarious, noisy, and has bare skin around the eye. Its diet consists almost entirely of fruit, primarily figs.*

Genera 2
Species 29

Australia & New Guinea to temperate Eurasia, sub-Saharan Africa

Family Artamidae Wood-swallows have strong, pointed wings, blue-gray bills with black tips, and short legs. They forage in all types of habitats, hawking on the wing. Sociable birds, they perch side by side and preen each other.

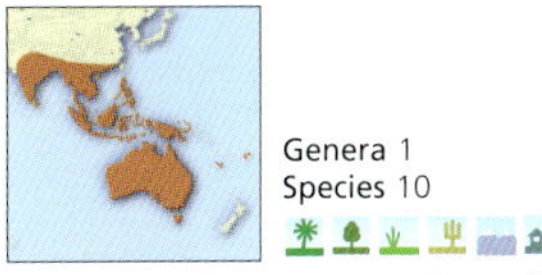

Genera 1
Species 10

Australia & Melanesia to S.E. Asia & India

SONG

Bird songs have inspired musicians, poets, and other artists through the ages. The syrinx, the resonating chamber in which bird sounds are produced, is a complex structure that is best developed in the group of passerines known as oscines. Different species can be distinguished by their songs, although some birds, such as the village indigobird, are very skilled at mimicking other birds. Songs are generally partly or entirely learned, and different dialects often exist among neighboring populations. In most species, it is the males that sing, to attract females or to warn other males to stay away from an established territory. Studies have shown that males can distinguish the songs of their neighbors from the songs of other males. Sometimes birds sing together in duets.

Volume matters *Male zebra finches (left) sing far more loudly when other birds are present, in order to prompt females (below) to respond to their advances and to ward off other males.*

Well-researched singers Zebra finches (*Taeniopygia guttata*), widespread in the dry regions of Australia, are among the best-studied songbirds. To make their songs heard far afield, birds may sing from the highest perches available. This is also one reason that many species prefer to sing in the morning, when the clear air helps carry the sound further. Birds can produce sounds with higher frequencies than the human voice.

Repertoires matter *Most male songbirds can sing two or more different songs. Female zebra finches have been shown to consistently prefer males with more complex songs.*

Stimulating sounds *Songs help coordinate the reproductive cycles of mates. They can prompt females to ovulate, build nests, and lay eggs.*

Common green magpie
Cissa chinensis,
family Corvidae

Red-billed blue-magpie
Urocissa erythrorhyncha,
family Corvidae

This species is split in its distribution between southwestern Europe and eastern Asia

Eurasian nutcracker
Nucifraga caryocatactes,
family Corvidae

Azure-winged magpie
Cyanopica cyanus,
family Corvidae

Characteristic long tail

Clark's nutcracker
Nucifraga columbiana,
family Corvidae

Red-billed chough
Pyrrhocorax pyrrhocorax,
family Corvidae

Alpine chough
Pyrrhocorax graculus,
family Corvidae

Distinctively massive bill

Rufous treepie
Dendrocitta vagabunda,
family Corvidae

White-necked raven
Corvus albicollis,
family Corvidae

FAMILY CORVIDAE

These medium to large birds have bristle-covered nostrils and fairly long legs. Coloring varies from the raven's deep black to the brilliant reds and greens of the Asian magpies. They may eat berries, insects or seeds, or scavenge meat from carcasses.

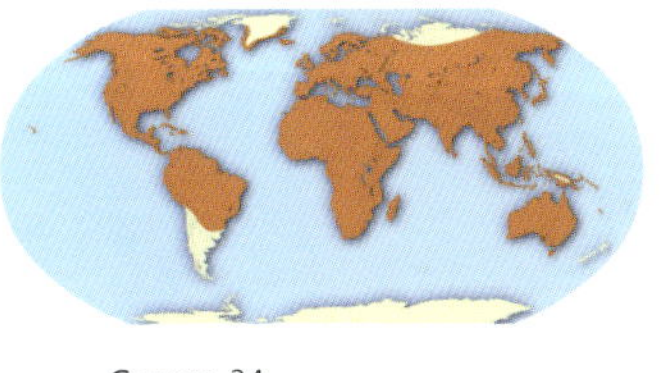

Genera 24
Species 117

Worldwide except S. South America & polar regions

Nest robber *The North American scrub jay is notorious for taking the eggs or young of many songbirds.*

FOOD STORAGE HABITS

Many corvids hide food in caches, to which they return long afterward. Their ability to recall the exact location of the caches has intrigued scientists, who believe they rely on landmarks and possess a well-evolved spatial memory.

Western parotia
Parotia sefilata,
family Paradisaeidae

Wire-like head plumes are twirled in display

Arfak astrapia
Astrapia nigra,
family Paradisaeidae

Velvet sheen gorget

Magnificent riflebird
Ptiloris magnificus,
family Paradisaeidae

Wallace's standardwing
Semioptera wallacii,
family Paradisaeidae

White plumes are raised in display

Great bowerbird
Chlamydera nuchalis,
family Ptilonorhynchidae

Twelve-wired bird-of-paradise
Seleucidis melanoleucus,
family Paradisaeidae

Satin bowerbird
Ptilonorhynchus violaceus,
family Ptilonorhynchidae

Always decorates his bower with blue or violet-colored objects

Golden bowerbird
Prionodura newtoniana,
family Ptilonorhynchidae

FAMILY PTILONORHYNCHIDAE

Most bowerbirds live in forested areas, although others prefer the more open woodlands of the Australian bush. Most species can mimic bird calls or other sounds, and several species are called catbirds because of their cat-like calls. They eat mostly fruit and other plant matter, although their nestlings are sometimes fed insects or even other birds' nestlings.

Bower on display *Male bowerbirds build elaborate bowers, which they decorate with various small objects.*

Family Paradisaeidae The birds of paradise are stout- or long-billed and strong-footed birds similar to crows in size and appearance. But they are distinguished by their gaudy plumage, which the males show off singly or in groups to attract females.

Genera 16
Species 40

New Guinea & satellite islands, E. Australia, N. Moluccas

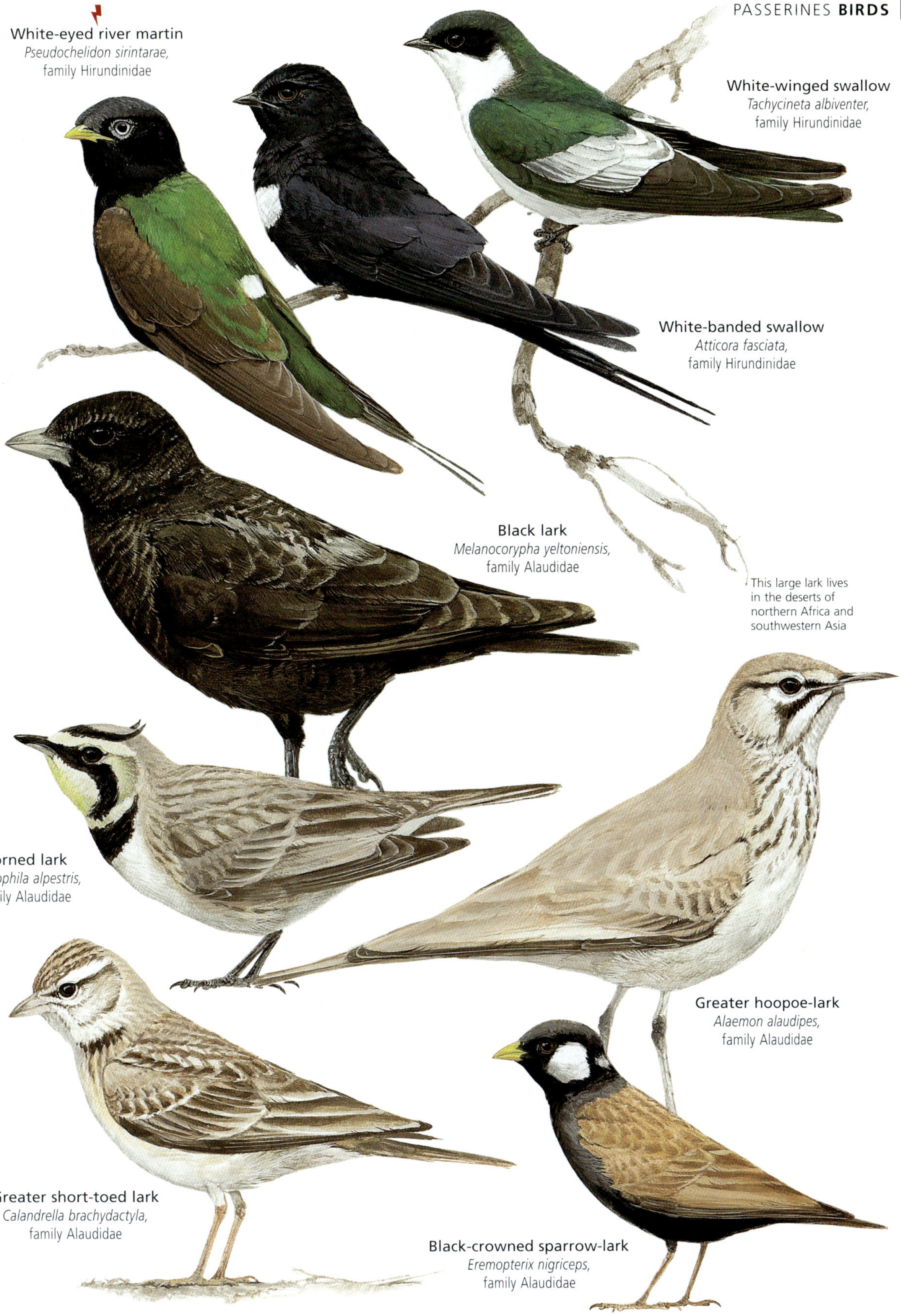

White-eyed river martin
Pseudochelidon sirintarae,
family Hirundinidae

White-winged swallow
Tachycineta albiventer,
family Hirundinidae

White-banded swallow
Atticora fasciata,
family Hirundinidae

Black lark
Melanocorypha yeltoniensis,
family Alaudidae

This large lark lives in the deserts of northern Africa and southwestern Asia

Horned lark
Eremophila alpestris,
family Alaudidae

Greater hoopoe-lark
Alaemon alaudipes,
family Alaudidae

Greater short-toed lark
Calandrella brachydactyla,
family Alaudidae

Black-crowned sparrow-lark
Eremopterix nigriceps,
family Alaudidae

Family Hirundinidae Typical swallows are often recognized by their outer tail-quills, which extend into long, narrow filaments. However, some species have squared tails. Many are strong migrants and are usually silent except near the nest. Like swallows, martins feed on insects taken during flight.

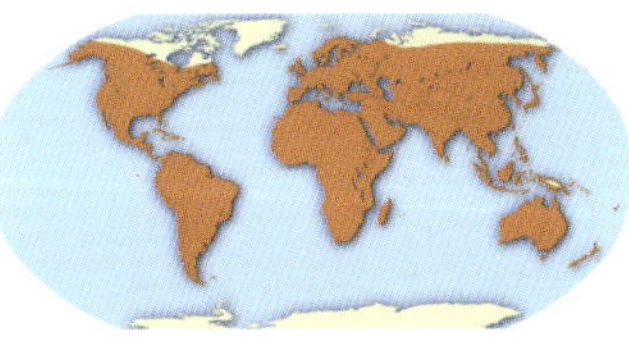

Genera 20
Species 84

Cosmopolitan except polar regions

COURTSHIP

The courtship of the male horned lark (*Eremophila alpestris*) involves some complicated acrobatics.

Scarlet minivet
Pericrocotus flammeus,
family Campephagidae

Bar-bellied cuckoo-shrike
Coracina striata,
family Campephagidae

Black-faced cuckoo-shrike
Coracina novaehollandiae,
family Campephagidae

Ranges nomadically through the woodlands of Australia

Varied triller
Lalage leucomela,
family Campephagidae

Ashy minivet
Pericrocotus divaricatus,
family Campephagidae

Madagascan wagtail
Motacilla flaviventris,
family Motacillidae

Rosy pipit
Anthus roseatus,
family Motacillidae

Wagtails are so called because of their habit of waving their tails up and down

Red-throated pipit
Anthus cervinus,
family Motacillidae

Family Campephagidae
Birds in this family are known as cuckooshrikes because their bustle of feathers on the rump and undulating flight feathers are reminiscent of the cuckoos. These mostly secretive birds forage for insects and fruit in trees, and build tiny, saucer nests high up in the branches.

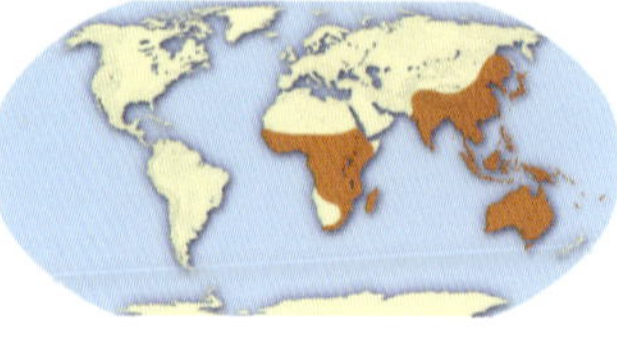

Genera 7
Species 81

Australia & Melanesia to S. & E. Asia, Madagascar, Africa

Family Motacillidae
These small, slender, often long-tailed birds are known as wagtails and pipits. They mostly nest on the ground and feed on insects. The typical wagtails are often found near running water, on riverbanks, and in moist grassland areas. Pipits are very idespread, and prefer open country.

Genera 5
Species 64

Cosmopolitan (only one species in North America)

Family Cinclidae
The only truly aquatic passerines, dippers are rarely seen far from the clear, swiftly flowing streams where they feed on aquatic insect larvae and occasionally small fishes. They usually catch their prey by diving into the water and swimming or walking along the bottom.

Genera 1
Species 5

W. North & N.W. South America, Eurasia

Family Bombycillidae
Waxwings get their name from red, wax-like droplets that form at the tip of each secondary wing feather; these serve no known function. These very gregarious, often fawn-colored birds have sleek silhouettes and vagrant tendencies, appearing suddenly in unlikely places.

Genera 5
Species 8

North America to N.W. South America, Eurasia

Black-throated accentor
Prunella atrogularis,
family Prunellidae

A shrubbery bird of western Europe

White-browed shortwing
Brachypteryx montana,
family Turdidae

European robin
Erithacus rubecula,
family Muscicapidae

Rufous scrub-robin
Cercotrichas galactotes,
family Muscicapidae

White-starred robin
Pogonocichla stellata,
family Muscicapidae

Siberian rubythroat
Luscinia calliope,
family Muscicapidae

Northern scrub-robin
Drymodes superciliaris,
family Petroicidae

This scrub-robin is superficially thrush-like

Bluethroat
Luscinia svecica,
family Muscicapidae

FAMILY PETROICIDAE

Although known as robins, these common Australasian birds are not related to European or American robins, but resemble them in shape, size, and nature. They live in forests and woods.

Red-capped robin
The red-capped robin (Petroica goodenovii) catches insects by pouncing down from perches.

Genera 13
Species 45
Australia, New Guinea, New Zealand, S.W. Pacific

Family Prunellidae These birds, known as accentors, look like slender-billed sparrows. They prefer high altitude habitats, and forage for seeds, insects, or berries. They insulate their ground-level nests with feathers.

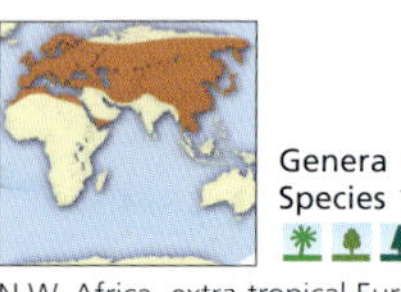

Genera 1
Species 13
N.W. Africa, extra-tropical Eurasia

FAMILY TURDIDAE

Thrushes live in most parts of the world. Familiar members of this family are the Eurasian blackbird and the American robin. They vary in color and habitat preferences. The sexes are usually alike.

Fine singer *The white-rumped shama (Copsychus malabaricus) has a fine voice that carries far and rivals that of the nightingale in its richness.*

Genera 24
Species 165

Cosmopolitan except W. & C. Australia, New Zealand

Family Pomatostomidae These birds are known as Australian babblers. Isolated in Australia and New Guinea, they have evolved convergently with the true babblers, imitating them in appearance and behavior. Most species are gray, white, and ruddy.

Genera 2
Species 5

Australia, lowland New Guinea

Icterine warbler
Hippolais icterina,
family Sylviidae

African yellow warbler
Chloropeta natalensis,
family Sylviidae

Gives clicking calls in display flights

Zitting cisticola
Cisticola juncidis,
family Cisticolidae

Sedge warbler
Acrocephalus schoenobaenus,
family Sylviidae

White-necked picathartes
Picathartes gymnocephalus,
family Picathartidae

Goldcrest
Regulus regulus,
family Regulidae

Gray-headed parrotbi
Paradoxornis gularis,
family Timaliidae

Tropical gnatcatcher
Polioptila plumbea,
family Polioptilidae

TINY SONGSTER

The golden-crowned kinglet (*Regulus satrapa*) inhabits coniferous forests. The tiny, plump warblers in its genus are characterized by a patch of vivid red or yellow on the crown.

Family Picathartidae The bald crows, or rockfowl, are unusual birds that look like long-legged thrushes. Their bare heads are colored either orange-yellow or blue and pink. They inhabit dense forests in West Africa, where they build bowl-like mud nests.

Genera 1
Species 2

Equatorial W. Africa (Guinea to Gabon)

Family Polioptilidae The insect-eating gnatcatchers are very small birds, even when their long tails are taken into account. Notable is the long-billed gnatwren, whose bill is equal to more than one-third of the bird's body length.

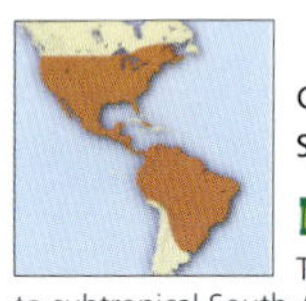

Genera 3
Species 14

Temperate North America to subtropical South America

Some colonies number in the millions

Red-billed quelea
Quelea quelea,
family Ploceidae

Southern red bishop
Euplectes orix,
family Ploceidae

Long-tailed widowbird
Euplectes progne,
family Ploceidae

Spot-winged starling
Saroglossa spiloptera,
family Sturnidae

Amethyst starling
Cinnyricinclus leucogaster,
family Sturnidae

Metallic starling
Aplonis metallica,
family Sturnidae

Red-billed buffalo weaver
Bubalornis niger,
family Ploceidae

Builds globular hanging nests in trees in large colonies

This weaver is a member of the Old World sparrow family

Gray-headed social weaver
Pseudonigrita arnaudi,
family Passeridae

Family Ploceidae The weaver birds are notable for their nests. The males of some species build their own nests by looping and knotting blades of grass in a process akin to basket-weaving. Other species build large communal nests of thorns or sticks.

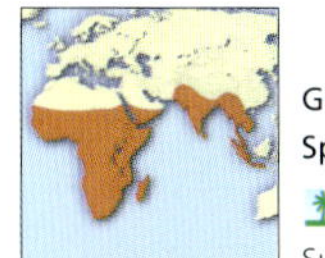

Genera 11
Species 108

Sub-Saharan Africa, S. & S.E. Asia, Gr. Sundas (except Borneo)

FAMILY STURNIDAE

The starling family includes the common starling, one of the most familiar of garden birds. This stumpy, relatively short-tailed bird has a confident strut and a pugnacious air. Most species live in rain forests and eat fruit.

Safari bird *The superb starling* (Lamprotornis superbus) *is often seen by safari-goers in East Africa. It commonly visits campsites. Its plumage makes it one of the more beautiful of the starlings.*

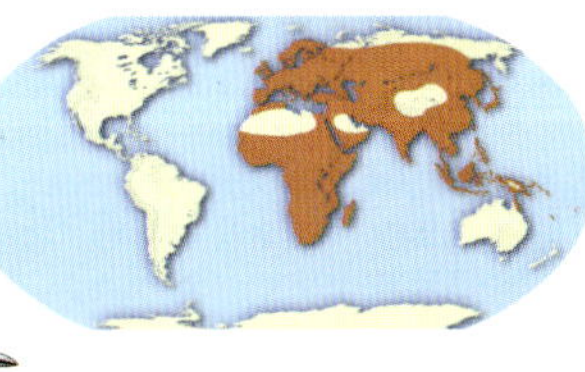

Genera 25
Species 115

Africa, temperate & tropical Eurasia to N.E. Australia & Oceania

European goldfinch
Carduelis carduelis,
family Fringillidae

Introduced to Australia and New Zealand

Common redpoll
Carduelis flammea,
family Fringillidae

Zebra finch
Taeniopygia guttata,
family Estrildidae

Widespread in dry regions of Australia

Pin-tailed whydah
Vidua macroura,
family Viduidae

Eastern paradise whydah
Vidua paradisaea,
family Viduidae

Straw-tailed whydah
Vidua fischeri,
family Viduidae

Gouldian finch
Erythrura gouldiae,
family Estrildidae

Blue-faced parrotfinch
Erythrura trichroa,
family Estrildidae

American goldfinch
Carduelis tristis,
family Fringillidae

Black-capped vireo
Vireo atricapilla,
family Vireonidae

FAMILY FRINGILLIDAE

Almost all finches within this family have red or yellow in their plumage. They are migratory birds that inhabit temperate climates. They have only nine large primary feathers in each wing; the tenth is vestigial.

Specialized bill
The parrot crossbill (Loxia pytyopsittacus) uses its large, cross-tipped bill to extract the nuts from hard pine cones.

Genera 42
Species 168

The Americas, Africa, Eurasia to S.E. Asia

Family Viduidae Whydahs get their name from the Portuguese word for a widow's veil. Each species is a brood parasite of a particular waxbill species, and their young even have the same begging calls as the foster species' own young.

Genera 2
Species 20

Sub-saharan Africa (except rainforested Congo & Namibian desert)

Scarlet tanager
Piranga olivacea,
family Thraupidae

A common, gregarious forest bird

Red-legged honeycreeper
Cyanerpes cyaneus,
family Thraupidae

Green honeycreeper
Chlorophanes spiza,
family Thraupidae

Giant conebill
Oreomanes fraseri,
family Thraupidae

Golden-hooded tanager
Tangara larvata,
family Thraupidae

Paradise tanager
Tangara chilensis,
family Thraupidae

Swallow-tanager
Tersina viridis,
family Thraupidae

Bananaquit
Coereba flaveola,
family Coerebidae

Their brilliant color makes tanagers attractive to the live bird trade

Family Coerebidae The bananaquit is the only member of this family. It lives in the tropics of South America, north to Mexico and the Caribbean. This very small bird has a slender, curved bill adapted for taking nectar. It is often seen in gardens.

Genera 1
Species 1
S. Mexico to W.C. South America

FAMILY THRAUPIDAE

The Andes are the center of radiation for the tanagers. A few species of tanager are drab and secretive, but the family also includes some of the most colorful of birds. The honeycreepers have long, thin, delicate bills for extracting nectar. The insect specialists have heavy, notched bills. Most tanagers build cup-shaped nests among moss or dead leaves. Females carry out most of the nesting tasks.

Genera 50
Species 202
North America to South America

Superb plumage
The seven-colored tanager (Tangara fastuosa) lives in a small sector of Brazil. Its numbers have declined greatly due to trapping for the pet trade and habitat loss.

Golden-winged warbler
Vermivora chrysoptera,
family Parulidae

Breeds in eastern United States and migrates as far as northern South America for winter

Northern parula
Parula americana,
family Parulidae

MacGillivray's warbler
Oporornis tolmiei,
family Parulidae

Unlike its relatives, may nest in areas without wooded vegetation

Most New World warblers build simple cup-shaped nests. Only the female builds and incubates

Canada warbler
Wilsonia canadensis,
family Parulidae

Common yellowthroat
Geothlypis trichas,
family Parulidae

Yellow-rumped warbler
Dendroica coronata,
family Parulidae

Elegant patterns of red and yellow on head and breast characterize wood warblers in breeding plumage

Collared redstart
Myioborus torquatus,
family Parulidae

Nests in cavities in trees

Prothonotary warbler
Protonotaria citrea,
family Parulidae

SEXUAL PLUMAGE

Most New World warblers are brightly colored and patterned. Among the North American species, the breeding males are often more brightly colored than females, but both sexes usually molt into more drab plumage before their autumn migrations.

Variety *The range of sexual plumage dimorphism in this group is exemplified by the similarity of the sexes in the orange-crowned warbler* (Vermivora celata)*; the slight overlap in the Magnolia warbler* (Dendroica magnolia)*; and the huge differences in the American redstart* (Setophaga ruticilla).

Family Parulidae These warblers are active foliage gleaners. They sometimes form multi-species "guilds" comprising species that each forage in a different way. Most build nests on trees, shrubs, or vines.

Genera 24
Species 112

North America to South America, West Indies

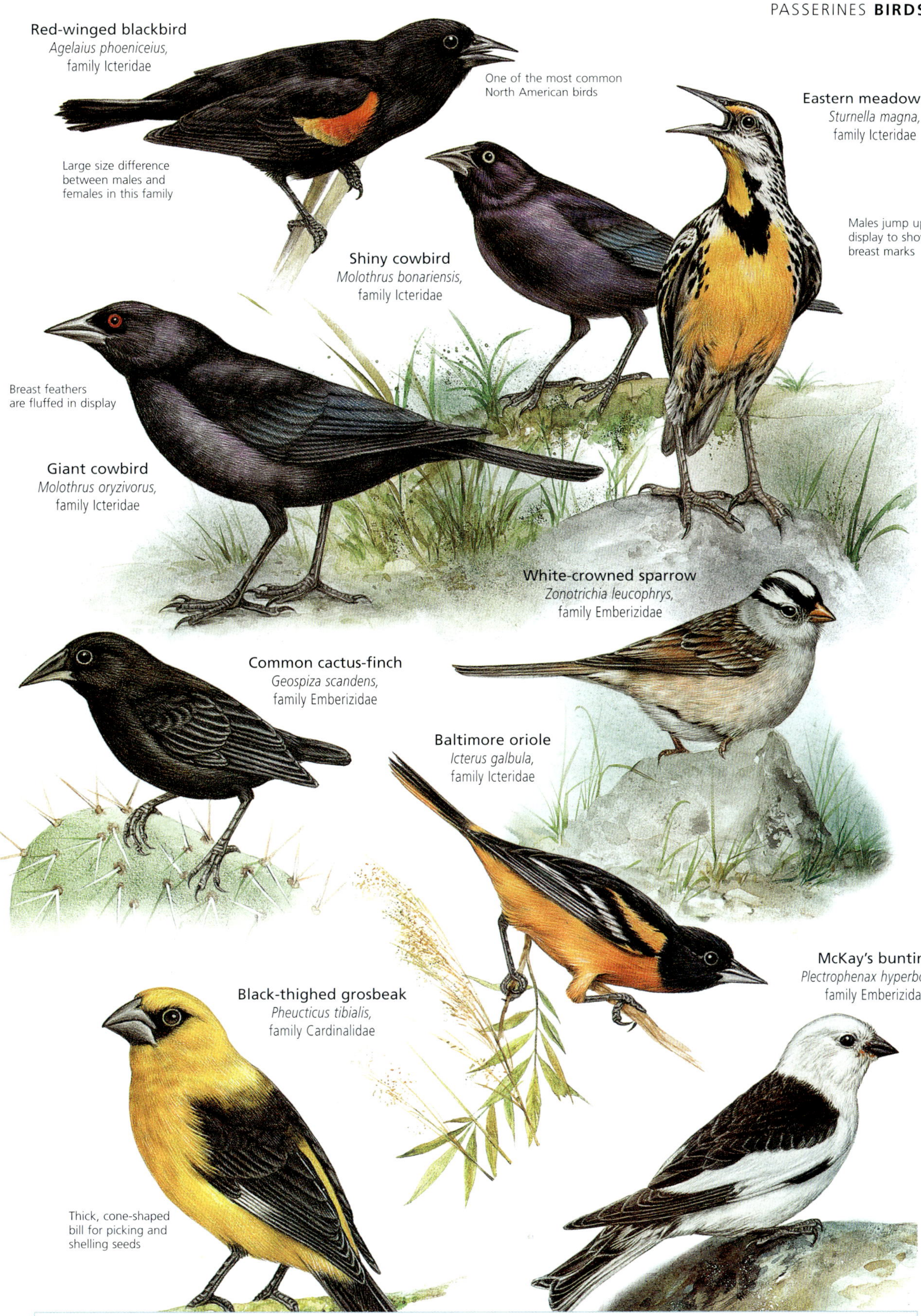

Red-winged blackbird
Agelaius phoeniceius,
family Icteridae

One of the most common North American birds

Large size difference between males and females in this family

Eastern meadowlark
Sturnella magna,
family Icteridae

Males jump up in display to show breast marks

Shiny cowbird
Molothrus bonariensis,
family Icteridae

Breast feathers are fluffed in display

Giant cowbird
Molothrus oryzivorus,
family Icteridae

White-crowned sparrow
Zonotrichia leucophrys,
family Emberizidae

Common cactus-finch
Geospiza scandens,
family Emberizidae

Baltimore oriole
Icterus galbula,
family Icteridae

McKay's bunting
Plectrophenax hyperboreus,
family Emberizidae

Black-thighed grosbeak
Pheucticus tibialis,
family Cardinalidae

Thick, cone-shaped bill for picking and shelling seeds

FAMILY EMBERIZIDAE

Included in this family are New World sparrows and finches and Old World buntings. These small birds all have short, conical bills adapted to eating seeds, but they tend to feed their young with insects. They are primarily terrestrial and most have brown-streaked plumage. The snow bunting has the distinction of breeding further north (in northern Greenland) than any other land bird. These birds tend to have small song repertoires.

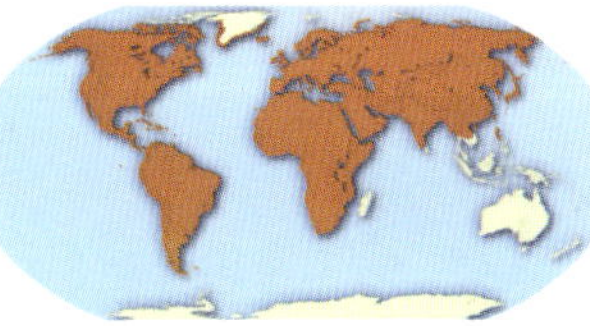

Genera 73
Species 308

Cosmopolitan except Madagascar, Indonesia & Australasia

A standout
Cardinals, such as the painted bunting (Passerina ciris), tend to be much more brightly colored than other buntings. The male in particular has notably multi-hued plumage.

EVOLUTION OF BILLS

The size and shape of birds' bills vary, adapted to their diets and food-gathering methods. It was Charles Darwin who first offered convincing proof that such physical features are not immutable, and that species change over time in response to their environments. He theorized that Galápagos Islands finches, which differed most noticeably in terms of bill size and shape, were all descended from a common ancestor and had evolved to exploit different ecological niches that were not being filled by other animals. Genetic mutations that helped certain birds thrive were likely to be passed on to successive generations. This adaptive radiation is most noticeable on remote islands, where outside influences and competitor species are almost absent.

A rainbow of differences Hawaiian honeycreepers have evolved different kinds of bills depending on their eating habits. The seed eaters have strong, stout bills while the nectar specialists have long, slender bills.

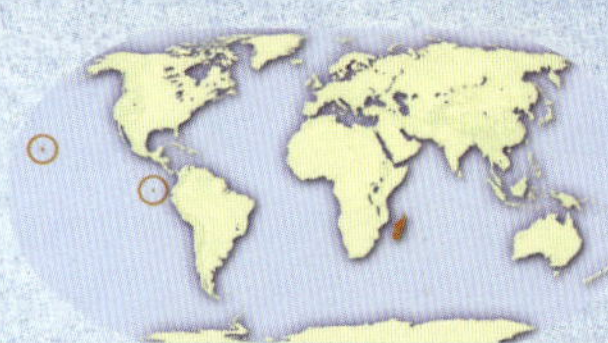

Isolated habitats Left to right: Hawaii; Galápagos Islands; Madagascar

G1

Blood-sucker *One finch (G5) is also known as the "vampire finch" due to its habit of pecking seabirds in order to drink their blood.*

Special skills *One Darwin finch species (G1) has learned to extract insects from under bark or cracks in wood using a cactus spine or twig. The largest of the ground finches (G2) uses its sturdy bill to eat large seeds.*

G3

G2

Diversity on a large island The vangas are a diverse family (called Vangidae) of passerines found only on the island of Madagascar. The 22 species all prey on invertebrates or small reptiles, but they have evolved to fill many niches on a long-isolated island whose very different habitats span a latitudinal range equal to that of California. The vangas pictured are: M1: Sickle-billed vanga (*Falculea palliata*); M2: Pollen's vanga (*Xenopirostris polleni*); M3: Helmet vanga (*Euryceros prevostii*); M4: Red-shouldered vanga (*Calicalicus rufocarpalis*); M5: Blue vanga (*Cyanolanius madagascarinus*).

A variety of roles
The vangas have specialized so much that there are almost as many genera as species in this group. The long bill of M1 is ideal for feeding in spiny forests. M2 probes in dead wood; M4 gleans insects from bushes and in flight.

Similarities to other birds
Individual vanga species have filled niches similar to those of birds elsewhere and have thus grown to resemble them. The M3, for example, recalls a small hornbill.

Versatile bird *One of the ground finches (G6) evolved a longer bill, which allows it to feed on cactus flowers and fruits as well as seeds. Diet specialization is especially important in lean times.*

Spawning a scientific revolution Modern genetic analyses have proven Darwin right, showing that the 14 species of Galápagos-based birds now known as Darwin's finches are indeed descended from one species, a seed-eating finch that arrived there from the South American mainland. The Galápagos birds pictured are: G1: woodpecker finch (*Camarhynchus pallidus*); G2: large ground finch (*Geospiza magnirostris*); G3: warbler finch (*Certhidea olivacea*); G4: masked booby (*Sula dactylatra*); G5: sharp-beaked ground finch (*Geospiza difficilis*); G6: common cactus-finch (*Geospiza scandens*).

REPTILES

PHYLUM	Chordata
CLASS	Reptilia
ORDERS	4
FAMILIES	60
GENERA	1,012
SPECIES	8,163

Reptiles are often thought of as relics of an age gone by, leftovers from the age of the dinosaurs. But in fact reptiles are continuously evolving. Modern species are the product of over 300 million years of evolution since the basal amniotes split into two lineages: one leading to mammals (Synapsida) and the other to birds and reptiles (Diapsida). The diapsid reptiles diverged into two main groups, the Lepidosaura (scaly reptiles—lizards, snakes, amphisbaenas, and tuataras) and the Archosauria (ruling reptiles—crocodiles, pterosaurs, dino-saurs, and ancestral birds) in the Triassic. Turtles appeared in the fossil record about 210 million years ago. Even though current skeletal shapes look similar to species living 200 million years ago, today's reptiles are highly specialized.

Swimming turtles Using their front limbs for propulsion and their back limbs for steering, sea turtles can swim at a speed of up to 18 miles per hour (29 km/h).

THE CLEDOIC EGG

The closed system of the egg—where water, nutrients, and waste are stored until hatching—allowed reptiles to be independent of water and invade land. All eggs need to respire; they have tiny pores that allow oxygen to enter. Some eggs are very permeable, and if they are not buried in a moist place will dehydrate. Other species have developed very thick-shelled eggs to avoid dehydration. Eggs are sometimes retained in many snakes and lizards, and development progresses in the female. Variations on this theme led to viviparity evolving in lizards and snakes, whereby large, yolked, shell-less eggs were maintained in the oviduct until hatching. Where there was nutrient exchange between the female and her embryos this evolved into the formation of a placenta.

Internal fertilization was another mechanism that freed reptiles from an aquatic life. Lizards and snakes have two functional hemipenes. Tuataras abut their cloacas, and turtles and crocodilians have a penis for internal sperm transmission.

Development of impermeable skin allowed reptiles the chance to live permanently out of water without dehydrating. The epidermis radiated in diverse directions—the skin beads on gila monsters, the crests on iguanas, the keeled scales in snakes, the rattle on rattlesnakes, the scutes on turtle shells, and the armored plates on crocodiles. Reptile skin varies in its permeability to water and gas exchange.

On land, temperatures fluctuate more rapidly than in the water, so reptiles had to develop behavioral mechanisms to control their core body temperatures so that enzyme processes would function. Since, unlike mammals, most reptiles do not burn calories to maintain body temperatures, they use behavioral means, sunning themselves in the early morning hours, remaining in the shade or underground in the heat of the day, or becoming nocturnal in hot environments. Many behavioral, structural, and physiological mechanisms evolved to allow reptiles to be more efficient in using energy than mammals.

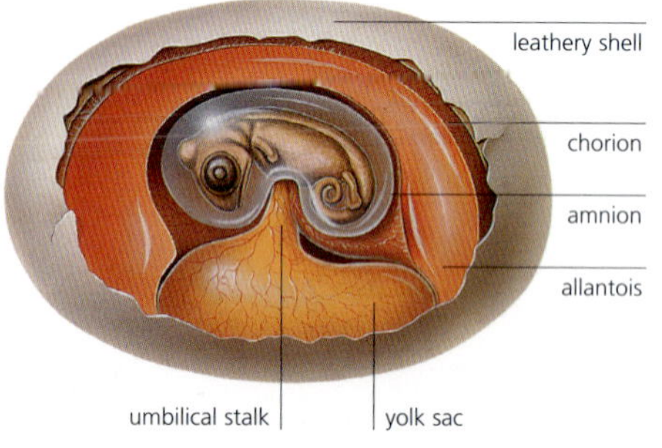

Reptile egg The embryo absorbs oxygen through the blood vessels lying adjacent to the pores of the shell. Waste products are stored in the allantois. The amnion maintains the water balance in the egg, and acts as a shock absorber. The embryo grows from the energy supplied in the yolk sac, the oxygen that filters through the shell, and the water absorbed through the leathery shell.

Parental care Some species of crocodiles guard their nests and carry their young to the water after they hatch. Some stay with their pod for over a year.

Mobile predators Snakes have evolved many ways of moving around, to compensate for not having limbs. They have also established themselves on all the world's major landmasses (with the exception of Antarctica) as well as on many islands.

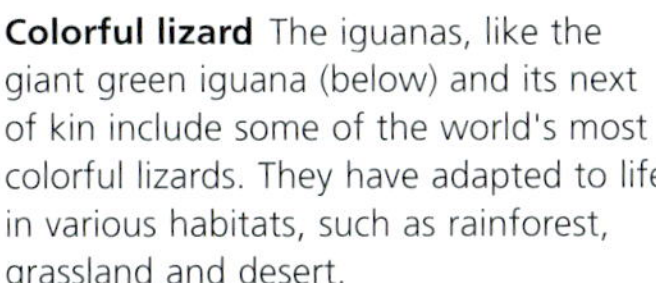

Colorful lizard The iguanas, like the giant green iguana (below) and its next of kin include some of the world's most colorful lizards. They have adapted to life in various habitats, such as rainforest, grassland and desert.

Tortoises and turtles

CLASS Reptilia	
ORDER Testudines	
FAMILIES 14	
GENERA 99	
SPECIES 293	

Turtles and tortoises are the only vertebrates that house the pelvic and pectoral girdles within a shell made of ribs fused with bone. The first fossils to show this trait appeared 220 million years ago in the Triassic period, when dinosaurs roamed the earth. The shell design has since undergone various modifications. The African pancake tortoise has a flexible flat shell allowing it to wedge into crevices; it then inflates, after which it is impossible to pull out. Softshell turtles lack a hard shell; instead, the bones are covered with a smooth leathery shell, giving them more speed. There is still controversy as to whether turtles are derived from a different lineage to other reptiles.

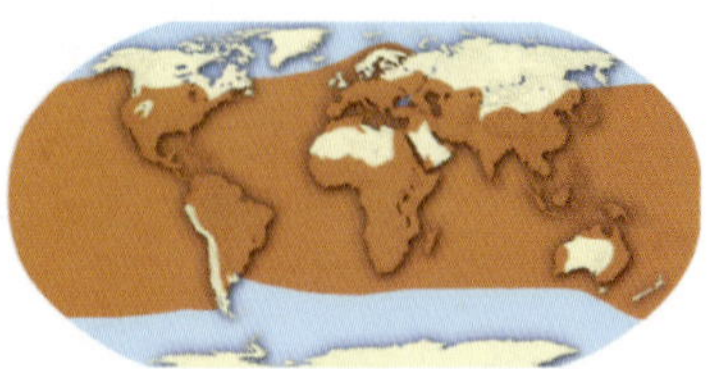

Distribution Turtles are found on all continents except Antarctica, and in all oceans: 241 species of turtles have adapted to freshwater rivers, lakes, and ponds; 45 tortoises are terrestrial; only seven species live in the oceans.

EVOLUTION SHELLED OUT

In addition to the shell, turtles and tortoises have evolved many unique features. Turtle muscles are able to withstand high levels of lactic acid, so they do not tire after swimming rapidly. They also fill their large urinary bladders with water for ballast to maintain buoyancy.

Turtles and tortoises occupy niches as predators, grazers, croppers, and scavengers. Some have gut flora to digest plant cells; others are seed dispersers. Wood turtles stomp on the ground with their front feet, then eat the earthworms that come out of the ground.

Some sea turtles and freshwater turtles nest within a few days in tremendous numbers on the same nesting beach. These mass nestings, known as arribadas, evolved to swamp local predators—it would be impossible for them to eat all of the eggs or kill all of the females. The success of these adaptations has helped turtles to survive over millions of years; now, as a result of human intervention, two-thirds of all turtle species are being carefully monitored by the IUCN.

Swimming sea turtle The forelimbs of sea turtles have been modified into flipper-shaped paddles that move synchronously, resembling aquatic flying rather than swimming. Sea turtles are so adept in the water that they rarely leave it except to nest. Freshwater turtles swim using all four limbs in turn.

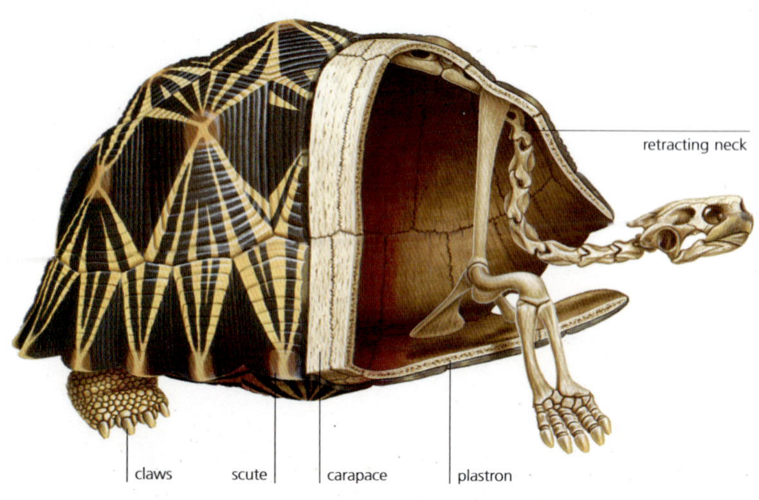

Turtle skeleton A turtle's shell has an outer layer of epidermal scutes (usually 38 on the carapace and 16 on the plastron). Under these, a shell of fused ribs (the plastron) supports the body. The vertebrae are joined to the inside of the carapace.

Matamata
Chelus fimbriatus

Twist-necked turtle
Platemys platycephala

Matamata characterized by its long snorkel nose, which allows it to lie in shallow water and reach the surface to respire without frightening its prey

Common snake-necked turtle
Chelodina longivollis

Twist-necked turtle is the only species of turtle known to have triploid chromosomes in some populations

Helmeted turtle
Pelomedusa subrufa

Giant Amazon river turtle
Podocnemis expansa

Chelidae Fitzroy turtle
Rheodytes leukops

Serrated turtle
Pelusios sinuatus

Hilaire's toadhead turtle
Phrynops hilarii

Victoria short-necked turtle
Emydura victoriae

Giant Amazon river turtle Selection for fast incubation was influenced by the unpredictability of rising river levels. The eggs hatch in 45 days, and nest temperatures often reach 104°F (40°C), the highest known for any turtle.

- Up to 42 in (107 cm)
- Aquatic
- TSD
- 60–150
- Conserv. dependent

Amazon & Orinoco basins (N. South America)

BEGINNING OF LIFE

Sea turtles lay their eggs on the same beaches that their ancestors have used for millennia. Once the nest has been covered with sand, the eggs are left to chance to go undetected by predators. Hatchlings must race to the sea, running the risk of predation by hungry mammals and birds, to the dubious safety of waters filled with even more predators. Survival rates for some species may be as low as 1 in 50,000.

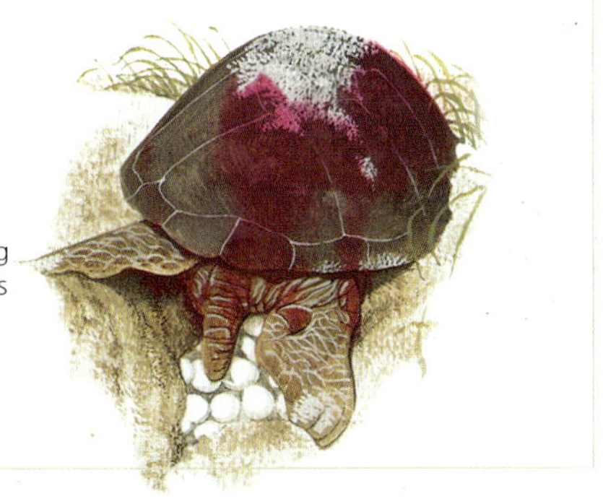

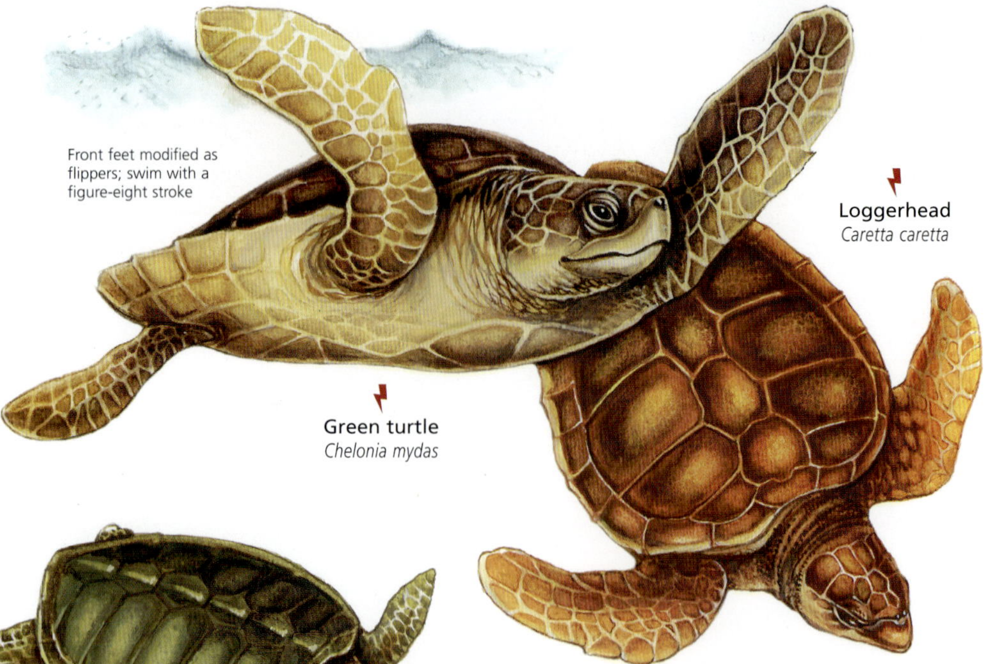

Front feet modified as
flippers; swim with a
figure-eight stroke

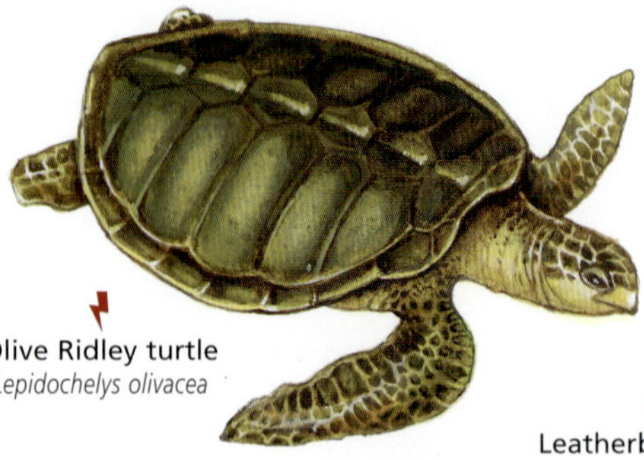

Green turtle
Chelonia mydas

Loggerhead
Caretta caretta

Olive Ridley turtle
Lepidochelys olivacea

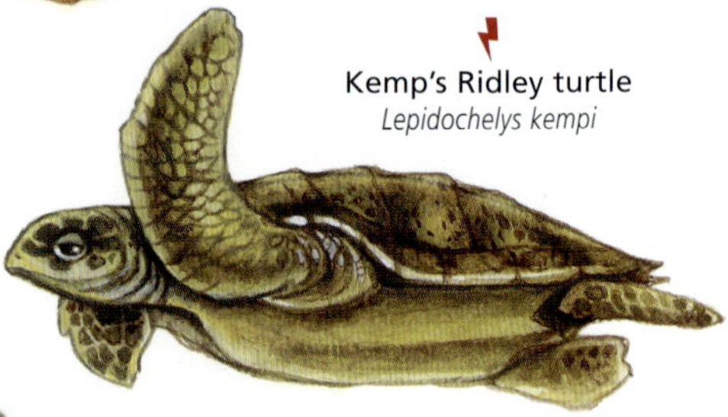

Kemp's Ridley turtle
Lepidochelys kempi

Leatherback turtle
Dermochelys coriacea

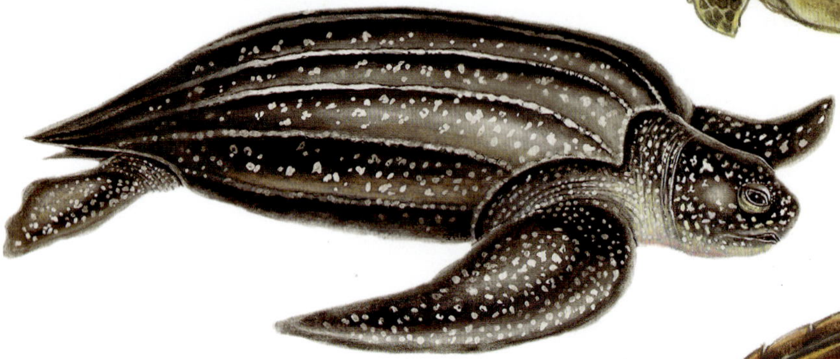

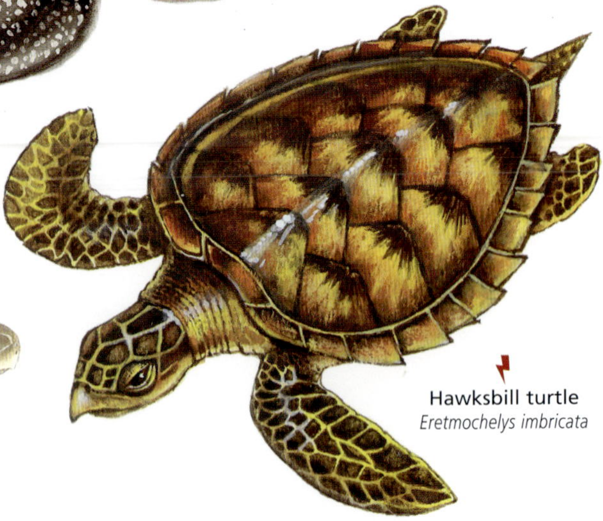

Carapace scutes used for making combs and hair ornaments

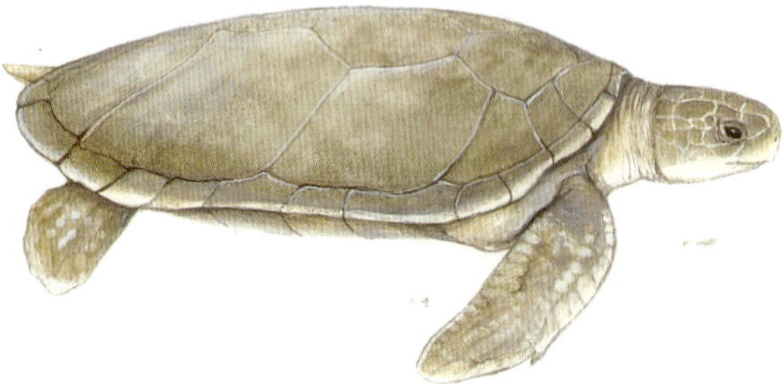

Hawksbill turtle
Eretmochelys imbricata

Flatback turtle
Natador depressus

SHELL SHAPES

Tortoises are high-domed to protect them from predators and store water. Pond turtles are more streamlined so that they can swim with less resistance. Semi-terrestrial turtles have higher domes for predator protection. Sea turtles are designed for least resistance to glide through the water.

Land tortoise

Semi-terrestrial

Sea turtle

Pond turtle

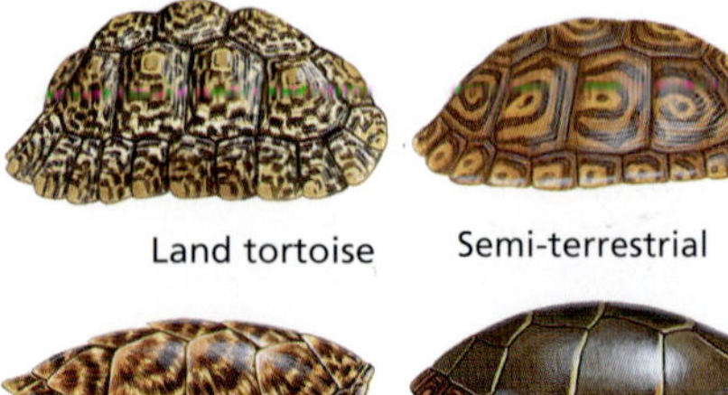

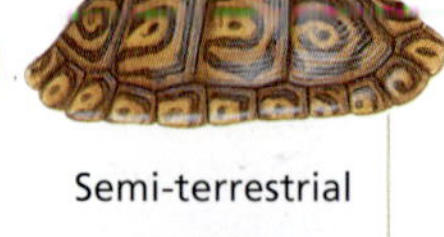

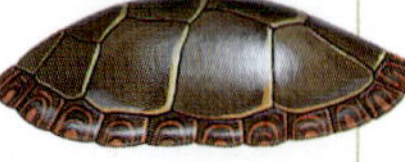

Green turtle Green turtles migrate huge distances across the open seas between their feeding grounds and nesting beaches.

- Up to 5 ft (1.5 m)
- TSD
- Endangered
- Aquatic
- 50–240

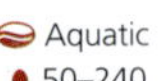

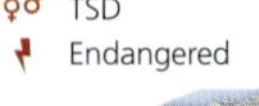

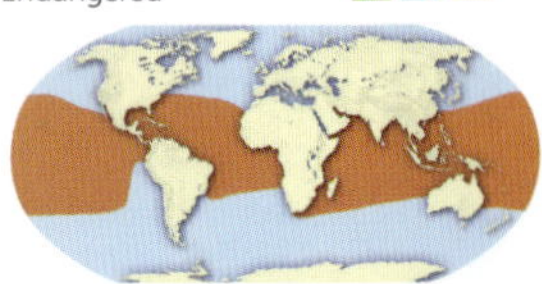

Tropical oceans worldwide & Mediterranean Sea

Ganges softshell turtle
Aspideretes gangenticus

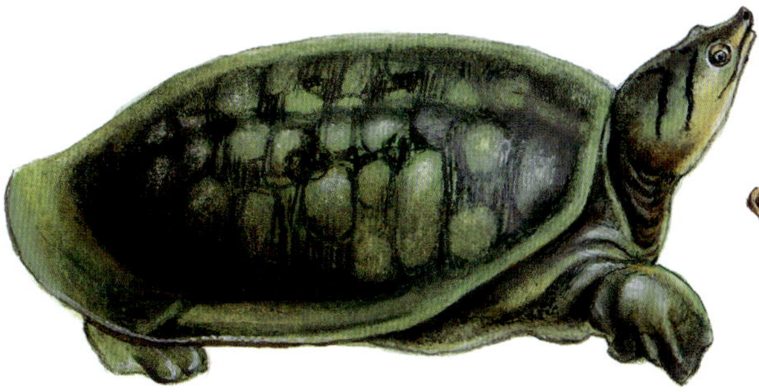

Indian flap-shelled turtle
Lissemys punctata

Spiny softshell turtle
Apalone spinifera

Australian pig-nosed turtle
Carettochelys insculpta

Nile softshell turtle
Trionyx triunguis

Nile softshell absorbs much of the oxygen it needs by pharyngeal respiration and by filtration through the skin

Smooth softshell turtle
Apalone mutica

Ganges softshell turtle An important scavenger, this species helps to lower pollution levels in the Ganges River by consuming partially cremated human corpses that are thrown into the river in traditional funeral rites.

Up to 28 in (71 cm)
Aquatic
Unknown
25–35
Vulnerable

N. India, N.W. Pakistan, Bangladesh & Nepal

HEAT AND SEX

Sex in some species of turtles, as in most vertebrates, is genetic (referred to as GSD, or genetic sex determination). In most turtles studied, some lizards, all crocodiles, and the tuatara, sex is controlled by the incubation temperature (referred to as TSD, or temperature sex determination).

Wood turtle hatching
Wood turtles are one of the few North American turtles in the family Emydidae to have GSD; most others have TSD.

Diamond-back terrapin
Malaclemys terrapin

Ringed sawback
Graptemys oculifera

Wood turtle
Clemmys insculpta

Eastern box turtle
Terrapene carolina

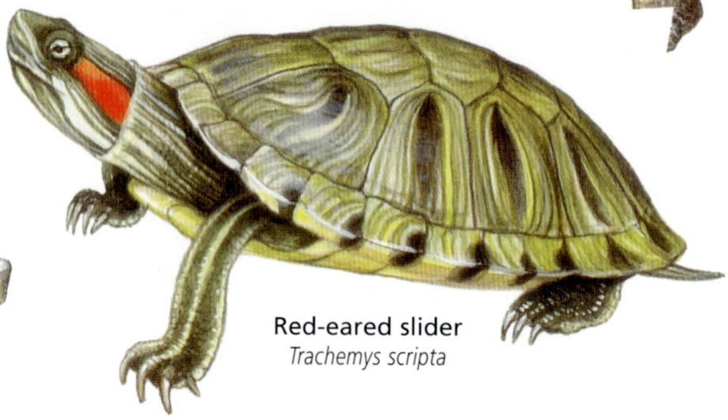

Cagle's map turtle
Graptemys caglei

Red-eared slider
Trachemys scripta

Red-bellied turtle
Pseudemys rubiventris

European pond turtle
Emys orbicularis

Scorpion mud turtle
Kinosternon scorpioides

Muhlenberg's turtle
Clemmys muhlenbergi

Painted turtle
Chrysemys picta

River cooter
Pseudemys concinna

Ringed sawback This species thrives in fast-moving water, where it forages for aquatic insect larvae. It is endangered due to pollution of the Pearl River. The pet trade and target practice have also reduced its numbers.

Up to 8½ in (21 cm)
Aquatic
TSD
4–8
Endangered

Pearl River, Mississippi (USA)

European pond turtle Only males hatch at incubation temperatures of 75–82°F (24–28°C), while at 86°F (30°C), 96 percent of the hatchlings are females. At higher temperatures, only females are produced.

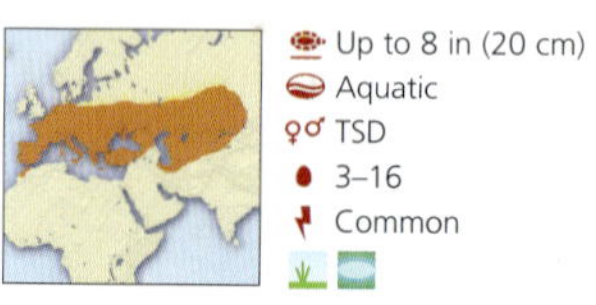

Up to 8 in (20 cm)
Aquatic
TSD
3–16
Common

S. Europe & W. Asia

River cooter A complete herbivore, this species is often seen basking by the dozens on floating logs in order to raise its body temperature to speed digestion. Basking is also important for ridding the shell and limbs of fungal and algal colonies.

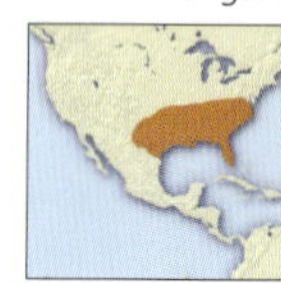

Up to 17 in (43 cm)
Aquatic
TSD
6–28
Common

S.E. USA

Furrowed wood turtle
Rhinoclemmys areolata

Indian roofed turtle
Kachuga tecta

Malayan box turtle
Cuora amboinensis

Caspian turtle
Mauremys caspica

Spotted pond turtle
Geoclemys hamiltoni

Malayan snail-eating turtle
Malayemys subtrijuga

One of the few freshwater turtles to nest on ocean beaches. The hatchlings can live for at least 2 weeks in sea water

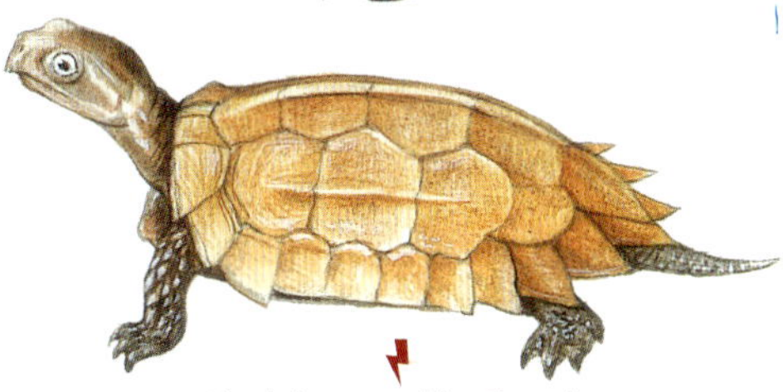

Black-breasted leaf turtle
Geoemyda spengleri

River terrapin
Batagur basca

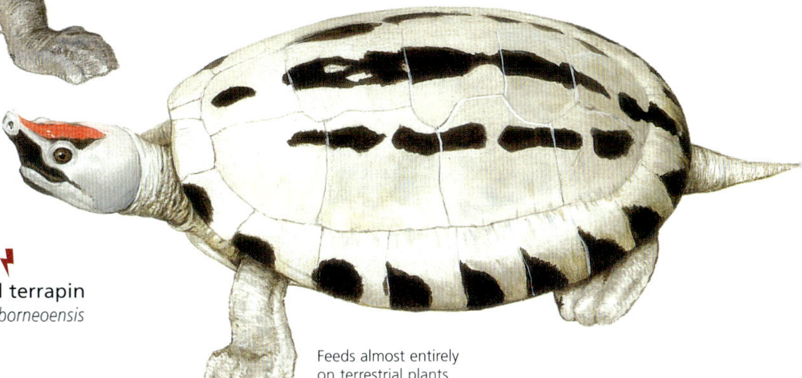

Painted terrapin
Callagur borneoensis

Feeds almost entirely on terrestrial plants

LEG ADAPTATIONS

From the shape of the leg, foot, and claws, it is possible to deduce what habitat the turtle was adapted for, but not necessarily where it is living today.

Sea turtle *The forelimbs of sea turtles are aerodynamically designed to fly through water. They are tapered from tip to base like the wings of an albatross. There is no webbing, as the toes are fused together within the flippers.*

Land tortoise *This animal is designed for walking with a load on land, not for swimming. The legs have armored plates, the unwebbed toes are elephantine, and the feet are flat to support the tortoise's weight on land.*

Pond turtle *The legs are only slightly paddle-shaped to navigate through aquatic vegetation and to walk on land. The toes are webbed, with long, claw-like toenails for traction when crawling up logs to bask.*

Dark blotches on carapace

Bell's hinge-back tortoise
Kinixis belliana

Leopard tortoise
Geochelone pardalis

Marginated tortoise
Testudo marginata

Radiated tortoise
Geochelone radiata

Dark lines on carapace

Speckled cape tortoise
Homopus signatus

Elongated tortoise
Indotestudo elongata

Bright red scales on legs

South American red-footed tortoise
Geochelone carbonaria

African tent tortoise
Psammobates tentorius

Texas tortoise
Gopherus polyphemus

African pancake tortoise
Malacochersus tornieri

Leopard tortoise During courtship, the male trails the female, butting her into submission. After mounting, he extends his neck and releases a gruntlike bellow. Females lay 5–7 clutches of 5–30 eggs.

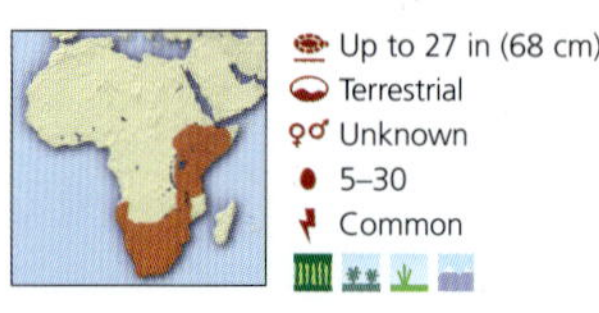

Up to 27 in (68 cm)
Terrestrial
Unknown
5–30
Common

S. Sudan & Ethiopia to Natal & South Africa

African tent tortoise Nesting takes place from September to December, when up to three ellipsoidal eggs are laid in a single yearly clutch. The eggs hatch between April and May.

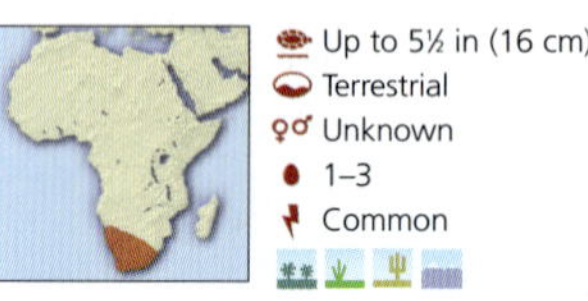

Up to 5½ in (16 cm)
Terrestrial
Unknown
1–3
Common

S.W. Africa to Cape (South Africa)

Texas tortoise This species feeds on the pads, flowers, and fruit of cactus. In the Chihuahuan Desert, they are active early in the morning, and rest in the shade or in burrows during the heat of the day.

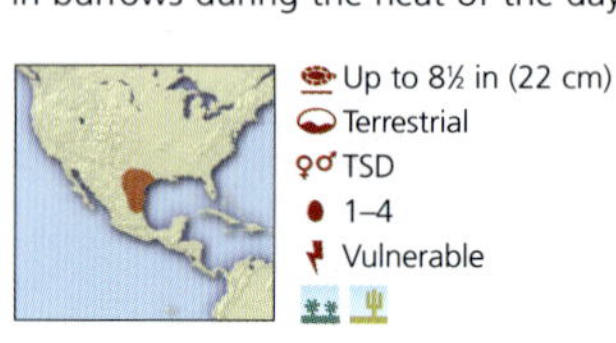

Up to 8½ in (22 cm)
Terrestrial
TSD
1–4
Vulnerable

S. Texas (USA) to N. Mexico

Razorback musk turtle
Kinosternon carinatus

Mexican giant musk turtle
Staurotypus triporcatus

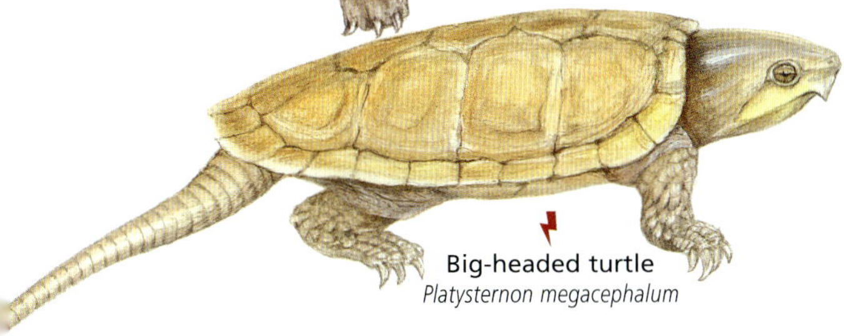

Big-headed turtle
Platysternon megacephalum

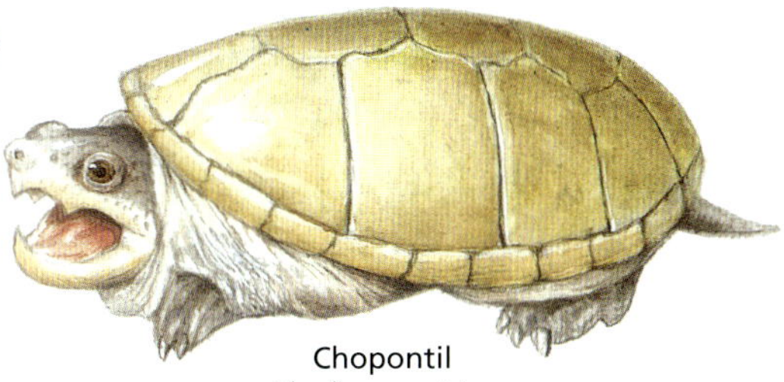

Chopontil
Claudius angustatus

Loggerhead musk turtle
Kinosternon minor

Common snapping turtle
Chelydra serpentina

Common mud turtle
Kinosternon subrubrum

Alligator snapping turtle
Macrochelys temmincki

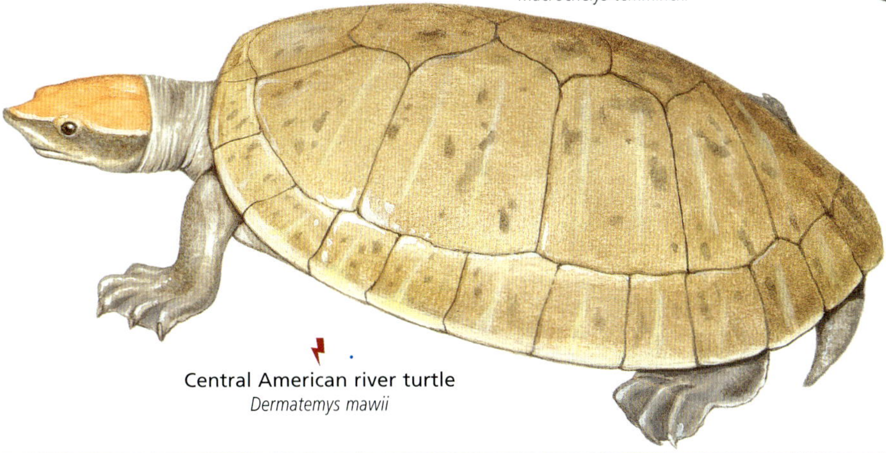

Central American river turtle
Dermatemys mawii

Common snapping turtle This species is known for its ability to lurch forward rapidly with snapping jaws, an adaptation for crayfish predation and for protection. These turtles are omnivores.

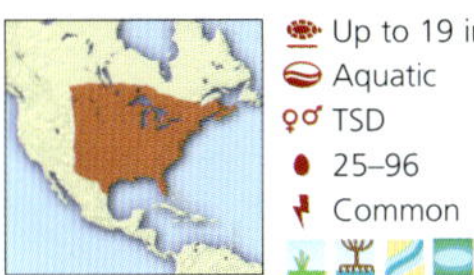

Up to 19 in (48 cm)
Aquatic
TSD
25–96
Common

S. Canada, E. & C. USA

Alligator snapping turtle The largest species of turtle in North America was being driven to extinction by the soup industry. Until recently, this rare beast could be found on American grocery store shelves.

Up to 31½ in (80 cm)
Aquatic
TSD
8–52
Vulnerable

S.E. USA

Central American river turtle This turtle seldom leaves the water even to nest. They often nest by digging into the river bank under the water level. The eggs do not begin to develop until the river is lower.

Up to 26 in (66 cm)
Aquatic
TSD
8–26
Endangered

S. Mexico, Guatemala, Belize

DEATH ROW

During the last decade, the decimation of turtle populations has accelerated rapidly. The Turtle Survival Alliance (TSA) has been formed to monitor and attempt to reverse this trend. TSA shortlisted 25 of the most critically endangered species to raise awareness of the problem. Most of these species occur in hotspots for high biodiversity—areas that are important for the critical habitat of a number of other taxa as well. The turtles are in trouble because of overexploitation for food or traditional remedies, the pet trade, restricted range, and habitat degradation. Only five of these are suffering due to habitat degradation—15 of the 25 are on the brink of extinction because of uncontrolled exploitation by humans.

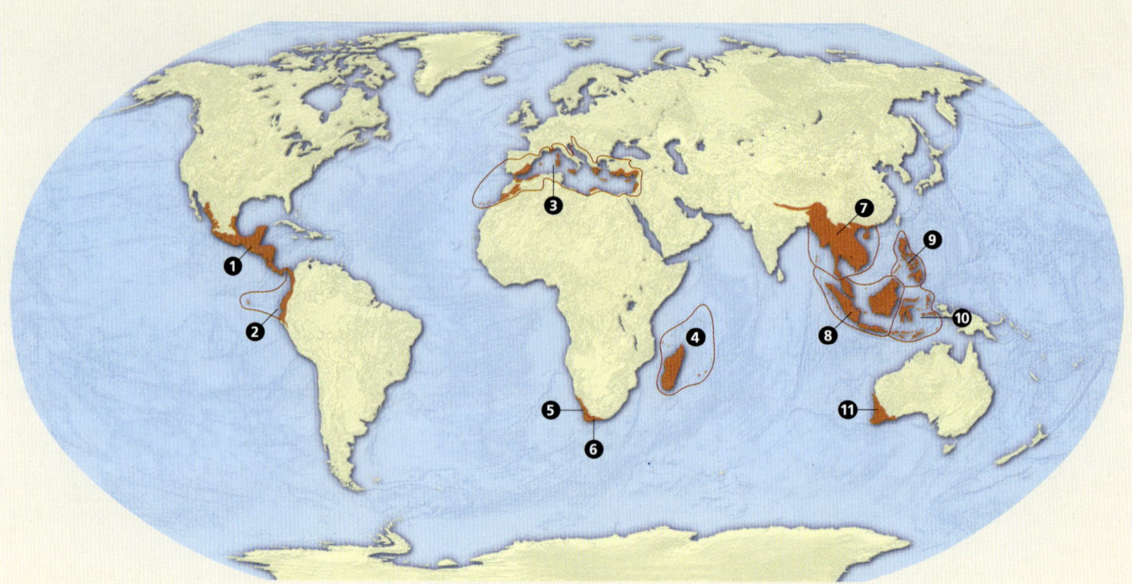

THE WORLD'S MOST ENDANGERED TURTLES

1. MESOAMERICA HOTSPOT
Central American river turtle
Dermatemys mawii

2. CHOCÓ-DARIÉN–WESTERN ECUADOR HOTSPOT
Dahl's toad-headed tortoise
Batrachemys dahli

3. MEDITERRANEAN BASIN HOTSPOT
Egyptian tortoise
Testudo kleinmanni

4. MADAGASCAR AND INDIAN OCEAN ISLANDS HOTSPOT
Madagascar big-headed turtle
Erymnochelys madagascariensis

Ploughshare tortoise
Geochelone yniphora

Flat-tailed tortoise
Pyxis planicauda

5. SUCCULENT KAROO HOTSPOT
Southern speckled padloper tortoise
Homopus signatus cafer

6. CAPE FLORISTIC REGION HOTSPOT
Geometric tortoise
Psammobates geometricus

7. INDO-BURMA HOTSPOT
Striped narrow-headed softshell turtle

Chinese three-striped box turtle
Cuora trifasciata

Arakan forest turtle
Heosemys depressa

Burmese star tortoise
Geochelone platynota

Burmese roofed turtle
Kachuga trivittata

Leaf turtle
Mauremys annamensis

Yangtze giant softshell turtle
Rafetus swinhoei

8. SUNDALAND HOTSPOT
River terrapin
Batagur baska

Painted terrapin
Callagur borneoensis

9. PHILIPPINES HOTSPOT
Philippine forest turtle
Heosemys leytensis

10. WALLACEA HOTSPOT
Roti snake-necked turtle
Chelodina mccordi

Sulawesi forest turtle
Leucocephalon yuwonoi

11. SOUTHWESTERN AUSTRALIA HOTSPOT
Western swamp turtle
Pseudemydura umbrina

Crocodilians

CLASS	Reptilia
ORDER	Crocodilia
FAMILIES	3
GENERA	8
SPECIES	23

Alligators, caimans, crocodiles, and gharials are all crocodilians, belonging to the lineage of archosaurs, which includes the dinosaurs and birds. Crocodilians are much more closely related to birds than they are to other reptiles, and have existed for 220 million years, since the Triassic period. Part of their success is a result of being the top aquatic predator in their domain. As the body form of crocodilians has stayed the same, they are often called living fossils, but these beasts have been evolving for millions of years and are very different from their ancestors in the age of the dinosaurs. Unlike most reptiles, crocodilians are very vocal, especially during courtship.

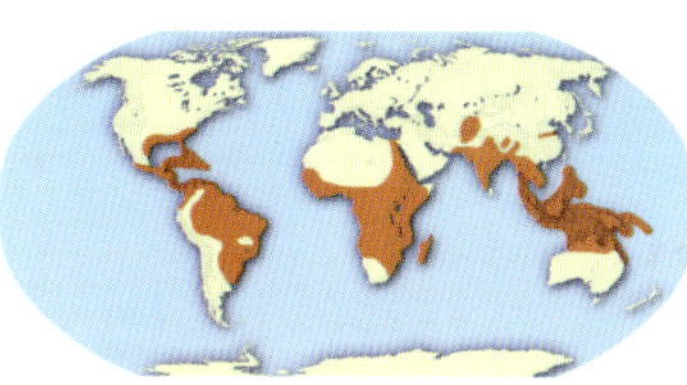

Distribution Crocodilians are found in tropical, subtropical, and temperate zones worldwide: Gavialidae in South Asia, and Alligatoridae in eastern North America, Central and South America, and eastern China. Crocodylidae inhabit estuaries and freshwater streams of Africa, India, Indonesia, Australia, northern South America, Central America, and the West Indies.

A fine figure of a crocodile The Nile crocodile grows to 19 feet (6 m) and can live for 40 years or more. Young Nile crocodiles eat insects, spiders, and frogs. Adults eat anything from monkeys and antelopes to zebras and humans. On land, many mammals could outrun a crocodile, so a preferred killing method is to drag animals underwater to drown them.

CANNIBALISM AND CARE

Crocodilians have an elongated, cylindrical body with short, muscular limbs and a laterally compressed tail. The massive skull is set on a short neck and features strongly toothed jaws. Crocodilians are aquatic, yet they bask and nest on shorelines.

Crocodilians are oviparous, with internal fertilization. Clutches usually contain 12–48 eggs; all species studied have temperature-controlled sex determination, in which females are produced at high and low incubation temperatures and males only at a narrow range of intermediate temperatures.

The eggs are laid in nests made of mounds of vegetation or in nests dug in the sand. Nest site selection can control the sex of the young. Both males and females of some species guard the nest. Females of some species are known to respond to the grunts of the nearly hatched embryos by opening the nests for them. The female will also carry the hatchlings in her mouth to the water, where she has bulldozed a nursery pond for them.

Females will protect their young for the first 2 months. However, males are also known to protect and maintain their territories and food supplies by killing and eating any young male crocodilian in their territory, so the female is often protecting her young from their own father. Female American alligators often stay with their young near the nesting site for 1–2 years.

American alligator
Alligator mississippiensis

Black caiman
Melanosuchus niger

Cuvier's dwarf caiman
Paleosuchus palpebrosus

African dwarf crocodil
Osteolaemus tetraspis

Chinese alligator
Alligator sinensis

False gharial
Tomistoma schlegeli

Mugger
Crocodylus palustri

Saltwater crocodile
Crocodylus porosus

Orinoco crocodile
Crocodylus intermedius

American alligator Populations of this species were low in the 1950s and protected as endangered in 1967. Populations rebounded in 20 years to over 800,000 animals.

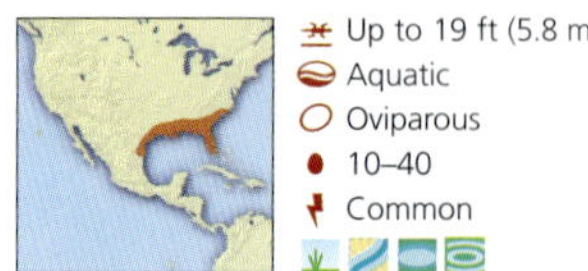

- Up to 19 ft (5.8 m)
- Aquatic
- Oviparous
- 10–40
- Common

S.E. USA

Black caiman This species has recently been reclassified from endangered in Brazil. Conservation efforts over more than 10 years were highly successful, with populations recuperating rapidly.

- Up to 20 ft (6 m)
- Aquatic
- Oviparous
- 35–50
- Conserv. dependent

Amazon Basin (N. South America)

Chinese alligator This species spends the majority of its life in a complex of underground burrows. The alligators build these systems with pools of water above and below ground as well as air holes.

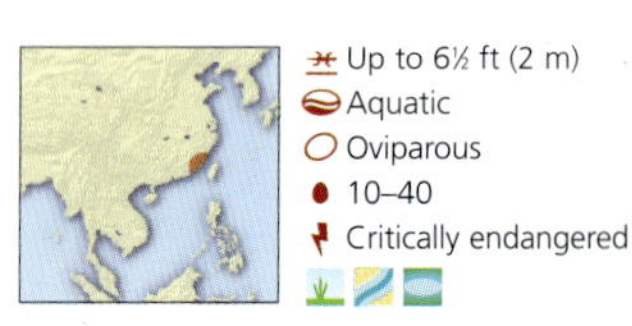

- Up to 6½ ft (2 m)
- Aquatic
- Oviparous
- 10–40
- Critically endangered

Yangtze Valley (China)

Spectacled caiman
Caiman crocodilus

Black tail bands are characteristic of the caiman

Siamese crocodile
Crocodylus siamensis

Five clawed digits are present on the forelimb

American crocodile
Crocodylus acutus

Gharial
Gavialis gangeticus

Only the male has a distinct boss on the tip of the snout

Nile crocodile
Crocodylus niloticus

American crocodile This species was once abundant and widespread but was extirpated in many regions due to the skin trade and habitat destruction. Populations have not recuperated in many areas.

- Up to 23 ft (7 m)
- Aquatic
- Oviparous
- 30–40
- Vulnerable

S. Florida (USA), Mexico to Colombia, Ecuador

Nile crocodile Maturity is reached in 12–15 years at 6–9 feet (1.8–2.8 m). The female guards the nest, opens it at hatching, and carries the hatchlings to the water. Both parents guard the young.

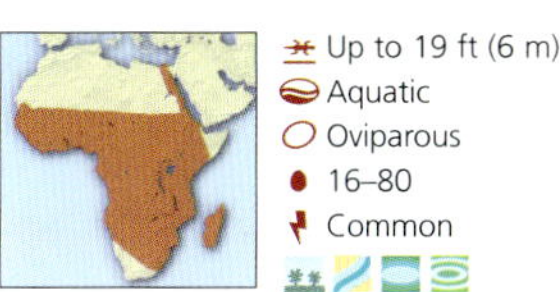

- Up to 19 ft (6 m)
- Aquatic
- Oviparous
- 16–80
- Common

Sub-Saharan Africa

ANATOMY

Crocodilians have various adaptations to an aquatic life. The position of the eyes, ear openings, and nostrils at the highest part of the body allows them to lie concealed just below the surface of the water when stalking prey. A secondary palate lets them breathe with the mouth closed. A flap of skin in the throat prevents water from entering the throat when the jaws are being opened to capture prey.

Tuatara

CLASS	Reptilia
ORDER	Rhynchocephalia
FAMILY	Sphenodontidae
GENUS	Sphenodon
SPECIES	2

The tuatara is often called a "living fossil" and is now only to be found on the islands of New Zealand. It is the only survivor of a large group of reptiles that roamed with the dinosaurs over 225 million years ago, the rest of which became extinct 60 million years ago. Its tooth arrangement is unique: a single row in the lower jaw fits between two rows of teeth in the upper jaw. Lizards have visible ear openings but tuatara do not.

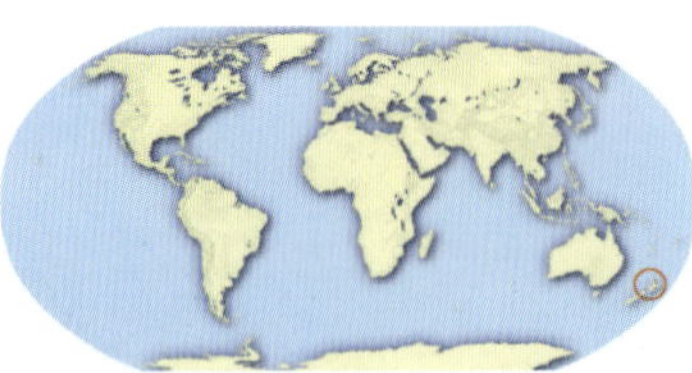

Distribution About 400 *Sphenodon guntheri* live on North Brother Island, New Zealand. More than 60,000 *S. punctatus* live on some 30 islands off the northeast coast of New Zealand's North Island.

Tuatara
Sphenodon punctatus

Amphisbaenians

CLASS	Reptilia
ORDER	Squamata
SUBORDER	Amphisbaenia
FAMILIES	4
GENERA	21
SPECIES	140

These legless squamates have reduced pectoral and pelvic girdles. They have an annulated pattern of scutes, and short tails. Amphisbaenians are built for burrowing, with heavily ossified skulls. Their brain is surrounded by the frontal bones. The right lung of amphisbaenians is reduced in size, while in other limbless lizards and snakes the left lung is smaller. Three of the four families of worm lizards have no limbs whatsoever, while the remaining family has enlarged forelimbs, which help it with locomotion and digging.

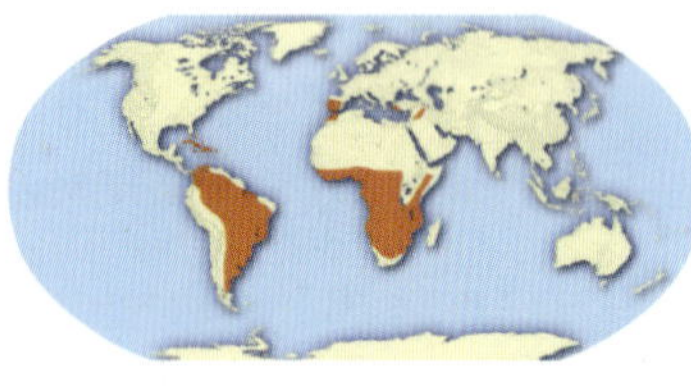

Distribution Amphisbaenians are found in tropical and subtropical regions of southern North America, South America to Patagonia, West Indies, Africa, the Iberian Penisula, Arabia, and western Asia.

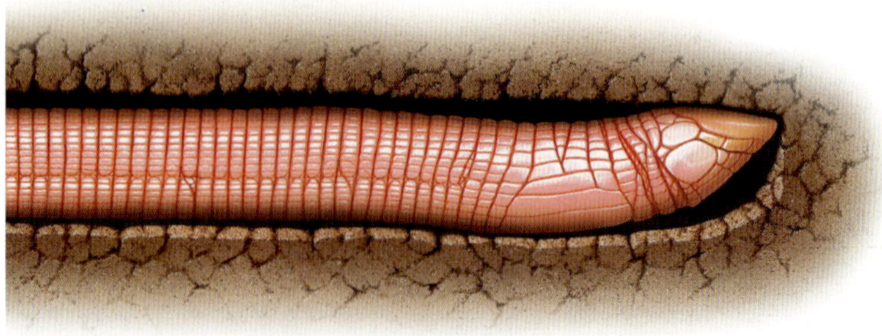

Shovel-snouted worm lizard The illustration below shows how the worm lizard pushes its head against the ceiling of the tunnel to widen it. Worm lizards have large, interlocking upper and lower teeth, which allow them to grasp prey and drag it into their tunnel.

LIZARDS

CLASS	Reptilia
ORDER	Squamata
SUBORDER	Sauria
FAMILIES	27
GENERA	442
SPECIES	4,560

Today, lizards occupy almost all landmasses except for Antarctica and some Arctic regions. At the end of the Cretaceous period, some 65 million years ago, lizards survived when the dinosaurs and other large reptiles died out. With more than 4,000 species, they are the largest group of living reptiles. Although the largest lizard, the Komodo dragon, reaches an impressive length of 10 feet (3 m), few lizards exceed 12 inches (30 cm) and this is one reason for their continued success. Lizards tend to be restricted to specific habitat niches, as mountains and water are significant barriers to their movement.

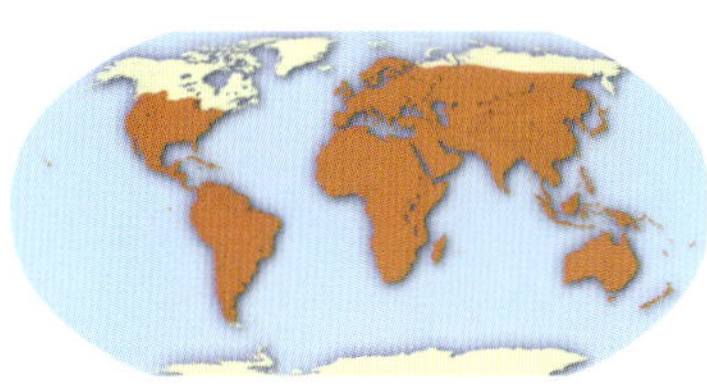

Distribution Lizards are found from New Zealand to Norway, and from southern Canada to Tierra del Fuego. They are also endemic to many islands in the world's oceans. The only continent they have not colonized is Antarctica.

Territorial defense A defined territory provides an area in which an adult male lizard can hunt for food and find a female for mating. If a rival invades its territory, a collared lizard does a series of "push-ups" to make itself look bigger and more threatening to the unwelcome interloper.

Predatory dragon The Indonesian Komodo dragon (below) tests the air for the scent of warm-blooded prey. These lizards grow to be more than 10 feet (3 m) in total length and feed on large mammals such as deer, pigs, goats, and even water buffalo.

DEFENSE AND ESCAPE

Lizards are preyed upon by spiders, scorpions, other lizards, snakes, birds, and mammals. The Gila monster and the Mexican beaded lizard are the only two venomous species, but even these will resort to scare tactics at the start of a confrontation. Many others have developed an impressive array of tactics to defend themselves or to escape an attacker.

Most lizards are extremely well camouflaged and may keep absolutely still until a predator passes by. Chameleons, in particular, are well known for their ability to change color to blend in with their environment. Other lizards surprise or distract a predator to give themselves a chance to escape. The Australian frilled lizard, for example, opens its mouth, hisses loudly, and flourishes its neck frill before scampering away.

Some species have sharp spines that can injure a predator's mouth, or slippery scales that make them hard to grip. The armadillo girdle-tailed lizard curls itself into a ball and protects itself with a prickly fence of spikes, while the basilisk escapes by skimming on water before diving in to safety.

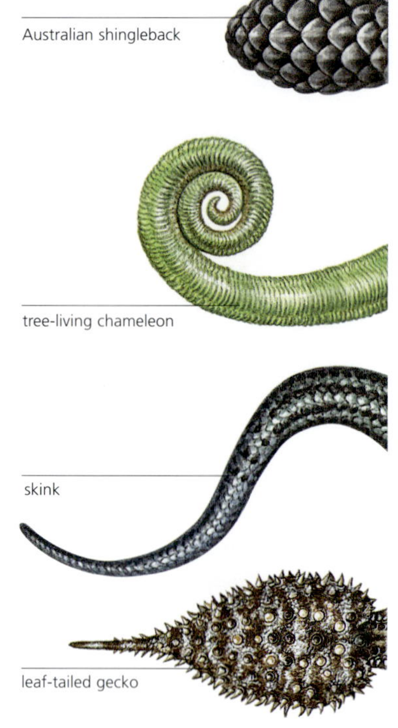

Cactus eaters Galápagos land iguanas are vegetarians, subsisting mostly on the fruit and pads of *Opuntia* cactus. These iguanas live in dry areas, and in the mornings they bask in the hot equatorial sun.

Lizard tails Some lizards have tails that mimic leaves; others mimic heads. Some are used to hold onto branches, and many are expendable. If a predator grabs a skink's tail, the tail stays in the predator's mouth and the lizard escapes.

Marine iguana The only seagoing lizard, the marine iguana develops its colors with age. Young iguanas are black, while adults can be shades of green, red, gray, or black, depending on their island home. Males favor sunny, rocky shores during the day, where sea breezes keep them cool.

Lizards use their tails for defense. Monitors and iguanas beat their attackers with their tails. Skinks and other small lizards may give up a little piece of tail in return for their lives. These lizards often have brightly colored tails that they will wave from side to side, inducing the predator to attack the tail and not the head. The tail continues to wriggle after the lizard has escaped. Shed tails grow back after time; the lizard loses some stored energy but remains alive to reproduce. In territorial bouts, some geckos attack and eat the tails of their opponents.

Horned lizards, when attacked by foxes and coyotes, will squirt a stream of bad tasting blood out of their eyes to distract and discourage potential predators. Some lizards also attempt to scare off predators by sticking out their tongues. Australian skinks hiss and thrust their brightly colored tongues out to startle predators. Even though the Gila monster is venomous and brightly colored to warn predators of the risk of attacking, if molested it will display its bright purple tongue and hiss at its aggressor.

Fiji banded iguana
Brachylophus fasciatus, family Iguanidae

West Indian iguana
Iguana delicatissima, family Iguanidae

Marine iguana
Amblyrhynchus cristatus, family Iguanidae

Galápagos land iguana
Conolophus subcristatus, family Iguanidae

Rhinoceros iguana
Cyclura cornuta, family Iguanidae

Black iguana
Ctenosaura similis, family Iguanidae

The warmer male black iguana (top) is lighter in color; the other male is colder and darker

Green iguana
Iguana iguana, family Iguanidae

Merrem's Madagascar swift
Oplurus cyclurus, family Opluridae

Collard lizard
Crotaphytus collaris, family Crotaphytidae

Side-blotched lizard
Uta stansburiana, family Phrynosomatidae

Chuckwala
Sauromalus obesus, family Phrynosomatidae

Leopard lizard
Gambelia sila, family Crotaphytidae

Regal horned lizard
Phrynosoma solare, family Phrynosomatidae

Short-horned lizard
Phrynosoma douglass, family Phrynosomatidae

Desert spiny lizard
Sceloporus magister, family Phrynosomatidae

Colorado desert fringe-toed lizard
Uma notata, family Phrynosomatidae

Family Crotaphytidae Collard and leopard lizards are medium-length diurnal lizards active in deserts and other rocky, arid areas. They feed on invertebrates, lizards, and other small vertebrates. They employ squealing vocalizations when stressed, and can use a form of bipedalism when moving among rocks.

S.W. North America

Genera 2
Species 12

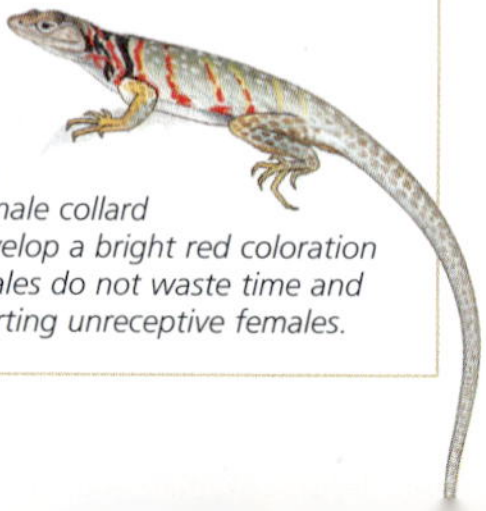

Red alert
Gravid female collard lizards develop a bright red coloration so that males do not waste time and effort courting unreceptive females.

Family Phrynosomatidae The spiny lizards, *Sceloporus*, are the most diverse group in this family, with 70 species. They have the generalized body form of a sit and wait predator. They predominate in desert regions, where there are both terrestrial and arboreal species. Horned lizards are terrestrial with a flattened body.

Genera 9
Species 110

Southern Canada & USA to Panama, Central and South America, Caribbean, USA

Defensive spines
The regal horned lizard has the largest crown of spines of any species.

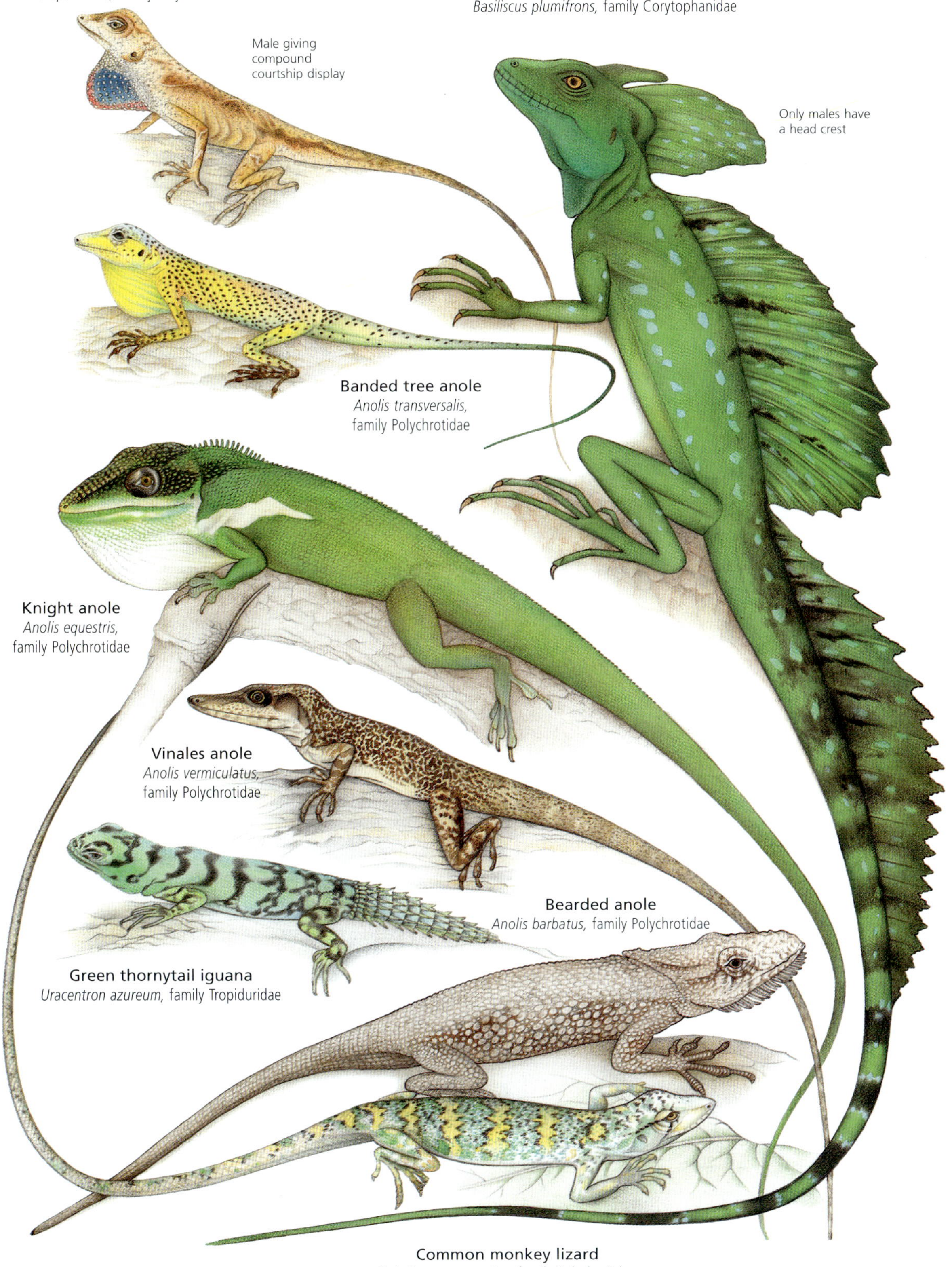
Blue-throated anole
Norops nitens, family Polychrotidae
Male giving compound courtship display
Green basilisk lizard
Basiliscus plumifrons, family Corytophanidae
Only males have a head crest
Banded tree anole
Anolis transversalis, family Polychrotidae
Knight anole
Anolis equestris, family Polychrotidae
Vinales anole
Anolis vermiculatus, family Polychrotidae
Green thornytail iguana
Uracentron azureum, family Tropiduridae
Bearded anole
Anolis barbatus, family Polychrotidae
Common monkey lizard
Polychrus marmoratus, family Polychrotidae

Family Corytophanidae All members of this family are carnivorous. Crested lizards are arboreal, living in rain forests or tropical scrub forests. Up to 8 inches (20 cm) in snout–vent length, they have a casque-shaped head, slender body, long legs, and long tails. Basilisk lizards forage mainly on the ground.

Mexico to Colombia, Venezuela

Genera 3
Species 9

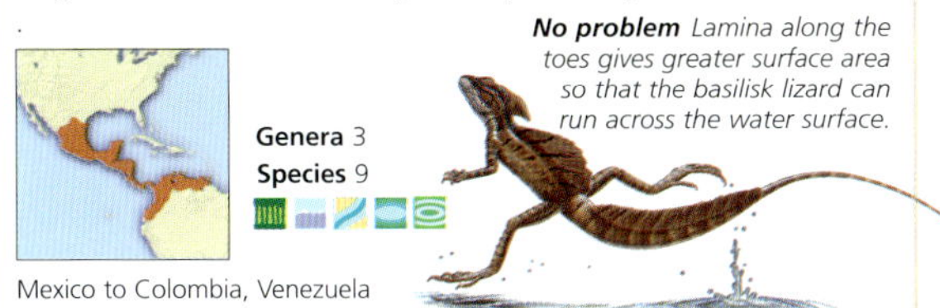

No problem *Lamina along the toes gives greater surface area so that the basilisk lizard can run across the water surface.*

Family Polychrotidae All anoles have subdigital lamellae for climbing and a colorful dewlap for intraspecific communication. Some species are terrestrial, others are aquatic. Many are arboreal. All will lose their tails if they are grasped, and are oviparous, laying one to two eggs per clutch. Most are entirely carnivorous.

Central and South America, Caribbean, USA

Genera 8
Species 395

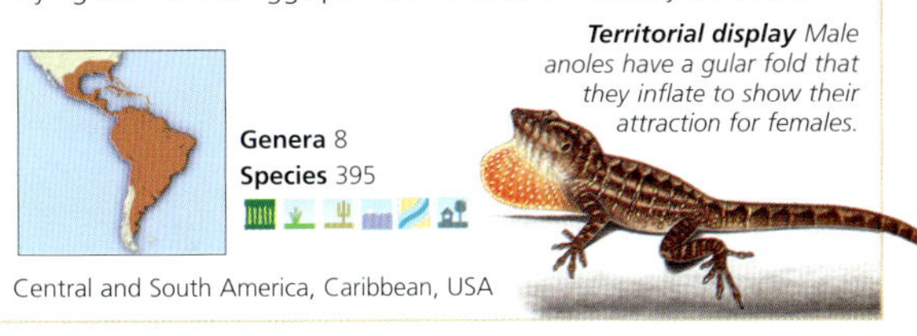

Territorial display *Male anoles have a gular fold that they inflate to show their attraction for females.*

1 **Frilled lizard**
Chlamydosaurus kingii, family Agamidae

2 **Banded agama**
Laudakia stellio, family Agamidae

3 **Thorny devil**
Moloch horridus, family Agamidae

4 **Eastern bearded dragon**
Pogona barbata, family Agamidae

5 **Ocellated mastigure**
Uromastyx ocellata, family Agamidae

6 **Sinai agama**
Pseudotrapelus sinaitus, family Agamidae

7 **Iranian toad-headed agama**
Phrynocephalus persicus, family Agamidae

8 **Persian agama**
Trapelus persicus, family Agamidae

9 **Moroccan dabb lizard**
Uromastyx acanthinura, family Agamidae

10 **Common agama**
Agama agama, family Agamidae

TREE DRAGON

The tree dragon (*Diporiphora superba*) is one of the most slender agamids. This semiarboreal lizard measures 3 inches (8 cm) in snout–vent length and its tail is up to four times its body length.

Frilled lizard This arboreal lizard is active diurnally in dry woodlands. When escaping, it runs bipedally. If threatened, it erects its huge frilled collar to appear much larger than it is.

Up to 11¼ in (28 cm)
Arboreal
Oviparous
8–23
Common

N. Australia & S. New Guinea

Thorny devil A hygroscopic system of grooves on this lizard's skin leads to the corners of its mouth, allowing it to drink the dew that falls on its back. It feeds only on ants.

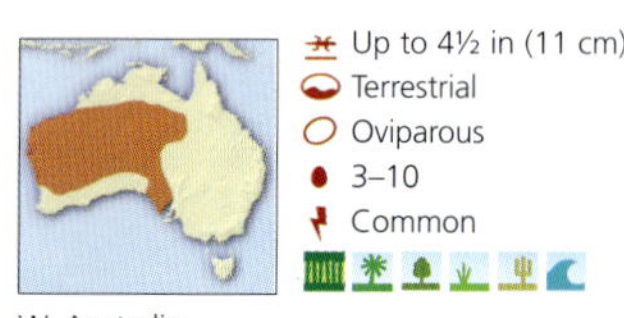

Up to 4½ in (11 cm)
Terrestrial
Oviparous
3–10
Common

W. Australia

1 Armored pricklenape
Acanthosaura armata, family Agamidae

2 Five-lined flying dragon
Draco quinquefasciatus, family Agamidae

3 Sailfin lizard
Hydrosaurus amboinensis, family Agamidae

4 Giant forest dragon
Gonocephalus grandis, family Agamidae

5 Chinese water dragon
Physignathus cocincinus, family Agamidae

6 Sumatra nose-horned lizard
Harpesaurus beccarii, family Agamidae

7 Indo-Chinese forest lizard
Calotes mystaceus, family Agamidae

8 Common green forest lizard
Calotes calotes, family Agamidae

9 Tropical forest dragon
Gonocephalus liogaster, family Agamidae

Sailfin lizard This semiaquatic lizard forages and basks along the edges of streams. When danger approaches, it runs bipedally across the surface of the water with the aid of a fringe on the toes of its hindlimbs.

Up to 40 in (100 cm)
Variable
Oviparous
6–12
Common

S.E. Asia, New Guinea

Chinese water dragon These lizards are semiaquatic, living along the shores of rivers, climbing into the lower branches of trees, and sharing burrows in colonies of one dominant male and several females.

Up to 12 in (30 cm)
Variable
Oviparous
7–12
Common

Thailand, E. Indochina, Vietnam, S. China

Common green forest lizard This arboreal lizard inhabits most forest areas to an altitude of 4,900 feet (1,500 m), beyond which they are replaced by other species. Males have very vivid coloration.

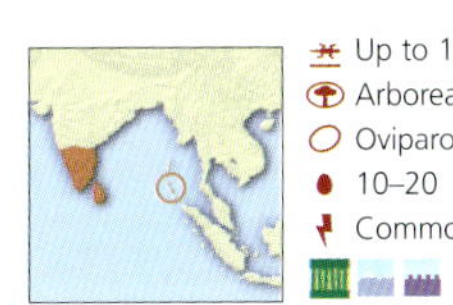

Up to 12 in (30 cm)
Arboreal
Oviparous
10–20
Common

India & Sri Lanka

LIZARD REPRODUCTION

Reproductive patterns in lizards vary widely. Some lizard species mature rapidly, are short lived, and are continuously producing an egg. Other species take several years to mature and then lay large clutches of large eggs for many years. Between these two extremes there are all possible variations of numbers of clutches per year and viviparity. Some species are egg layers at low altitudes but populations at high altitudes or latitudes are viviparous. The female acts as an incubator, moving with the developing embryos into optimum incubation temperatures. Equally amazing are some species that are stuck in evolutionary inertia and produce no more than two eggs per clutch regardless of their size or the amount of stored energy.

Clutch size Geckos (above), regardless of size or nutrient condition, produce two eggs per clutch. Some species can vary clutch size and egg size (center). Large species (right) have large clutches of large eggs and the number and size of the eggs is proportionate to the size and physical condition of the female.

Stuck together Each female gecko lays two eggs at a time. When first laid, the eggs are sticky. Occasionally, females will share a good nesting sight (below). They glue their eggs to well-protected, vertical surfaces, such as tree hollows, away from predators.

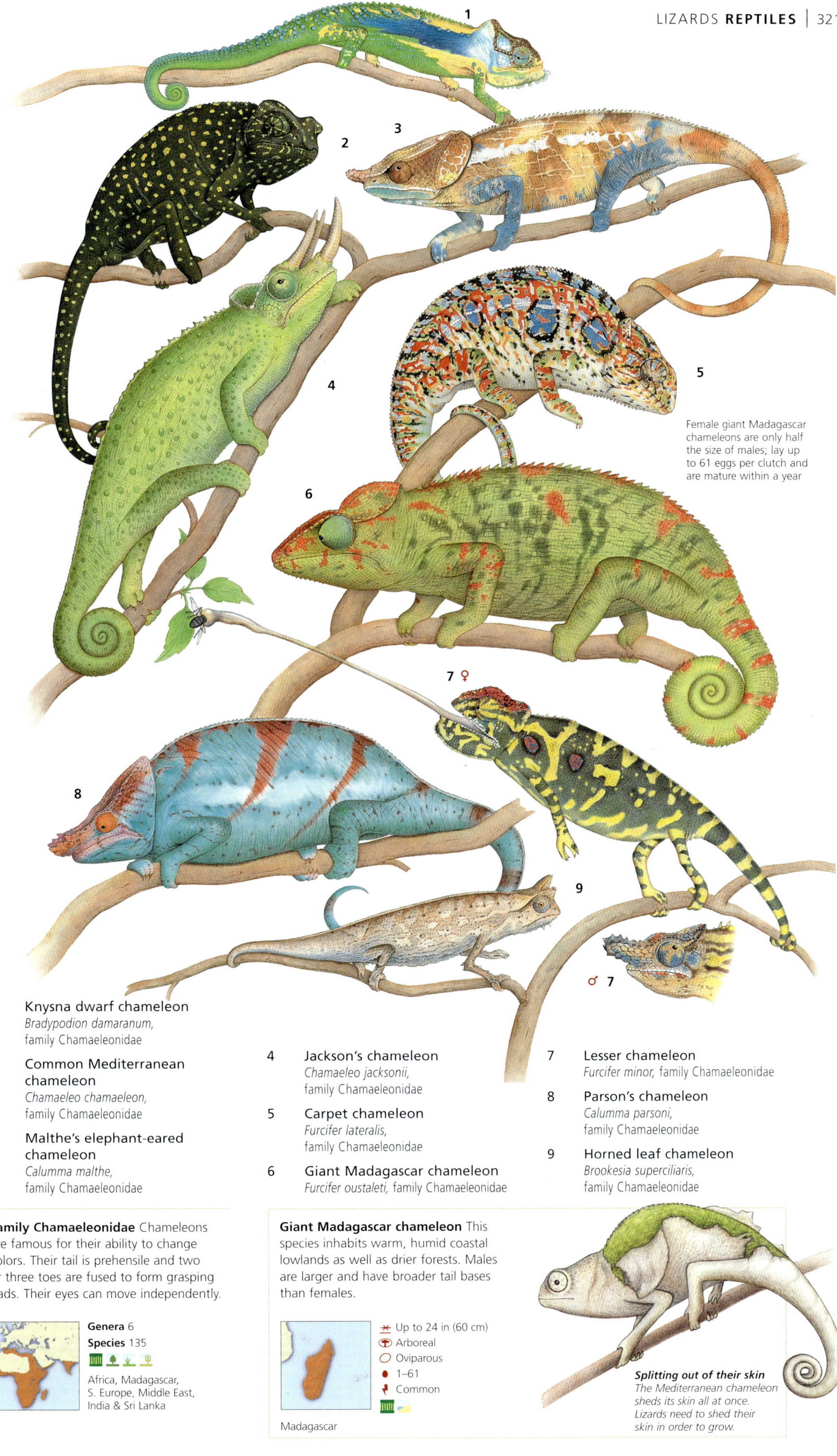

Female giant Madagascar chameleons are only half the size of males; lay up to 61 eggs per clutch and are mature within a year

1 Knysna dwarf chameleon
Bradypodion damaranum, family Chamaeleonidae

2 Common Mediterranean chameleon
Chamaeleo chamaeleon, family Chamaeleonidae

3 Malthe's elephant-eared chameleon
Calumma malthe, family Chamaeleonidae

4 Jackson's chameleon
Chamaeleo jacksonii, family Chamaeleonidae

5 Carpet chameleon
Furcifer lateralis, family Chamaeleonidae

6 Giant Madagascar chameleon
Furcifer oustaleti, family Chamaeleonidae

7 Lesser chameleon
Furcifer minor, family Chamaeleonidae

8 Parson's chameleon
Calumma parsoni, family Chamaeleonidae

9 Horned leaf chameleon
Brookesia superciliaris, family Chamaeleonidae

Family Chamaeleonidae Chameleons are famous for their ability to change colors. Their tail is prehensile and two or three toes are fused to form grasping pads. Their eyes can move independently.

Genera 6
Species 135

Africa, Madagascar, S. Europe, Middle East, India & Sri Lanka

Giant Madagascar chameleon This species inhabits warm, humid coastal lowlands as well as drier forests. Males are larger and have broader tail bases than females.

Up to 24 in (60 cm)
Arboreal
Oviparous
1–61
Common

Madagascar

Splitting out of their skin
The Mediterranean chameleon sheds its skin all at once. Lizards need to shed their skin in order to grow.

1

2

3

4

One of the few lizards native to New Zealand

5

6

7

8

9

Tail mimics head to distract predators

1 **Tokashiki gecko**
Goniurosaurus kuroiwae,
family Gekkonidae

2 **Common leopard gecko**
Eublepharis macularius,
family Gekkonidae

3 **Prehensile tailed gecko**
Aeluroscalabotes felinus,
family Gekkonidae

4 **Banded gecko**
Coleonyx variegatus,
family Gekkonidae

5 **Green tree gecko**
Naultinus elegans,
family Gekkonidae

6 **Thomas's sticky-toed gecko**
Hoplodactylus rakiurae,
family Gekkonidae

7 **New Caledonia bumpy gecko**
Rhacodactylus auriculatus,
family Gekkonidae

8 **Stellate knob-tail**
Nephrurus stellatus,
family Gekkonidae

9 **Northern leaf-tailed gecko**
Saltuarius cornutus,
family Gekkonidae

ADAPTED FOR SAND

Fine free-flowing sand requires specialized feet to efficiently traverse the terrain. The African web-footed gecko (*Pallmatogecko rangeri*) uses its feet like snow shoes on the sand to keep it on the surface.

Sand shoes *Some lizards have fringes along the digits to allow them better traction on sand; others have paddle-like webbed feet.*

Common leopard gecko Leopard geckos have moveable eyelids and lack toe pads. The sex is determined by incubation temperature.

Up to 10 in (25 cm)
Terrestrial
Oviparous
2
Common

Afghanistan, Pakistan, W. India, Iraq, Iran

Banded gecko This nocturnal gecko has large, moveable eyelids, no toe pads and pre-anal pores in a continuous row. Tails are easily lost.

Up to 4 in (10 cm)
Terrestrial
Oviparous
2
Common

S.W. USA to Mexico & Panama

Tail is flattened

Two-thirds of length is tail

Enormous eyes, bulbous head, short tail

1 **Lined gecko**
Gekko vittatus, family Gekkonidae

2 **Henkel's flat-tailed gecko**
Uroplatus henkeli, family Gekkonidae

3 **Tokay gecko**
Gekko gecko, family Gekkonidae

4 **Northern spiny-tailed gecko**
Diplodactylus ciliaris, family Gekkonidae

5 **Marbled velvet gecko**
Oedura marmorata, family Gekkonidae

6 **Marble-faced worm lizard**
Delma australus, family Pygopodidae

7 **Burton's snake-lizard**
Lialis burtonis, family Pygopodidae

8 **Gray's bow-fingered gecko**
Cyrtodactylus pulchellus, family Gekkonidae

9 **Common wonder gecko**
Teratoscincus scincus, family Gekkonidae

Lined gecko This gekko has a distinct white dorsal stripe. The tail is cross-banded. They are nocturnal and have large distinct laminae on the toes for climbing.

Up to 10 in (25 cm)
Terrestrial
Oviparous
2
Common

Indo-Australian archipelago

Gray's bow-fingered gecko This gecko has a flattened, compact body. The toes are long and thin, bending upward and at the ultimate phalange turning downwards, terminating with small lamellae.

Up to 8 in (20 cm)
Terrestrial
Oviparous
2
Common

Central Asia, S.E. Asia

Family Pygopodidae This family is closely related to the geckos. Forelimbs are lacking, and the hindlimbs are represented by a scaly flap just anterior to the cloaca. Their tails are fragile and easily broken. The eyes are snake-like, as they lack lids.

Rudimentary feet
The hindlimb of the Burton's snake lizard has been reduced to a flap-like scale, useful for traction in flowing sand.

Broad-tailed day gecko is native to Madagascar but is well established on Hawaii

1 **Madagascar day gecko**
Phelsuma madagascariensis, family Gekkonidae

2 **Broad-tailed day gecko**
Phelsuma laticauda, family Gekkonidae

3 **Mourning gecko**
Lepidodactylus lugubris, family Gekkonidae

4 **Common wall gecko**
Tarentola mauritanica, family Gekkonidae

5 **Israeli fan-fingered gecko**
Ptyodactylus puiseuxi, family Gekkonidae

6 **Kuhl's flying gecko**
Ptychozoon kuhli, family Gekkonidae

7 **Anderson's short-fingered gecko**
Stenodactylus petrii, family Gekkonidae

8 **Ruppell's leaf-toed gecko**
Hemidactylus flaviviridis, family Gekkonidae

9 **Cradock thick-toed gecko**
Pachydactylus geitje, family Gekkonidae

10 **Wiegmann's striped gecko**
Gonatodes vittatus, family Gekkonidae

Mourning gecko This arboreal gecko is covered with small scales. The tail is long and depressed with a lateral fringe of small, spinose scales.

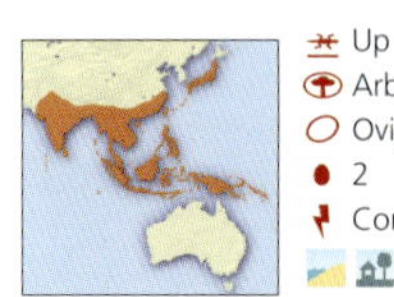

Up to 2 in (5 cm)
Arboreal
Oviparous
2
Common

N.E. Australia, Malaysia to Oceania

Common wall gecko This gecko is strong and heavily built, with rows of keeled, tubercular scales. Males are territorial. The eggs take 10 weeks to hatch and the young 2 years to mature.

Up to 6 in (15 cm)
Arboreal
Oviparous
2
Common

Mediterranean & S. European coast

Kuhl's flying gecko This terrestrial species lays clutches of two eggs about 30 days apart. The eggs attach to bark or rocks and hatch in about 60 days.

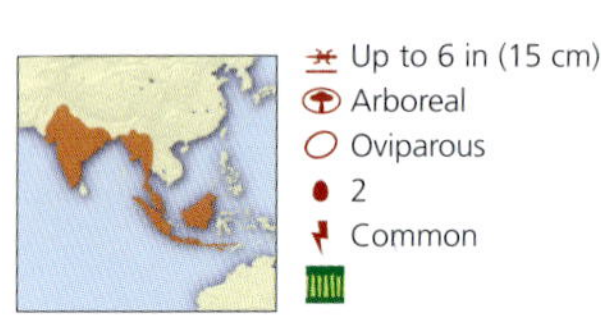

Up to 6 in (15 cm)
Arboreal
Oviparous
2
Common

S.E. Asia

1

2

Long-tail whip lizard is more than two-thirds tail; the tail will break off, if grasped by a predator, so that the lizard can escape

4

3

5

6

7

8

9

1 Rough-scaled plated lizard
Gerrhosaurus major, family Gerrhosauridae

2 Madagascar girdled lizard
Zonosaurus madagascariensis, family Gerrhosauridae

3 Long-tail whip lizard
Tetradactylus tetradactylus, family Gerrhosauridae

4 Lesser flat lizard
Platysaurus guttatus, family Cordylidae

5 Karoo girdled lizard
Cordylus polyzonus, family Cordylidae

6 Black crag lizard
Pseudocordylus melanotus, family Cordylidae

7 Mole skink
Eumeces egregius, family Scincidae

8 Schneider's skink
Novoeumeces schneideri, family Scincidae

9 Sand fish
Scincus scincus, family Scincidae

Family Gerrhosauridae Plated lizards have large, symmetrical shields with bony plates, and the scales on the body are rectangular and overlapping. They are terrestrial and omnivorous, consuming insects and plants.

Genera 6
Species 32

S. Africa & Madagascar

Family Cordylidae Girdled lizards have large symmetrical scales with bony plates on the head. The rectangular, overlapping body scales are usually strongly keeled but are granular in flat lizards.

Genera 4
Species 52

S. Africa

Rolled up in a ball
The armadillo girdled lizard (Cordylus cataphractus) bites its own tail to present an unswallowable form to predators.

Solomon Island skink can grasp a branch and hang by its prehensile tail

1 **Solomon Island skink**
Corucia zebrata, family Scincidae

2 **Emerald skink**
Dasia smaragdina, family Scincidae

3 **Hosmer's spiny-tailed skink**
Egernia hosmeri, family Scincidae

4 **Centralian blue-tongued lizard**
Tiliqua multifasciata, family Scincidae

5 **Shingleback lizard**
Tiliqua rugosa, family Scincidae

6 **Three-lined burrowing skink**
Androngo trivittatus, family Scincidae

7 **Otago skink**
Oligosoma otagense, family Scincidae

TELLTALE TAILS

Skinks tails are long to moderately long. Territorial male five-lined skinks (*Eumeces fasciatus*) recognize females and juveniles by their bright blue tails and allow them access to their territories.

Hosmer's spiny-tailed skink The base of this skink's tail lacks enlarged scales, and long, rugose ear lobules almost conceal the ear. This diurnal rock-dwelling lizard found in rock outcrops and stony hillsides.

N.E. Australia

- Up to 7 in (18 cm)
- Terrestrial
- Oviparous
- 2
- Common

Centralian blue-tongued lizard This lizard has a bright blue tongue, relatively short five-toed limbs, and a short tail. It thrives in stony areas in desert and semidesert areas.

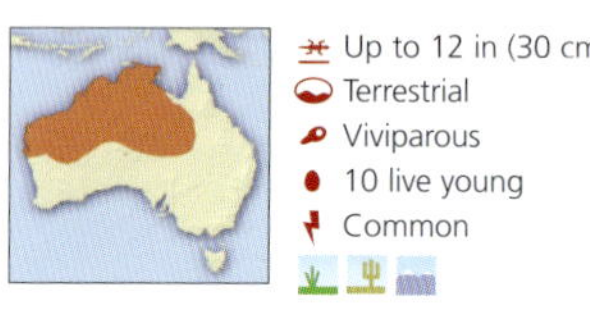

N. & N.W. Australia

- Up to 12 in (30 cm)
- Terrestrial
- Viviparous
- 10 live young
- Common

1 **Bridled mabuya**
Mabuya vittata, family Scincidae

2 **Christmas Island grass skink**
Lygosoma bowringii, family Scincidae

3 **Boulenger's legless skink**
Typhlosaurus vermis, family Scincidae

4 **Red-sided ctenotus**
Ctenotus pulchellus, family Scincidae

5 **Northwestern sandslider**
Lerista bipes, family Scincidae

6 **Cape York mulch skink**
Glaphyromorphus crassicaudum, family Scincidae

7 **Lined fire-tailed skink**
Morethia ruficauda, family Scincidae

8 **Juniper skink**
Ablepharus kitaibelii, family Scincidae

9 **Desert rainbow skink**
Carlia triacantha, family Scincidae

10 **Six-lined burrowing skink**
Scelotes sexlineatus, family Scincidae

Northwestern sandslider This lizard slides through sand without any front limbs; the hindlimbs are reduced and have only two digits. It is found on both sandy plains and coastal dunes.

Up to 2½ in (6 cm)
Terrestrial
Oviparous
2
Common

N.W. Australia

Cape York mulch skink This smooth-scaled skink lives in forest and coastal dune communities where it forages under logs, stones, or in leaf litter.

Up to 2 in (5 cm)
Terrestrial
Oviparous
2
Common

N.E. Australia

BLUE TONGUE PYGMY SKINK

Australia's *Tiliqua adelaidensis* was thought to be extinct, but in 1992 one was found in a snake's stomach. Since that time, a recovery plan has been in operation and numbers have now reached 5,500.

LIZARD COURTSHIP

Courtship displays have evolved in lizards to ensure that gametes are not wasted on mating with the wrong species or inferior individuals. Males often maintain territories in resource-rich areas and will fight other males to maintain them. Females are allowed to enter males' territories and are encouraged to mate by energetic courtship displays. When there are many similar species or species in the same genus in the same habitat the courtship behavioral patterns are more complex. Courtship displays are most complex and have been most intensively studied in the family Polychrotidae. Courtship in monitor lizards (Varanidae) involves tongue flicks by males along different parts of the female's body during the initiation of courtship. Communication in skinks is primarily chemical.

Expanding frill Dominant male frilled lizards show off different display sequences during courtship of females or territoriality displays toward other males. Over 75 different sequences have been recorded, including head bobbing, push-ups, beard erection, color changes, body inflations, head licking, and jaw gaping. Apart from the changes in shape and size of the lizard, changes in color and flashes of color as they open the mouth are a major part of their repertoire.

Anole courtship
The display of the lichen anole (N. pentaprion) *(right) is complex. The dewlap has red basal color with blue lines; the number of blue lines exposed varies according to the extension of the dewlap. Its slow bobbing sequence is synchronized with the dewlap's extension—bob and extend, retract, bob and extend.*

1 European green lizard
Ablepharus kitaibelii, family Lacertidae

2 Giant Canary Island lizard
Gallotia stehlini, family Lacertidae

3 Zagrosian lizard
Timon princeps, family Lacertidae

4 Tiger lizard
Nucras tessellata, family Lacertidae

5 Jeweled lizard
Timon lepidus, family Lacertidae

6 Sand lizard
Lacerta agilis, family Lacertidae

7 Gallot's lizard
Gallotia galloti, family Lacertidae

8 Balkan emerald lizard
Lacerta trilineata, family Lacertidae

Family Lacertidae Adult lacertids are 1½–10 inches (4–25 cm) in body length. The scales are variable, from large, overlapping smooth or keeled to small and granular. All species have limbs.

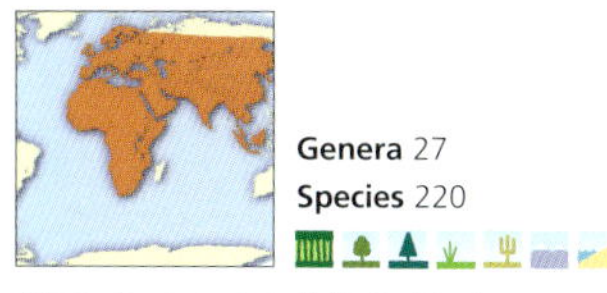

Genera 27
Species 220

Africa, Europe, Asia & N. East Indies

Giant Canary Island lizard Hatchlings take 3 years to reach maturity and live for about 12 years. Adults are primarily herbivorous and are important in seed dispersal.

Up to 10½ in (27 cm)
Terrestrial
Oviparous
10
Common

Grand Canary Island (Canary Is.)

Tiger lizard These lizards actively hunt scorpions, which are dormant during the day and sparsely distributed. Tiger lizards are active during the heat of the day at a body temperature of 102°F (39°C).

Up to 8 in (20 cm)
Terrestrial
Oviparous
3–8
Common

S. Namibia, S.W. Botswana, South Africa

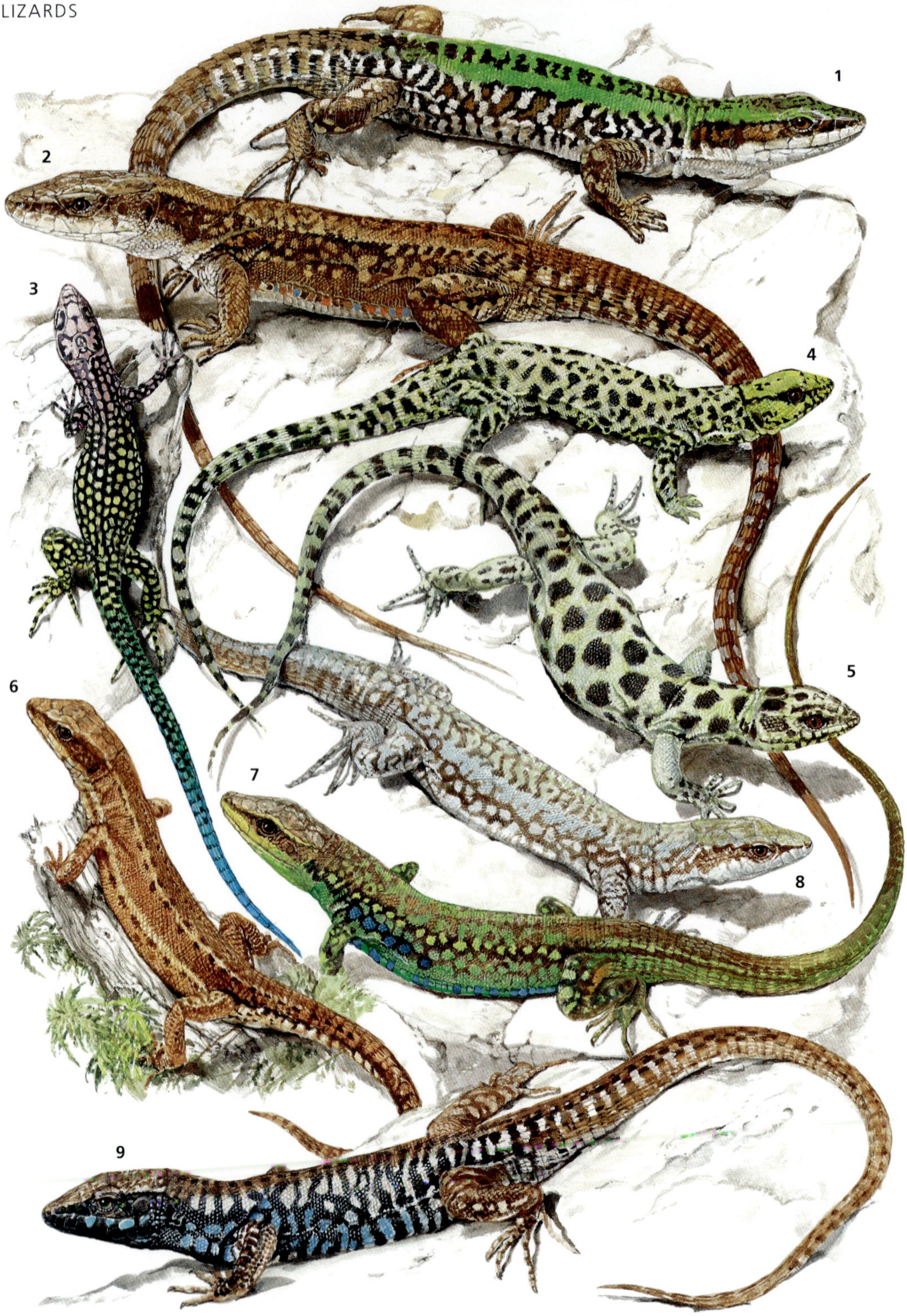

1 **Italian wall lizard**
Podarcis sicula, family Lacertidae

2 **Common wall lizard**
Podarcis muralis, family Lacertidae

3 **Menorca wall lizard**
Podarcis perspicillata, family Lacertidae

4 **Desert night lizard**
Xantusia vigilis, family Xantusiidae

5 **Granite night lizard**
Xantusia henshawi, family Xantusiidae

6 **Viviporus lizard**
Lacerta vivipara, family Xantusiidae

7 **Radd's rock lizard**
Lacerta raddei, family Xantusiidae

8 **Uzzell's rock lizard**
Lacerta uzzeli, family Xantusiidae

9 **Milo's wall lizard**
Podarcis milensis, family Xantusiidae

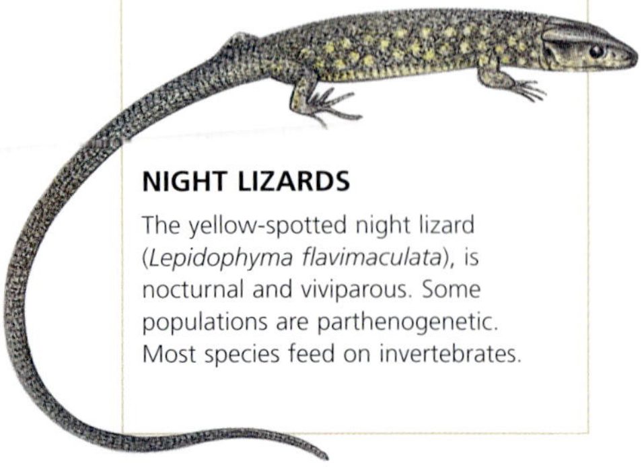

NIGHT LIZARDS

The yellow-spotted night lizard (*Lepidophyma flavimaculata*), is nocturnal and viviparous. Some populations are parthenogenetic. Most species feed on invertebrates.

Family Xantusiidae Even though Xantusidae are called night lizards, many species are diurnal. They are small, up to 4 inches (10 cm) long, with small granular dorsal scales and large ventral scales.

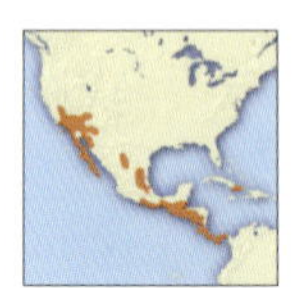

Genera 3
Species 20

W. USA & E. Mexico, Central America & N. South America

Common wall lizard This is a colonial species. It is a good climber and basks on rocks. Wall lizards feed on invertebrates: beetles, flies, butterflies, and spiders. They are prey items for raptors and snakes.

Up to 4 in (10 cm)
Terrestrial, arboreal
Oviparous
2–8
Common

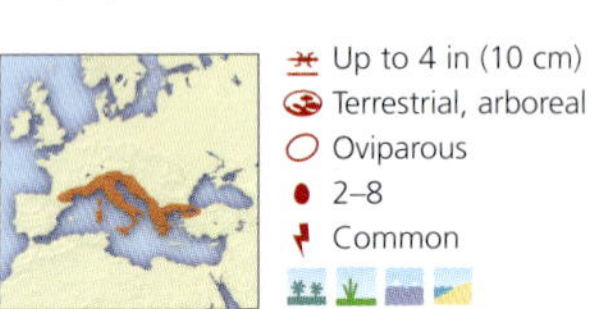

Southern Europe & Balkans

1 Egyptian fringe-fingered lizard
Acanthodactylus pardalis, family Lacertidae

2 Rapid racerunner
Eremias velox, family Lacertidae

3 Algerian psammodromus
Psammodromus algirus, family Lacertidae

4 Blue-throated keeled-lizard
Algyroides nigropunctatus, family Lacertidae

5 Small-spotted lizard
Mesalina guttulata, family Lacertidae

6 Asian grass lizard
Takydromus sexlineatus, family Lacertidae

7 Sawtail lizard
Holaspis guentheri, family Lacertidae

8 Mourning racerunner
Heliobolus lugubris, family Lacertidae

9 Slender sand lizard
Meroles anchietae, family Lacertidae

10 Snake-eyed lizard
Ophisops elegans, family Lacertidae

Egyptian fringe-fingered lizard The toes on these lizards have a fringe enabling better traction over windblown sand dunes. They also have shovel noses and countersunk lower jaws to plow through sand.

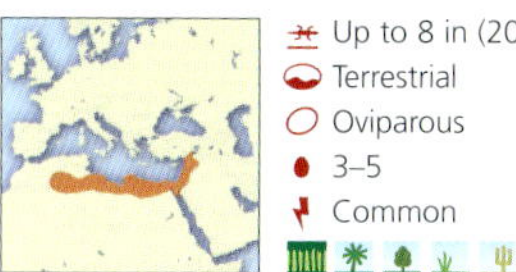

Up to 8 in (20 cm)
Terrestrial
Oviparous
3–5
Common

Algeria, Egypt, Israel, Jordan & Libya

Blue-throated keeled-lizard These diurnal lizards are the smallest lacertids in Europe, and forage for insects in vineyards and buildings. They emerge from hibernation and breed in April.

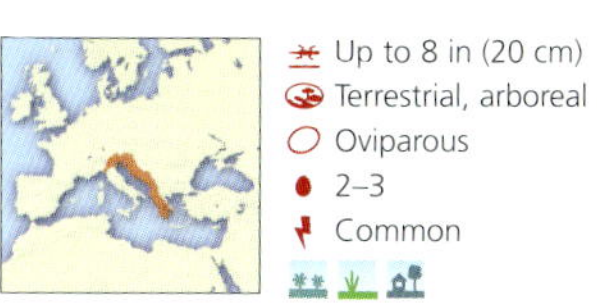

Up to 8 in (20 cm)
Terrestrial, arboreal
Oviparous
2–3
Common

N.E. Italy to Gulf of Corinth & Ionian islands

Snake-eyed lizard This lizard gets its name from the large transparent disks that cover its eyes, so large that the eyelids no longer cover the eye. When escaping from a predator, they scamper from bush to bush.

Up to 2 in (5 cm)
Terrestrial
Oviparous
4–5
Common

N. Africa, S.E. Europe to Middle East & India

Heliothermic species that feeds on ants and spiders in the leaf litter

1 **Ocellated tegu**
Cercosaura ocellata, family Gymnophthalmidae

2 **Thomas's bachia**
Bachia panoplia, family Gymnophthalmidae

3 **Two-ridged neusticurus**
Neusticurus bicarinatus, family Gymnophthalmidae

4 **Rainbow lizard**
Cnemidophorus lemniscatus, family Teiidae

5 **Green calango**
Kentropyx calcarata, family Teiidae

6 **Four-toed tegu**
Teius teyou, family Teiidae

7 **Slow worm**
Anguis fragilis, family Anguidae

8 **Common worm lizard**
Ophiodes intermedius, family Anguidae

9 **Southern alligator lizard**
Elgaria multicarinata, family Anguidae

Family Gymnophthalmidae The microtiids are small, oviparous lizards, up to 2½ inches (6 cm) long, with a wide variation of scale patterns. Most of the lizards in this diverse family live within the forest-floor litter.

Genera 36
Species 160

S. Central America & N.W. South America

Family Teiidae Whiptail lizards and tegus, which are oviparous, vary in adult size from 2 to 16 inches (5 to 40 cm) in snout–vent length. Larger tegus are omnivorous. The smaller whiptails prey on invertebrates.

Genera 9
Species 118

N. USA to South America

Golden tegu This heavy-bodied lizard is one of the main predators of turtle eggs in the Amazon Basin. To protect their own eggs from predation, they lay them in arborial termite nests.

Up to 12 in (30 cm)
Terrestrial
Oviparous
4–32
Common

N. South America

1
2
3
4
5
6
7
8

Lives in the low branches of trees above the water; the tail is high and flattened dorso-laterally for swimming

1 Giant ameiva
Ameiva ameiva, family Teiidae

2 Banded galliwasp
Diploglossus fasciatus, family Anguidae

3 Chinese crocodile lizard
Shinisaurus crocodilurus, family Xenosauridae

4 Golden tegu
Tupinambis teguixin, family Teiidae

5 European glass lizard
Pseudopus apodus, family Anguidae

6 Slender glass lizard
Ophisaurus attenuatus family Xenosauridae

7 Crocodile tegu
Crocodilurus lacertinus, family Teiidae

8 Paraguay Caiman lizard
Dracaena paraguayensis

Family Anguidae Anguids have heavily armored scales with underlying osteoderms. Most have limbs, but a large number of them are limbless with long tails that are two-thirds of their total length. They are commonly called glass snakes, because when attacked by a predator the tail breaks into several pieces.

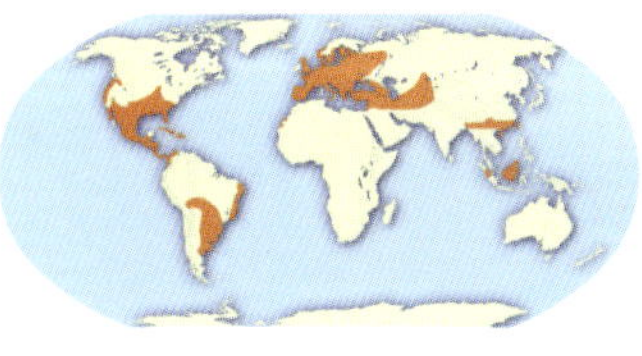

Genera 13
Species 101
S. Asia, S.W. Asia, Europe & Americas

Prehensile tails
Cope's arboreal alligator lizard (Abronia aurita) *is adapted for living high in the canopy of tropical rain forests; they hold on with their prehensile tails.*

Tail is round in cross section, with no evidence of a dorsal keel

1 **Mexican beaded lizard**
Heloderma horridum, family Helodermatidae

2 **Spiny-tailed monitor**
Varanus acanthurus, family Varanidae

3 **Roughneck monitor**
Varanus rudicollis, family Varanidae

4 **Emerald monitor**
Varanus prasinus, family Varanidae

5 **Gila monster**
Heloderma suspectum, family Helodermatidae

6 **Borneo earless monitor**
Lanthanotus borneensis, family Lanthanotidae

7 **Pygmy Mulga monitor**
Varanus gilleni, family Varanidae

SHEDDING SKIN

The outer layer of a lizard's skin (the epidermis) is composed of keratin; scales are thickenings of this layer. As the lizard grows, the outer layer of keratin is shed in large flakes.

Splitting out
All lizards need to shed their skin periodically in order to grow.

Family Helodermatidae The Gila monster and the Mexican beaded lizard are the only known venomous lizards. They have broad heads, and stocky bodies, strong limbs, and thick tails that are used for fat storage.

Genera 2
Species 2

S.W. USA to Guatemala

Family Lanthanotidae These earless monitors have pterygoid teeth and lack a parietal eye. They are related to the Varanidae and Helodermatidae. The species is nocturnal and semi-aquatic.

Genera 1
Species 1

Borneo

Komodo dragons have an extremely virulent symbiotic bacteria living in their mouths; within a few hours of being bitten, wounded prey will lose energy due to a fever caused by the bacteria, then drop to the ground. The dragon trails the animal and eats it when it is down.

1 **Crocodile monitor**
Varanus salvadorii, family Varanidae

2 **Perentie**
Varanus giganteus, family Varanidae

3 **Gray's monitor**
Varanus olivaceus, family Varanidae

4 **Desert monitor**
Varanus griseus, family Varanidae

5 **Sand monitor**
Varanus gouldii, family Varanidae

6 **Nile monitor**
Varanus niloticus, family Varanidae

7 **Komodo dragon**
Varanus komodoensis, family Varanidae

Family Varanidae Monitors are large, long-necked lizards with thick skin and many rows of small, rounded scales circling the body. Their tails are long and lack caudal autonomy. Their tongues are long and forked.The largest of the monitors is the komodo dragon.

Genera 1
Species 50

Africa, Asia, Australia & Pacific Is.

JACOBSON'S ORGAN

A large number of lizards sense the air around them with their tongues, picking up chemical cues regarding food, likely mates, and predators. The Jacobson's organ has a pair of cavities lined with sensory cells on the roof of the mouth. The particles collected on the lizard's tongue are transported to ducts leading to these cavities by the tongue, explaining why the tongue is forked. A nerve then transmits the information to the brain.

Tongue testing
Monitors and many other lizards pick up airborne scents on their forked tongues.

SNAKES

CLASS	Reptilia
ORDERS	Squamata
FAMILIES	17
GENERA	438
SPECIES	2,955

At nearly 3,000 species, the variety in snakes is enormous. They range from the tiny burrowing blind snakes of a few inches (10 cm) long to huge constrictors over 33 feet (10 m) in length. Snakes have evolved special methods of locomotion, different ways of gathering environmental cues, and diverse venom delivery systems. Lizards appeared in the fossil record before snakes, and the presence of vestigial pelvic girdles and spurs in some primitive snakes indicates that snakes have evolved from them; the left lung is reduced or absent in snakes and legless lizards, but the right lung is reduced in amphisbaenians. Most of the organs of snakes are reduced in girth and elongated.

Tropic lovers Snakes are found on all continents except Antarctica, and ranges of some even pass above the Arctic circle. Some islands are also devoid of snakes, such as New Zealand, Ireland, and Iceland, and sea snakes are absent from the Atlantic. Snake biodiversity is highest in the tropics.

Snake anatomy As in the example of the Western diamondback rattlesnake (below), most snake organs have been reduced in diameter and have increased in length. Lungs in most species have been reduced to one, but males have two functional reproductive organs.

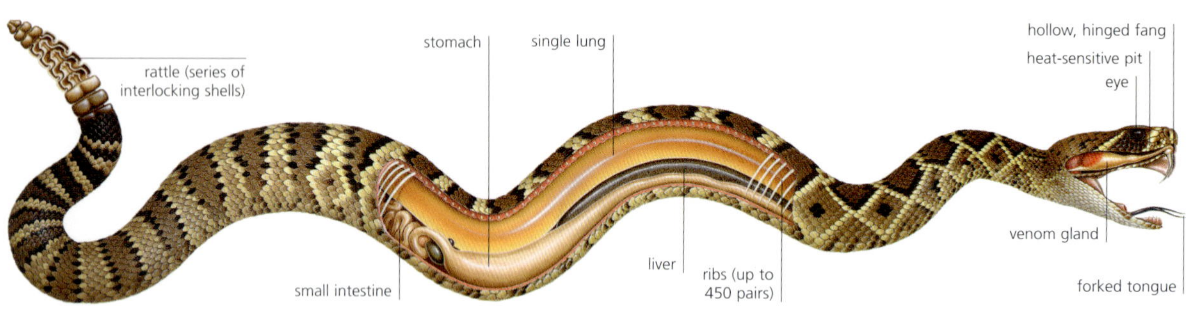

Green constrictors The green tree python is an arboreal predator, feeding on birds, mammals, and lizards. Grasping a bird or bat on the wing with its long teeth, it simultaneously throws a coil around the prey and constricts it. The hatchlings of the green tree python are 11–14 inches (28–35 cm) long and are colored differently from the adults. The juvenile color pattern, often yellow (above), transforms to a vivid green in 6–8 months.

AN EAR TO THE GROUND

Snakes are very close to their environment. In fact, their bellies are in contact with the substrate most of the time—only when they are gliding between trees, swimming, or climbing trees are they free from this close contact with the ground. It is through this earthy contact that they pick up sound vibrations and follow the scent trails of prey and receptive females.

A snake's shape determines where it will live and what it will do. Shovel-nosed snakes and worm snakes remain within the ground with modified snouts for burrowing and feeding on their subterranean prey. The slender, thin-necked, arboreal snakes are adapted for gliding through the branches and stretching out to reach the next branch only a nose away. They have small clutches of elongate eggs and eat thin, elongate prey, such as lizards or frogs. Short, thick-bodied snakes, like some vipers, are sit-and-wait predators and often gorge on a meal that weighs more than they do. Their awkward shape does not affect them, as they do not move much anyway, and many are venomous. Long-bodied, agile, fast-moving snakes are the athletes in the group, chasing down lizards and other snakes, and using their speed for escape. It is amazing how fast mambas and whip snakes can move through the forest canopy: they do not slither, but flow like a stream of water. Sea snakes have modified their shape to have a paddle-like tail, useful for a pelagic life. Many species never touch land, but feed, breed, and give birth at sea.

All male snakes have two hemipenes in the base of the tail; they are usually used alternately, each receiving sperm from only one of the testicles. This system perhaps evolved from the selective advantage of having a fast recovery time when participating in mass spring orgies following emergence from hibernation. Sperm can be stored for months or more than a year before it is utilized to fertilize eggs.

Females have evolved a range of modified behavior patterns to

Granular scales

Smooth scales

Keeled scales

SNAKE SKIN TYPES

Snake skin is made up of scales, which are not separate items, but a thickened part of the skin connected to other scales by thinner areas of elastic skin. Fine, granular scales are common in boas, the smooth, glossy over-lapping scales are common in racers, and strongly keeled scales are characteristic of rattlesnakes.

Tight squeeze The boa constrictor is a large, solitary predator with powerful muscles that allow the snake to squeeze its prey until it suffocates. The snake can open its jaws wide and swallow its prey whole, head-first. Strong acids in the stomach help the constrictor digest its meal.

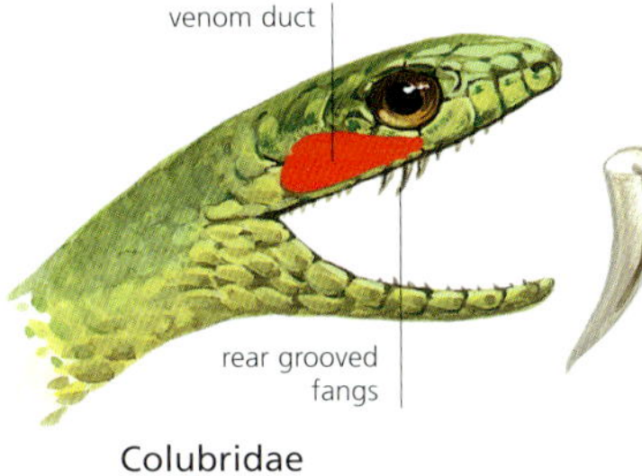

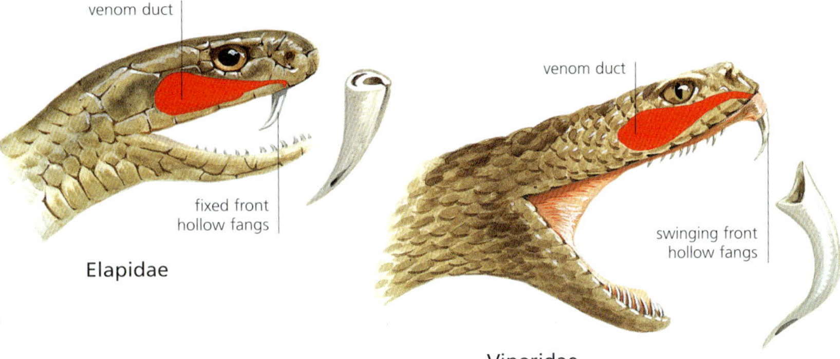

Fang varieties Rear grooved fangs are associated with Duvernoy's gland secretions in Colubridae. Elapidae (cobras, coral snakes, sea snakes, and taipans) have fixed short front fangs. The vipers all have hinged front fangs.

enhance their genes. Most snakes lay eggs, but many of the more recent snakes have evolved viviparity and some parental care. Through muscle contractions, pythons can raise their body temperatures when brooding their eggs. Female cobras guard their nest of eggs in a loose construction of leaves built over the nest.

Some of the most fantastic adaptations of modern snakes are the mechanisms they have developed for finding, capturing, killing, and swallowing prey. Highly developed chemical receptors on the roof of the mouth help snakes to identify mates, enemies, and prey. Infrared receptors possessed by pit vipers, boas, and pythons allow them to visualize warm-blooded prey in terms of heat. The venom produced by Viperidae, Elapidae, and Colubridae allow them to stop prey in its tracks and help in swallowing and digesting the meal. Obviously some snakes have evolved more than others, and many are now losing ground—humans are changing the environment faster than snakes can evolve.

Tasting Snakes, such as Australian rough-scaled snakes, flick their tongue to collect chemicals from the air. When a snake retracts its tongue, it wipes the forked tips over tubes that lead to its Jacobson's organ, a cluster of cells on the roof of the mouth that analyzes the particles. These chemicals can help them locate food, avoid predators, or follow the trail of a possible mate.

Male Mexican rosy boas have well-developed spurs; those of the female are smaller

Tatar sand boas are sold for high prices in central Asia for medicinal purposes

1 **Mexican rosy boa**
Charina trivirgata, family Boidae

2 **Wood snake**
Tropidophis melanurus, family Tropidophiidae

3 **New Guinea ground boa**
Candoia aspera, family Boidae

4 **Calabar ground python**
Calabaria reinhardtii, family Boidae

5 **Rubber boa**
Charina bottae, family Boidae

6 **East African sand boa**
Gongylophis colubrinu, family Boidae

7 **Schaefer's dwarf boa**
Xenophidion schaeferi, family Tropidophiidae

8 **Isthmian dwarf boa**
Ungaliophis continentalis, family Tropidophiidae

9 **Tatar sand boa**
Eryx tataricus, family Boidae

10 **Feick's dwarf boa**
Tropidophis feicki, family Tropidophiidae

BOA AND PYTHON SKULLS

The kinetic jaws of snakes are designed to swallow objects up to five times their diameter. Teeth are curved so that by first moving one side and then the other, the food is pushed down the throat.

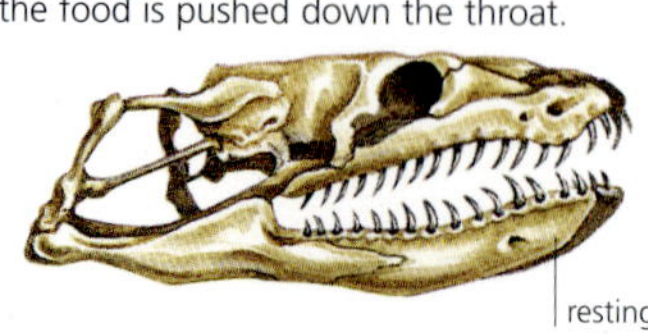

resting jaw

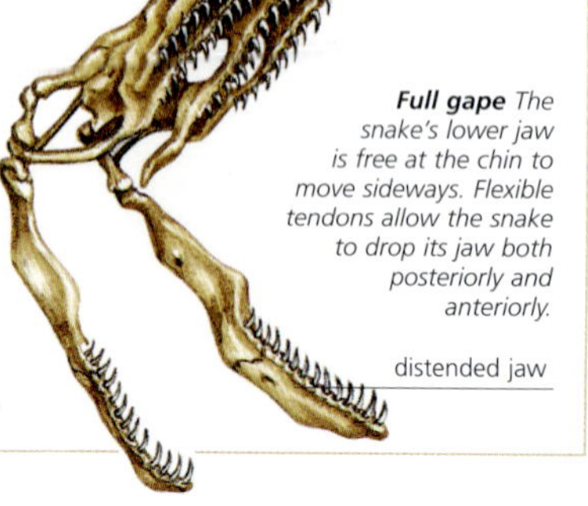

Full gape *The snake's lower jaw is free at the chin to move sideways. Flexible tendons allow the snake to drop its jaw both posteriorly and anteriorly.*

distended jaw

Family Tropidophiidae When disturbed, these snakes bleed from the mouth and eyes and roll up into a tight ball as a defense mechanism.

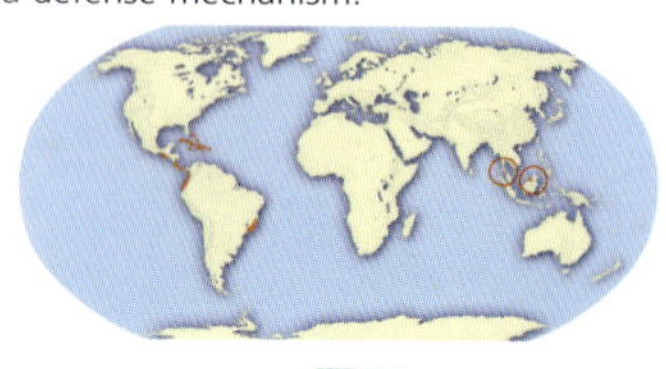

Genera 2 **Species** 31
West Indies to Ecuador, S.E. Brazil, S.E. Asia

Because of its gentle disposition and beautiful pattern, the Indian python has been prized in the pet trade for decades; now most individuals sold are captive-born

Heat-sensing pits are located in the labial scales

Tongue is used to collect airborne scents

1 Common boa constrictor
Boa constrictor, family Boidae

2 Indian python
Python molarus, family Boidae

3 Madagascar ground boa
Acrantophis madagascariensis, family Boidae

4 African rock python
Python sebae, family Boidae

5 Reticulate python
Python reticulatus, family Boidae

6 Anaconda
Eunectes murinus, family Boidae

7 Emerald tree boa
Corallus caninus, family Boidae

Common boa constrictor Heavy-bodied, arboreal snakes, boas kill their prey by constriction. They have vestigial legs as spurs on either side of the cloaca, which suggests that they are of ancient origin.

Up to 14 ft (4.2 m)
Terrestrial, arboreal
Viviparous
30–50 live young
Common

S. Mexico to Argentina

Reticulated python One of the two largest snakes in the world, this python is bigger than the anaconda. Its prey includes large reptiles, such as lizards and crocodilians, and medium to large mammals, even humans.

Up to 33 ft (10 m)
Terrestrial
Oviparous
80–100
Common

S.E. Asia

EGG BROODING

The children's python *(Antaresia childreni)* coils around its eggs to conceal and protect them from predators. It increases its body temperature with gently vibrating muscle contractions.

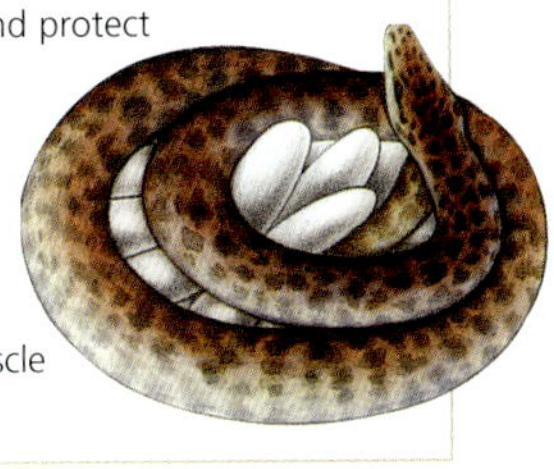

Green tree python
Morelia viridis, family Boidae

Cook's tree boa
Corallus cookii, family Boidae

Carpet python
Morelia spilota, family Boidae

The ventral scales are a yellowish cream color with gray streaking

White-lipped python
Leiopython albertisii, family Boidae

Cuban boa
Epicrates angulifer, family Boidae

Black-headed python
Aspidites melanocephalus, family Boidae

Ball python
Python regalis, family Boidae

Blood python
Python curtus, family Boidae

Cook's tree boa This boa feeds on small lizards when young, progressing to rodents and birds as it grows. Prey is ambushed: in one motion, the boa bites and wraps several coils of its body around its victim while suspended from a branch by its tail.

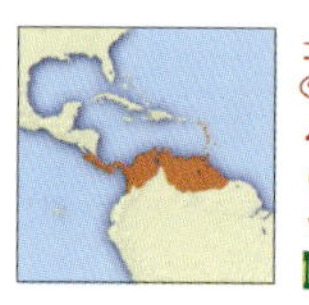

- Up to 6¼ ft (1.9 m)
- Arboreal
- Viviparous
- 30–80 live young
- Common

Panama, N. South America, West Indies

Black-headed python This python is nocturnal in warmer months and diurnal in cooler months. Its diet consists of monitor lizards, frogs, birds, mammals, and snakes—even death adders.

- Up to 8½ ft (2.6 m)
- Terrestrial
- Oviparous
- 5–10
- Common

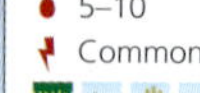

N. Australia

Blood python This species is semiaquatic, spending much time underwater in swamps or streams. They feed primarily on mammals and small birds. Blood python populations are diminishing due to the skin trade: the colorful skin is prized for luxury leather.

- Up to 10 ft (3 m)
- Terrestrial, aquatic
- Oviparous
- 18–30
- Common

S.E. Asia

Coral cylinder snake
Anilius scytale, family Aniliidae

Mexican burrowing python
Loxocemus bicolor, family Loxocemidae

Sunbeam snake
Xenopeltis unicolor, family Xenopeltidae

Drummond Hay's earth snake
Rhinophis drummondhayi, family Uropeltidae

Natal black snake
Macrelaps microlepidotus, family Atractaspidae

Mole viper
Atractaspis engaddensis, family Atractaspidae

Beaked burrowing asp
Atractaspis duerdeni, family Atractaspidae

Red cylinder snake
Cylindrophis ruffus, family Cylindrophiidae

Arafura file snake
Acrochordus arafurae, family Acrochordidae

Mexican burrowing python A nocturnal python, it forages for small mammals and reptiles, and sea turtle and lizard eggs. The large plates on the head suggest those of more recent colubrid snakes, rather than the closely related boas.

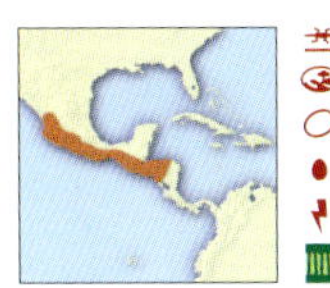

Up to 4¾ ft (1.4 m)
Terrestrial, burrowing
Oviparous
2–4
Rare

S. Mexico to Costa Rica

Family Aniliidae These brightly colored coral-snake mimics live in leaf litter and feed on earthworms, caecilians, eels, amphisbaenians, and snakes. They are on occasion found in the water, but are probably burrowers.

Genera 1
Species 1

Amazonia (South America)

Family Acrochordidae These snakes have small, strongly keeled scales. File snakes are one of the most aquatic snakes, and are almost incapable of moving on land. They are nocturnal, and feed on fishes and crustaceans.

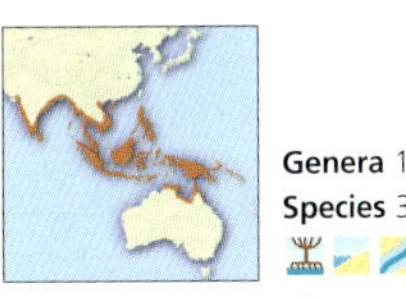

Genera 1
Species 3

India, S.E. Asia, Australia

Chinese rat snake
Ptyas karros, family Colubridae

Large whip snake
Coluber jugularis, family Colubridae

Aesculapian snake
Elaphe longissima, family Colubridae

Desert whip snake
Masticophis flagellum, family Colubridae

Green whip snake
Coluber viridiflavus, family Colubridae

Black-banded trinket snake
Elaphe porphyracea, family Colubridae

Red-tailed green ratsnake
Gonyosoma oxycephalum, family Colubridae

Dahl's whip snake
Coluber najadum, family Colubridae

Beauty snake
Elaphe taeniura, family Colubridae

Tropical ratsnake
Spilotes pullatus, family Colubridae

Family Colubridae About 63 percent of all species of snakes are in this family, utilizing all available reproductive styles and habitats. There are six groups: Natricinae, small aquatic or terrestrial snakes; the largest group, Colubrinae, exploit all snake habitats except marine; Xenodontinae contains small, New World tropical and terrestrial species; Dipsadinae, diversified in Central and South America; Homalopsinae, aquatic snakes; and the small snakes of the Aparallactinae group.

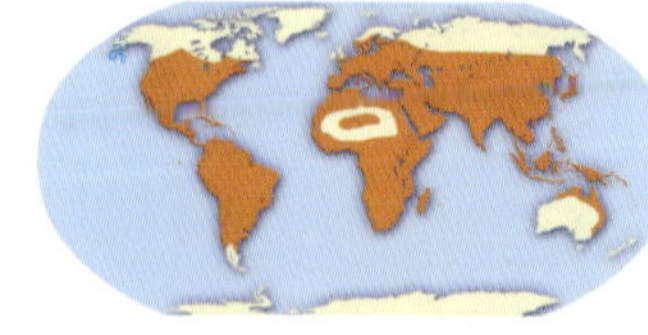

Genera 320
Species 1,800

Worldwide, except Antarctica & polar regions

Snake trade
*The brightly colored corn snake (*Elaphe guttata*) has been sought after in the pet trade for over 50 years because of its attractive colors and mild disposition. Captive breeding colonies are now providing most pet snakes sold today.*

Indigo snake
Drymarchon corais,
family Colubridae

Spotted bush snake
Philothamnus semivariegatus,
family Colubridae

Chinese slug snake
Pareas chinensis,
family Colubridae

Gray-banded king snake
Lampropeltis alterna,
family Colubridae

Milk snake
Lampropeltis triangulum,
family Colubridae

Aesculapian false coral snake
Erythrolamprus aesculapii,
family Colubridae

Milk snake
(juvenile color pattern)
Lampropeltis triangulum,
family Colubridae

Rhombic egg-eating snake
Dasypeltis scabra,
family Colubridae

Montpelier snake is rear-fanged and has venom potentially dangerous to humans

Schokari sand racer
Psammophis schokar,
family Colubridae

Montpelier snake
Malpolon monspessulanus,
family Colubridae

Indigo snake This large diurnal snake is prized by the pet trade for its dark, blue-black coloration and docile temperament. Indigos feed on turtle eggs, reptiles, birds, amphibians, small mammals, and snakes.

Up to 8¾ ft (2.7 m)
Terrestrial
Oviparous
15–26
Uncommon

S.E. USA, Mexico to Paraguay

Milk snake This wide-ranging species is divided into 24 diverse subspecies. Milk snakes often mimic the color pattern of coral snakes for defense.

Up to 4 ft (1.2 m)
Terrestrial
Oviparous
8–12
Common

E. USA, Mexico to Venezuela & Ecuador

Schokari sand racer This slender, fast-moving, diurnal predator searches its desert habitat for prey. They have rear fangs and are mildly venomous.

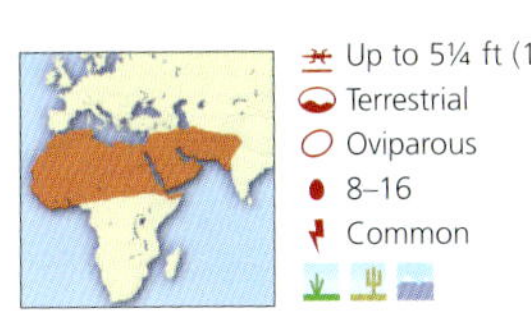

Up to 5¼ ft (1.6 m)
Terrestrial
Oviparous
8–16
Common

N.W. India, Afghanistan, Pakistan to N. Africa

Striped kukri snake
Oligodon octolineatus, family Colubridae

Western ground snake
Sonora semiannulata, family Colubridae

Striped kukri snake is so-named because the shape of its rear fangs resembles the kukri knife used by Gurkha soldiers

Common bronze-back snake
Dendrelaphis pictus, family Colubridae

Red-sided garter snake
Thamnophis sirtalis, family Colubridae

Laotian wolf snake
Lycodon laoensis, family Colubridae

Ring-neck snake
Diadophis punctatus, family Colubridae

Hong Kong dwarf snake
Calamaria septemtrionalis, family Colubridae

Asia Minor dwarf racer
Eirenis modestus, family Colubridae

Crowned leaf-nosed snake
Lytorhynchus diadema, family Colubridae

Western ground snake This snake is secretive, inhabiting areas with loose alluvial sand in dry washes, creosote bushes, desert flats, and rocky hillsides. It prefers to hunt at night and consumes centipedes, spiders, crickets, grasshoppers, and insect larvae.

- Up to 20 in (50 cm)
- Terrestrial
- Oviparous
- Unknown
- Uncommon

S.W. USA & N.W. Mexico

Common bronze-back snake Bronze-backs are primarily arboreal in rain forest, coconut plantations, and urban areas. Frogs and lizards, including flying lizards (Draco), are consumed.

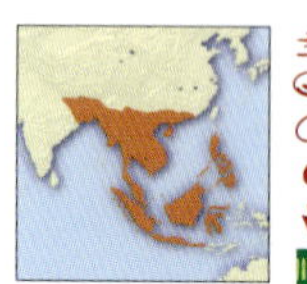

- Up to 3¼ ft (1 m)
- Terrestrial, arboreal
- Oviparous
- Unknown
- Common

India, Burma, W. Malaysia, Indonesia, S. China

Ring-neck snake When attacked, these small woodland snakes hide their head and raise a coiled tail exposing the bright orange ventral surface. They feed on beetles, slugs, frogs, salamanders, and small snakes.

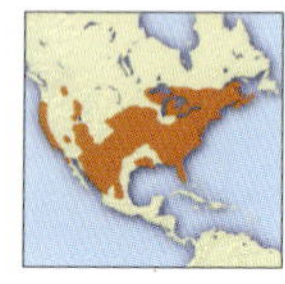

- Up to 28 in (71 cm)
- Terrestrial
- Oviparous
- 1–7
- Common

S.E. Canada, USA, N. Mexico

1 **Blue-necked keelback**
Macropisthodon rhodomelas, family Colubridae

2 **Red-necked keelback**
Rhabdophis subminiatus, family Colubridae

3 **European grass snake**
Natrix natrix, family Colubridae

4 **Northern water snake**
Nerodia sipedon, family Colubridae

5 **Queen snake**
Regina septemvittata, family Colubridae

6 **Masked water snake**
Homalopsis buccata, family Colubridae

7 **New Guinea bockadam**
Cerberus rynchops, family Colubridae

8 **Tentacle snake**
Erpeton tentaculatum, family Colubridae

9 **Rainbow water snake**
Enhydris enhydris, family Colubridae

European grass snake This is one of the few reptiles to live inside the arctic circle and above 7,000 feet (2,121 m). Frogs are one of their main dietary items. Declining frog populations mean grass snake numbers are falling.

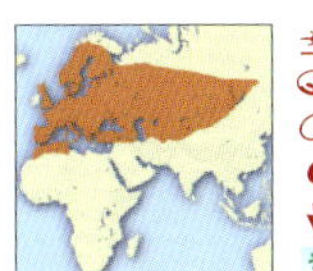

- Up to 6½ ft (2 m)
- Terrestrial, aquatic
- Oviparous
- 15–35
- Common

Europe, W. Asia, N.W. Africa

Queen snake The queen snake is endangered in parts of its range due to habitat destruction and pollution. This water snake inhabits cool, clear streams, where it forages for newly molted crayfish under flat rocks.

- Up to 36½ in (93 cm)
- Aquatic
- Ovoviviparous
- 5–23 live young
- Status

S.E. Canada, E. USA

Masked water snake This nocturnal aquatic snake feeds on fishes and frogs. It has well-developed Duvernoy's glands and grooved, posterior maxillary teeth. Duvernoy's secretions are used to incapacitate prey animals.

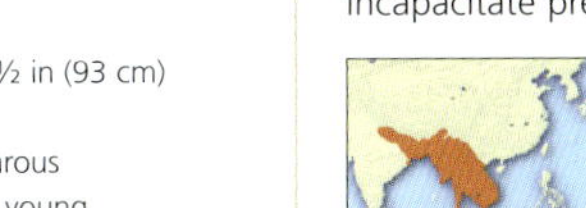

- Up to 4 ft (1.2 m)
- Aquatic
- Viviparous
- Unknown
- Common

S.E. Asia

Rear-fanged nocturnal predator of lizards and birds; mildly venomous

1 **Mangrove snake**
Boiga dendrophila, family Colubridae

2 **Brown vine snake**
Oxybelis aeneus, family Colubridae

3 **Long-nosed tree snake**
Ahaetulla nasuta, family Colubridae

4 **Boomslang**
Dispholidus typus, family Colubridae

5 **Paradise flying snake**
Chrysopelea paradisi, family Colubridae

6 **Cape twig snake**
Thelotornis capensis, family Colubridae

7 **Mediterranean cat snake**
Telescopus fallax, family Colubridae

8 **Forest flame snake**
Oxyrhopus petola, family Colubridae

Mangrove snake This rear-fanged snake has mildly potent venom, and is capable of killing humans. They are semiarboreal, feeding on small mammals, tree shrews, birds, and other snakes in mangrove swamps.

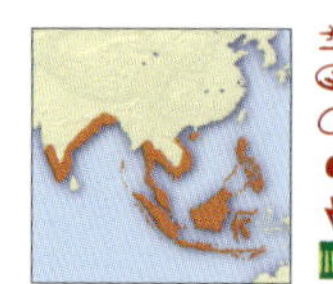

Up to 8¼ ft (2.5 m)
Aquatic, arboreal
Oviparous
7–14
Common

S.E. Asia

Boomslang This venomous snake is the most dangerous colubrid. It is a rear-fanged snake, but its venom is more potent than that of cobras or vipers and causes internal bleeding.

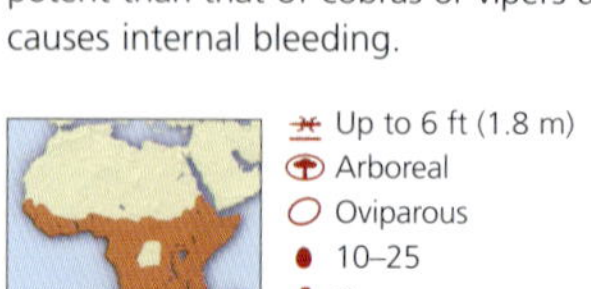

Up to 6 ft (1.8 m)
Arboreal
Oviparous
10–25
Common

Sub-Saharan Africa

Cape twig snake This is a venomous rear-fanged colubrid, capable of inflicting fatal bites on humans. The venom causes internal bleeding. The twig snake puffs its neck up in a threat display.

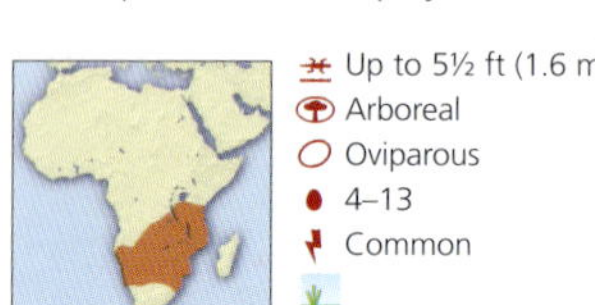

Up to 5½ ft (1.6 m)
Arboreal
Oviparous
4–13
Common

Sub-Saharan Africa

SNAKES AS PREDATORS

All snakes are carnivorous. The variety of prey sizes and taxa ranges from termites to crocodiles. Small snakes that feed on invertebrates search for their prey by sight or smell, often attacking them in their nests. Snakes use their tongues to collect scent data on where their prey has been and often trail them or wait for them to return; many of the large vipers and constricting snakes wait along rodent or game trails for prey to come to them. The large-eyed diurnal racers and whipsnakes are visual predators and chase down lizards and other snakes. The venom that vipers, elapids, and some colubrids inject to kill their prey also helps in the digestion of the animal, since venom is an enzyme. Many large snakes eat a meal only a few times a year, but sometimes this meal has a greater mass than they do.

Strike force A rattlesnake's heat-sensing organs are very accurate: it can even strike in total darkness. Its fangs, which normally lie flat against the roof of its mouth, swing forward to inject fast-acting venom into prey.

Swing action *The rattlesnake's fangs rotate forward to strike. The thin fleshy sheath covering each fang slides back as the snake bites.*

Quick lunge *Ambush predators move with lightning-fast speed when they strike. A rattlesnake's muscular body provides a solid launching pad for the head and neck to shoot forward.*

Catesby's snail eater
Dipsas catesbyi, family Colubridae

Mountain sipo
Chironius monticola,
family Colubridae

Amazon false ferdelance
Xenodon severus,
family Colubridae

Blunthead tree snake
Imantodes cenchoa, family Colubridae

Mussurana
Clelia clelia, family Colubridae

Ringed hognose snake
Lystrophis semicinctus,
family Colubridae

Eastern hognose snake
Heterodon platyrhinos,
family Colubridae

Rainbow snake
Farancia erytrogramma, family Colubridae

False water cobra
Hydrodynastes gigas, family Colubridae

Catesby's snail eater These nocturnal snakes stalk snails and slugs. They extract the snail meat from the shell with their toothed mandible.

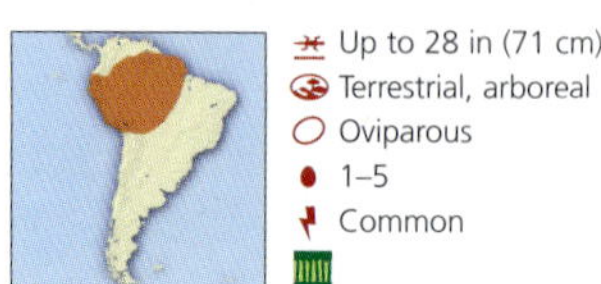

Up to 28 in (71 cm)
Terrestrial, arboreal
Oviparous
1–5
Common

Amazonia (N. South America)

Amazon false ferdelance When alarmed, this snake will flatten its head and neck and distend its neck to make a hood. It is a rear-fanged snake, so the bite is potentially dangerous.

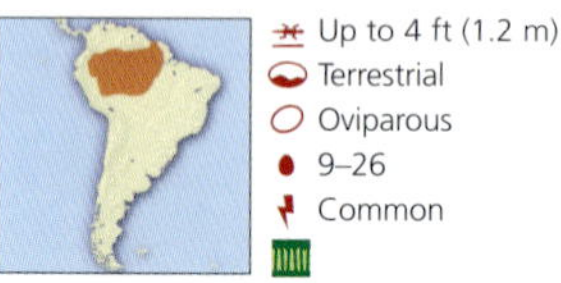

Up to 4 ft (1.2 m)
Terrestrial
Oviparous
9–26
Common

Amazonia (N. South America)

False water cobra These snakes are not highly dangerous, but their bites have caused serious side-effects in humans. Adults both bite and constrict mammals and birds, but swallow frogs and fishes directly.

Up to 9 ft (2.8 m)
Aquatic
Oviparous
20–36
Common

N. South America

Egyptian cobra
Naja haje, family Elapidae

King cobra
Ophiophagus hannah, family Elapidae

Ringed water cobra
Boulengerina annulata, family Elapidae

Monocled cobra
Naja kaouthia, family Elapidae

Venom can cause blindness if not treated

Spitting cobra
Naja nigricollis, family Elapidae

Western green mamba
Debdroaspis viridis, family Elapidae

Fastest snake known: clocked at up to 12½ miles per hour (20 km/h)

Black mamba
Dendroaspis polylepis, family Elapidae

Blue Malaysian coral snake
Maticora bivirgata, family Elapidae

Banded krait
Bungarus fasciatus, family Elapidae

Spitting cobra Although this species usually does not bite, it can squirt its venom accurately up to 9 feet (2.6 m), aiming for its attacker's eyes.

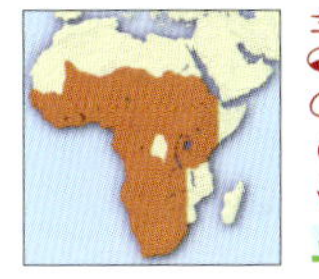

Up to 7¼ ft (2.2 m)
Terrestrial
Oviparous
10–22
Common

Sub-Saharan Africa

Black mamba This large snake carries enough neuro-toxic venom to kill up to 10 adult humans. It is one of the fastest and most fearless snakes known. It can strike out to 40 percent of its length.

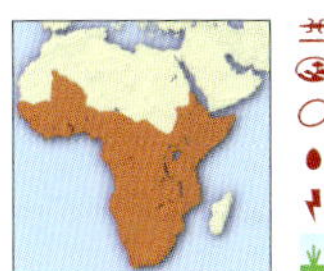

Up to 14 ft (4.2 m)
Terrestrial, arboreal
Oviparous
12–14
Common

Africa

SPITTING COBRA

Spitting cobras have an opening on the anterior surface of the fang above the tip. The inside of the fangs have spiral grooves that put a spin on the projected venom.

Eastern coral snake
Micrurus fulvius, family Elapidae

Arizona coral snake
Micruroides euryxanthus, family Elapidae

Redtail coral snake
Micrurus mipartitus, family Elapidae

Many-banded coral snake
Micrurus multifasciatus, family Elapidae

Loveridge's garter snake
Elapsoidea loveridgei, family Elapidae

Loveridge's garter snake is nocturnal and secretive and feeds on lizards; although venomous, there are few recorded bites on humans

Cape coral snake
Aspidelaps lubricus, family Elapidae

Common death adder
Acanthophis antarcticus, family Elapidae

Curl snake
Suta suta, family Elapidae

Black-striped burrowing snake
Simoselaps calonotus, family Elapidae

Eastern coral snake Even though it has neuro-toxic venom, this snake is not very dangerous to humans because of its small size. Several non-venomous snakes mimic the color pattern.

Up to 4 ft (1.2 m)
Terrestrial
Oviparous
3–13
Common

S.E. USA & E. Mexico

Redtail coral snake Eggs are laid in January and hatch in March. Hatchlings are 6¾ inches (17 cm) long. Cope's false coral snake (*Pliocercus euryzona*) mimics the color pattern in Costa Rica.

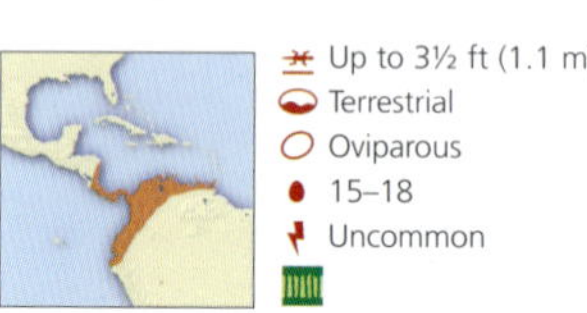

Up to 3½ ft (1.1 m)
Terrestrial
Oviparous
15–18
Uncommon

S. Central America & N.W. South America

Common death adder These nocturnal snakes spend the day buried in sand or leaf litter, often at the base of trees or shrubs. Death adders have large fangs and toxic venom.

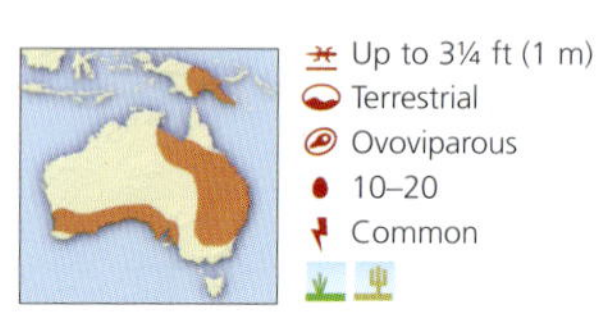

Up to 3¼ ft (1 m)
Terrestrial
Ovoviparous
10–20
Common

Australia & Papua New Guinea

Taipan
Oxyuranus scutellatus, family Elapidae

Black tiger snake
Notechis ater, family Elapidae

Mulga
Pseudechis australis, family Elapidae

Western brown snake
Pseudonaja nuchalis, family Elapidae

Blue-lipped sea krait
Laticauda laticauda, family Elapidae

Olive-brown sea snake
Aipysurus laevis, family Elapidae

Yellow-belly sea snake
Pelamis platurus, family Elapidae

Turtle-headed sea snake
Emydocephalus annulatus, family Elapidae

Ornate reef snake
Hydrophis ornatus, family Elapidae

Scales at center of ornate reef snake's body are hexagonal in shape

Taipan The taipan is the world's deadliest snake, due to the toxicity of its venom. It inhabits wet tropical forests as well as dry forests and open savanna, and feeds on small mammals.

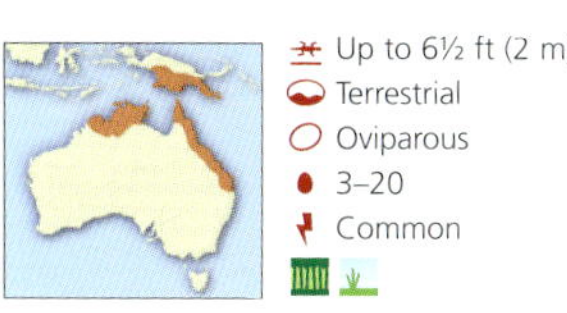

S. New Guinea, Indonesia & Australia

SEA SNAKES

Sea snakes were once grouped with Hydrophiidae, but are now part of the family Elapidae. They have many similar characteristics: all are venomous, and all except the sea krait (*Laticauda*) are viviparous and give birth at sea. All of these snakes have laterally compressed bodies, paddle-like tails, reduced ventral scales, and are incapable of moving well on land. There are 15 genera and 70 species.

Stunners *Sea snakes evolved a very potent neuro-toxic venom so that they could stun fishes rapidly, and then easily manipulate them into their mouths without dropping them to the ocean floor.*

Lichtenstein's night adder
Causus lichtensteini, family Viperidae

Adder
Vipera berus, family Viperidae

Persian horned viper
Pseudocerastes persicus, family Viperidae

Wagner's viper
Vipera wagneri, family Viperidae

Green bush viper
Atheris squamigera, family Viperidae

Russel's viper
Daboia russelii, family Viperidae

Saw-scaled viper
Echis carinatus, family Viperidae

Horned viper
Cerastes cerastes, family Viperidae

Fea viper
Azemiops faae, family Viperidae

Family Viperidae All vipers are venomous, and possess movable front fangs. The pit vipers (rattlesnakes and allies) have an external loreal pit for infrared detection. Adults range from 1 to 12½ feet (0.3 to 3.75 m) in length. The vipers lack the loreal pit; adults range from 1 to 6½ feet (0.3 to 2 m) in length. Most vipers have thick bodies and triangular-shaped heads. Young often eat amphibians, lizards, or snakes and change to a diet of mammals and birds as adults.

Retractable fangs
Vipers are characterized by large retractable front fangs in the upper jaw, which are used to inject a potent venom to prey and predators alike.

Genera 32
Species 221

Central & South America, Africa, Europe, Asia

Levantine viper
Macrovipera lebetina, family Viperidae

Rhinoceros viper
Bitis nasicornis, family Viperidae

Mauritanic viper
Macrovipera mauritanica, family Viperidae

Tropical rattlesnake
Crotalus durissus, family Viperidae

Gaboon adder
Bitis gabonica, family Viperidae

Pygmy rattlesnake
Sistrurus miliarius, family Viperidae

Banded rock rattlesnake
Crotalus lepidus, family Viperidae

Copperhead
Agkistrodon contortix, family Viperidae

Cantil
Agkistrodon bilineatus, family Viperidae

Western diamondback rattlesnake
Crotalus atrox
family Viperidae

Rhinoceros viper This heavy-bodied viper has two to three horns on the tip of its snout and is venomous but not aggressive. It is a sit-and-wait predator and mostly preys on mammals.

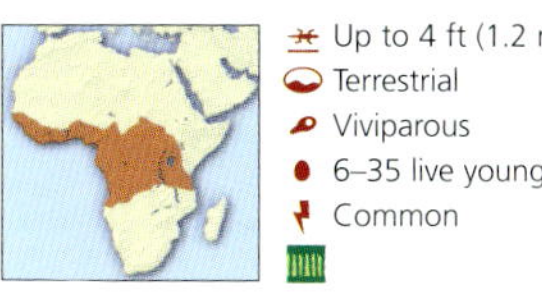

- Up to 4 ft (1.2 m)
- Terrestrial
- Viviparous
- 6–35 live young
- Common

C. & W. Africa

Pygmy rattlesnake The genus *Sistrurus* is distinguished by having nine large scales on the dorsal surface of the head. Young pygmy rattlesnakes have a bright yellow tail tip, which they wiggle as a lure to attract small prey.

- Up to 31½ in (80 cm)
- Terrestrial
- Viviparous
- 6–10 live young
- Common

S.E. USA

Western diamondback rattlesnake One of the most abundant rattlesnakes in the United States, it is responsible for more snake bites there than any other snake. Diet is primarily rodents, but the young prey on lizards.

- Up to 7½ ft (2.3 m)
- Terrestrial
- Viviparous
- 4–25 live young
- Common

S.W. USA & N.W. Mexico

Tiger rattlesnake
Crotalus tigris, family Viperidae

Named for its solid black unbanded tail

Massasauga
Sistrurus catenatus, family Viperidae

Black-tailed rattlesnake
Crotalus molossus, family Viperidae

Twin-spotted rattlesnake
Crotalus pricei, family Viperidae

Ridge-nosed rattlesnake
Crotalus willardi, family Viperidae

Mojave rattlesnake
Crotalus scutulatus, family Viperidae

Speckled rattlesnake
Crotalus mitchelli, family Viperidae

Prairie rattlesnake
Crotalus viridis, family Viperidae

Timber rattlesnake
Crotalus horridus, family Viperidae

Tiger rattlesnake This small nocturnal rattlesnake can often be found resting in pack rat (*Neotoma*) mounds in the daytime. At night it feeds on rodents and lizards. It has long fangs and potent venom.

Up to 36 in (92 cm)
Terrestrial
Viviparous
2–5 live young
Common

Sonoran Desert (S.W. USA & N.W. Mexico)

Massasauga The preferred habitat for this species is river-bottom forest in the floodplains of large rivers. They hibernate in cray-fish burrows after foraging for rodents and frogs during the summer.

Up to 30 in (76 cm)
Terrestrial
Viviparous
8–20 live young
Rare

S.E. Canada, N.E. USA to N.W. USA, N. Mexico

Black-tailed rattlesnake In the spring, males follow the chemical trails of females. Courtship and mating often last several days. Females give birth in August and stay with their young until they have shed.

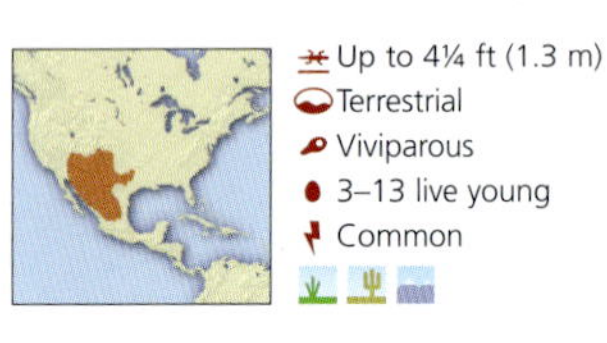

Up to 4¼ ft (1.3 m)
Terrestrial
Viviparous
3–13 live young
Common

S.W. USA & N.W. Mexico

Wagler's palm viper
Tropidolaemus wagleri, family Viperidae

White-lipped tree viper
Trimeresurus albolabris, family Viperidae

Yellow-blotched palm pit viper
Bothriechis aurifer, family Viperidae

Speckled forest pit viper
Bothriopsis taeniata, family Viperidae

Malayan pit viper
Calloselasma rhodostoma, family Viperidae

Hump-nosed moccasin
Hypnale hypnale, family Viperidae

Brazil's lancehead
Bothrops brazil, family Viperidae

Bushmaster
Lachesis muta, family Viperidae

Jararacussu
Bothrops jararacussu, family Viperidae

Hognosed pit viper
Porthidium nasutum, family Viperidae

Wagler's palm viper This viper can be identified by its keeled chin scales. It has long fangs and the venom is hemotoxic, causing cell and tissue destruction.

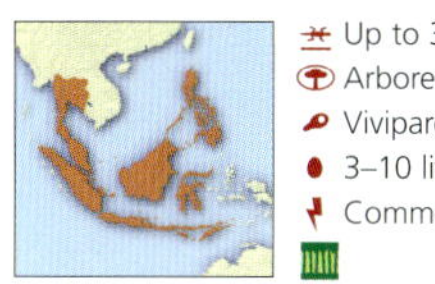

Up to 3¼ ft (1 m)
Arboreal
Viviparous
3–10 live young
Common

S.E. Asia

Bushmaster This is the largest venomous snake in the Americas and the largest viper in the world. They are sit-and-wait predators, traveling short distances between sites then waiting for days or weeks for a meal.

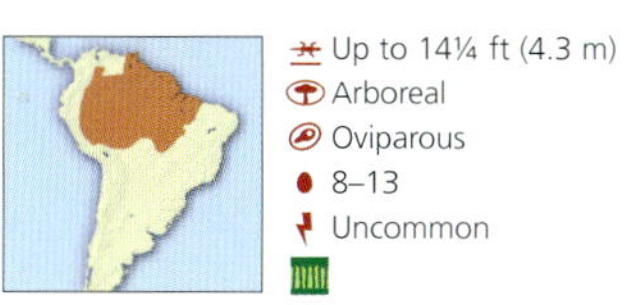

Up to 14¼ ft (4.3 m)
Arboreal
Oviparous
8–13
Uncommon

Amazonia (N. South America)

Hognosed pit viper These stocky snakes are called jumping vipers, as their whole body appears to jump forward when they strike. In reality they can only strike about half their length.

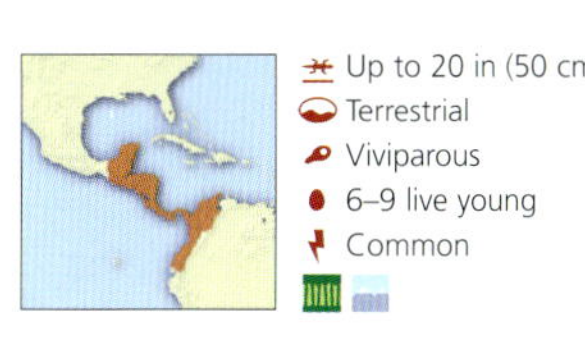

Up to 20 in (50 cm)
Terrestrial
Viviparous
6–9 live young
Common

Central America to Colombia & Ecuador

AMPHIBIANS

PHYLUM	Chordata
CLASS	Amphibia
ORDERS	3
FAMILIES	44
GENERA	434
SPECIES	5,400

Amphibians appeared 360 million years ago. Descended directly from early lobe-finned fishes, they were the first vertebrates to become established on land, where they gave rise to the reptiles. Amphibians are cold-blooded, with moist skin and no scales or claws. There are three orders. Caecilians resemble earthworms, being limbless with bullet-shaped heads and tails. Salamanders (Caudata, meaning "with tail") have cylindrical bodies, long tails, distinct heads and necks, and well-developed limbs. Frogs and toads (Anura, meaning "lacking tail") differ from all other vertebrates in having stout, tailless bodies with a continuous head and body and well-developed limbs.

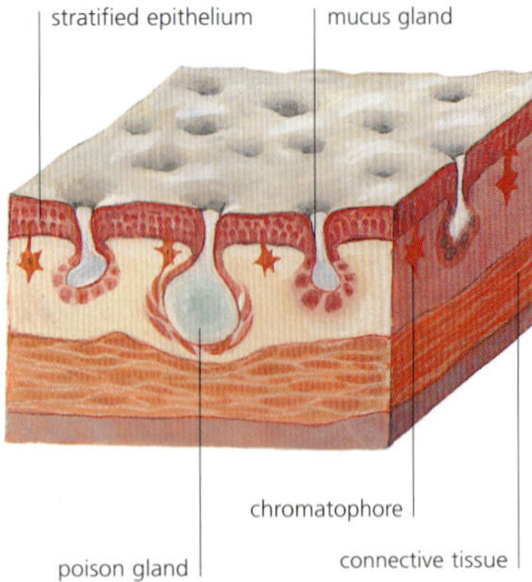

Amphibian skin The skin is moistened by mucus gland secretions. It maintains water balance, respires, and protects the body. Many species possess antibiotic or protective substances, including poisons, in their skin secretions. These secretions are antipredatory mechanisms, being distasteful and often deadly.

Healthy environment Frogs thrive in unpolluted environments where the water supply is pure (far right). Worldwide, frog populations have been on the decline for many years, largely due, it is thought, to environmental pollution.

A DOUBLE LIFE

The word amphibian is from the Greek *amphibios*, meaning "a being with a double life." This refers to the reproductive cycle of most amphibians, in which they have an aquatic larval form and a terrestrial adult life. Most species breed and lay eggs in the water. In these cases, the young have a free-swimming larval stage with external gills, and metamorphose later into a miniature replica of the terrestrial adult. There are many exceptions to this. Some species have direct development on land and no larval stage. Others are entirely aquatic and viviparous. Still others are neotenic.

Some of the distinctive features that amphibians have developed to allow them to invade terrestrial habitats are: the tongue, which allows them to moisten and move food; eyelids, which along with adjacent glands wet and protect the cornea; an outer layer of dead skin cells that is regularly sloughed; ears; a sound-producing structure called the larynx; and the Jacobson's chemosensory organ in the nasal cavity, which gives them a sense of smell and taste.

An amphibian's skin is active in controlling water balance. This can be illustrated by looking more closely at frogs. The permeability of the skin is dynamic, changing with the activity cycle of the frog; when the frog is foraging away from water, the skin is highly permeable to absorb water. When the animal is aquatic, the permeability of the skin is reduced. Some terrestrial frogs have a special patch of skin in the pelvic region that is highly vascularized, allowing them to absorb water from any moist area. When frogs enter hibernation in the water, they must change their ionic balance and skin permeability to lower the absorption of water by osmosis. Some desert toads are able to absorb water from dry soils by retaining urea in their urine, thus creating an osmotic gradient across their skin.

Adults of all amphibians are carnivores, although a few species of frogs have been reported to eat some fruits. The larvae of salamanders are carnivorous, while those of most anurans are herbivores. The tadpoles of some species of frogs are cannibalistic, as are the adults of many species.

Axolotl life Amphibians like the axolotl (right) usually undergo metamorphosis, transforming from a larva to an adult. Axolotls, however, are the exception to this rule. Babies are identical in every respect to their parents, simply smaller.

SALAMANDERS AND NEWTS

CLASS	Amphibia
ORDER	Caudata
FAMILIES	10
GENERA	60
SPECIES	472

Salamanders, newts, mudpuppies, waterdogs, and sirens belong to the order Caudata. All have tails and most have four legs. Most species are very secretive, living terrestrially in leaf litter or rotting logs, underground, in arboreal bromeliads, or underwater. Some live terrestrially but return to the water to reproduce. Although seldom seen, they are not uncommon; astoundingly, in some deciduous forests in North America, the biomass of salamander populations is greater than that of the birds and mammals. All salamanders and their larvae are carnivorous, and they in turn become prey for many larger carnivores, such as snakes, birds, and mammals.

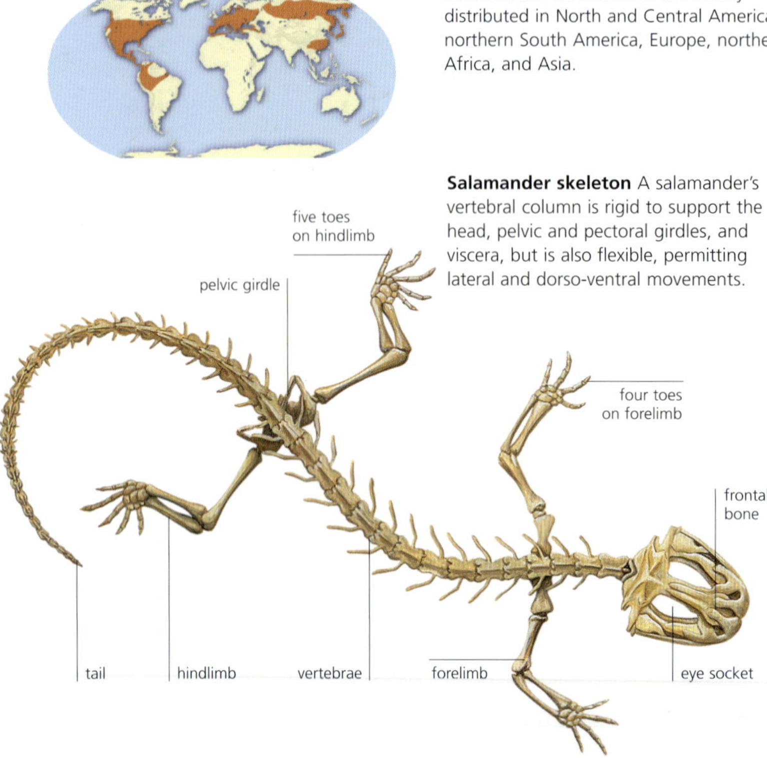

Distribution Salamanders are widely distributed in North and Central America, northern South America, Europe, northern Africa, and Asia.

Salamander skeleton A salamander's vertebral column is rigid to support the head, pelvic and pectoral girdles, and viscera, but is also flexible, permitting lateral and dorso-ventral movements.

SILENT CARNIVORES

Salamanders evolved to live on land, with four legs and a tail. Some species have secondarily returned to live permanently in the water and have reduced limbs; these are known as sirens.

Salamanders more closely resemble early fossil amphibians than do frogs and toads. Unlike frogs, they do not emphasize vocal communications during breeding. A salamander's skin is generally smooth, moist, and flexible, with oxygen exchange taking place through the skin. Many species have lungs as well, but some families lack lungs and respire entirely cutaneously. As the salamander grows, the skin is regularly shed. Because it is made up of the salamander's own cells and is easy to digest, the animal usually devours its own skin as it is sloughed.

Salamanders and their larvae are completely carnivorous. Adults eat insects, spiders, snails, slugs, worms, and other invertebrates. The larvae devour mosquito larvae and other insects. As they grow, they consume larger prey, including tadpoles. Some species are able to extend their tongues rapidly for half the length of their body to capture prey.

Newts, both in their aquatic and terrestrial phases, have rough, dry skin. Some species of newts live on land as adults, coming to the water only to breed, while the adults of some species are aquatic. Red-spotted newts are equally at home in the water or on land. The aquatic larvae metamorphose into a terrestrial red eft, and after 1–2 years on land return to the water to breed.

Gills Neoteny is the developmental stage where the aquatic larval form with gills is maintained throughout adulthood. This usually occurs in species that live in oxygen-deficient water. Some species are permanently neotenic. The axolotl (left) is neotenic if iodine is absent in its habitat.

Hellbender
Cryptobranchus alleganiensis

Greater siren
Siren lacertina

Chinese giant salamander
Andrias davidianus

Lesser siren
Siren intermedia

Lacks eyelids

Two-toed amphiuma
Amphiuma means

Dwarf siren
Pseudobranchus striatus

Feathery protrusions near head are external gills

Hellbender Hellbenders live under rocks and thrive in streams, where they forage for crayfish, mollusks, and small fishes. They are harmless but extremely slimy; in order to pick them up, they must be grasped firmly around the head.

Up to 29½ in (74 cm)
Aquatic
Fall
Common

E.C. USA

Greater siren Although most salamanders are silent, greater sirens when first caught will sometimes yelp. They also make clicking sounds, which perhaps may be used in intraspecific communication.

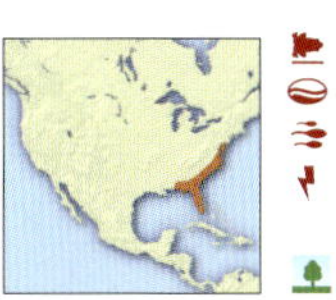

Up to 38½ in (98 cm)
Aquatic
Unknown
Common

S.E. USA

Chinese giant salamander Breeding takes place in an underwater cavity where the female lays up to 500 eggs in a cylindrical string. The male fertilizes the eggs externally and guards them until they hatch, in 50–60 days.

Up to 6 ft (1.8 m)
Aquatic
Fall
Data deficient

E.C. China

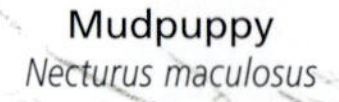
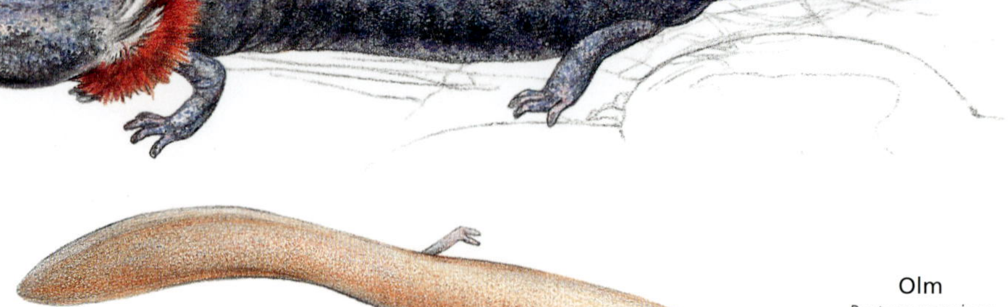

Mudpuppy
Necturus maculosus

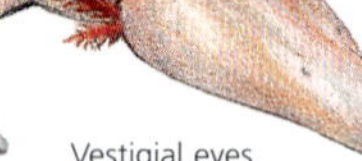

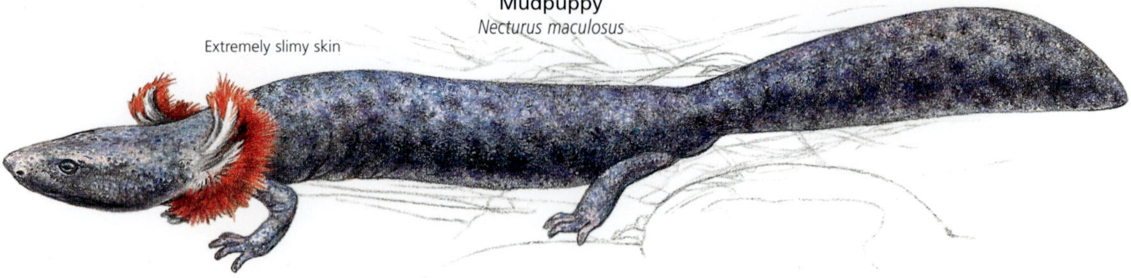

Olm
Proteus anguinus

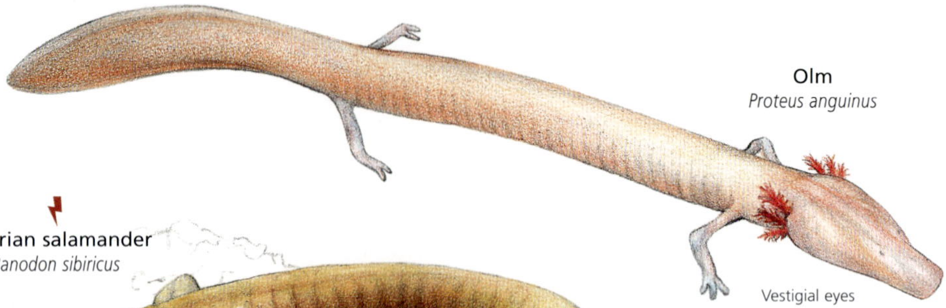

Siberian salamander
Ranodon sibiricus

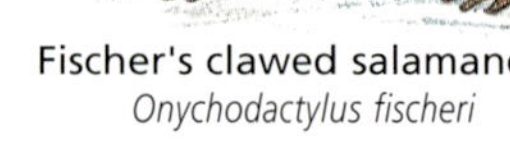

Fischer's clawed salamander
Onychodactylus fischeri

Dybowski's salamander
Salamandrella keyserlingii

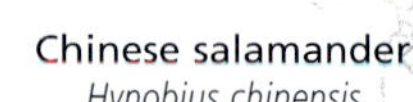
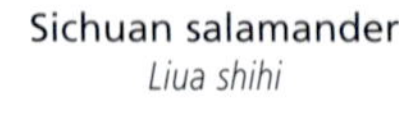

Sichuan salamander
Liua shihi

Chinese salamander
Hynobius chinensis

Paghman mountain salamander
Batrachuperus mustersi

Mudpuppy These neotenic salamanders are characterized by their large, red external gills. They feed nocturnally on crustaceans, insect larvae, fishes, worms, mollusks, and other amphibians. Females guard and aerate their eggs.

- Up to 19 in (48 cm)
- Aquatic
- Fall
- Common

N.C. USA

Olm Olms live in flooded underground caves and are blind, with only vestigial eyes occurring beneath the pigmentless skin. They are neotenic. They lay up to 70 eggs, or may retain eggs and produce fewer living young.

- Up to 12 in (30 cm)
- Aquatic
- Spring
- Vulnerable

Adriatic seaboard & N.E. Italy

Chinese salamander This terrestrial salamander with short limbs migrates to streams or ponds for breeding. Males are attracted to females by chemical and visual cues.

- Up to 4 in (10 cm)
- Terrestrial
- Spring
- Common

Hubei Province (S. China)

Luschan's salamander
Mertensiella luschani

European fire salamander
Salamandra salamandra

Olympic torrent salamander
Rhyacotriton olympicus

Golden striped salamander
Chioglossa lusitanica

Pyrenees mountain salamander
Euproctus asper

Californian giant salamander
Dicamptodon ensatus

Red-spotted newt
Notophthalmus viridescens

Common newt
Triturus vulgaris

Male in breeding condition

Luristan newt
Neurergus kaiseri

Banded newt
Triturus vittatus

European fire salamander This oviparous salamander breeds on land. Once the embryos are mature, the female releases 20–30 membranous eggs into the water, where they hatch immediately. The larvae metamorphose in 2–6 months.

Up to 10 in (25 cm)
Terrest., aquatic larvae
Spring
Common

Europe & W. Asia

Californian giant salamander Females guard their 50 or more eggs for up to 6 months in streams beneath rocks. Terrestrial adults often climb into low shrubs up to 8 feet (2.4 m) above the ground. The males make a low rattling sound when disturbed.

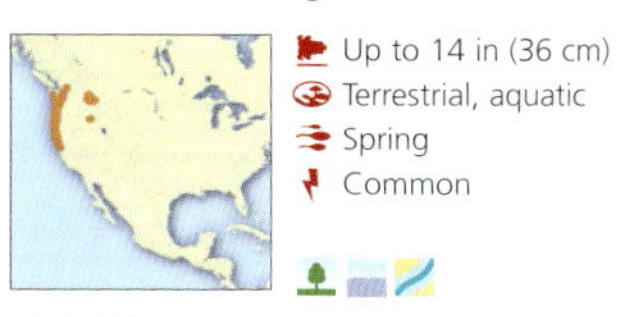

Up to 14 in (36 cm)
Terrestrial, aquatic
Spring
Common

N.W. USA

Common newt Males are smaller than females. In the breeding season, males develop a wavy crest from head to tail, as well as a fringe on the hind toes. They remain in the breeding ponds from March to July, after which they are terrestrial.

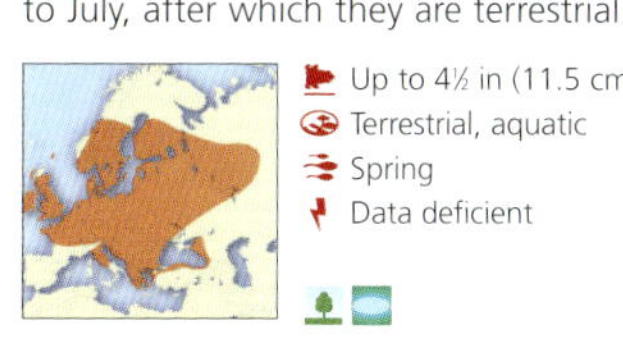

Up to 4½ in (11.5 cm)
Terrestrial, aquatic
Spring
Data deficient

Europe & W. Asia

Ringed salamander
Ambystoma annulatum

Kweichow crocodile newt
Tylototriton kweichowensis

Tiger salamander
Ambystoma tigrinum

California newt
Taricha torosa

Blue-spotted salamander
Ambystoma laterale

Marbled salamander
Ambystoma opacum

Japanese firebelly newt
Cynops pyrrhogaster

Shau Tau Kok newt
Pleurodeles poireti

Markings on the belly are variable, and may take the form of spots, blotches, wavy lines, or reticulations

Vietnam warty newt
Paramesotriton deloustali

Unusually large head

Marbled salamander This species gathers to lay up to 230 eggs in dry depressions from September to December. Females stay wrapped around the eggs, protecting them from predation and desiccation, until fall rains fill the pools.

S.E. USA

- Up to 5 in (13 cm)
- Terrestrial, aquatic
- Fall
- Common

Japanese firebelly newt Males of this species differ from females by having swollen cloacas. The tails of breeding males develop a bluish iridescent sheen and a small filament at the tip.

Japan

- Up to 5¼ in (13.2 cm)
- Aquatic
- Spring
- Common

Vietnam warty newt This, the largest species of the genus, is characterized by a distinct belly pattern of orange-yellow or reddish blotches, each with black or dark brown mottling.

N. Vietnam

- Up to 8 in (20 cm)
- Aquatic
- Spring
- Vulnerable

Arboreal salamander This aggressive species has a heavily ossified skull, strong jaws, hypertrophied jaw muscles, and enlarged, flattened, blade-like, monocuspid teeth. Scarred individuals are often found.

Up to 4 in (10 cm)
Terrestrial, arboreal
Spring
Common, declining

Coast Ranges of California (USA)

LUNGLESS SALAMANDERS

Plethodontidae is the most diverse family of salamanders, with 269 species. Jordan's salamander (*Plethodon jordani*) inhabits the leaf litter and rotting logs on the humid forest floor of the Appalachian Mountains, in the eastern United States. It is adapted to cool, moist habitats up to 6,643 feet (1,993 m). Highly vascularized blood vessels near the skin's surface mean it can respire through the skin and lining of the mouth.

Desirable colors
The Jordan's salamander has a variety of color morphs. There are other species that mimic the color pattern of this red-cheeked morph, presumably because it has a bad taste to predators.

AMPHIBIAN LIFE-CYCLES

Most amphibians have a dual life-cycle involving a courtship ritual, the deposition of sperm and eggs, external fertilization, an aquatic larval stage, and transformation into the adult life form. Modifications to this cycle involve internal fertilization, direct development, viviparity, neoteny, or parental care. There is a trade-off between multitudes of small eggs with no parental care and a few large eggs with parental care.

Frog and toad life-cycle Males call to attract females, and shed sperm over the eggs as the female deposits them in a selected site. The eggs absorb water and in a few days develop into free-swimming tadpoles. After a few weeks, months, or years (depending on the species) the tadpoles metamorphose into identical small replicas of the adults. It then takes months or years of a carnivorous diet before sexual maturity is reached.

1. Mating *The male clasps the female around the middle. The female carries the male until she is ready to lay her eggs. The male then sheds his sperm over the eggs as they are laid.*

2. Egg development *The eggs float in large masses while the embryos develop inside. The eggs are covered in a jelly-like substance that swells when it comes into contact with water.*

3. Tadpoles emerge *The eggs hatch into tadpoles after 2 weeks. When the mouth develops, the tadpoles feed on algae. They breathe through external gills.*

4. Tadpole development *Lungs begin to form after 3 weeks, and legs after 4 weeks, the hindlimbs developing first. The tadpole's tail shrinks and its gills are reabsorbed.*

5. Change complete *After 6 weeks, the metamorphosis from tadpole to adult frog is complete, and the young frog begins to hunt insects.*

Caecilians

CLASS Amphibia
ORDER Gymnophiona
FAMILIES 6
GENERA 33
SPECIES 149

Caecilians have radiated to a variety of aquatic and terrestrial habitats, with specialized characteristics for a fossorial life not present in any other vertebrates—chemo-sensing tentacles, an extra set of jaw-closing muscles, and skin that is ossified to the skull. All species are chacterized by distinct annuli, which have caused some people to confuse them with earthworms. Adults range in size from just under 3 inches (7 cm) to 5¼ feet (1.6 m). Fertilization is internal. Some species are oviparous; others have direct development; and about half of the known species are viviparous. Caecilians have a retractable tentacle on each side of the head, which helps them to search for prey by transmitting chemical clues to the nasal cavity. They feed on earthworms, beetle larvae, termites, and crickets.

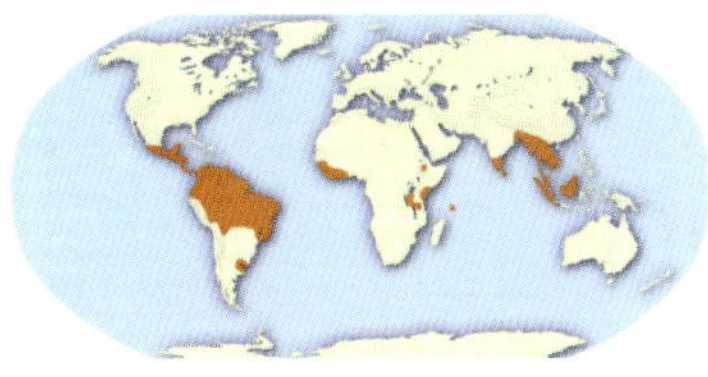

Distribution Caecilians have a limited pantropical distribution in India, southern China, Malaysia, the Philippines, Africa, Central America, and South America. No species are known from Europe, Australia, or Antarctica.

Young caecilians The São Tomé caecilian (*Schistometopum thomense*) (below) is viviparous. The eggs develop in safety in the oviducts. Upon hatching, the larvae feed on nutrient-rich fluids secreted by glands in the oviducts for 7–10 months. The bright coloration of this species is a warning to potential predators that its skin secretions make it unpalatable.

Distinct, widely spaced annuli

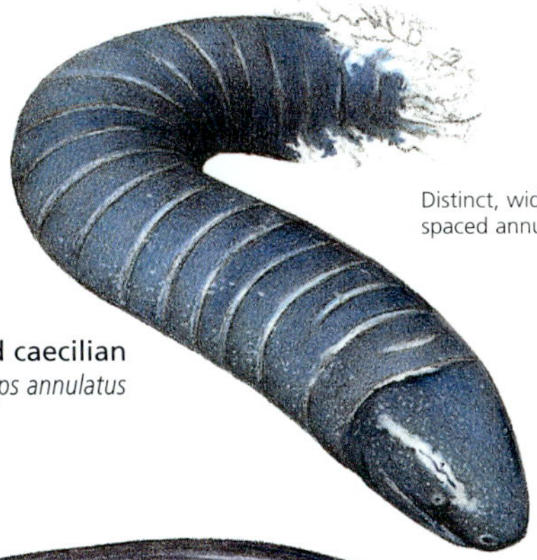

Ringed caecilian
Siphonops annulatus

Mengla County caecilian
Ichthyophis bannanicus

Trade-off
Viviparous caecilians produce fewer young than oviparous species but provide better care.

Cayenne caecilian
Typhlonectes compressicauda

Laterally compressed tail

Epicrionops petersi

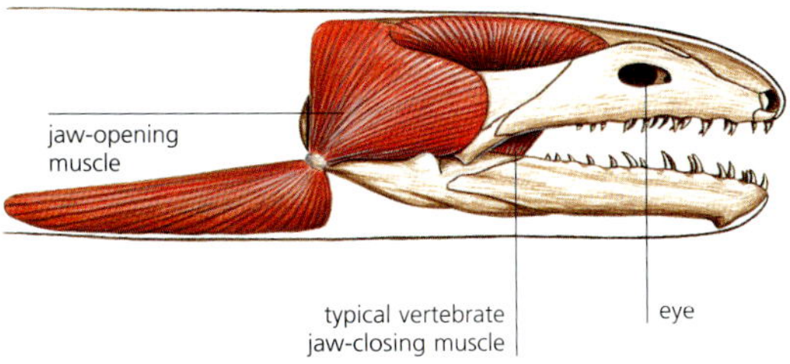

jaw-opening muscle

typical vertebrate jaw-closing muscle

eye

Icthyophis glutinous

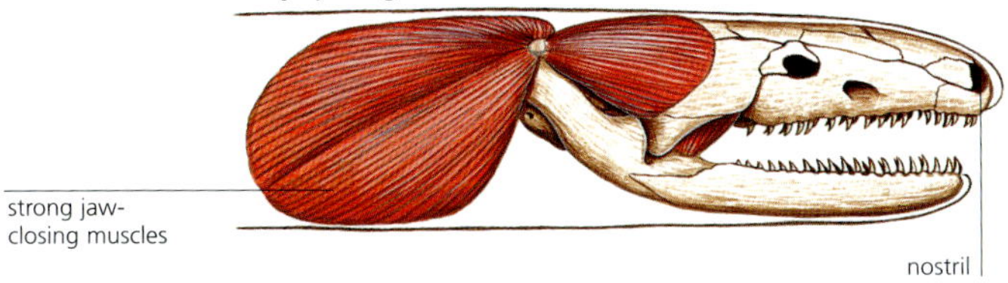

strong jaw-closing muscles

nostril

Caecilian skull Caecilians are built for burrowing, with massive, bullet-shaped skulls. They are streamlined, but reinforced for burrowing through soil or protruding into the substrate in search of invertebrate prey. They move through soil by undulating the body: the muscles move in a wave from the head to the tail. The body curves resist soil or water, and forward motion is achieved.

FROGS AND TOADS

CLASS	Amphibia
ORDER	Anura
FAMILIES	28
GENERA	338
SPECIES	4,937

Anuran is the collective term for this group of tailless amphibians. This is the largest order of amphibians, with nearly 5,000 known species. Anurans come in all shapes and sizes, from the tiny Brazilian Izecksohn's toad (*Psyllophryne didactyla*), which is less than half an inch (1 cm) long, to the African goliath frog (Conraua goliath), which is 12 inches (30 cm) long and weighs 7⅓ pounds (3.3 kg). Frogs and toads are easily identified by their long hindlegs, short trunk, moist skin, and lack of tail. Anurans are the only amphibians that are extremely vocal when breeding. The largest genus of frogs—Eleutherodactylus (rain frogs)—is also the largest genus of all vertebrates.

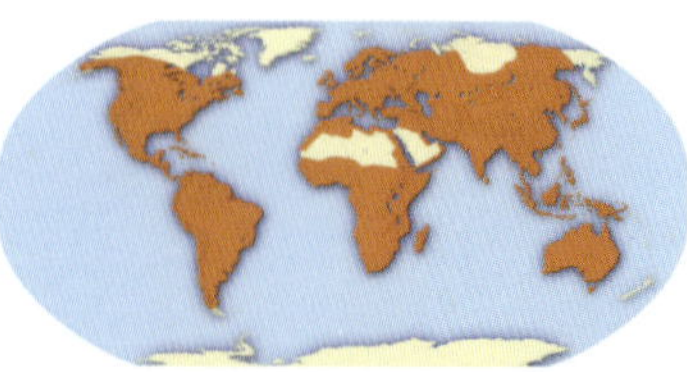

Distribution Frogs and toads occur on all continents except Antarctica. More than 80 percent of species are tropical; at one site alone, 67 species have been recorded. Two species extend north of the Arctic Circle.

SINGING FOR THEIR LIVES

A notable feature of frogs and toads is the voices of the males calling for mates. These calls herald spring in the temperate zone and the onset of the rainy season in the tropics. Each species of frog has a specific call that distinguishes it from all other species. Females can assess the size of the male, and thus his fitness, by the tone of his call. Only males call, and females come to the calling male to mate. These calls also have the disadvantage of attracting some predators, notably bats.

In some species, calling sites are at a premium and males defend them against other calling males. To avoid confrontations, satellite males wait silently near the calling male in hopes of intercepting a gravid female before she reaches him. This lessens their chance of being eaten by a bat, and reduces the energy lost by calling or by fighting with the dominant male.

The above are characteristics of frogs with long breeding seasons. Many other species are explosive breeders, coming to the breeding ponds with the first heavy rains, breeding by the thousands for 2–3 days, and then retreating until the following year. These species usually have loud, carrying calls and lay thousands of small eggs. To avoid predation of their eggs by aquatic predators, many species of tree frogs (*Hylidae*) lay their eggs in vegetation overhanging their breeding ponds. When the eggs hatch, the tadpoles fall into the water, where they complete their development. Other species

Handsome but toxic Poison frogs are some of the most colorful and interesting frogs. Their bright gaudy hues are a warning that their skin glands secrete poisons that can act on the nervous system, causing paralysis or even death.

Frog skeleton Anurans have a shortened vertebral column of nine or fewer vertebrae. The epipodial elements of both fore and hindlimbs are fused; also, the ankle is elongated and made up of two fused elements. The postsacral vertebrae are uniquely fused into a rod-shaped urostyle. The long hindlimbs give frogs the propulsive force to propel themselves forward. A short, compact body is easier to move forward. The reinforced pectoral girdle and forelimbs are built to absorb the shock of landing.

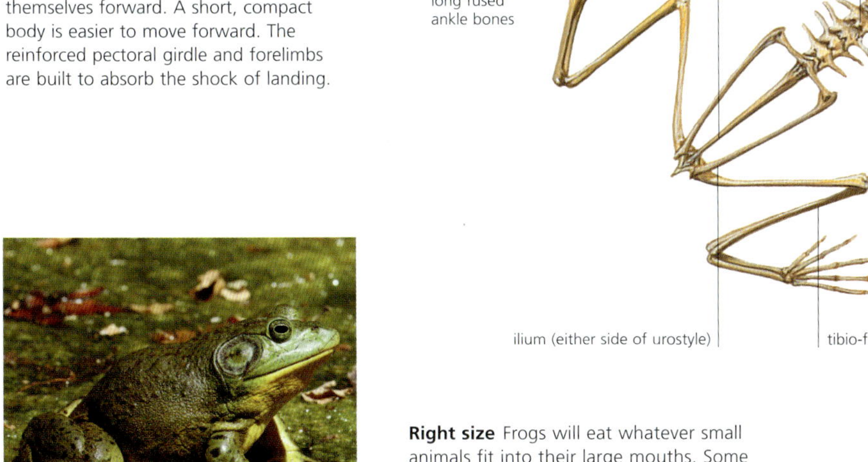

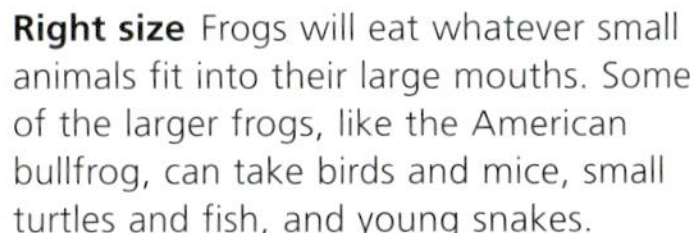

Right size Frogs will eat whatever small animals fit into their large mouths. Some of the larger frogs, like the American bullfrog, can take birds and mice, small turtles and fish, and young snakes.

of casque-headed tree frogs (*Osteocephalus*) lay a few large fertilized eggs in tree holes. The female returns regularly to deposit unfertilized eggs in the nest to feed her tadpoles. Many species of poison frogs (*Dendrobatidae*) guard their eggs in terrestrial nests and, upon hatching, carry their tadpoles to the water, or in some cases up trees and into bromeliads. Many species of *Eleutherodactylus* lay 15–25 large yolked eggs in terrestrial nests, the eggs having direct development. Midwife toads, Surinam toads, and gastric-brooding frog females carry the eggs with them until they hatch into small frogs. Other *Leptodactylid* and *Rhacophorid* frogs build foam nests to deposit their eggs and rear their tadpoles.

Frogs are indicators of environmental pollution. Since the 1970s, there has been growing worldwide evidence that frog populations are declining on a massive scale and that many species are going extinct, even though their populations appear to be within pristine areas. The decline may be the result of a number of factors, including pollution, pesticides (such as DDT), herbicides (such as Atrizine), road salts, UV radiation, global warming, lower rainfall, or lower humidity. At present, the fungus chytridiomycosis seems to be the greatest threat. The fungus, which probably originated in Africa, invades the skin and lowers the frogs' immune system so they are more susceptible to diseases.

Energy transfers The green and golden bell frog leaps into the water to escape a predator. Frogs and tadpoles transform the energy in their food (algae, detritus, and insects) into larger packets of energy that are then available to bigger predators.

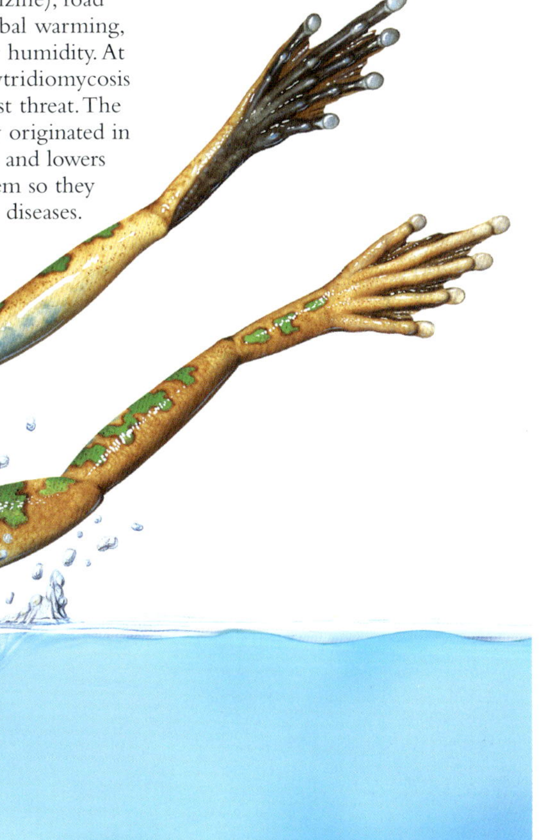

Brown midwife toad
Alytes cisternasii

Yellowbelly toad
Bombina variegata

Brown New Zealand frog
Leiopelma hamiltoni

Olive midwife toad
Alytes obstetricans

Hochstetter's New Zealand frog
Leiopelma hochstetteri

Oriental firebelly toad
Bombina orientalis

Belly and underside of limbs are flashed as a warning to predators

Painted frog
Discoglossus pictus

Corsica painted frog
Discoglossus montalentii

Tailed frog
Ascaphus truei

ANURAN FERTILIZATION

Fertilization of the eggs usually occurs while the frogs are in amplexus. As the female deposits her eggs, males release sperm above them in the water. The coqui and some other terrestrial species achieve internal fertilization by cloacal apposition; only the tailed frog has an intermittent organ.

Frogs in amplexus
The female carries the male to an appropriate spot to deposit the eggs.

Tailed frog This frog has no breeding call. Its "tail"—used for internal fertilization in fast-moving streams—is unique among anurans. Eggs are deposited in strings under rocks in streams. Tadpoles take 1–4 years to metamorphose to juveniles, and then 7–8 years to first breeding.

N.W. USA

- Up to 2 in (5 cm)
- Terrestrial, aquatic
- Summer
- Common

Malayan horned frog
Megophrys nasuta

Shaping frog
Atympanophrys shapingensis

Common parsley frog
Pelodytes punctatus

Surinam toad
Pipa pipa

Eggs embedded in back

Mexican spadefoot toad
Spea multiplicata

Mid-dorsal red-orange stripe

Mueller's clawed frog
Xenopus muelleri

Mexican burrowing toad
Rhinophrynus dorsalis

Cape clawed frog
Xenopus gilli

Syrian spadefoot toad
Pelobates syriacus

Common parsley frog Male parsley frogs can be heard calling from below the surface of the water. The tadpoles are larger than the adult frogs, at up to 2½ in (6 cm) in length. The eggs are laid in strings across aquatic plants.

Up to 2 in (5 cm)
Terrestrial, aquatic
Spring
Common

Iberia, France, W. Belgium, N.W. Italy

Surinam toad During amplexus, the female releases eggs one at a time. The male fertilizes and embeds each egg in the skin of her back, where the eggs sink in and form a honeycomb of pockets. After 3–4 months, froglets emerge from her back.

Up to 8 in (20 cm)
Aquatic
Spring
Common

N. South America

Mexican spadefoot toad Spadefoots are nocturnal during the summer rainy season, when they are breeding. The rest of the year, they occupy burrows that they dig in the soft earth.

Up to 2 in (5 cm)
Terrestrial
Summer
Common

S.C. USA, N. Mexico

Red-crowned toadlet
Pseudophryne australis

Skeleton ghost frog
Heleophryne rosei

Spatulate toes used for climbing rocks

Thomasset's frog
Nesomantis thomasseti

Turtle frog
Myobatrachus gouldii

Crucifix toad
Notaden bennettii

Natal ghost frog
Heleophryne natalensis

Giant banjo frog
Limnodynastes interioris

Differs from the giant burrowing frog (*Heliopórus australicacus*) as lacks black spines on back and throat

Western marsh frog
Heleioporus barycragus

Turtle frog This frog burrows head-first with its strong forelimbs, often into termite nests to eat the inhabitants. It lives in sandy soil. Males call from the soil surface or with only their heads uncovered. Up to 40 large eggs are laid underground.

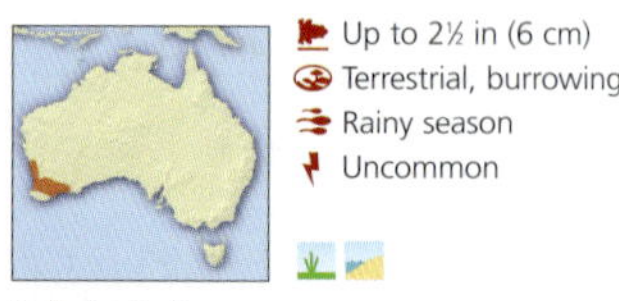

Up to 2½ in (6 cm)
Terrestrial, burrowing
Rainy season
Uncommon

S.W. Australia

Crucifix toad This fossorial toad has a cross-shaped pattern of spots on its back. It spends most of its life underground and only emerges to breed in temporary pools after heavy rains. Small eggs are deposited in the water.

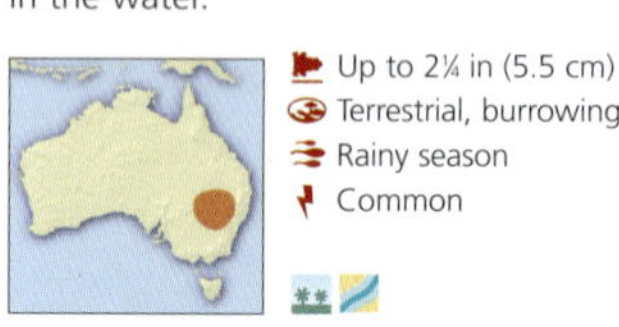

Up to 2¼ in (5.5 cm)
Terrestrial, burrowing
Rainy season
Common

S.E. Australia

Giant banjo frog Adults spend daylight hours and drier months buried beneath the surface. Males call while floating on vegetation or concealed in burrows at the water's edge. The call sounds like a banjo being plucked.

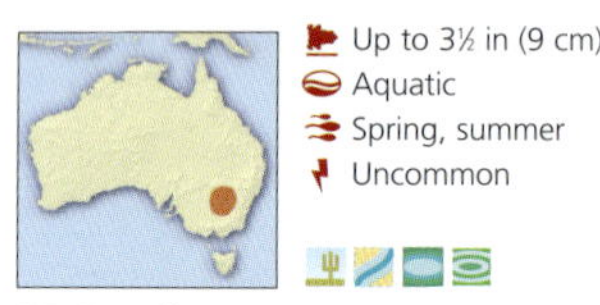

Up to 3½ in (9 cm)
Aquatic
Spring, summer
Uncommon

S.E. Australia

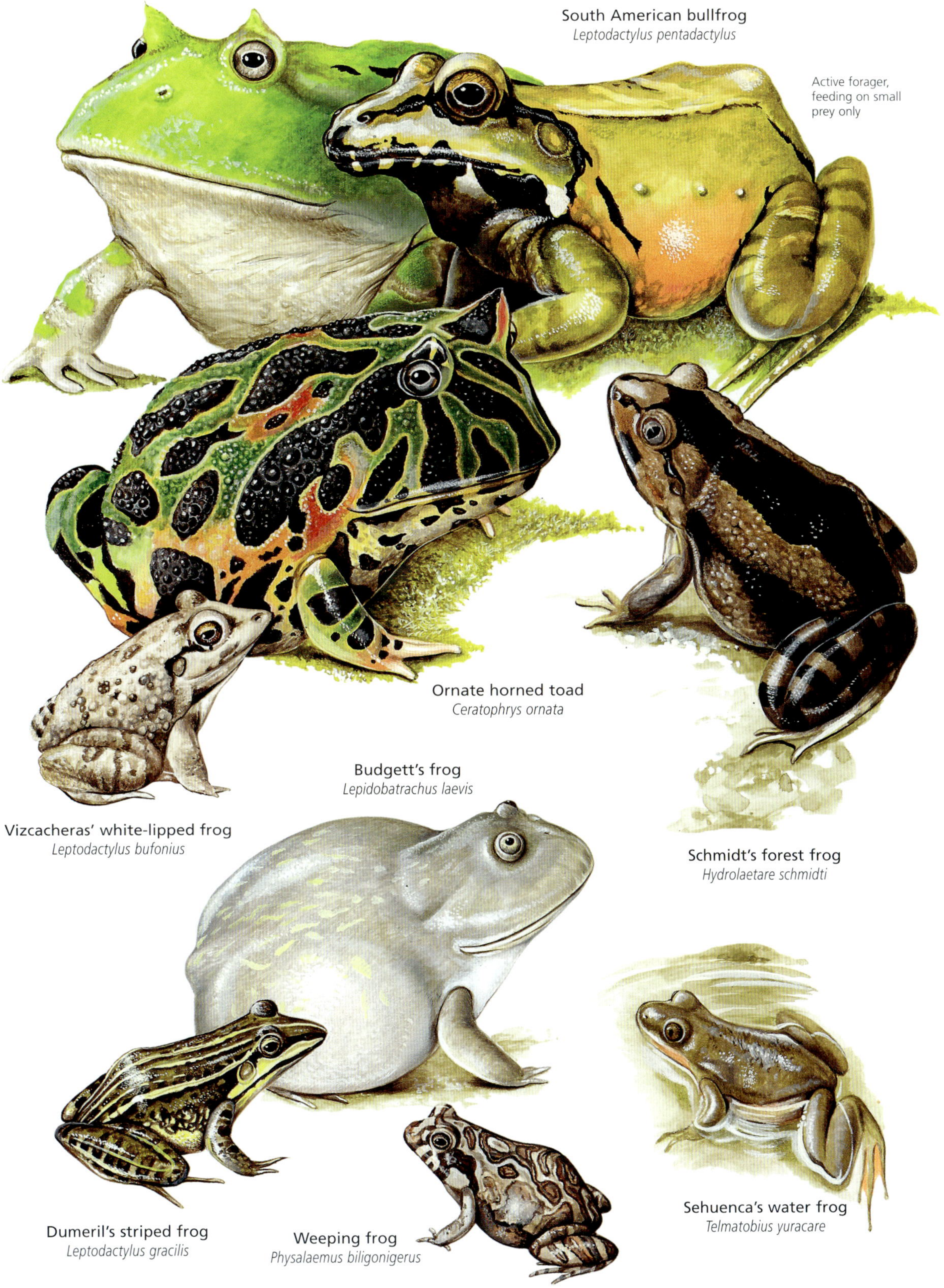

Surinam horned frog
Ceratophrys cornuta

South American bullfrog
Leptodactylus pentadactylus

Active forager, feeding on small prey only

Ornate horned toad
Ceratophrys ornata

Vizcacheras' white-lipped frog
Leptodactylus bufonius

Budgett's frog
Lepidobatrachus laevis

Schmidt's forest frog
Hydrolaetare schmidti

Sehuenca's water frog
Telmatobius yuracare

Dumeril's striped frog
Leptodactylus gracilis

Weeping frog
Physalaemus biligonigerus

Surinam horned frog When prey ventures nearby, this frog lurches forward, opening the mouth that dominates the entire top of its head, and swallows the prey whole.

Up to 4¾ in (12 cm)
Terrestrial
Fall rainy season
Common

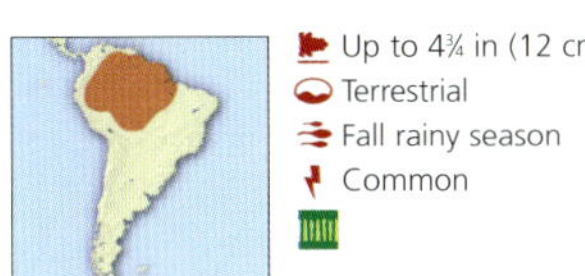

Amazonia (South America)

DROUGHT RESISTANCE

During the dry winter months, ornate horned toads are inactive underground, encased in a hard shell composed of unshed skin. This "cocoon" protects the animal from excessive water loss and allows it to survive until the rains arrive. These signal the beginning of the wet summer months in South America, which last from October to February.

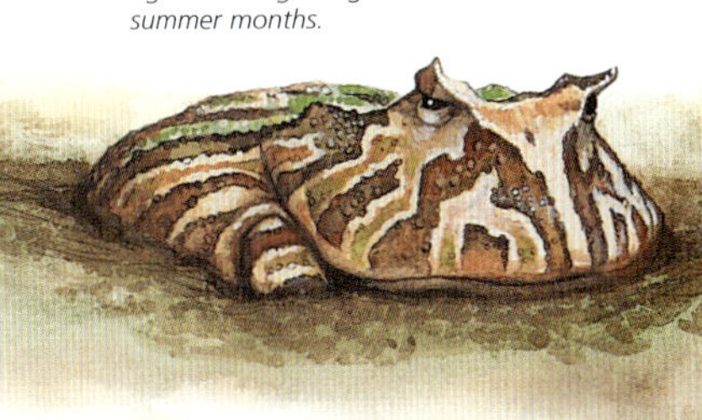

Digging in *Ornate horned toads will remain buried until rains signal the beginning of the wet summer months.*

Coqui
Eleutherodactylus coqui

Calls can reach 100 decibels; recently introduced to Hawaii

Barking frog
Eleutherodactylus augusti

Seldom venture out from their limestone cave haunts

No webbing between the toes

Jimenez robber frog
Eleutherodactylus lacrimosus

Carabaya robber frog
Eleutherodactylus ockendeni

Peru robber frog
Eleutherodactylus peruvianus

Variable robber frog
Eleutherodactylus variabilis

Lowland tropical frog
Adenomera andreae

Red spots on thighs mimic those of a poisonous frog

Gold-striped frog
Lithodytes lineatus

Darwin's mouthbreeder
Rhinoderma darwinii

Coqui This species gets its name from its high-pitched chirp, which sounds like "co qui." Eggs are laid on land with direct development in the egg. Coquis have been introduced into Florida and Hawaii, United States.

Up to 2¼ in (5.5 cm)
Terrestrial, arboreal
Year round
Common

Puerto Rico (Caribbean)

Barking frog The breeding call of this species sounds like a dog barking when heard at a distance; close up, it is a guttural "whurr." Females make a screech when grasped. When captured, their bodies puff up immensely.

Up to 3¾ in (9.5 cm)
Terrestrial
Spring
Common

S.W. USA to C.W. Mexico

Peru robber frog This species has direct development. The eggs are deposited in moist areas in the leaf litter. The males, which are considerably smaller than the females, call day and night from the forest floor.

Up to 1¼ in (3 cm)
Terrestrial
Fall rainy season
Common

W. Amazonia (Brazil, Bolivia, Peru, Ecuador)

ADVANCE OF THE CANE TOAD

Successful stories of the cane toad (*Bufo marinos*) (also known as the marine toad) being introduced into the sugarcane fields of Puerto Rico to control insect pests convinced the sugarcane growers in Queensland, Australia, to import and breed 100 adult toads. In 1935, they released 62,000 subadults in Queensland to control the grayback cane beetle (*Dermolepida albohirtum*). Since the neotropical cane toad has no natural predators in Australia, populations have reached plague proportions. The toads adapted well to feeding on all animals smaller than themselves, even competing with dogs for the rations in their bowls. Studies in Brazil are yet to come up with a solution.

Cane toad poison When cane toads are threatened, they inflate their bodies and begin sweating a milky, latex-like liquid from the paratoid glands. In extreme cases, the secretions can be projected up to 3 feet (92 cm) toward the aggressive attacker, but are not toxic unless ingested. The poison has been used as a hallucinogenic drug.

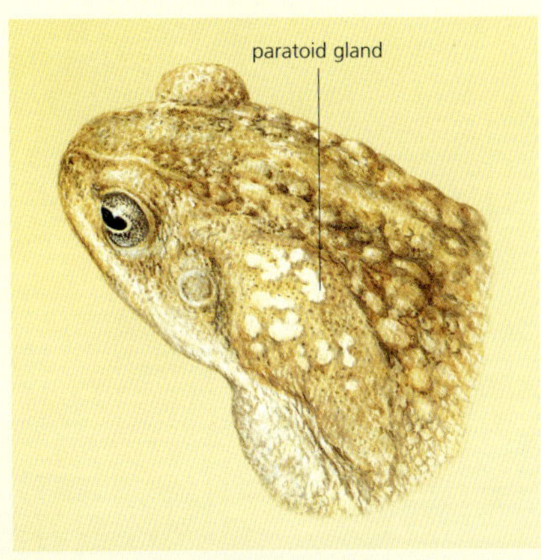

Breeding The cane toad breeds year-round in Australia in temporary pools, stock tanks, lakes, and streams. The prolific toads can produce up to 13,000 eggs at a time. Their tadpoles transform rapidly and have minimal habitat requirements: algae and water. Although cane toads take several years to mature, they can live for over 20 years. Calling cane toads interfere with the breeding choruses of native frogs.

1 2 3

4 5

White chest and abdomen

6 7 8

Raised ridges of warts

9

1 Cururu toad *Bufo paracnemis*

2 Colombian giant toad *Bufo blombergi*

3 Cane toad *Bufo marinus*

4 Cameroon toad *Bufo superciliaris*

5 Square-marked toad *Bufo regularis*

6 Tschudi's Caribbean toad *Bufo pelticephalus*

7 Colorado river toad *Bufo alvarius*

8 Berber toad *Bufo mauritanicus*

9 Common Asian toad *Bufo melanostictus*

AMPHIBIAN EGG

Amphibian embryos have gelatinous membranes around them, but lack the protective amnion present in reptiles. Since eggs also lack shells, they must be protected from desiccation until tadpoles emerge as free-swimming larvae. Eggs of some species contain enough yolk to nourish embryos until they hatch as miniature adults.

Colorado river toad The skin glands of these toads produce a toxin that can cause intoxication and hallucination in humans, and is a controlled substance in the United States, where people have been arrested for milking toads.

Up to 8 in (20 cm)
Terrestrial
Summer
Common

S.W. USA & N.W. Mexico

Common Asian toad Once common, this species is declining due to habitat alteration and loss, caused by drought, deforestation, drainage of habitat, pesticides, fertilizers, and pollutants.

Up to 6 in (15 cm)
Terrestrial
Summer
Declining

S.E. Asia

Everett's Asian tree toad
Pedostibes everetti

Sonoran green toad
Bufo retiformis

Horned toad
Bufo ceratophrys

South American common toad
Bufo margaritifer

Red toad
Schismaderma carens

Oak toad
Bufo quercinus

Red-spotted toad
Bufo punctatus

Two color morphs of harlequin frogs

Green toad
Bufo viridis

Harlequin frog
Atelopus varius

Green toad Green toads form dense populations in areas altered by humans, often in higher numbers than in nearby natural habitats. Abundance can reach more than 100 individuals per 1,075 square feet (100 sq. m).

Europe to Asia

Up to 4¾ in (12 cm)

Terrestrial
Spring
Common

TOADS VERSUS FROGS

In Europe and North America, toads (family Bufonidae) have short legs for hopping, dry warty skin, and are terrestrial. Frogs (family Ranidae) have long, slender legs for leaping great distances, moist skin, and are aquatic. However, the use of the terms "frog" and "toad" depends on the region of the world. In Africa, the smooth and moist-skinned aquatic Cape clawed frog (*Xenopus gilli*) is called a clawed toad.

To kiss a toad
Even though South Africans call Bufo pardalis *(above) the leopard frog, it is part the toad family Bufonidae.*

Australian lace-lid frog
Nyctimystes dayi

Dainty green tree frog
Litoria gracilenta

Southern bell frog
Litoria raniformis

Blue Mountains tree frog
Litoria citropa

Green and golden bell frog
Litoria aurea

White's tree frog
Litoria caerulea

Short-footed frog
Cyclorana brevipes

Paradox frog
Pseudis paradoxa

Water-holding frog
Cyclorana platycephala

Extensively webbed feet

Short-footed frog This is a robust, burrowing frog with striking, highly variable markings. The back is dark brown with silver-brown blotches. The frog usually also has a silver-brown stripe on the back and a white belly.

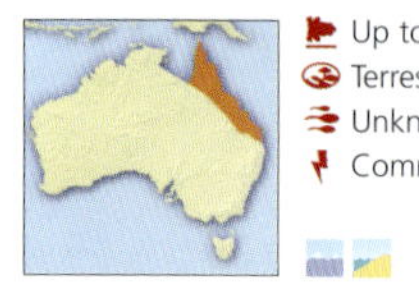

Up to 2 in (5 cm)
Terrestrial, burrowing
Unknown
Common

N.E. Australia

Paradox frog This frog has large webbed feet and extremely slippery skin. The tadpoles are three times as large as the adults, up to 10 inches (25 cm) in length.

Up to 3 in (7.5 cm)
Aquatic
Rainy season
Common

Amazonia (South America) to N. Argentina

Water-holding frog This species spends the long dry periods, which can be up to several years long, in burrows deep underground. It surrounds itself with a cocoon-like chamber, formed with sloughed dead layers of its skin.

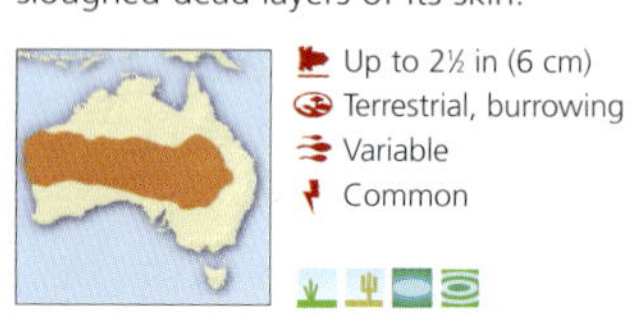

Up to 2½ in (6 cm)
Terrestrial, burrowing
Variable
Common

Central Australia

Fringed leaf frog
Agalychnis craspedopus

Painted-belly leaf frog
Phyllomedusa sauvagii

Shovel-headed tree frog
Triprion spatulatus

Jaguar leaf frog
Phyllomedusa palliate

Cayenne slender-legged tree frog
Osteocephalus leprieurii

Map tree frog
Hyla geographica

Red-eyed tree frog
Agalychnis callidryas

Veined tree frog
Phrynohyas venulosa

Rusty tree frog
Hyla boans

Jordan's casque-headed tree frog
Trachycephalus jordani

Red-eyed tree frog This species lays its green eggs on vegetation or rocks overhanging temporary ponds. The tadpoles hatch and fall into the water to complete development. The males become sexually mature after a year.

Up to 3 in (7.5 cm)
Arboreal
Summer
Common

S. Mexico to Colombia

FROG FEET

Frogs have four digits on the forelimbs and five on the hindlimbs. The shapes vary depending on their habits: modified claws with extra tubercules for digging, or toe pads for climbing, or fully webbed feet ideal for swimming.

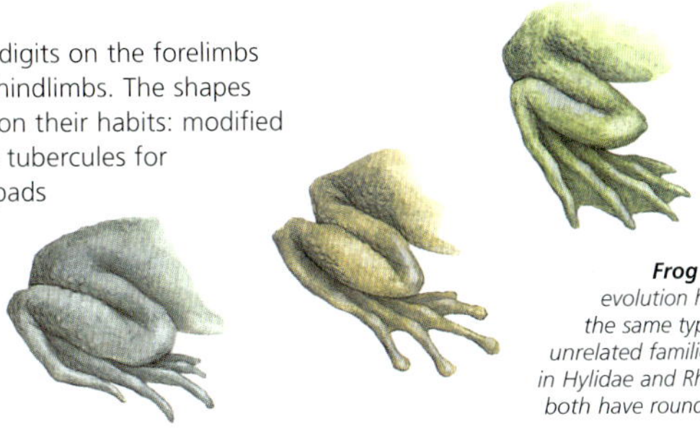

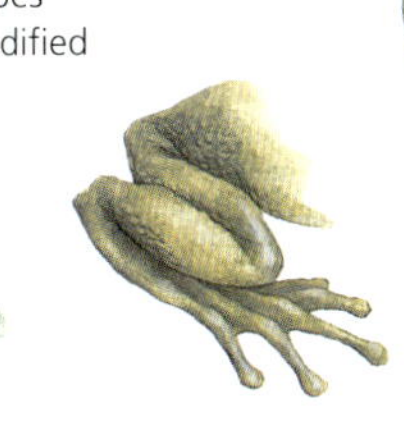

Frog toes *Parallel evolution has produced the same types of toes in unrelated families: tree frogs in Hylidae and Rhacophoridae both have rounded toe pads.*

Nicaragua giant glass frog
Centrolene prosoblepon

Glass frogs named for their translucent undersides

Hourglass tree frog
Hyla ebraccata

Bereis' tree frog
Hyla leucophyllata

Barking tree frog
Hyla gratiosa

Equally at home in treetops or underground burrows

Green tree frog
Hyla cinera

Sumaco horned tree frog
Hemiphractus proboscideus

Mexican tree frog
Smilisca baudini

European tree frog
Hyla arborea

Marsupial frog
Gastrotheca marsupiata

Female frog without eggs in pouch

Ornate chorus frog
Pseudacris ornata

Nicaragua giant glass frog This species calls—using a three-note series of high-pitched "ticks"—from above streams. It attaches eggs to leaves that overhang the streams. Tadpoles attach to rocks with their suction mouth.

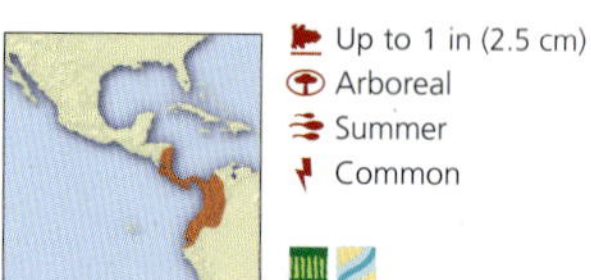

Up to 1 in (2.5 cm)
Arboreal
Summer
Common

N. Nicaragua to Colombia & Ecuador

Barking tree frog This frog gets its name from its loud raucous call of nine or ten notes, possibly a territorial call. Its breeding call is a simple "doonk" or "toonk." The female deposits eggs singly on the bottom of the breeding pond.

Up to 2¾ in (7 cm)
Terrestrial, arboreal
Spring, summer
Common

S.E. USA

European tree frog This tree frog species is declining in western and central Europe due to loss of habitat, habitat isolation, pollution, and climate change. Flying insects make up the majority of its diet.

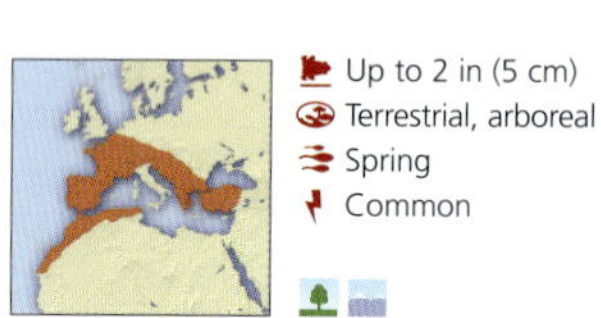

Up to 2 in (5 cm)
Terrestrial, arboreal
Spring
Common

Europe, W. Asia, N.W. Africa

TREE FROGS

The family Hylidae—tree frogs—comprises 855 species in 42 genera, distributed primarily in North America, South America, and Australia, and to a lesser extent in Europe and Asia. Tree frogs range in size from the half an inch (12 mm) adult javelin frog (*Litoria microbelos*) to the 5½ inch (14 cm) Hispaniola tree frog (*Hyla vasta*). Most hylids are arboreal, but some are terrestrial, aquatic, or burrowing. Most possess adhesive toe pads, which allow them to cling to any surface. Almost all are very flattened and streamlined, with long legs that are useful for leaping from branch to branch. Tree frogs have horizontally elliptical pupils, with the exception of the Phyllomedusids, which have vertical pupils.

Burrowing habits The burrowing tree frog (*Pternohyla fodiens*) is found in many habitats subject to seasonal inundation. They are well adapted for burrowing and spend the dry season in a dormant state beneath the surface. They will only breed after heavy rain.

Hanging on *Glass frogs typically call from the upper side of leaves overhanging streams and attach their eggs to the underside of the same leaves.*

Red-eyed tree frog Climbing easily with sucker-padded toes, these frogs spends most of their lives in the trees of the always-humid tropical rainforests of Central America. When these frogs mate, the males come down from the canopy to the low-growing foliage and, in a chorus of raucous croaking, call the females out of the treetops.

Ornate rice frog
Microhyla ornata

Nosy Be giant tree frog
Platypelis milloti

Red rain frog
Scaphiophryne gottlebei

Dotted humming frog
Chiasmocleis ventrimaculata

Red-banded frog
Phrynomantis bifasciata

Shown in defensive posture, with the head lowered and the rear legs extended

Muller's termite frog
Dermatonotus muelleri

Great Plains narrow-mouthed toad
Gastrophryne olivacea

Malaysian narrow-mouthed toad
Kaloula pulchra

Males are patterned like females, but are slightly smaller in size

Tomato frog
Dyscophus antongilii

RAIN FROGS OF AFRICA

Rain frogs (genus *Breviceps*) have short heads and stout bodies. They appear above ground only during torrential rain. Males have loud, bellowing calls that can be heard from great distances, and sticky skin secretions that attach them to the females during amplexus. They lay their eggs underground, and the young frogs develop without any need for water.

Puffed up like a balloon *Desert rain frog males (*Breviceps macrops*) inflate their bodies enormously when calling for mates.*

Tomato frog The bright red coloration of this frog acts as a clear warning to potential predators that it is a toxic meal. A white, sticky fluid secreted from the skin deters snakes and can produce an allergic reaction in humans.

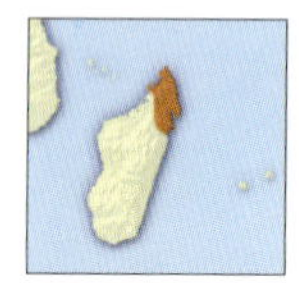

N.E. Madagascar

Up to 4 in (10 cm)
Terrestrial, aquatic
Summer
Vulnerable

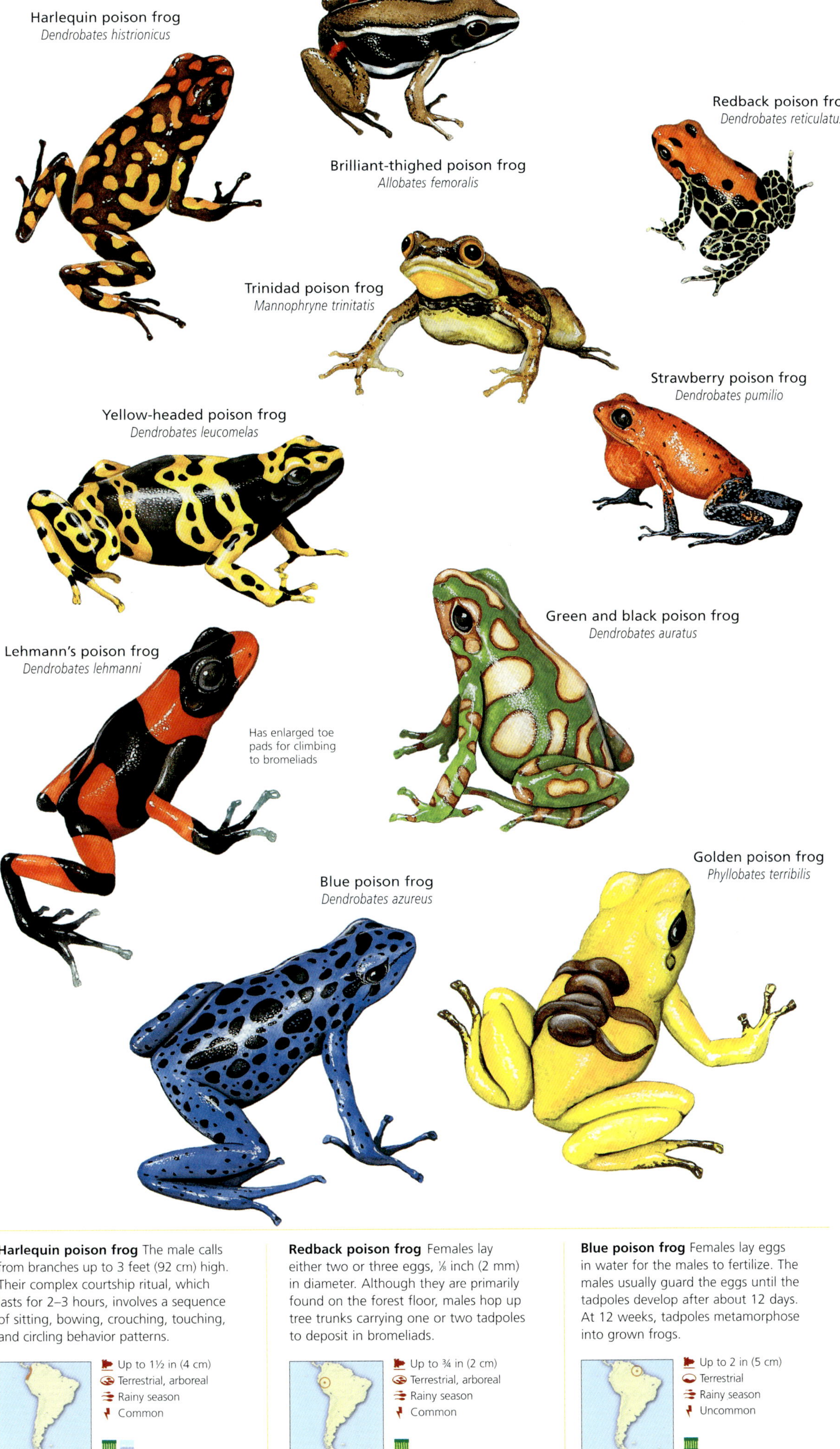

Harlequin poison frog
Dendrobates histrionicus

Brilliant-thighed poison frog
Allobates femoralis

Redback poison frog
Dendrobates reticulatus

Trinidad poison frog
Mannophryne trinitatis

Strawberry poison frog
Dendrobates pumilio

Yellow-headed poison frog
Dendrobates leucomelas

Green and black poison frog
Dendrobates auratus

Lehmann's poison frog
Dendrobates lehmanni

Has enlarged toe pads for climbing to bromeliads

Blue poison frog
Dendrobates azureus

Golden poison frog
Phyllobates terribilis

Harlequin poison frog The male calls from branches up to 3 feet (92 cm) high. Their complex courtship ritual, which lasts for 2–3 hours, involves a sequence of sitting, bowing, crouching, touching, and circling behavior patterns.

Colombia, Ecuador

- Up to 1½ in (4 cm)
- Terrestrial, arboreal
- Rainy season
- Common

Redback poison frog Females lay either two or three eggs, ⅛ inch (2 mm) in diameter. Although they are primarily found on the forest floor, males hop up tree trunks carrying one or two tadpoles to deposit in bromeliads.

N.E. Peru, W. Brazil

- Up to ¾ in (2 cm)
- Terrestrial, arboreal
- Rainy season
- Common

Blue poison frog Females lay eggs in water for the males to fertilize. The males usually guard the eggs until the tadpoles develop after about 12 days. At 12 weeks, tadpoles metamorphose into grown frogs.

Suriname

- Up to 2 in (5 cm)
- Terrestrial
- Rainy season
- Uncommon

Madagascar reed frog
Heterixalus madagascariensis

Striped spiny reed frog
Afrixalus dorsalis

Distinctive yellow stripe on body, bordered on both sides with black

Yellow-striped reed frog
Hyperolius semidiscus

Natal forest tree frog
Leptopelis natalensis

Seychelles reed frog
Tachycnemis seychellensis

Marbled reed frog
Hyperolius marmoratus

Weal's running frog
Semnodactylus wealii

Runs instead of hopping, jumping, or crawling

Tulear golden frog
Mantella expectata

Madagascar golden frog
Mantella madagascariensis

Red-legged kassina
Kassina maculata

Madagascar reed frog Common around dunes, savanna, and deforested habitats of the east coast of Madagascar, these frogs are active at night. During the day they take refuge near pools, submerging when disturbed by a potential predator.

Up to 1½ in (4 cm)
Terrestrial
Year round
Common

Madagascar

Tulear golden frog The species is subject to intense collecting for the pet trade at the beginning of the rainy season (October to December). During the first rains, these frogs, which breed explosively, come out from their refuges and are easy to capture.

Up to 1¼ in (3.2 cm)
Terrestrial
Rainy season
Uncommon

Madagascar

Weal's running frog Males call from elevated positions in vegetation on the banks of or partly submerged in a pond. The voice is a coarse, loud rattle. One call, lasting about half a second, is emitted every 3–5 seconds.

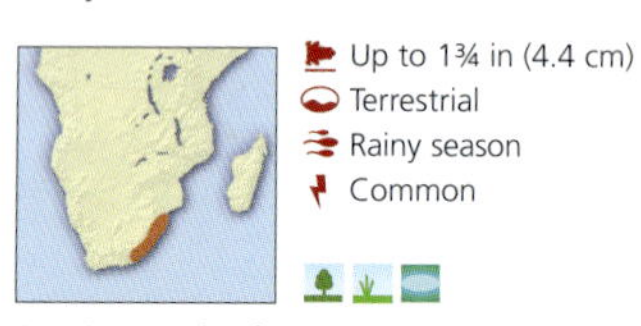

Up to 1¾ in (4.4 cm)
Terrestrial
Rainy season
Common

E. & S.E. South Africa

Red-eared frog
Rana erythraea

African bullfrog
Pyxicephalus adspersus

European common frog
Rana temporaria

Pickerel frog
Rana palustris

Malaysian frog
Rana luctuosa

Marsh frog
Rana ridibunda

Pig frog
Rana grylio

Dorsum has two rows of square chocolate-colored blotches, while the leopard frog, for which it has been mistaken, usually has round blotches

Amazon river frog
Rana palmipes

Bullfrog
Rana catesbeiana

Pickerel frog These frogs produce toxic skin secretions that irritate human skin and can be fatal to small animals, especially other amphibians. Many frog-eating snakes avoid these frogs.

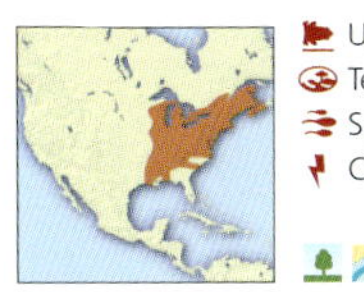

Up to 3 in (7.5 cm)
Terrestrial, aquatic
Spring
Common

E. USA

Amazon river frog This aquatic species forages day and night along the edges of rivers, permanent ponds, and lakes. They consume invertebrates and vertebrates, such as insects, fishes, other frogs, and small birds, both in and out of the water.

Up to 4½ in (11.5 cm)
Aquatic
Rainy season
Common

C. America to Peru & Brazil

Bullfrog This large frog eats anything that moves that it can swallow. This species was introduced into the western United States, where it reduced or extirpated local species of amphibians and reptiles. Tadpoles take 2 years to develop.

Up to 8 in (20 cm)
Aquatic
Summer
Common

E. USA

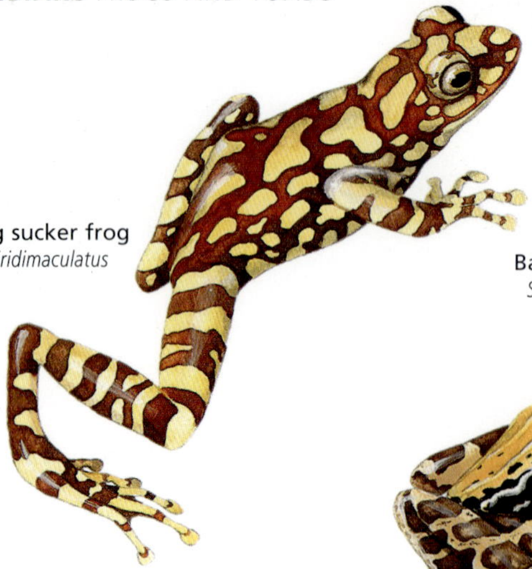

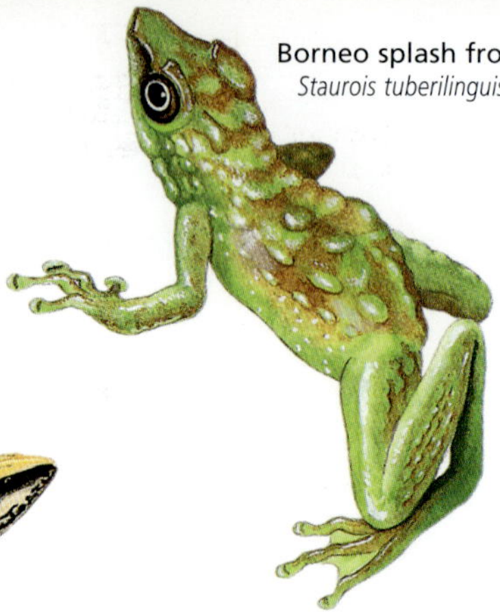

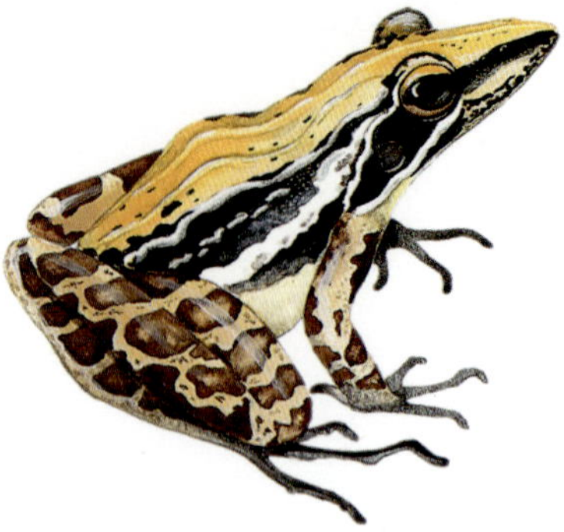

Dahaoping sucker frog
Amolops viridimaculatus

Borneo splash frog
Staurois tuberilinguis

Banded stream frog
Strongylopus bonaspei

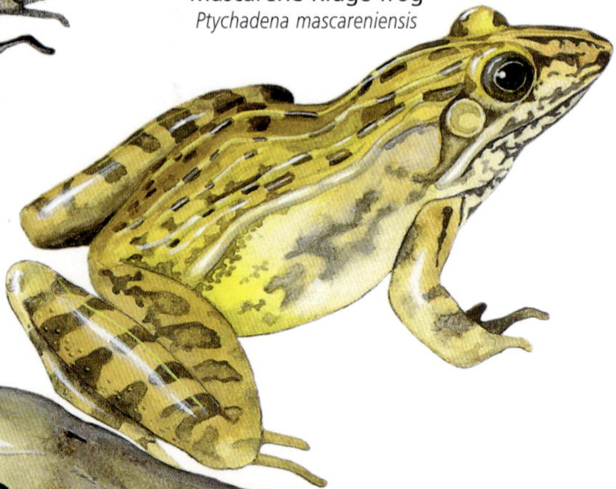

Mascarene Ridge frog
Ptychadena mascareniensis

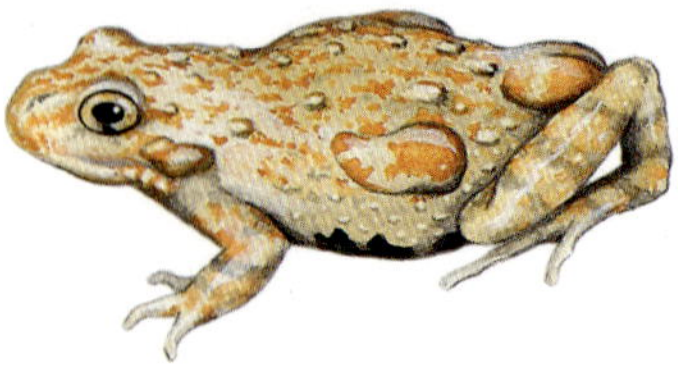

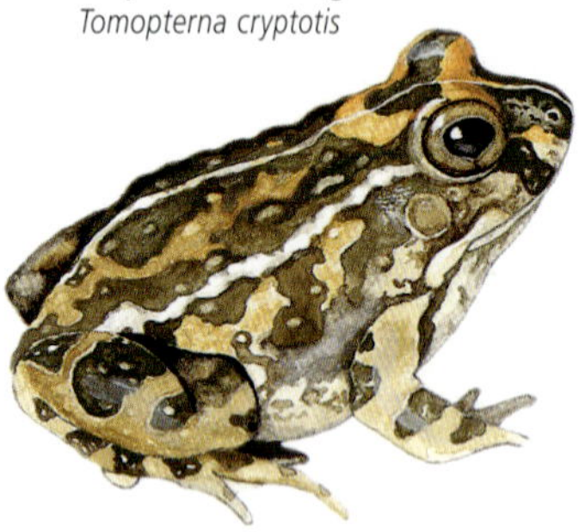

Bulbous glands on the back and sides

Cape dainty frog
Cacosternum capense

Striped sand frog
Tomopterna cryptotis

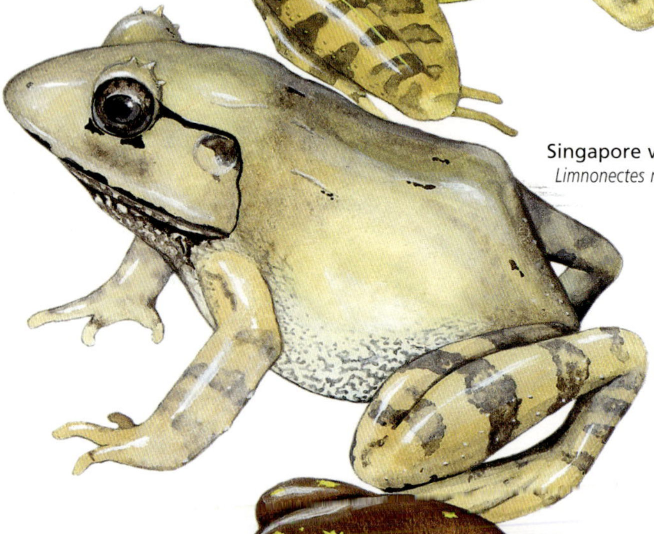

Singapore wart frog
Limnonectes malesianus

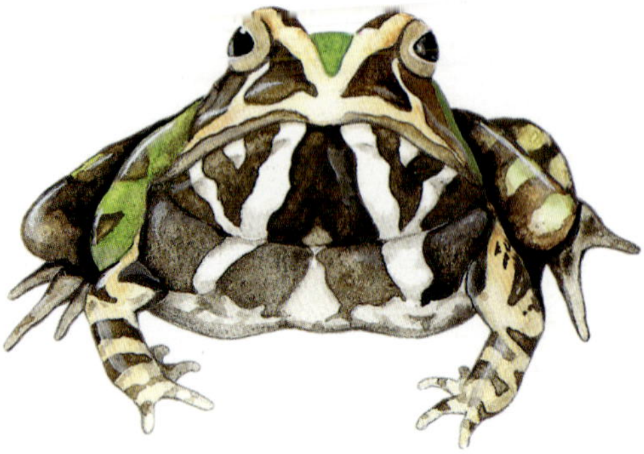

African ornate frog
Hildebrandtia ornata

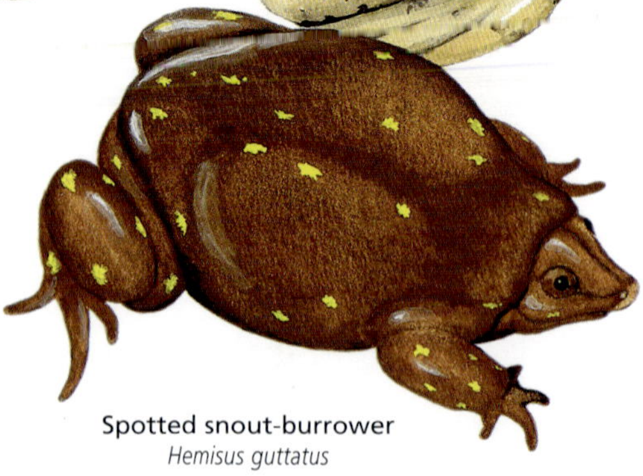

Spotted snout-burrower
Hemisus guttatus

ANURAN DEFORMITIES

Frogs abnormalities are a warning about chemical pollution when they begin to appear with grossly deformed bodies. Part of the chemical pollution is from the progesterone used in birth control pills not being filtered out at sewage-treatment plants.

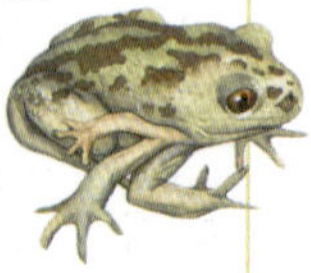

Multi-legged frog
Chemical pollution in breeding ponds may cause frogs to develop with extra legs.

Singapore wart frog In the field, this frog can be distinguished from Blyth's giant frog (*Limnonectes blythii*) by the sharp angle of the dark line behind the eye, and by the clearly defined black patch on the upper tympanum.

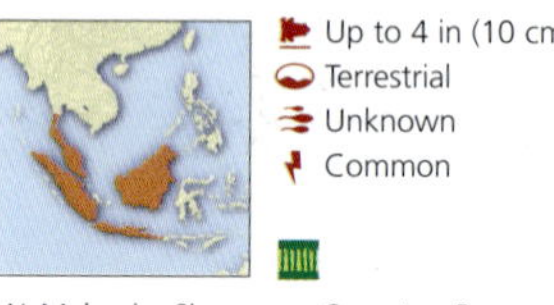

W. Malaysia, Singapore, Sumatra, Borneo, Java

- Up to 4 in (10 cm)
- Terrestrial
- Unknown
- Common

Cape dainty frog This tiny South African frog lives in inundated grasslands and depressions in dunes and on cultivated lands on poorly drained clays.

South Africa

- Up to 1½ in (4 cm)
- Terrestrial
- Winter
- Near threatened

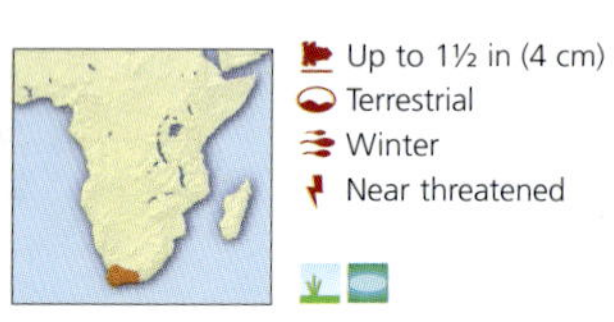

Bush squeaker
Arthroleptis wahlbergii

Java whipping frog
Polypedates leucomystax

Painted Indonesian tree frog
Nyctixalus pictus

Warty tree frog
Theloderma asperum

Distinctive black membranes between the heavily webbed toes

Abah River flying frog
Rhacophorus nigropalmatus

Madagascar bright-eyed frog
Boophis madagascariensis

Hairy frog
Trichobatrachus robustus

African gray tree frog
Chiromantis xerampelina

Bongon whipping frog
Polypedates otilophus

Highly vascularized papillae on the hindlimbs

Java whipping frog One of the more successful frogs of modern times, it adapts well to human environments. Their foam nests are found inside bathrooms, water tanks, drains, or any damp spot.

Up to 3¼ in (8 cm)
Arboreal
April to December
Common

S.E. Asia

Abah River flying frog The female beats a body fluid into a foam in a nest of twigs and leaves above water. When the fertilized embryos hatch into tadpoles, the nest deteriorates and they drop into the water.

Up to 4 in (10 cm)
Arboreal
Rainy season
Common

Indonesia, Malaysia, Thailand

CONTROLLED FALLING

Frogs with expanded webs of the feet are able to fall in a controlled fashion to avoid predation or move about. They can adjust the impact and control the direction of their fall.

Expanded webbing
*The Java flying frog (*Rhacophorus reinwardtii*) uses its expanded webbed feet to guide its leaps from tree to tree.*

FISHES

PHYLUM	Chordata
SUBPHYLUM	Vertebrata
CLASSES	5
ORDERS	62
FAMILIES	504
SPECIES	25,777

With more than 25,000 described so far, and perhaps several thousand more still to be identified, fishes comprise over half of all living vertebrate species. They first evolved about 500 million years ago in fresh water and now exploit almost every aquatic habitat from polar seas to tropical ponds. Some even live briefly on land. This diverse group includes tiny gobies ⅖ inch (1 cm) long and massive 40 foot (12 m) whale sharks, drab wobbegongs and vibrantly colored butterfly fish, fiercely predatory great white sharks and placid algae-grazing parrotfish. Five classes survive today: hagfishes, lampreys, cartilaginous fishes, lobe-finned fishes, and ray-finned fishes.

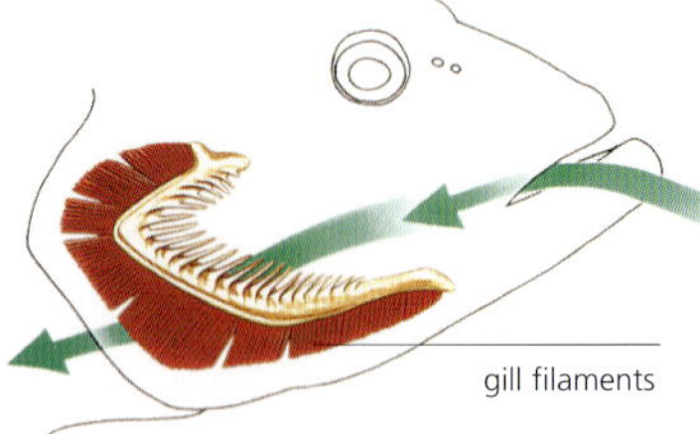

Gill-breathing Water enters the mouth and passes over feather-like gill filaments that are richly endowed with capillaries. Each filament has many folds, creating an enormous area for gas exchange. As blood and water pass closely in opposite directions, oxygen diffuses into the capillaries as carbon dioxide is expelled.

AQUATIC ADAPTATIONS

Many adaptations in the fishes have been driven by water's physical and chemical properties. The basic streamlined torpedo fish body shape, for example, has evolved to move through this medium, which is 800 times denser than air.

Forward motion in most species is created by oscillations of the tail fin and body, the other fins providing stability and maneuverability. The demands of such movements mean swimming muscles make up about half the body weight of most fishes. Maneuverability is enhanced in water by neutral buoyancy (weightlessness), achieved in most fishes by a swim bladder, an internal gas-filled sac.

The aquatic medium has also propelled the evolution of specialist sensory equipment, such as the lateral line organs. These usually run the length of the body and detect tiny changes in the surrounding water pressure, assisting in both prey location and avoiding obstacles.

Some fishes reproduce by means of internal fertilization and produce live young. Water, however, provides for the ready mixing of sex cells and dispersal of offspring and so most fishes lay eggs that are fertilized and hatched outside the female.

Fishes are the only vertebrates with true fins and most use them to swim. In some, however, these appendages allow for "walking," while in others they accommodate brief bouts of "flight."

One of the most important early evolutionary events in fishes, and indeed all vertebrates, was the development of jaws. This is thought to have first occurred in the fishes about 450 million years ago. Jaws probably evolved from gill arches. It is thought one of the anterior arches became fused to the skull, its upper section developing into the top jaw, the lower section developing into the bottom jaw.

The earliest fishes were filter feeders, but the evolution of jaws expanded options considerably and underpinned the group's enormous diversification that followed.

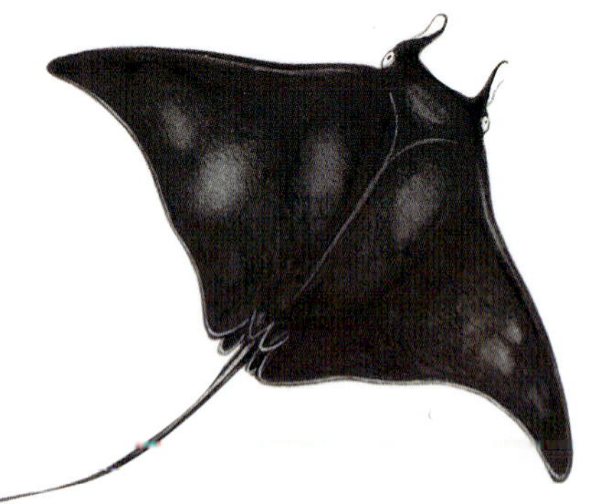

Leaping devil The acrobatic lesser devil ray *(Mobula hypostoma)* (above) is a small Atlantic species of up to 4 feet (1.2 m) wide.

Fish family As a group, seahorses (right) compose the fish genus *Hippocampus*. The name *"hippocampus"* comes from the Ancient Greek *hippos,* meaning "horse" and *kampos* meaning "sea monster."

Ancestral fish Scientists think that the first fish appeared around 500 million years ago and probably evolved from soft-bodied, filter-feeding invertebrates.

Ocean swimmers Found in most oceans of the world, butterflyfish are a group of distinctively patterned fish. Many have eyespots (right) and are striking in color.

HERMAPHRODITISM

Most fishes are dioecious; they are either male or female from an early age and remain that way throughout life. Hermaphroditism (where individuals possess both male and female sex organs during their lives) is, however, widespread in the group, much more so than among other vertebrates. In some species, individuals are sequential hermaphrodites, changing gender once they reach a certain size or when there is a shortfall of one sex or the other. Less common are simultaneous hermaphrodites, which can perform as either sex at the same time. Hermaphroditism tends to be more prevalent among fishes of lower latitudes. The fish life on tropical coral reefs, in particular, is noted for the phenomenon.

Replaceable males Hermaphroditism is common among parrotfishes (right). A dominant male often has a harem of females but, if he disappears, the largest and most aggressive female undergoes a sex change accompanied by a dramatic color change, all in a matter of a few weeks.

Juvenile: not sexually active

Initial phase: usually female

Terminal phase: always a mature male

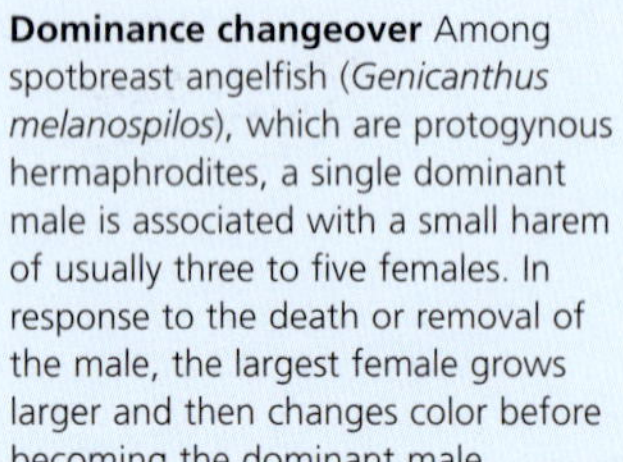

Dominance changeover Among spotbreast angelfish (*Genicanthus melanospilos*), which are protogynous hermaphrodites, a single dominant male is associated with a small harem of usually three to five females. In response to the death or removal of the male, the largest female grows larger and then changes color before becoming the dominant male.

1. Harem life *Female spotbreast angelfish are yellow on their upper surface and pale blue below, with strong black lines defining their tails.*

2. Changing colors *When the dominant male disappears, the harem's largest female begins to grow and develop the distinctive coloration of a male: a pale blue body with black stripes.*

3. Girl power *As her sexual transformation continues, her body reabsorbs its eggs and begins sperm production. She starts behaving more aggressively and displays courtship behavior toward the harem's other females.*

4. Quick change *The sexual transformation is completed within about 14 days and the new male begins spawning with members of his harem.*

JAWLESS FISHES

SUPERCLASS	Agnatha
CLASSES	2
ORDERS	2
FAMILIES	2
SPECIES	105

These were the first fishes and most became extinct by about 360 million years ago. The two surviving groups—hagfishes and lampreys—are probably only distantly related to each other. They comprise 105 species. All lack scales and jaws and have cartilaginous skeletons. True fins are either absent or poorly developed. Hagfishes are eel-like in appearance and produce copious quantities of mucus from slime glands along the length of the body, possibly for defensive reasons. These scavengers of dead and dying fishes and invertebrates have photoreceptors but do not have true eyes or a larval phase. Lampreys have functional eyes and a long larval phase. Larval lampreys are filter feeders but most adults are external parasites on other fishes.

Widespread curiosities Hagfishes and lampreys are found in temperate waters, and in cool, deep waters in parts of the tropics. Although lampreys are found in both fresh and salt water, hagfishes are exclusively marine. Larval lampreys burrow into soft stream and river substrates where they feed by filtering algae, detritus, and microorganisms from the water.

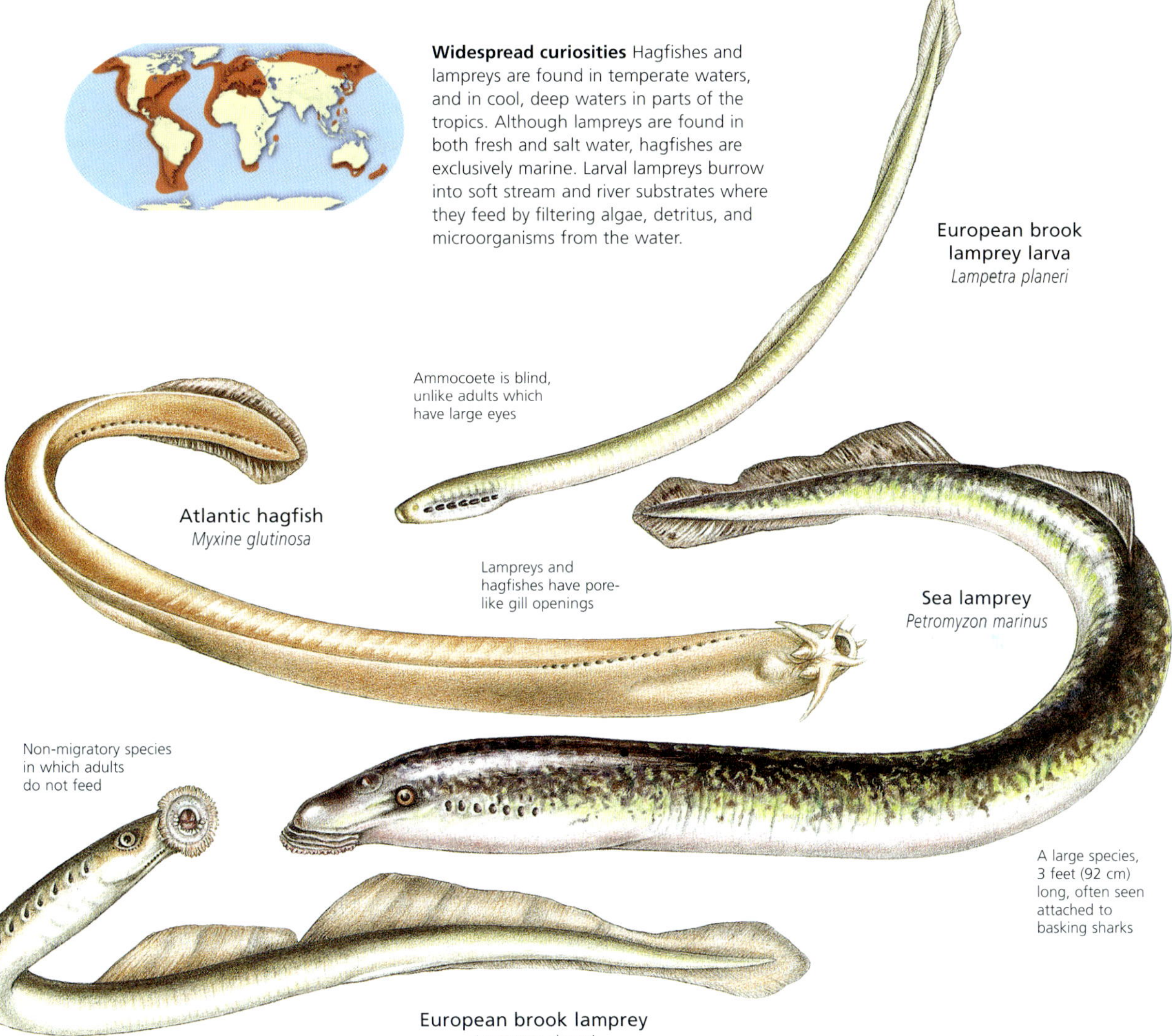

REPRODUCTION

Hagfish reproduction remains largely a mystery. These fishes begin life as hermaphrodites, later becoming either female or male. They are thought to spawn repeatedly during their lives, each time producing a small number of large eggs—about 1 inch (2.5 cm) long—in toughened cases.

Lampreys spawn only once, producing large numbers of much smaller eggs before dying. They spend their first few years as filter-feeding larvae called ammocoetes. Metamorphosis into young adults usually occurs when they are 3–6½ inches (7.5–16.5 cm) long and takes 3–6 months. Some migrate downstream to the sea.

SHARKS, RAYS, AND ALLIES

CLASS	Chondrichthyes
SUBCLASSES	2
ORDERS	12
FAMILIES	47
SPECIES	999

Having skeletons of cartilage, not bone, sharks and rays are known collectively as cartilaginous fishes. Shark evolutionary history extends back some 400 million years, while rays probably first appeared 200 million years ago. All feed on other animals and most are marine. Teeth are embedded in connective tissue and replaced throughout life; they usually have five, but sometimes six or seven, external gill slits on each side of the pharynx; and all have internal fertilization. Together with skates (a type of ray) and chimaeras (a mainly deepwater group, including spookfishes, ghost sharks, and elephant-fishes), sharks and rays today comprise about 1,000 species.

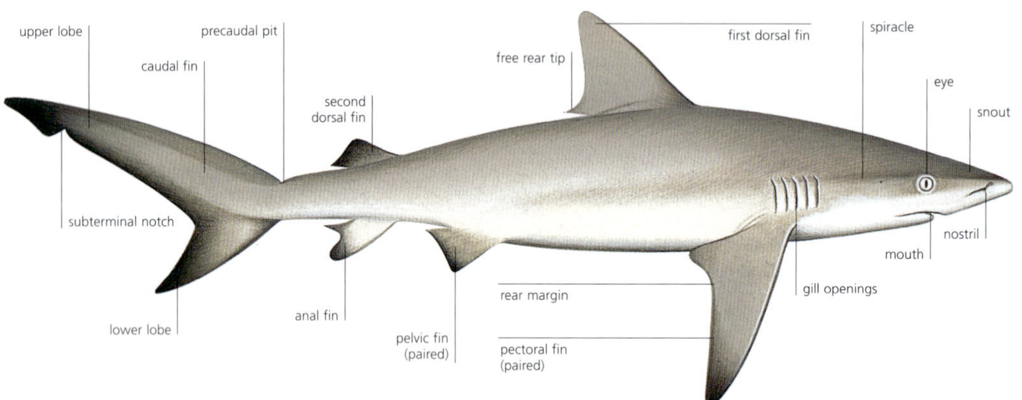

ANCIENT FEATURES

Unlike jawless fishes (which also have skeletons of cartilage), sharks, rays, and their close relatives have well-developed jaws, paired nostrils, and paired pectoral and pelvic fins. They also differ from the bony fishes in having dermal denticles and teeth that are replaced throughout their lives or which are fused into continuously growing bony plates.

The class Chondrichthyes is usually divided into two subclasses. Elasmobranchii is a large group that includes sharks, skates, and rays. The smaller Holocephali group contains the relatively "primitive" chimaeras. These bizarre-looking and mostly bottom-dwelling fishes differ from the elasmobranchs in having only a single gill cleft and four gills on each side of the head, mainly naked skin, teeth fused into plates, and the upper jaw fused to the skull.

Shark anatomy Basic shark attributes (above) have changed little since the group's early evolution. Body shape is designed for hydrodynamic efficiency and to suit a predatory lifestyle. Fins are thick, stiff, and usually lack spines. All sharks have paired pelvic and pectoral fins and between two and four unpaired anal, caudal, and dorsal fins. Gill openings are visible externally.

Powerful swimmers Rays are highly social fish that frequently frolic in groups near the ocean's surface. They move through the water with great strength and grace and have been reported to escape predators by leaping clear of the water. Often, feeding remoras catch a ride.

Ray anatomy Rays have flattened, disk-like bodies with eyes on the dorsal surface and a ventrally located mouth. They also have five or six pairs of gill slits located on the underside of the body, just behind the mouth. Rays take in water for respiration through large openings on the upper surface of the head called spiracles, as well as or instead of through the mouth. The long slender tail is often equipped with sharp spines to deter predators and is not used for swimming. Instead, rays propel themselves using their greatly enlarged wing-like pectoral fins. The caudal and dorsal fins are often reduced or lacking.

REPRODUCTION

All of the cartilaginous fishes have internal fertilization, and elaborate courtship rituals are believed to be widespread in the group. Males do not possess a penis but have claspers instead. These stiffened, fin-like rods, which extend parallel to the body from behind the pelvic fins, are inserted into the female's cloaca to guide sperm. Pup numbers range from 2–300 per pregnancy, depending on the species.

Most fishes disperse sometimes millions of tiny eggs at a time so that at least some reach adulthood. In contrast, sharks and rays produce very few offspring but invest far more energy in their early development.

Some sharks and rays lay large yolky eggs that nourish embryos for months before hatching. In most sharks and rays, however, mothers retain young internally to give birth after a lengthy gestation. Either way, baby sharks and rays are born as miniature versions of their parents.

Parental devotion almost always ends with birth. In fact, the pups of many shark species probably separate quickly from their mothers to avoid ending up as prey.

Gentle giant The largest living fish, the whale shark (left) is about 25 years old and 30 feet (9 m) long before it reaches sexual maturity. It roams the Indian, Pacific, and Atlantic oceans, mouth agape, filtering mainly plankton from surface waters.

UPPERSIDE VIEW

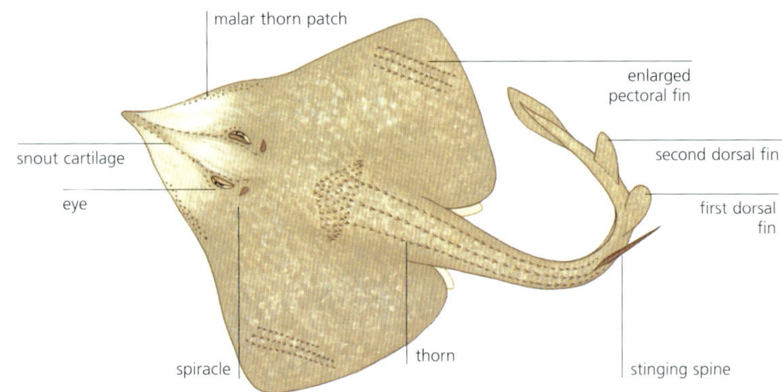

UNDERSIDE VIEW

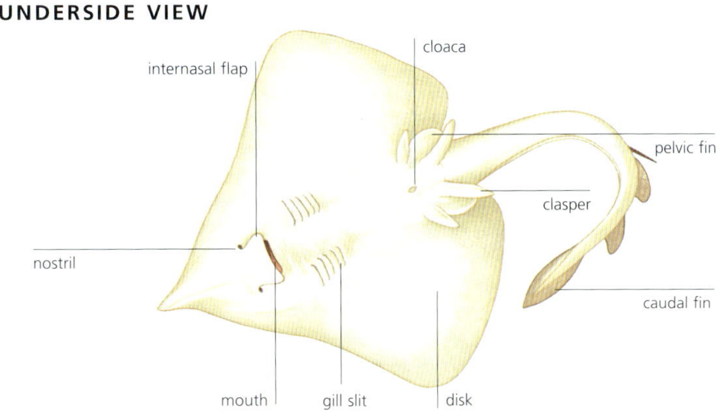

SHARKS

CLASS	Chondrichthyes
SUBCLASS	Elasmobranchii
ORDERS	8
FAMILIES	31
SPECIES	415

Although less than 2 percent of living fish species are sharks, they are of crucial ecological importance in marine ecosystems because they are apex predators. As a result they occur at naturally low levels of abundance compared with most bony fishes. This, combined with low reproductive rates, makes sharks particularly vulnerable to overexploitation. Most only reach sexual maturity after 6 years, and some not until 18 years or more. They produce few young and embryos undergo long periods of development before birth. Chimaeras are related to the sharks but belong to the subclass Holocephali, which contains a single order with 3 families and 37 species.

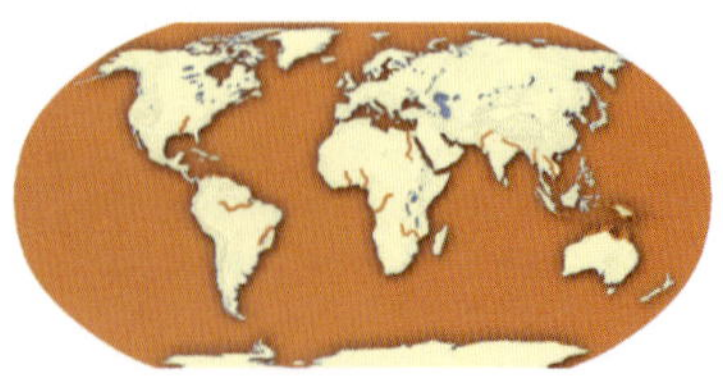

Widespread distribution Sharks are found throughout the world's oceans, although very few species can tolerate polar waters. Most prefer shallow marine habitats but a small number of species, such as dogfish sharks, dwell at great depths.

PREDATORY LIVES

Because most sharks are highly active predators, they tend to be strong, agile swimmers, and some species make long migrations covering many hundreds of miles (km) in search of food. As they have slower metabolisms than most fishes, sharks do not need to feed as frequently and, despite their reputation as irrepressible killers, only hunt when necessary. Prey typically includes small fishes and invertebrates but the larger sharks also hunt sea turtles and marine mammals.

They have no internal swim bladder but other adaptations help to increase buoyancy and enhance their swimming capabilities. The cartilaginous skeleton, which is lighter than true bone, contributes. And large livers with high levels of oil also help keep many afloat. They do need, however, to keep swimming to avoid dropping to the ocean floor.

Sharks reduce water loss by retaining urea in their tissues.

Wasteful harvest The great white shark is legally protected in many areas, but it is still among the 250,00 sharks killed daily for sport, for their fins (for Chinese soup), and as accidental bycatch of other fisheries.

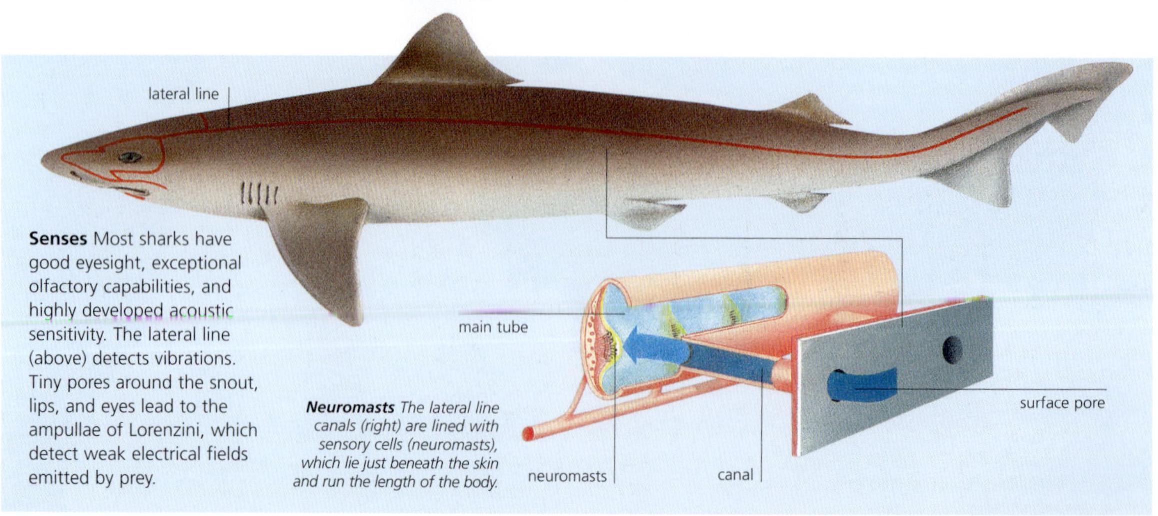

Senses Most sharks have good eyesight, exceptional olfactory capabilities, and highly developed acoustic sensitivity. The lateral line (above) detects vibrations. Tiny pores around the snout, lips, and eyes lead to the ampullae of Lorenzini, which detect weak electrical fields emitted by prey.

Neuromasts *The lateral line canals (right) are lined with sensory cells (neuromasts), which lie just beneath the skin and run the length of the body.*

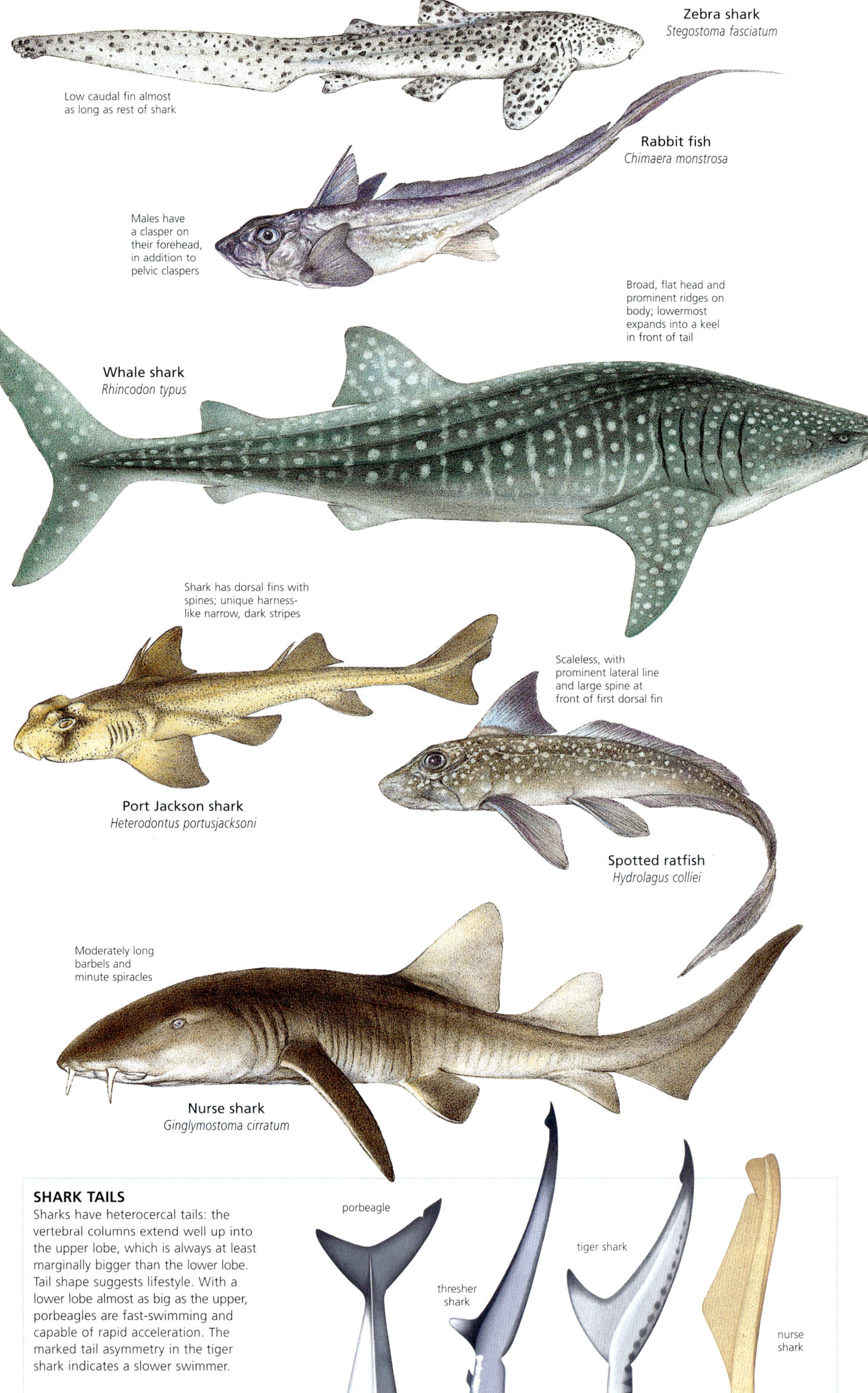

SHARK TAILS

Sharks have heterocercal tails: the vertebral columns extend well up into the upper lobe, which is always at least marginally bigger than the lower lobe. Tail shape suggests lifestyle. With a lower lobe almost as big as the upper, porbeagles are fast-swimming and capable of rapid acceleration. The marked tail asymmetry in the tiger shark indicates a slower swimmer.

Lower caudal lobe strong; posterior margin of anal fin deeply notched

Hammer-shaped head enhances maneuverability and prey capture and might also improve sensory capabilities

Smooth hammerhead
Sphyrna zygaena

Tiger shark
Galeocerdo cuvier

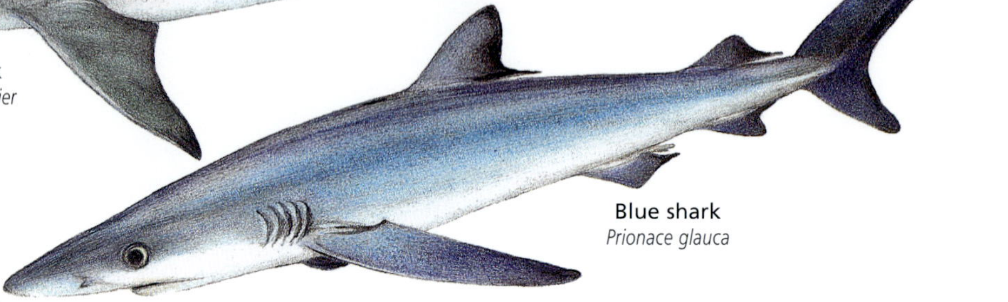

Blue shark
Prionace glauca

Long snout and large eyes

Bull shark
Carcharhinus leucas

One of the few species that can enter fresh water, bull sharks have been found as far as 2,600 miles (4,200 km) up the Amazon River

Smooth hound
Mustelus mustelus

Compressed precaudal tail and long anal fin

Blackmouth catshark
Galeus melastomus

AGONISTIC DISPLAY

When threatened, grey reef sharks indicate their readiness to attack by raising the snout, dropping the pectoral fins, and holding the tail sideways as the back is arched and flexed. Then, they swim in a figure-of-eight pattern with increasing intensity until they either make a rapid attack or retreat.

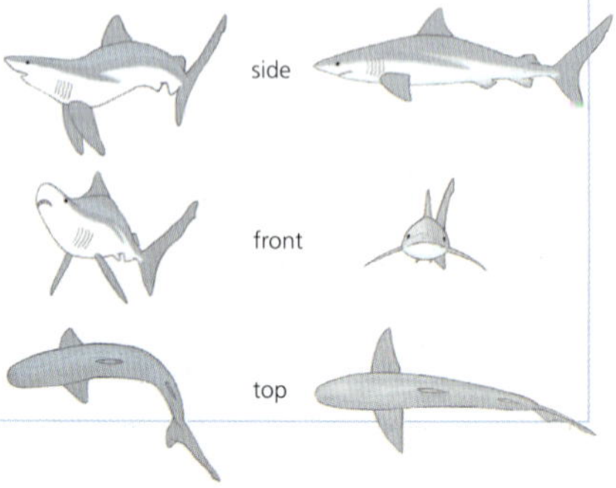

aggressive behavior

non-aggressive behavior

side

front

top

Smooth hammerhead This mostly temperate species feeds on bony fishes, small sharks, rays, crustaceans, and squid.

Up to 16½ ft (5 m) Up to 880 lb (400 kg)
Viviparous Male & female
Vulnerable
Widespread in temperate and tropical seas

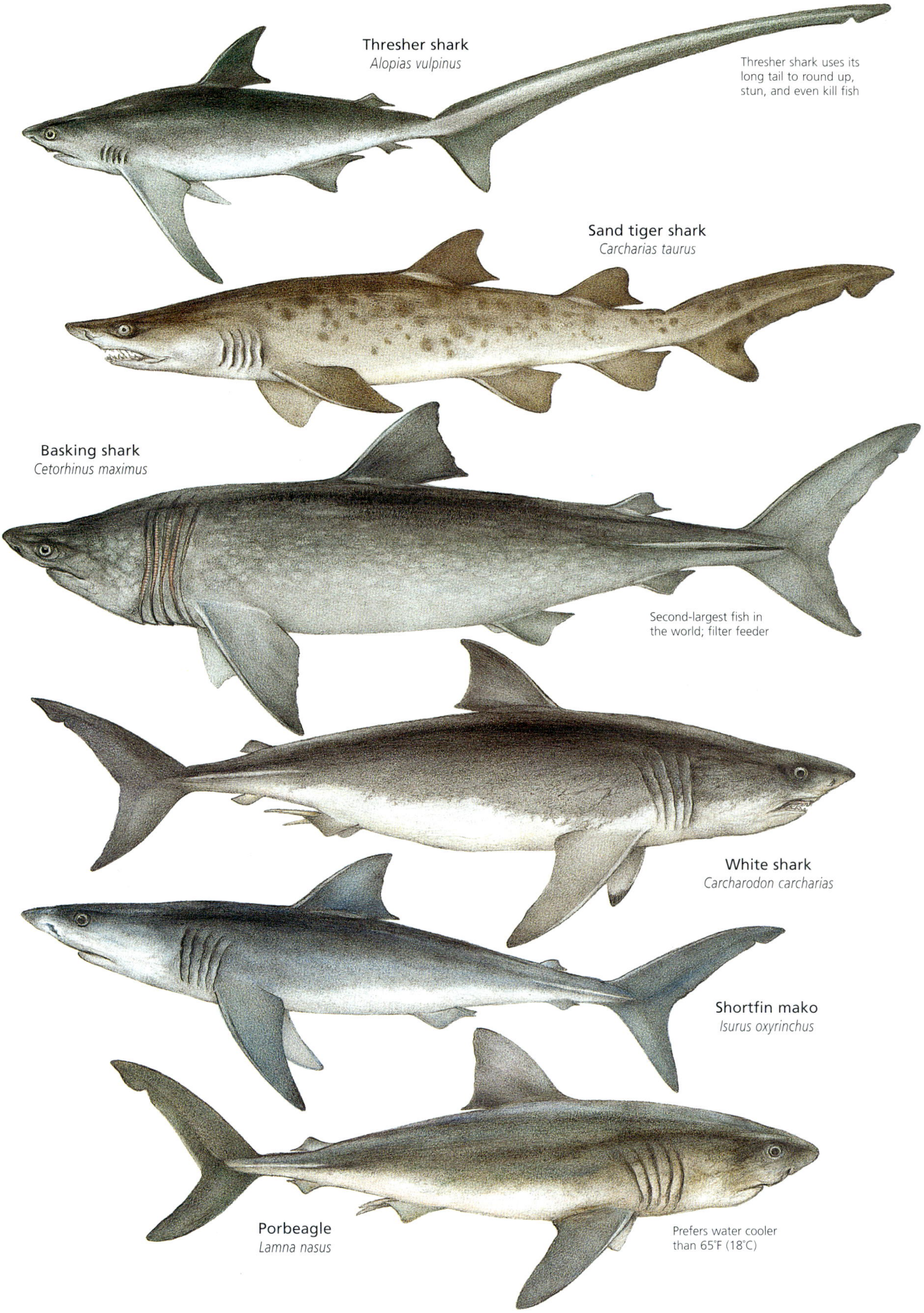

Thresher shark
Alopias vulpinus
Thresher shark uses its long tail to round up, stun, and even kill fish
Sand tiger shark
Carcharias taurus
Basking shark
Cetorhinus maximus
Second-largest fish in the world; filter feeder
White shark
Carcharodon carcharias
Shortfin mako
Isurus oxyrinchus
Porbeagle
Lamna nasus
Prefers water cooler than 65°F (18°C)

Sand tiger shark The embryos of sand tiger sharks cannibalize each other until only one in each uterus survives.

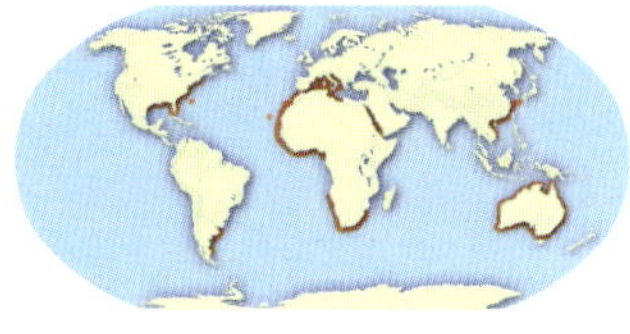

Up to 10½ ft (3.2 m) Up to 348 lb (158 kg)
Viviparous ♀♂ Male & female
Vulnerable
Widespread in warm seas except E. Pacific

PLACENTAL VIVIPARITY

Developing young in some sharks—including blue and hammerhead sharks—are nourished, like mammals, by maternal nutrients. Embryos hatch in the oviducts and live initially off egg sac yolk. Yolk stalks become "umbilical cords" extending from between the embryos' pectoral fins. Depending on the species, sharks with placental viviparity can produce up to a hundred offspring every pregnancy.

ostium
pelvic fins
cloaca
ovary
shell gland
uterus with pup

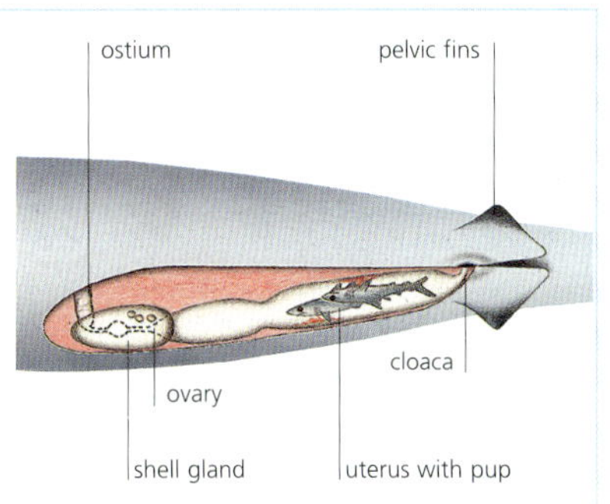

THE BIG BITE

A daytime predator of fish, squid, stingrays, sea turtles, and marine mammals, the white shark is believed to have exceptional eyesight and to see in color. This huge carnivore has evolved to consume large meals infrequently and is thought to undertake long migrations to favored hunting grounds. The white shark is one of just 27 shark species that is known to have attacked either people or boats.

Shark teeth Teeth reveal much about shark prey preferences: white sharks have large, triangular-shaped teeth with serrated edges for tearing flesh; the flattened hind teeth of the horn shark crush hard-shelled invertebrates; blue shark teeth are finely serrated for catching fishes and squid; and the needle-like mako teeth are adept at grasping large, slippery prey.

Protrusible jaws As in other sharks, white shark jaws are connected to each other at their outer corners but hang loosely beneath the skull. The snout lifts as the bite begins, the lower jaw drops and, as the mouth opens, the whole jaw arrangement thrusts forward, providing extra reach.

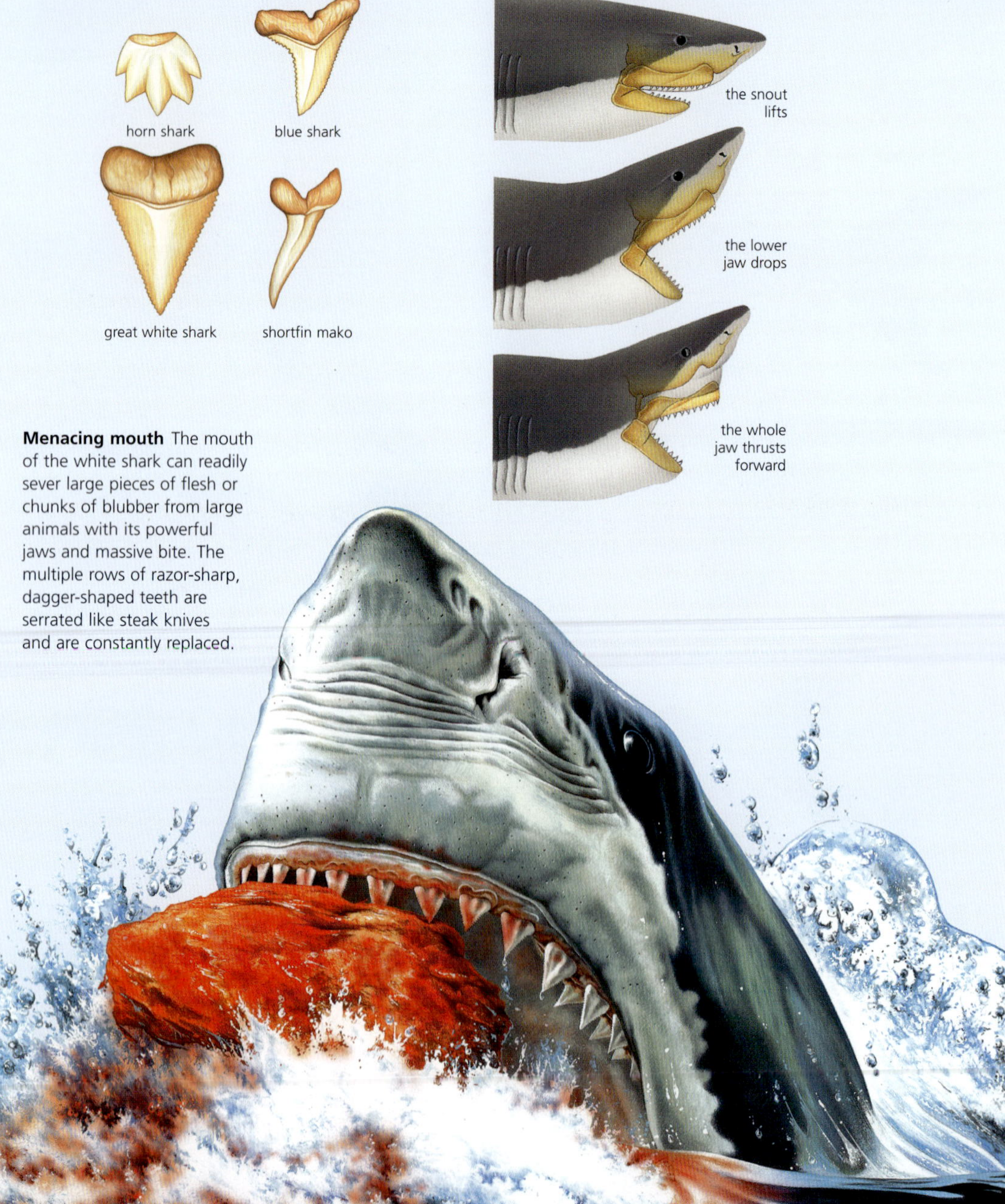

Menacing mouth The mouth of the white shark can readily sever large pieces of flesh or chunks of blubber from large animals with its powerful jaws and massive bite. The multiple rows of razor-sharp, dagger-shaped teeth are serrated like steak knives and are constantly replaced.

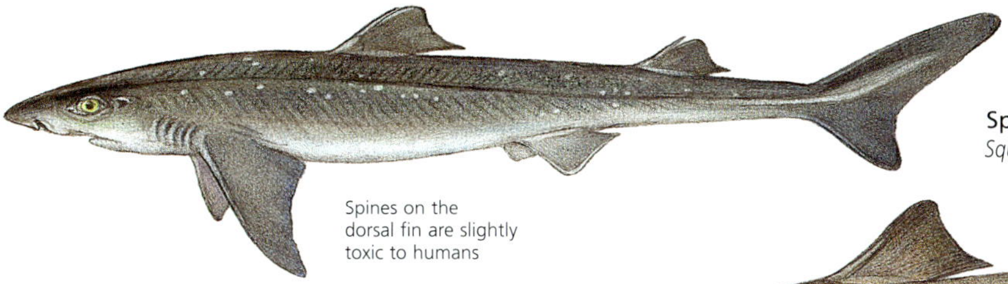

Spiny dogfish
Squalus acanthius

Spines on the dorsal fin are slightly toxic to humans

Six gill slits give this shark its name

Sixgill shark
Hexanchus griseus

Sharpnose sevengill shark
Heptranchias perlo

Frill shark
Chlamydoselachus anguineus

A slender eel-like shark with six gill slits

Bramble shark
Echinorhinus brucus

Common name comes from the unusual and prominent thorn-like scales covering the body

Angel shark
Squatina squatina

Tentacle-like sensory barbels hanging from snout are used to detect prey buried in ocean sediments

Longnose sawshark
Pristiophorus cirratus

Angel shark By day, this ambush predator lies buried in sediment with only its eyes uncovered. It bursts rapidly up to capture unsuspecting prey such as bony fishes, squid, skates, and crustaceans as they swim overhead.

- Up to 8 ft (2.4 m)
- Up to 175 lb (80 kg)
- Ovoviviparous
- Male & female
- Vulnerable

E. North Atlantic & Mediterranean Sea

COOKIE-CUTTER SHARKS

Cookie-cutters attach to prey with their suctorial lips, then spin around to carve deep round plugs of flesh using large triangular-shaped lower teeth. They remain attached with small hook-like upper teeth. It is thought that light-emitting organs on their undersides make them appear to be much smaller fishes when viewed from beneath. The illusion is revealed when would-be predators launch attacks and become prey instead.

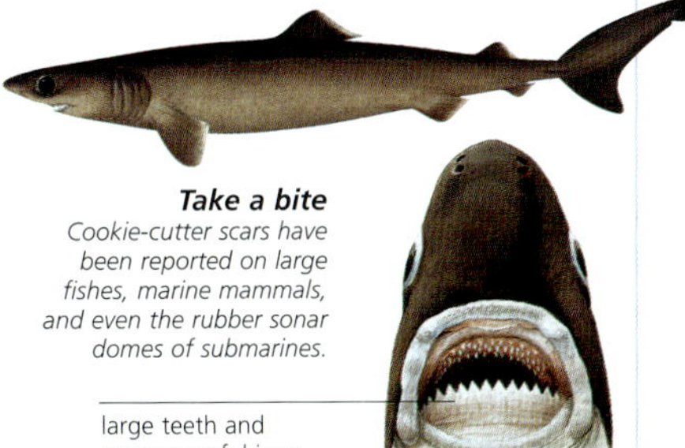

Take a bite
Cookie-cutter scars have been reported on large fishes, marine mammals, and even the rubber sonar domes of submarines.

large teeth and very powerful jaws

RAYS AND ALLIES

CLASS Chondrichthyes
SUBCLASS Elasmobranchii
ORDERS 3
FAMILIES 13
SPECIES 547

Body shape is the principal difference between these fishes and other cartilaginous fishes. Known as batoids, skates and rays are dorsoventrally flattened, an adaptation for a bottom-dwelling lifestyle. Their greatly enlarged pectoral fins extend from near the snout to the base of the tail. Together with the body, and often the head, they form the "disk," which can be triangular, round, or diamond-shaped. Most batoids take in water for gill ventilation through spiracles, openings located on the tops of their heads that are often mistaken for the eyes. Teeth are often plate-like and used for crushing prey, which ranges from bottom-dwelling invertebrates to pelagic fishes.

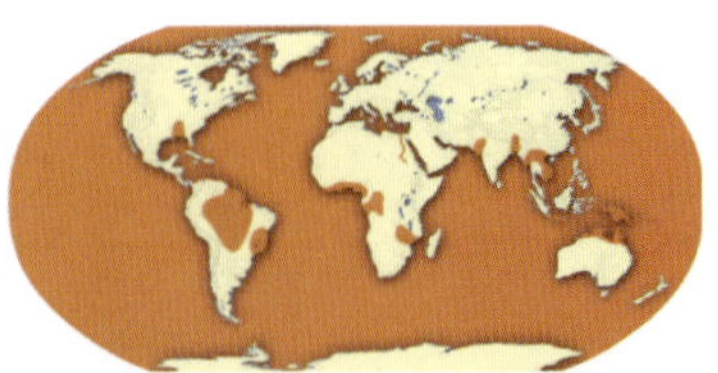

Mostly benthic Skates and rays can be found in most benthic marine communities. A small number of ray families live in the open ocean and a few stingray and sawfish species survive in brackish estuaries and freshwater rivers and lakes. Found in the temperate and tropical waters of the Atlantic, Pacific, and Indian oceans, guitarfishes are mostly marine.

Clever fishes Rays are inquisitive, often sociable animals with complex behaviors. Although they are usually seen alone, many species will aggregate into loosely formed groups, particularly for the purposes of mating or migration.

BATOID DIVERSITY

Almost half the living batoid species are skates, all of which are bottom-dwelling egg-layers with large flat disks and small tails. The dorsal surface usually bears "thorns" for defence against predators and these are also used by males to grip females during mating. Although some species can attain lengths over 8 feet (2.4 m), most are less than 3¼ feet (1 m). Most skates prefer shallow water but some are found to depths of 9,000 feet (2,750 m).

The rays are often categorized into four major groups: electric rays, sawfishes, stingrays and their allies, and guitarfishes.

Sawfishes have distinctive flat elongated snouts with conspicuous "teeth" along their edges. These can account for as much as a third of the body length in adults. They are thrashed about in schools of fishes to stun and kill prey. Some of the seven known living species attain lengths of over 23 feet (7 m).

Electric rays stun prey using electric organs located behind the eyes. There are more than 40 species worldwide, all of which tend to be rather lethargic bottom dwellers.

Many of the 150-plus species of stingrays and their allies have slender tails armed with serrated spines and are powerful swimmers. They tend to be bottom dwellers but three families have adopted an open ocean lifestyle. The largest, the manta, attains widths in excess of 21 feet (6.4 m).

Guitarfishes comprise about 50 species and have well-developed tails, either a poorly developed or absent disk and are all ovoviviparous.

15–20 pairs of rostral teeth on saw

Largetooth sawfish
Pristis pristis

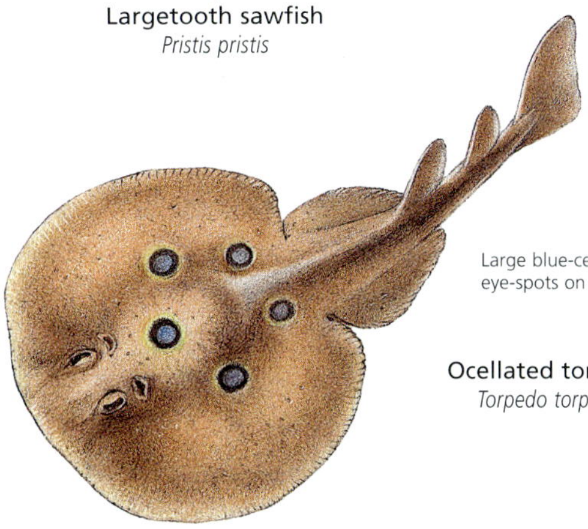

Large blue-centered eye-spots on disk

Ocellated torpedo
Torpedo torpedo

Marbled electric ray
Torpedo marmorata

Origin of the first dorsal fin is behind the pelvic fins

Atlantic guitarfish
Rhinobatos lentiginosus

Head and pectoral fins form a spade-like disk

Oda's skate
Rhinoraja odai

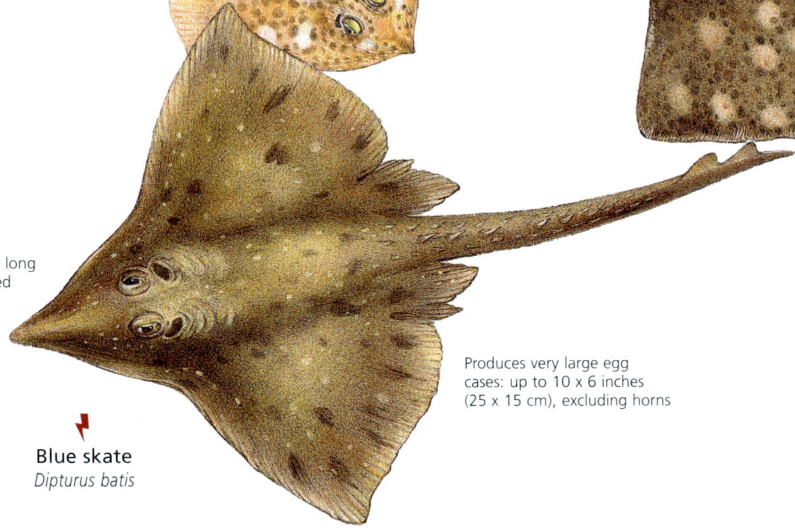

Snout very long and pointed

Thornback ray
Raja clavata

Produces very large egg cases: up to 10 x 6 inches (25 x 15 cm), excluding horns

Blue skate
Dipturus batis

Largetooth sawfish Critically endangered, this ray uses the flattened, tooth-studded "saw" that extends from the front of its head, like other sawfishes, to slash at prey as well as for defence against predators.

- Up to 15 ft (4.5 m)
- Up to 1,000 lb (454 kg)
- Ovoviviparous
- Male & female
- Critically endangered

W. & E. Atlantic, Amazon River

Marbled electric ray A nocturnal feeder that uses shock tactics to paralyze its prey, the marbled electric ray produces electrical discharges of up to 200 volts from two large electric organs on its head.

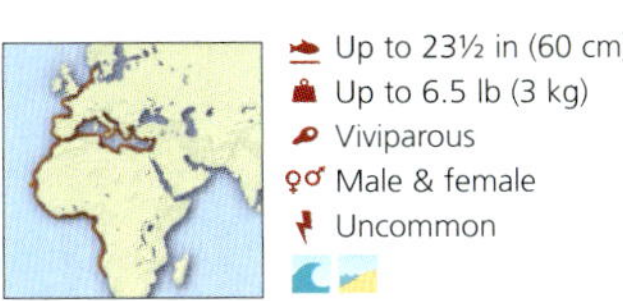

- Up to 23½ in (60 cm)
- Up to 6.5 lb (3 kg)
- Viviparous
- Male & female
- Uncommon

E. Atlantic & Mediterranean Sea

Blue skate The eggs of the blue skate, like those of other skate species, are encased in rectangular capsules of collagen, toughened with keratin and imbued with antibacterial sulphur.

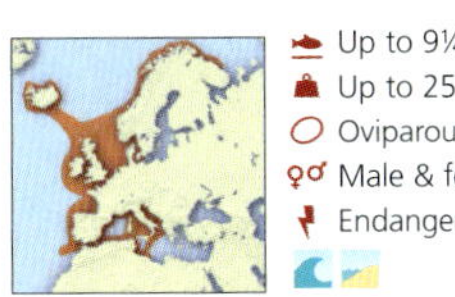

- Up to 9¼ ft (2.85 m)
- Up to 250 lb (113 kg)
- Oviparous
- Male & female
- Endangered

E. North Atlantic & W. Mediterranean Sea

Atlantic stingray
Dasyatis sabina

Length of disk equals width

Common stingray
Dasyatis pastinaca

Most stingrays have thin whip-like tails without a caudal fin

Round stingray
Urobatis halleri

Huge stingray with conspicuous dark spots on a light brown disk

Honeycomb stingray
Himantura uarnak

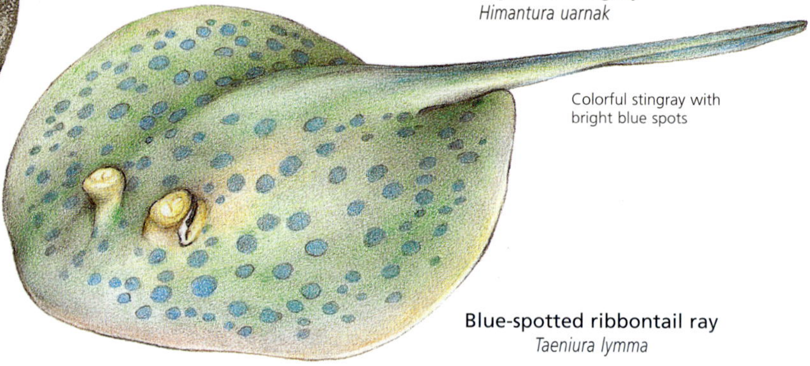

Colorful stingray with bright blue spots

Blue-spotted ribbontail ray
Taeniura lymma

One of 19 species of freshwater stingrays

Japanese butterfly ray
Gymnura japonica

Ocellate river stingray
Potamotrygon motoro

STINGRAY TAIL

Stingrays are the largest fishes with venom. Many have tails equipped with at least one, and often two, cartilaginous barbs or spines. These are usually notched with sharp, backward-facing teeth and encased in a thin sheath of tissue that keeps them bathed in a layer of mucus laced with venom.

Atlantic stingray Very few sharks, skates, and rays can tolerate fresh water but the Atlantic stingray can be found in marine, estuarine, and riverine environments. Some populations even breed in freshwater lakes.

- Up to 24 in (60 cm)
- Up to 10½ lb (4.7 kg)
- Ovoviviparous
- Male & female
- Locally common

W. Atlantic & Gulf of Mexico

Common stingray The poisonous barb on its tail can grow to 14 inches (35 cm). Occasionally it is shed, but a new one grows to replace it. Like most rays, this fish is not aggressive, preferring to flee rather than confront aggressors.

- Up to 4½ ft (1.4 m)
- Up to 56 lb (25.4 kg)
- Ovoviviparous
- Male & female
- Locally common

E. Atlantic & Mediterranean Sea

Manta
Manta birostris

Manta rays can have a 22 foot (6.7 m) wingspan

Cownose ray
Rhinoptera bonasus

Spotted eagle ray
Aetobatus narinari

An eagle ray with a long snout and thick head

Devil fish
Mobula mobular

Bat ray
Myliobatis californica

Plain eagle ray with a short rounded snout

Common eagle ray
Myliobatis aquila

Bat ray During mating, the male bat ray pokes thorns around his eyes into his mate's underbelly. This helps the pair remain together as the male maneuvers to insert a clasper into the female.

Up to 5 ft (1.5 m)
Up to 180 lb (82 kg)
Ovoviviparous
Male & female
Common

E. Pacific: Oregon to Gulf of California

FILTER FEEDING

Feeding mantas are often seen swimming in large loops in surface waters. These vertical flips may concentrate the tiny animals filtered from the plankton.

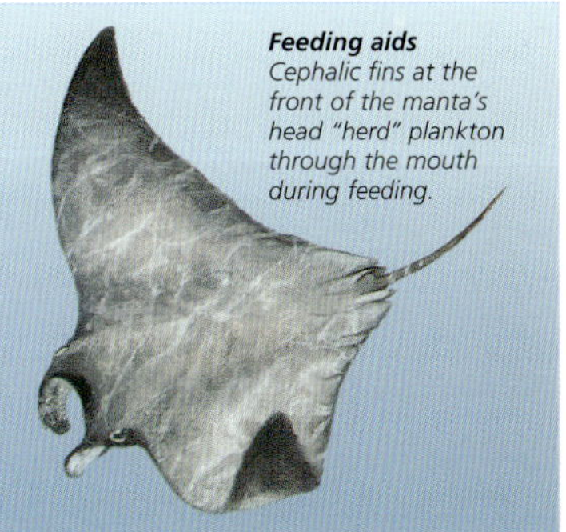

Feeding aids
Cephalic fins at the front of the manta's head "herd" plankton through the mouth during feeding.

BONY FISHES

SUPERCLASS	Gnathostomata
CLASSES	2
ORDERS	48
FAMILIES	455
SPECIES	24,673

In terms of both species and total numbers, this is by far the most successful living vertebrate group. Bony fishes first appeared about 395 million years ago, the fossil record indicating that the earliest forms inhabited fresh water. There are two distinct evolutionary lineages. The lobe-finned fishes (the sarcopterygians) are now represented by only a small number of living species. They are critically important in evolutionary terms because their ancestors gave rise to the earliest tetrapods—four-limbed land vertebrates—which, in turn, ultimately led to all other vertebrates. The overwhelming majority of living bony fishes, however, are the ray-finned fishes—the actinopterygians.

Successful radiation The bony fishes are found in virtually every available marine, freshwater, and brackish habitat throughout the world, and occasionally even desiccated environments. Species diversity increases nearer to the tropics and decreases toward the poles. Diversity tends to be highest close to coastlines and lowest in the open ocean.

EFFECTIVE EVOLUTION

Bony fishes are distinguished, as the name suggests, by a lightweight internal skeleton that is strengthened entirely or partially by true bone.

The fins of bony fishes are supported by more complex skeletal and muscle arrangements than those of cartilaginous fishes, giving bony fishes much finer control of swimming movements. As a result, many can move backward and even hover mid-water.

Maneuverability is also enhanced by the capacity to precisely and immediately adjust their buoyancy. This is achieved with a gas-filled sac known as a swim bladder. In some species, this is connected to the gullet and emptied and filled via the mouth, requiring fishes to gulp air at the water's surface. Mostly, however, there is no external connection and the swim bladder's content is controlled by the transfer of gases between adjacent blood vessels.

As they have a flap covering the gills and extra bony supports in the gill chamber known as branchiostegal rays, bony fishes can pump water over their gills and do not need to move forward to breathe.

About 90 percent of all bony fishes expel their reproductive cells from the body, making use of the watery environments in which they live for fertilization and the distribution of young.

Although cartilaginous fishes sometimes form aggregations, they don't school in the same highly coordinated way as many bony fishes, behavior made possible by a well-developed lateral line system and exceptional hearing and vision.

Safety in numbers Many fusiliers have elongated bullet-shaped bodies and iridescent coloring. Like all members of the family Caesionidae, these fish form huge, fast-moving, mid-water schools that feed on zooplankton by day but shelter by night on the outer slopes of reefs.

Deadly stalker Venom glands at the base of the fin spines of red lionfish (*Pterois volitans*) (right) deliver a poison that is potentially fatal to humans. This highly aggressive and mostly solitary fish tends to hide by day and hunt by night, stalking prey such as small fishes and crustaceans, which it corners by expanding its fan-like pectoral fins.

SUCCESSFUL MOVEMENTS

Migration is a another significant behavioral adaptation among bony fishes that has helped underpin the group's success. Mass, predictable movements are undertaken by many species to exploit changing food resources, avoid predators, or for the purposes of mating and spawning.

Such movements can be vertical, from deep to shallower waters and back, and measured in just meters. Horizontal migrations, however, can cover many thousands of miles (kilometers). When this involves migration from fresh water to ocean and back, and most of a species' life is spent at sea, it is termed anadromy. This is typical of the salmon family. The reverse—ocean to fresh water and back—is termed catadromy and is typical of freshwater eels.

Spawning migrations are common among the bony fishes because they allow adults and young to exploit different niches or even habitats.

Reef dweller The golden moray eel hides in crevices and nooks on coral reefs. Bold color patterns, which help to camouflage them, are common among the morays. There are more than 200 species of this largely nocturnal family of carnivores.

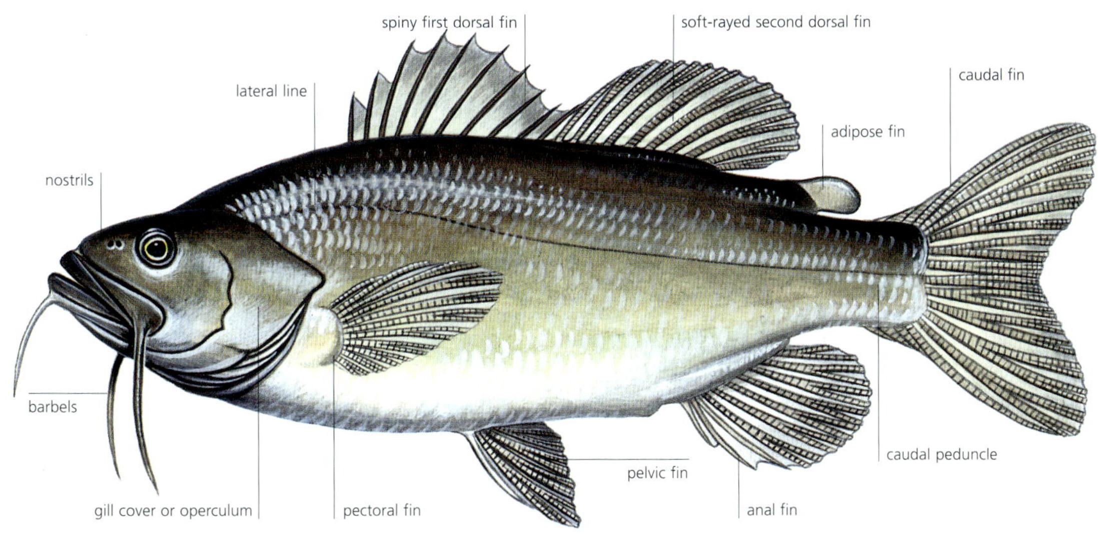

Bony fish anatomy The gills of nearly all adult bony fishes are covered by an operculum. Most species have thin, flexible scales covered by a thin mucus-secreting skin layer. Teeth are fixed to the upper jaw. The upper and lower lobes of the tail are usually symmetrical and the vertebral column stops before the fin. Most bony fishes have at least one dorsal fin, one anal fin, and paired pectoral fins. Except for a number of viviparous species, males do not have external organs to aid reproduction.

LUNGFISHES AND ALLIES

CLASS	Sarcopterygii
SUBCLASSES	2
ORDERS	3
FAMILIES	4
SPECIES	11

Long-extinct ancestors of this relict group, known as the Sarcopterygii or "lobe-finned fishes," gave rise to the earliest land animals. Today, just nine lungfish and two coelacanth species survive. All have fleshy fins with skeletal structures and musculature more reminiscent of tetrapod limbs than they are of the fan-like fins of most other living fishes. In fact, 19th-century scientists mistakenly classified lungfishes as both reptiles and amphibians before realizing that they were fishes. Larval lungfishes breathe through gills but, in all but one species, adults rely on lungs to survive, allowing them to exist in waters with low oxygen levels. They occur in tropical freshwater rivers, lakes, and flood-plains, while the gill-breathing coelacanths are exclusively marine.

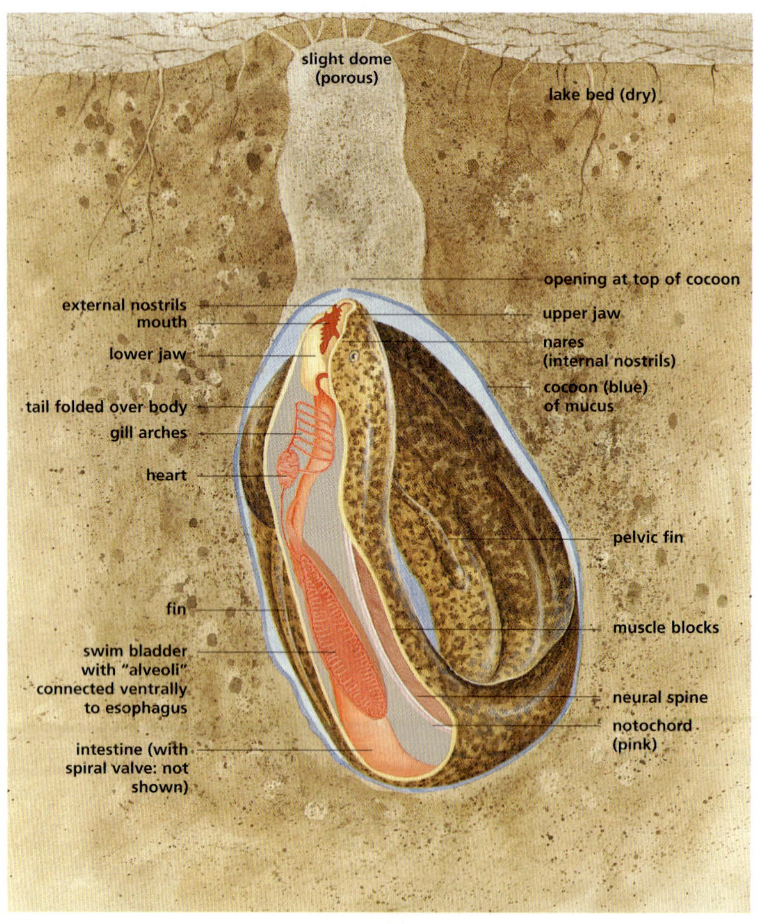

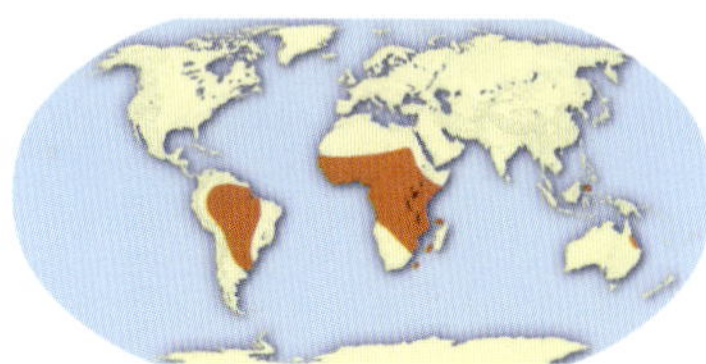

Restricted distribution Lungfishes survive only in Africa, South America, and Australia. One of the two coelacanth species (*Latimeria chalumnae*) is known from a small number of specimens caught off coastlines in the Indian Ocean around the Comoros Islands. The other (*L. menadoensis*) was discovered off Sulawesi in Indonesian waters in 1999. These are the only two known sites that support living coelacanth populations.

AIR-BREATHERS

African and American lungfishes have two lungs and thread-like paired pectoral and pelvic fins. Spawning occurs at the onset of the rainy season to aid survival of the gill-breathing larvae. Adults endure protracted dry spells by estivating in burrows, extracting oxygen from air via their lungs.

The Australian lungfish has a single lung and paddle-shaped fins. Although it can breathe air, it retains functional gills and cannot survive complete desiccation.

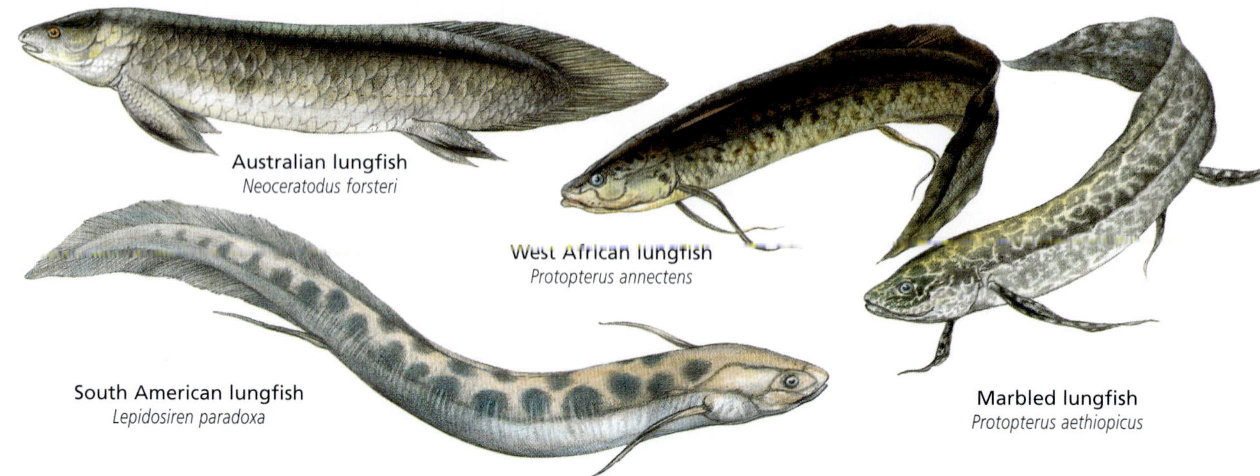

Australian lungfish *Neoceratodus forsteri*

West African lungfish *Protopterus annectens*

South American lungfish *Lepidosiren paradoxa*

Marbled lungfish *Protopterus aethiopicus*

Bichirs and Allies

CLASS	Actinopterygii
SUBCLASS	Chondrostei
ORDERS	2
FAMILIES	3
SPECIES	47

During the Permian period (285–245 million years ago) the direct ancestors of fishes such as the bichirs, reedfish, sturgeons, and paddlefishes—known collectively as chondostreans—were widespread and numerous both in species and total numbers. Today, these most primitive of the living ray-finned fishes have restricted distributions and are largely considered relics with archaic features. Most have rhombic-shaped scales hardened with a coating of an enamel-like substance called ganoine, not seen in modern species. Spiracles and intestinal spiral valves, characteristics more typical of cartilaginous fishes, are other common features. Most also have a single dorsal fin and a swim bladder, used for breathing air, connected to the gut.

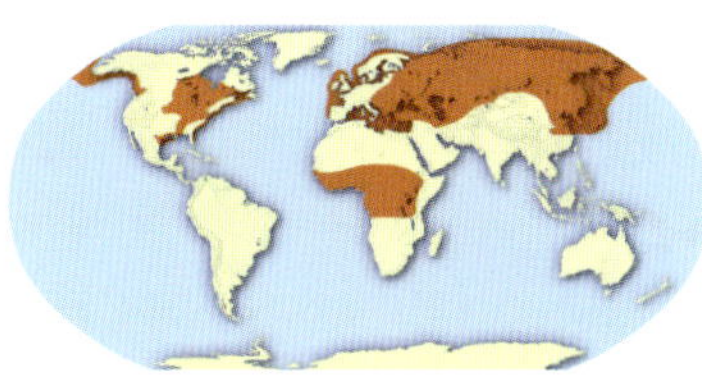

Relict distribution Although fossil records reveal this group once occurred worldwide, its distribution is now restricted to Europe, Asia, Africa, and North America. One paddlefish species occurs in the eastern United States, the other in China. Fossils of the freshwater bichir and reedfish family (Polypteridae) have been uncovered in North Africa but it is now confined to tropical Africa and the Nile River system.

STURGEON FEATURES

Sturgeons have been better studied than the other chondostreans, because of their high economic value. Found in both fresh and coastal marine waters, they are very large, long-lived fishes that take many years to reach sexual maturity. Females only spawn every few years.

Like the paddlefishes, sturgeons have several shark-like features, such as partly cartilaginous skeletons and heterocercal tails, where the vertebral column extends well into the tail's upper lobe.

They are unique among the living actinopterygians in having five rows of bony plates running along the body. They also have long flattened snouts and several tentacle-like barbels surrounding a ventrally located protractile mouth.

Reedfish
Erpetoichthys calabaricus

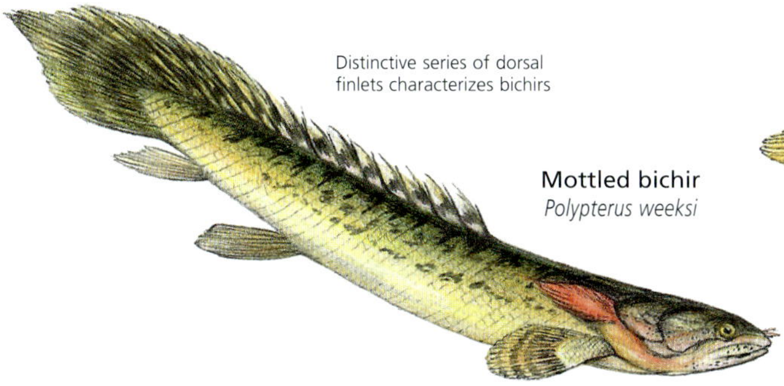

Distinctive series of dorsal finlets characterizes bichirs

Mottled bichir
Polypterus weeksi

Paddlefish
Polyodon spathula

Huge paddle-shaped snout covered in taste buds

China's largest freshwater fish, which grows to 10 feet (3 m) in length

Chinese swordfish
Psephurus gladius

Five rows of bony scutes or plates on the body

White sturgeon
Acipenser transmontanus

Snout is long and sharply V-shaped

Atlantic sturgeon
Acipenser oxyrinchus

European sturgeon
Acipenser sturio

Mouth is small and tubular

Pelvic fins are set well back

Stellate sturgeon
Acipenser stellatus

Beluga
Huso huso

Probably the largest fish to enter fresh water

White sturgeon The largest of the freshwater fishes in North America, the white sturgeon grows to 20 feet (6 m) in length and attains weights in excess of 1,500 pounds (680 kg). This species may have a life expectancy of at least a century.

- Up to 20 ft (6 m)
- Up to 1,800 lb (820 kg)
- Oviparous
- Male & female
- Near threatened

N.W. North America

Stellate sturgeon This fish has all the characteristic sturgeon features: five rows of bony plates on its dorsal surface, a flattened snout, a protractile toothless mouth, and sensitive, fleshy, tentacle-like projections known as barbels.

- Up to 7¼ ft (2.2 m)
- Up to 175 lb (80 kg)
- Oviparous
- Male & female
- Endangered

Basins of Black, Azov, Caspian seas; Adriatic Sea

Beluga Often described as the "most expensive fish in the world," the beluga is the most sought-after of the caviar sturgeons because of both the quality and quantity of its roe. A 13 foot (4 m) female can yield 400 pounds (180 kg).

- Up to 13 ft (4 m)
- Up to 1,760 lb (800 kg)
- Oviparous
- Male & female
- Endangered

Basins of Black & Caspian seas; Adriatic Sea

Primitive Neopterygii

CLASS	Actinopterygii
SUBCLASS	Neopterygii
ORDERS	2
FAMILIES	2
SPECIES	8

The Neopterygii are the second of the two living ray-finned fish groups. They arose from primitive fishes—probably early chondostreans—some 250 million years ago and, compared with their predecessors, had more mobile mouth parts, more symmetrical compact tails, and a simpler fin structure. Such advances enhanced feeding and swimming capabilities and the group ultimately gave rise to the extraordinarily successful teleosts, which comprise the majority of living fish species. Features of the earliest Neopterygii can still be seen today in a few surviving primitive representatives of the group: the bowfin and the seven species of gars, all of which are agile and voracious predators.

Northern Hemisphere inhabitants The bowfin is mostly restricted to still bodies of fresh water in temperate eastern North America. Five gar species are found in eastern North America. The remaining two species occur in Central America.

ANCIENT FISHES

Gars and the bowfin have elongated body shapes, abbreviated shark-like heterocercal tails, and numerous sharp teeth. They inhabit mostly still-water swamps and backwaters where they can survive low levels of dissolved oxygen by using lung-like gas bladders to breathe air. Gars have primitive, interlocking ganoid scales, while the bowfin has cycloid scales, the type common in more modern fishes.

Prominent black spot with yellow or orange halo that is bright in young fish but fades in adults

Bowfin
Amia calva

Shortnose gar
Lepisosteus platostomus

Longnose gar
Lepisosteus osseus

Non-overlapping diamond-shaped ganoid scales

Spotted gar
Lepisosteus oculatus

Long jaws with sharp teeth

BONYTONGUES AND ALLIES

CLASS	Actinopterygii
SUBDIVISION	Osteoglossomorpha
ORDER	Osteoglossiformes
FAMILIES	6
SPECIES	221

These fishes, the osteoglossomorphs, are considered the most primitive of the modern teleosts. The term bonytongue describes a trait shared by all group members: well-developed, tooth-like tongue bones that bite against teeth on the roof of the mouth. Beyond this unifying trait—and the fact that all are exclusively freshwater, even though the fossil record indicates that some past forms may have endured brackish water—these fishes exhibit an extraordinary diversity of form and behaviors. They are found on all continents except Europe and Antarctica, with most species occurring in Africa and just one family containing two species in North America.

Tropical endemics True bonytongues occur in South America, Africa, Australia, Malaysia, Borneo, Sumatra, Thailand, and New Guinea. Featherbacks are restricted to Asia and Africa. All elephantfish species are found only in Africa.

FRESHWATER CURIOSITIES

There are three main groups in the osteoglossomorph subdivision. True bonytongues range from a small butterfly fish that seemingly glides in air using its well-developed pectoral fins, to the arapaima, one of the largest freshwater fishes.

Featherbacks have almost no tail fin but their anal fin runs two-thirds of their body length.

The elephantfishes, the third main group, includes species with elongated snouts that look like elephant trunks. Anal fin shape differs between the sexes. It is thought these fins are brought together in spawning pairs to form a cup into which eggs and sperm are ejected and mixed.

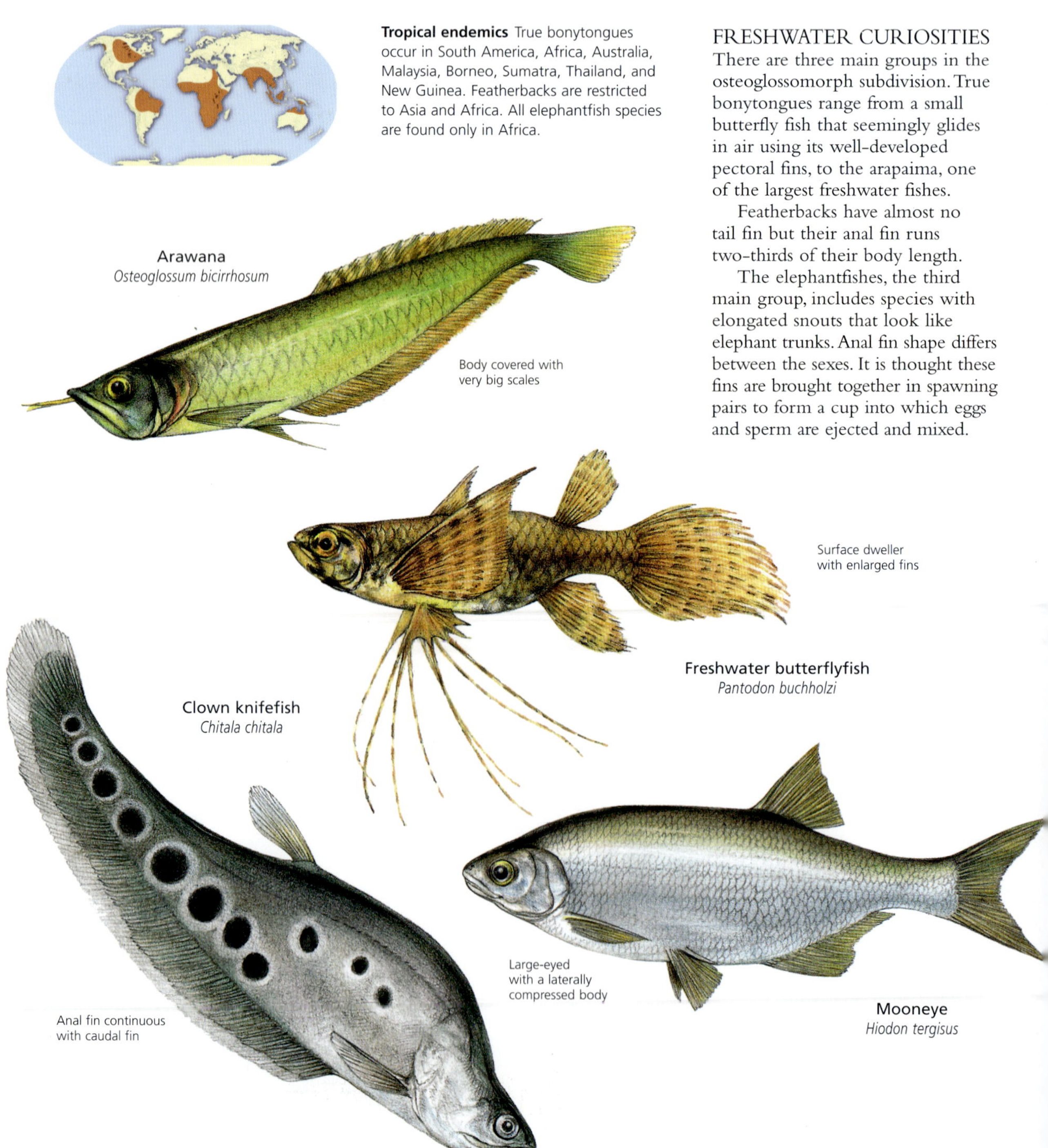

Arapaima
Arapaima gigas

Dorsal-band whale
Petrocephalus simus

Homocercal tail

Dorsal fin set well back on body

Bulldog
Marcusenius macrolepidotus

Aba
Gymnarchus niloticus

Long dorsal fin

Eastern bottlenose mormyrid
Mormyrus longirostris

Discharges weak electric currents

Blunt-jaw elephantnose
Campylomormyrus elephas

Arapaima Also known as the pirarucu, this species is one of the largest freshwater fishes. Parental care by both males and females begins with the making of nests for up to 50,000 eggs. Parents guard the developing embryos, then protect the fry.

N. South America

- Up to 15 ft (4.5 m)
- Up to 440 lb (200 kg)
- Oviparous
- Male & female
- Data deficient

Aba This species is in the same super-family as the elephantfishes, but differs from all other members of that group in that it has no caudal, anal, or pelvic fins. It does, however, have a very long dorsal fin that runs much of its body length.

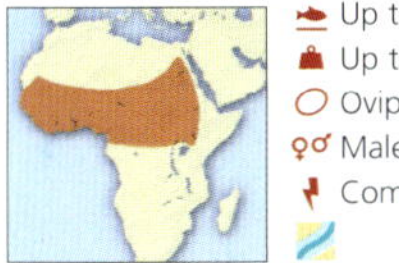

N. & N.W. Central Africa

- Up to 5½ ft (1.7 m)
- Up to 40 lb (18 kg)
- Oviparous, guarders
- Male & female
- Common

BIG BRAINS

Elephantfishes have a large brain with a well-developed cerebellum. Using electroreception, they can detect and produce weak electric currents—useful for navigating murky water and at night.

enlarged "chin" used to probe muddy substrates for food

Eels and allies

CLASS Actinopterygii	
SUBDIVISION Elopomorpha	
ORDERS 5	
FAMILIES 24	
SPECIES 911	

Many of the more than 900 species grouped together among these fishes, the elopomorphs, appear at first to have little to do with the elongated snake-like creatures most people know as eels. All, however, are linked because they invariably begin life as leptocephalus larvae. These translucent ribbon-shaped organisms drift in ocean currents for up to 3 years before metamorphosing into juvenile versions of their adult forms. This diverse assemblage of mainly marine and estuarine species includes three major groups: "true eels," such as moray, conger, and freshwater eels; the tarpons, ladyfishes, bonefishes, halosaurs, and spiny eels; and the bizarre deep-sea gulper eels.

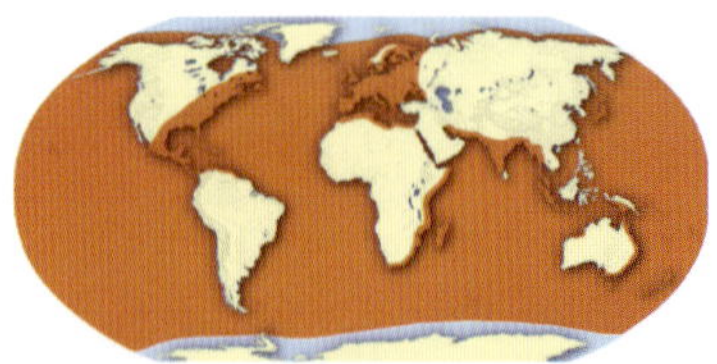

Widespread distribution Most true eels are found in tropical and subtropical seas and oceans, although members of the family Anguillidae spend most of their lives in temperate fresh water. Tarpons and their closest relatives occur mainly in warm coastal and estuarine waters. Spiny eels are found throughout the world's oceans down to depths of 16,000 feet (4,900 m). All four families of deep-sea gulper eels also occur worldwide in deep ocean waters.

VARIED FORMS

With more than 700 species, true eels are by far the largest group of elopomorphs. These have an elongated shape and almost all lack pelvic and pectoral fins. This streamlined body design suits a burrowing lifestyle and allows ease of movement into and out of holes around coral reefs.

Adult tarpons, bonefishes, and ladyfishes have large metallic scales, forked tails, and a typical fish shape. The former two groups are highly regarded as sportfish.

The spiny eels are long, scaled, and feed on slow-moving, bottom-dwelling invertebrates. Deep-sea gulper eels have elongate, scaleless bodies. Enormous mouths and highly distensible stomachs take advantage of infrequent large meals.

Dragon moray
Enchelycore pardalis

Geometric moray
Gymnothorax griseus

Laced moray
Gymnothorax favagineus

European eel
Anguilla anguilla

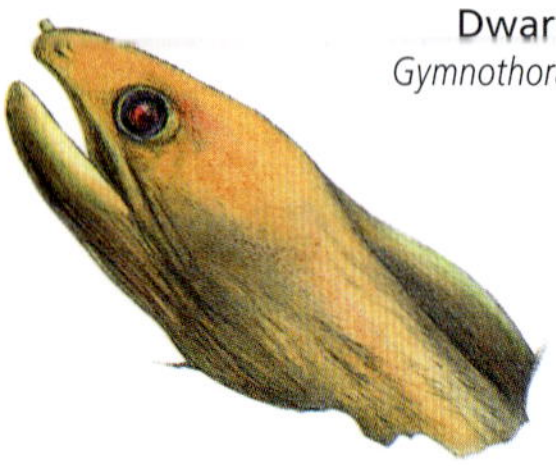

Dwarf moray
Gymnothorax melatremus

Ribbon moray
Rhinomuraena quaesita

Tarpon
Megalops atlanticus

Pectoral and pelvic fins with axillary scales

Oxeye
Megalops cyprinoides

Lower jaw projects beyond snout

Ladyfish
Elops saurus

Bonefish
Albula vulpes

Blunt snout extends beyond inferior mouth

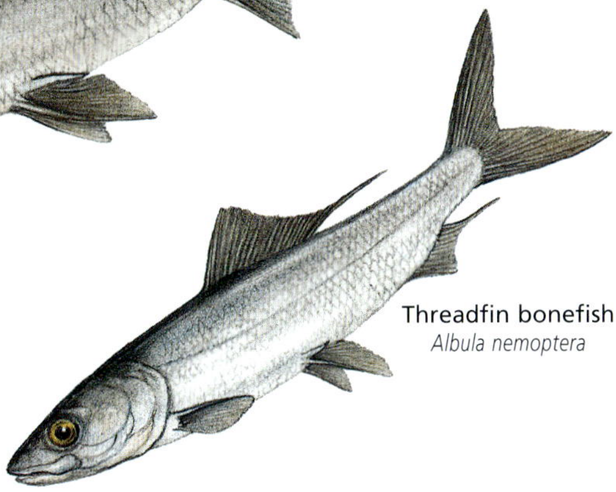

Threadfin bonefish
Albula nemoptera

Tarpon The Atlantic tarpon is sometimes called the "greatest gamefish in the world." They are hard to hook and land and will fight fiercely for many hours. During the struggle, they can leap explosively from the water.

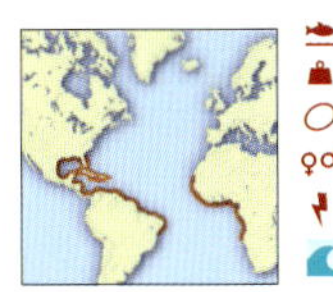

- Up to 8 ft (2.4 m)
- Up to 300 lb (135 kg)
- Oviparous
- Male & female
- Common

Caribbean Sea, W. & E. Atlantic

Oxeye Like its close relative the tarpon, the oxeye is able to breathe atmospheric air in oxygen-poor environments and is often seen gulping air at the water's surface. These fishes are intolerant of low water temperatures.

- Up to 5 ft (1.5 m)
- Up to 40 lb (18 kg)
- Oviparous
- Male & female
- Common

Indo-West Pacific

TOOTHED TURKEY

Like other morays, the turkey moray (*Gymnothorax meleagris*) has large canine teeth and powerful jaws. Morays have been known to bite divers when scared or threatened but usually reserve their aggression for hunting prey.

morays have small eyes and poor vision

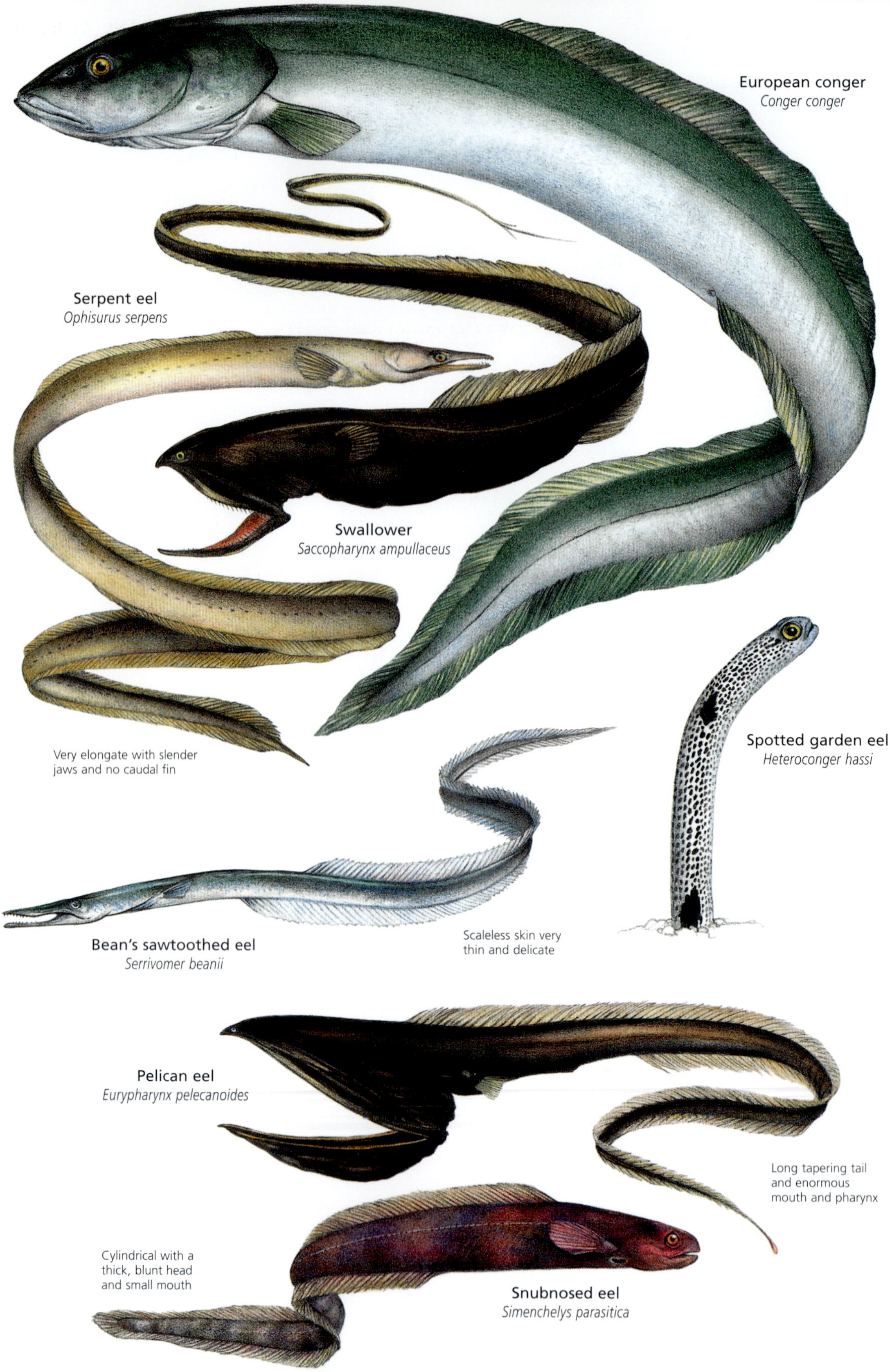

European conger
Conger conger

Serpent eel
Ophisurus serpens

Swallower
Saccopharynx ampullaceus

Very elongate with slender jaws and no caudal fin

Spotted garden eel
Heteroconger hassi

Bean's sawtoothed eel
Serrivomer beanii

Scaleless skin very thin and delicate

Pelican eel
Eurypharynx pelecanoides

Long tapering tail and enormous mouth and pharynx

Cylindrical with a thick, blunt head and small mouth

Snubnosed eel
Simenchelys parasitica

EEL GARDENS

At least 20 conger eel species spend their entire juvenile and adult lives in colonies—sometimes comprising thousands of individuals—permanently embedded in substrate. With tails down and heads waving, they feed on plankton. Known as garden eels, they occur at depths of up to 1,000 feet (300 m), retreating into their mucus-lined burrows when threatened.

European conger Juveniles occur around rocks and on sandy substrates near coasts but adults live in deeper offshore waters. Like other congers, reproduction is a once-in-a-lifetime event, during which a female can produce up to 8 million eggs.

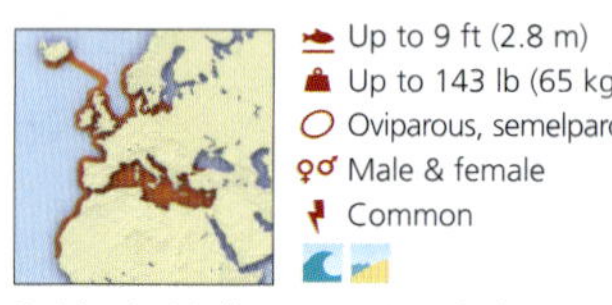

- Up to 9 ft (2.8 m)
- Up to 143 lb (65 kg)
- Oviparous, semelparous
- Male & female
- Common

E. Atlantic, Mediterranean Sea, Black Sea

THE MYSTERIOUS LIVES OF FRESHWATER EELS

It was not until late in the 19th century that scientists began uncovering the truth about the life-cycles of the European eel *(Anguilla anguilla)* and its relative the American eel *(Anguilla rostrata)*. Both species spawn in the Sargasso Sea, but precisely where remains a mystery. Their lives take decades and up to 7,000 miles (11,250 km) to complete. European eels head east, their American cousins west, both spending most of their lives in fresh water before returning to the Sargasso Sea to spawn once and die.

Parallel lives Adult European eels dwell in fresh water throughout Europe and parts of northern Africa. Their life history (below) mirrors that of the American species, in which adults occupy fresh water along North America's east coast. Research has shown that both species have suffered massive declines since the 1970s. Some investigations indicate adult European eel numbers may have fallen by 90 percent or more as a result of overfishing, pollution, disease, and habitat destruction.

1. The journey starts
Like American eels, European eels spawn in the salty waters of the western Atlantic's Sargasso Sea, somewhere off the Bermudan coast. Leaf-like larvae, known as leptocephali drift northeast on Gulf Stream currents for up to 3 years

2. See-through change
Larvae metamorphose into almost transparent juveniles known as glass eels when they arrive in Europe's coastal waters. As they accumulate pigment they become elvers, which swim up into the freshwater rivers and lakes where they will grow to maturity as yellow eels.

3. Long lives *Yellow eels (right) spend 6–20 years in fresh water, their growth rates depending on water temperature and food availability. At maturity, their eyes enlarge, their undersides turn shiny, and they become known as silver eels (left).*

4. Mature migration
Adult eels make their way downriver to the Atlantic Ocean to return to their birth-place, their sex cells maturing along the way. They do not eat during the journey. Males can grow to a length of 27½ inches (70 cm), and females to 51 inches (130 cm).

Sardines and Allies

DIVISION	Teleostei
SUBDIVISION	Clupeomorpha
ORDER	Clupeiformes
FAMILIES	5
SPECIES	378

Known as the clupeoids, this group of 378 species comprises some of the world's most commercially important fishes, including herrings, sardines, anchovies, shads, and pilchards. Most are marine filter feeders of zooplankton that form large schools and lead pelagic lifestyles, although their distributions tend to be more coastal than open water. Spawning is often in near-shore surface waters and usually seasonal. Females mostly produce large numbers of small eggs which develop into larvae that can drift in surface currents for many months before metamorphosing into juveniles. Some species undertake extraordinary migrations as adults, covering thousands of miles (kilometers) and several years.

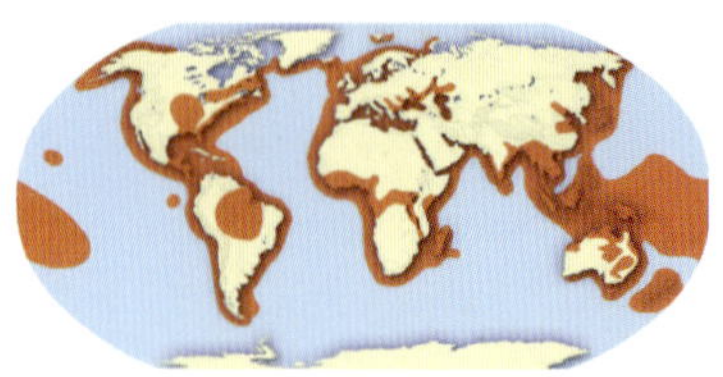

Coastal preferences The clupeoids are largely a Northern Hemisphere group. The adults of most species feed along the coasts of tropical and temperate seas, but more than 70 species inhabit freshwater rivers and lakes. Very few species occur far from shore in the open ocean and none are found in polar seas or at oceanic depths. Seasonal migrations to spawning grounds are common in the group.

BOOM-BUST POPULATIONS

Renowned for natural boom-bust population cycles, clupeoids suffer from extremely high mortality rates—frequently up to 99 percent—partly because they fall victim to a wide range of predators at all stages of their life cycle. They are also very vulnerable as adults to variations in food availability, a result of increasingly fluctuating environmental conditions.

Most species, however, have very high reproductive rates and reach sexual maturity at early ages (rarely later than 3 years of age). As a result, populations tend to bounce back quickly when good conditions prevail.

The clupeoid group contributes enormously to the total biomass of the world's aquatic environments and is a crucial lower link in many aquatic food chains.

Round herring
Etrumeus teres

Wolf-herring has fang-like teeth

Blackfin wolf-herring
Chirocentrus dorab

Gizzard shad
Dorosoma cepedianum

Atlantic thread herring
Opisthonema oglinum

Spanish sardine
Sardinella aurita

Deeply forked caudal fin

Allis shad
Alosa alosa

American shad
Alosa sapidissima

Lower jaw does not strongly project when mouth is closed

Cycloid scales are easily lost (deciduous)

Atlantic herring
Clupea harengus

Distinct ridges (bony striae) on operculum

European pilchard
Sardina pilchardus

Median notch in upper jaw and ridges on operculum

South American pilchard
Sardinops sagax

Fatty ridge with modified scales in front of dorsal fin

Ventral scutes; paired fins with axillary process; caudal fin deeply forked

European sprat
Sprattus sprattus

Atlantic menhaden
Brevoortia tyrannus

Anchoveta
Engraulis ringens

American shad Adults spend most of their lives in coastal marine and brackish waters but migrate without feeding to freshwater streams and rivers to spawn. Each female releases up to 600,000 eggs during one evening.

North America

- Up to 30 in (76 cm)
- Up to 12 lb (5.4 kg)
- Oviparous
- Male & female
- Common

Atlantic herring This plentiful fish species has been observed in 17 mile (27 km) long schools containing millions of individuals. Schools usually undertake daily vertical migrations, typically dropping to the ocean floor in the day, then rising to feed at night.

North Atlantic

- Up to 17 in (43 cm)
- Up to 24 oz (680 g)
- Oviparous
- Male & female
- Common

Anchoveta Perhaps the most commercially exploited fish in history, it was once taken at an annual rate of over 11 million tons (10 million tonnes). The natural boom-bust population cycles of the species appear to be exacerbated periodically by overfishing.

W. South American coast

- Up to 8 in (20 cm)
- Up to 2 oz (60 g)
- Oviparous
- Male & female
- Common

CATFISHES AND ALLIES

CLASS	Actinopterygii
SUPERORDER	Ostariophysi
ORDERS	5
FAMILIES	62
SPECIES	7,023

These fishes, the ostariophysans, dominate the world's freshwater habitats. They form a massive group of over 7,000 species united by two key characteristics. A unique set of bones, the Weberian apparatus, connects the gas bladder and inner ear in an arrangement that enhances hearing. The second trait is an alarm response involving the release of distress chemicals from special skin cells. Many ostariophysans not only produce these substances but can also perceive and respond to them by fleeing. Some species that eat other ostariophysans, however, are not capable of the latter as most of their prey produce these alarm chemicals and reacting would thus inhibit feeding.

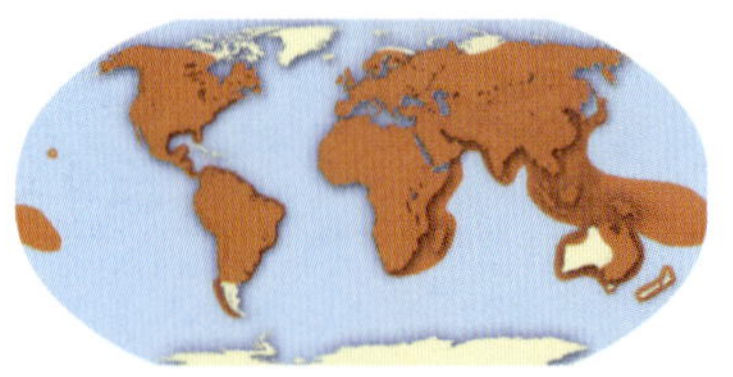

Freshwater dominance Most characins are tropical. Cyprinids are naturally found in North America and Africa, but most are found in Eurasia. Catfishes are found on all continents and the electric knifefishes are restricted to Central and South America.

Sensing Catfishes have elongated, fleshy sensitive barbels around the mouth, which function as taste and touch receptors. All species have at least one pair known as the maxillary barbels. Catfishes posses a wide variety of body forms and sizes within their diverse group.

ENORMOUS DIVERSITY

The most primitive ostariophysans are a tropical group with incomplete Weberian apparatus. The milk fish, an important protein source for people in Southeast Asia, is in this group.

The largest ostariophysan group, the cyprinids, includes the carps, minnows, and many of the world's most popular aquarium fishes, from goldfish to rasboras. These lack jaw teeth but grind food between a pharyngeal tooth-bearing bone and a hard keratinized pad at the base of the skull. This arrangement has been modified repeatedly to provide many different feeding strategies.

The mostly South American characins include species as diverse and well known as the piranhas and the aquarium favorites, the tetras.

There are well over 2,000 catfish species, all readily recognized by tentacle-like barbels around the mouth. All are substrate feeders.

The electric knifefishes make up a small, specialized, and poorly studied group that includes the legendary electric eel.

Many ostariophysans are now found well beyond their original range, partly as a consequence of the aquarium trade.

Milk fish
Chanos chanos

Cameroon shellear
Parakneria cameronensis

Slender, with a protractile upper jaw and subterminal mouth

Hingemouth
Phractolaemus ansorgii

Harlequin rasbora
Rasbora heteromorpha

Zebra danio
Danio rerio

Anal fin striped; lateral line absent

White cloud mountain fish
Tanichthys albonubes

Common carp
Cyprinus carpio

Flying fox
Epalzeorhynchos kalopterus

Slender fish with a gold band above a wide dark band

Bitterling
Rhodeus sericeus

Dorsal, caudal, anal, and pelvic fins with black margins

Tricolor sharkminnow
Balantiocheilos melanopterus

Milk fish Larval milkfish develop in brackish coastal wetlands and do occasionally enter fresh water but return to the sea to breed.

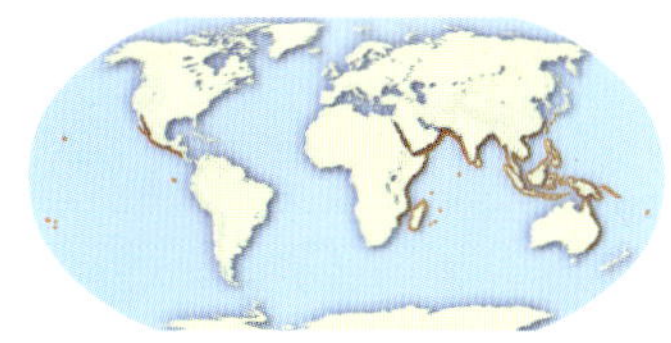

Up to 6 ft (1.8 m) Up to 30 lb (14.5 kg)
Oviparous Male & female
Common
E. Africa, S.E. Asia, Oceania & E. Pacific

BITTERLING

To breed, the female bitterling develops a long ovipositor down which she passes eggs into the gill chamber of a freshwater mussel. Eggs are fertilized when sperm, released by a male near the bivalve's inhalant aperture, is drawn in and mixes with the eggs. Bitterling larvae develop inside the mussel for up to a month. Female bitterlings were once used to test for pregnancy in women.

Pregnancy
Urine carrying traces of pregnancy hormones encourages ovipositor development when injected into these fish.

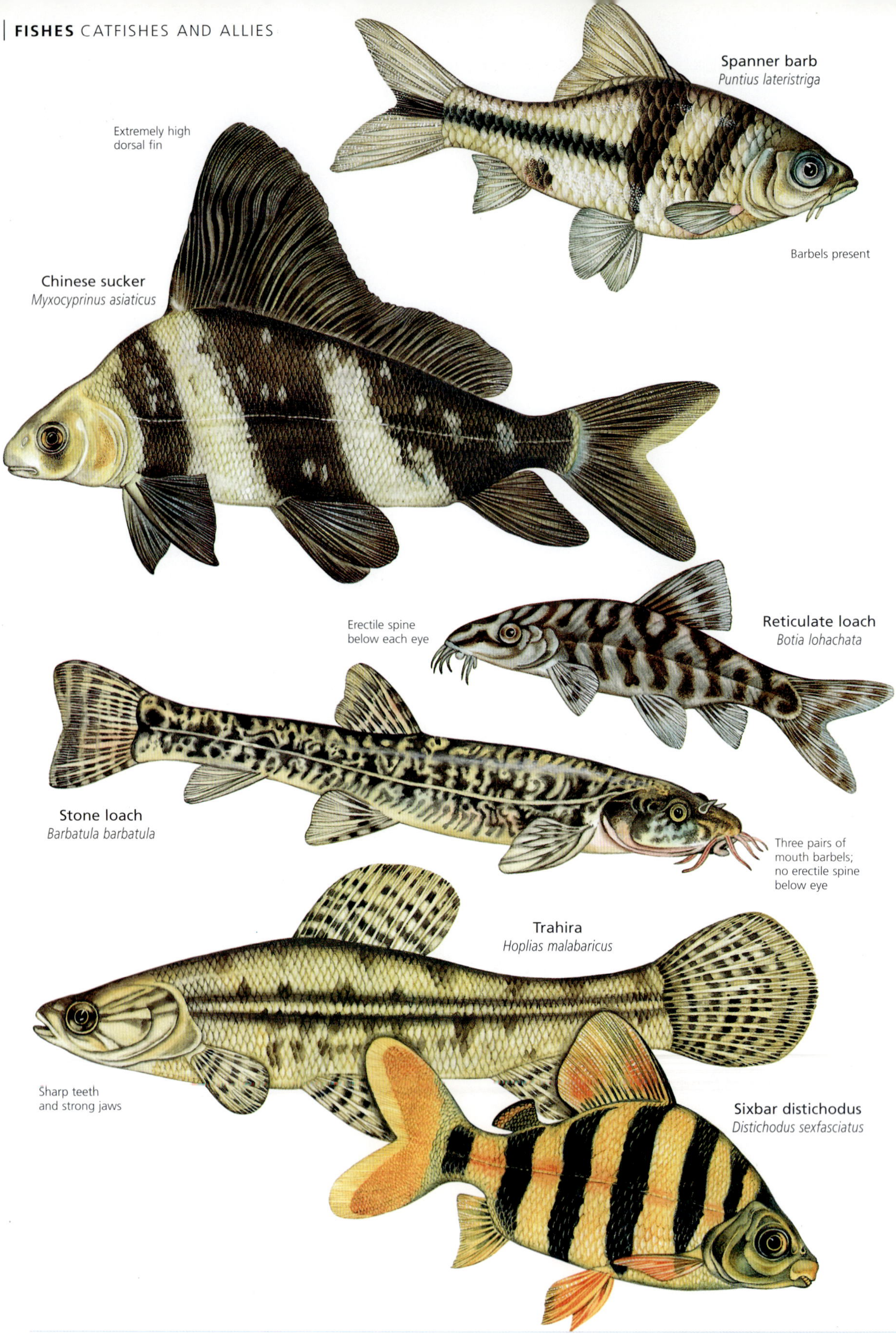

Spanner barb
Puntius lateristriga

Barbels present

Extremely high dorsal fin

Chinese sucker
Myxocyprinus asiaticus

Erectile spine below each eye

Reticulate loach
Botia lohachata

Stone loach
Barbatula barbatula

Three pairs of mouth barbels; no erectile spine below eye

Trahira
Hoplias malabaricus

Sharp teeth and strong jaws

Sixbar distichodus
Distichodus sexfasciatus

Spanner barb Many of the barbs are popular within the aquarium trade, although the spanner barb tends to grow a little too large to be a very highly sought after species.

- Up to 7 in (18 cm)
- Up to 8 oz (225 g)
- Oviparous
- Male & female
- Common

Malay Peninsula to Borneo

Chinese sucker Native to the Yangtze River basin, this species' common name stems from its thickened lips. In some specimens, the dorsal fin can be almost as tall as the body is long.

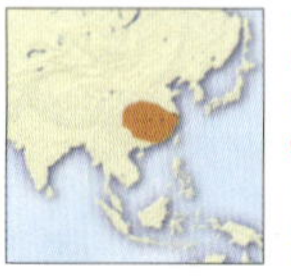

- Up to 23½ in (60 cm)
- Up to 8 lb (3.6 kg)
- Oviparous
- Male & female
- Locally common

China (Yangtze River basin)

Reticulate loach A tropical freshwater species, it is a nocturnal feeder with an omnivorous diet of mainly bottom-dwelling invertebrates, such as snails and worms, as well as algae.

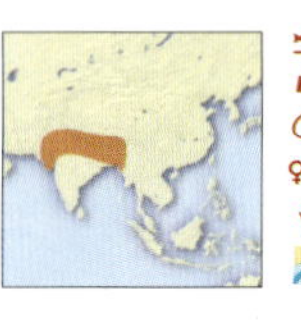

- Up to 4¾ in (12 cm)
- Up to 4 oz (115 g)
- Oviparous, hiders
- Male & female
- Common

Pakistan, India, Bangladesh, Nepal

Gar characin
Ctenolucius hujeta

Attenuate body; dorsal fin set well back with adipose fin behind

Spotted pike-characin
Boulengerella maculata

Splash tetra
Copella arnoldi

Capable of capturing small aerial insects

Marbled hatchetfish
Carnegiella strigata

River hatchetfish
Gasteropelecus sternicla

Giant tigerfish
Hydrocynus goliath

Redeye piranha
Serrasalmus rhombeus

Timid but possesses powerful dentition

Red piranha
Pygocentrus nattereri

Emperor tetra
Nematobrycon palmeri

Red piranha The most notorious of South America's piranhas, it forms the feeding frenzies for which piranhas are famous but only in the dry season when numbers are concentrated in shrinking water holes.

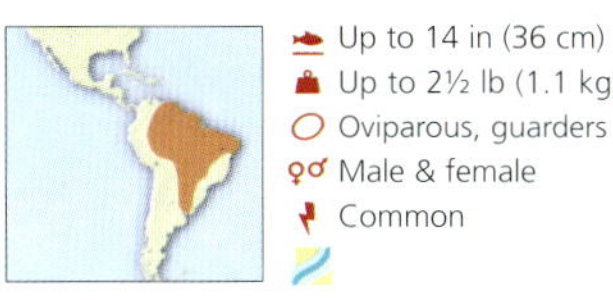

Up to 14 in (36 cm)
Up to 2½ lb (1.1 kg)
Oviparous, guarders
Male & female
Common

N.E. South America

SPLASH TETRA

Splash tetras avoid aquatic egg-eaters by spawning out of water. Females leap into the air and, as their wet bodies stick to overhanging leaves, deposit up to eight eggs at a time before falling. Males follow quickly and the pair interlock fins to hang from the leaves as the male deposits his sperm. The procedure is repeated until several hundred fertilized eggs are laid. Tail flicks from males keep the eggs moist until fry hatch and fall into the water.

Channel catfish
Ictalurus punctatus

Single dorsal fin with an adipose fin near the tail

Wels catfish
Silurus glanis

Striped eel-catfish
Plotosus lineatus

Glass catfish
Kryptopterus bicirrhis

Black walking catfish
Clarias angolensis

Capable of "walking" short distances on land

Stinging catfish possesses pectoral spines with associated poison gland

Stinging catfish
Heteropneustes fossilis

PARASITIC BROODERS

Africa's cuckoo catfish is the only fish species known to parasitize the reproductive brooding of another fish. Pairs of these catfish spawn near pairs of mouth-brooding cichlids doing the same, and consume most of the cichlid eggs. As the female cichlid collects what eggs she can into her mouth, she also gathers up the catfish eggs. The catfish fry hatch first, grow strong on their yolk sacs, and often consume hatching cichlids.

after hatching, catfish fry continue to return to the safety of their surrogate mother's mouth for protection

cichlid eggs are protected by tough skin

Glass catfish This species is a popular aquarium fish because of its transparent flesh, through which the skeleton and internal organs are visible. Glass catfish are a key ingredient in the salty fish sauce that is a flavoring staple in many Asian cuisines.

S.E. Asia

Blotched upsidedown catfish
Synodontis nigriventris

Electric catfish
Malapterurus electricus

Adipose fin with ossified rays

Angel squeaker
Synodontis angelicus

Spotted hoplo
Megalechis thoracata

A long-whiskered catfish

Masked corydoras
Corydoras metae

Pictus cat
Pimelodus pictus

Eye small and no teeth; pulsed electric discharge

Barred knifefish
Steatogenys elegans

Electric eel
Electrophorus electricus

Black ghost
Apteronotus albifrons

Small caudal fin; continuous weak electric discharge

Electric catfish With modified muscle around its body capable of discharging about 400 volts, this tropical African catfish generates electric shocks to stun prey and deter predators.

Nile River & Central Africa

- Up to 4 ft (1.2 m)
- Up to 44 lb (20 kg)
- Oviparous, guarders
- Male & female
- Common

Electric eel Unrelated to true eels, the electric eel is a relative of piranhas and tetras. It is unable to absorb sufficient oxygen through gills and gulps air to breathe by using its mouth, which is richly lined with blood vessels, as a lung.

N. South America

- Up to 8 ft (2.4 m)
- Up to 44 lb (20 kg)
- Oviparous, guarders
- Male & female
- Common

CANDIRU

The candiru, a catfish that grows to just 1 inch (2.5 cm) in length, leads a uniquely parasitic lifestyle. Found in the fresh water of the Amazon, it swims into gill cavities of larger fish to feed on blood. It has been known to enter and lodge in the urethras of urinating human bathers, causing hemorrhaging and sometimes even death.

SALMONS AND ALLIES

CLASS	Actinopterygii
SUPERORDER	Protacanthopterygii
ORDERS	3
FAMILIES	15
SPECIES	502

Some of the world's most sought-after angling and table fishes are among this group, the protacanthopterygians, which has ancient origins stemming back to the cretaceous (144–65 million years ago). It is usually divided into 15 families comprising more than 500 species arranged into three groups: esociforms, osmeriforms, and salmoniforms. These fishes are mostly carnivores equipped with large mouths and sharp teeth. Many are powerful and agile swimmers with elongate, streamlined bodies and well-developed tail fins. This swimming prowess is particularly evident in the salmons and trouts, many of which make extraordinarily long and arduous spawning migrations, journeys that have perplexed and fascinated scientists for centuries.

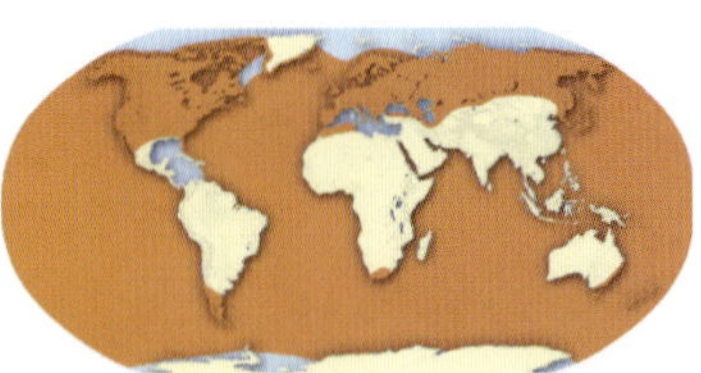

Expanding distribution Esociforms are found only in temperate North America, Europe, and Asia. Osmeriforms are also temperate fishes, with members in both hemispheres. The salmoniforms are Northern Hemisphere natives but many species have been introduced into suitable temperate-water ponds, streams, and rivers worldwide for recreational angling and aquaculture.

DIVERSE PREDATORS

The esociform group includes the pikes, muskellunge, pickerels, and mudminnows, all of which lead exclusively freshwater lives. Most are formidable ambush predators and the location of the median fins back toward the tail is an adaptation to this lifestyle.

There are more than 230 species of osmeriforms. Among the most bizarre are the barreleyes, which have upward-directed tubular eyes and other specialist adaptations for life in the high-pressure darkness of oceanic depths. Commercially important osmeriforms include the Northern Hemisphere's smelts—small, streamlined fishes that are particularly abundant in temperate coastal waters. In the Southern Hemisphere, the best-known osmeriforms are the galaxiids, a scale-less family with complex life cycles in which juvenile stages move between fresh and salt water.

The salmoniforms include the whitefishes, ciscoes, graylings, chars, salmons, and trouts. Most are commercially important but it is the homing and migratory capabilities of the salmons that attract particular attention.

The six species of Pacific salmons spend most of their lives at sea but all attempt to return to the freshwater streams in which they were spawned when they reach sexual maturity. The sense of smell is believed to be critical to these fishes in locating their birthplaces.

Migratory marvels Each stream has a distinct odor created by the soil and vegetation in its drainage. It is believed that this may become imprinted on young salmon and help them to find their way home during spawning migrations.

Muskellunge
Esox masquinongy

Large, aggressive North American species that preys on birds, mammals, reptiles, and also other fishes

Northern pike
Esox lucius

Chain pickerel
Esox niger

Greater argentine
Argentina silus

California smoothtongue
Bathylagus stilbius

Mudminnow
Umbra krameri

Alaska blackfish
Dallia pectoralis

Alaska blackfish can endure parts of its body being frozen for many days and thawed without ill effects

Specially adapted kidneys tolerant to salinity changes allow migration between marine and freshwater environments

Eastern mudminnow
Umbra pygmaea

Barreleye
Macropinna microstoma

Ayu
Plecoglossus altivelis

California slickhead
Alepocephalus tenebrosus

Mudminnow Habitat destruction, pollution, and competition from introduced fishes are threatening the future of Eastern Europe's mudminnow. This fish will gulp atmospheric air when dissolved oxygen levels drop in surrounding waters.

E. Europe

- Up to 5 in (13 cm)
- Up to 5 oz (140 g)
- Oviparous, guarders
- Male & female
- Vulnerable

PIKE TEETH

A pike's lower jaw is studded with a single row of numerous, needle-like teeth. Their acute points and razor-sharp edges pierce and cut prey as the fish lunges at speeds of up to 30 miles per hour (48 km/h). Backward-curved teeth on the roof of the mouth hold prey in place to prevent escape. Pike frequently attack prey as big as themselves and may be seen swimming with half-digested meals protruding from their mouths.

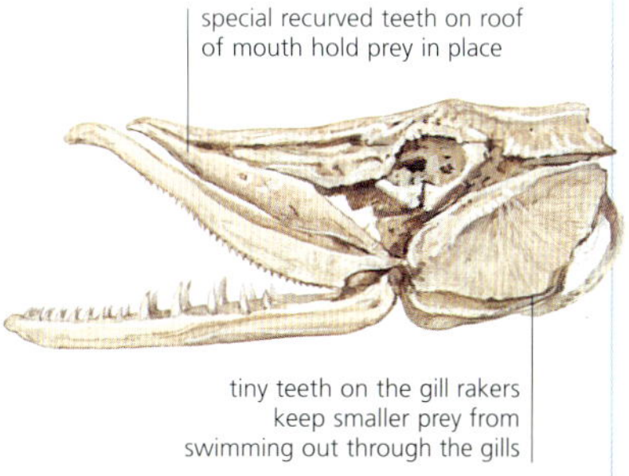

special recurved teeth on roof of mouth hold prey in place

tiny teeth on the gill rakers keep smaller prey from swimming out through the gills

Atlantic salmon
Salmo salar

Sea trout
Salmo trutta trutta

Cutthroat trout
Oncorhynchus clarki

Red "cutthroat" mark under lower jaw

Brown trout
Salmo trutta fario

Apache trout
Oncorhynchus apache

Large black spots on head, body, and dorsal, adipose, and caudal fins

European smelt
Osmerus eperlanus

Eulachon
Thaleichthys pacificus

Capelin
Mallotus villosus

Long-based anal fin, strongly arched in males

Inanga
Galaxias maculatus

Atlantic salmon Most young Atlantic salmon remain in fresh water for up to 4 years before migrating to the North Atlantic Ocean. Between 1 and 4 years later, they return to their birthplaces to spawn, often surviving to repeat the migration.

Up to 5 ft (1.5 m)
Up to 79 lb (36 kg)
Oviparous, hiders
Male & female
Common

N. Atlantic, N.W. Europe, N.E. North America

Brown trout A freshwater non-migratory form of the sea trout, this fish is prized by anglers and gourmet alike. As in other salmonids, females create nests, known as "redds," with their tail in clean sediments.

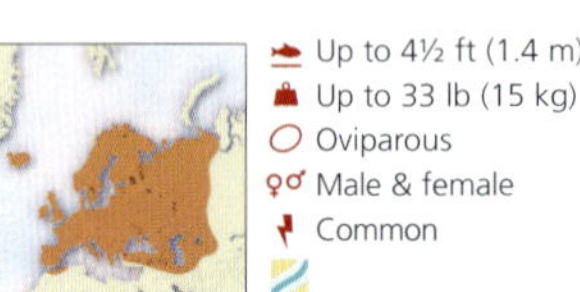

Up to 4½ ft (1.4 m)
Up to 33 lb (15 kg)
Oviparous
Male & female
Common

Europe, W. Asia, N.W. Africa; introd. worldwide

European smelt Like their close relatives the salmons and trouts, smelts have an adipose fin. This small fleshy lump on the dorsal side near the tail fin is considered a primitive feature.

Up to 12 in (30 cm)
Up to 7 oz (200 g)
Oviparous
Male & female
Data deficient

N.W. Europe

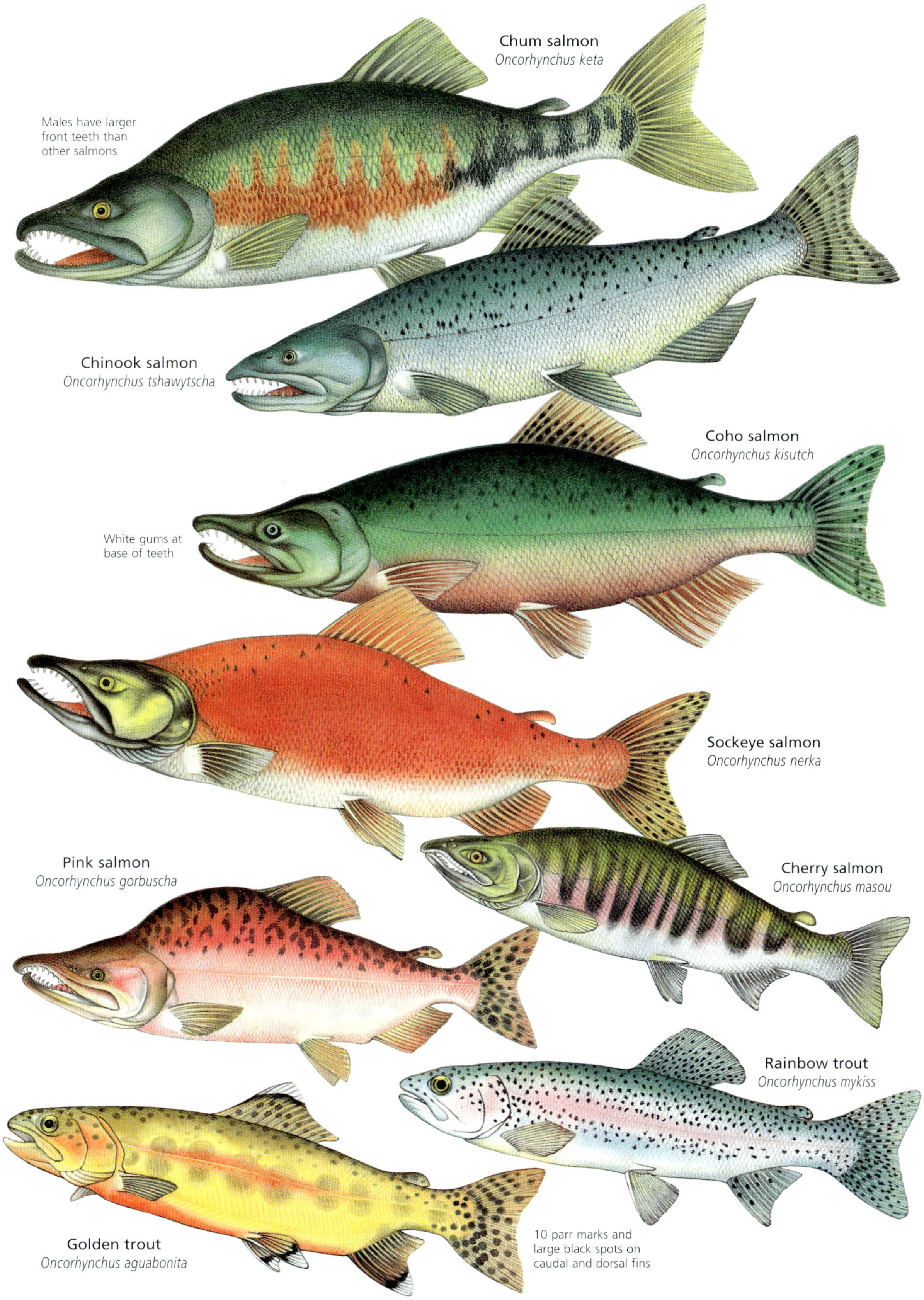

Chum salmon
Oncorhynchus keta
Males have larger front teeth than other salmons
Chinook salmon
Oncorhynchus tshawytscha
Coho salmon
Oncorhynchus kisutch
White gums at base of teeth
Sockeye salmon
Oncorhynchus nerka
Pink salmon
Oncorhynchus gorbuscha
Cherry salmon
Oncorhynchus masou
Rainbow trout
Oncorhynchus mykiss
Golden trout
Oncorhynchus aguabonita
10 parr marks and large black spots on caudal and dorsal fins

Rainbow trout This fish has become a freshwater favorite of anglers in the many parts of the world where the species has been introduced. Some North American rainbow trout stocks migrate to the sea as adults and return to fresh water to spawn.

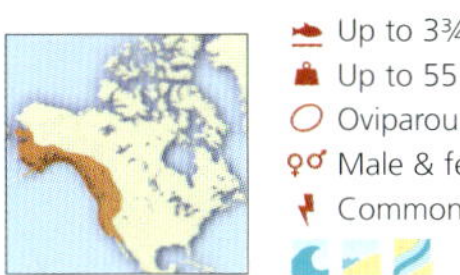

- Up to 3¾ ft (114 cm)
- Up to 55 lb (25 kg)
- Oviparous, hiders
- Male & female
- Common

N.W. North America; introd. widely

LIFE CYCLE OF SOCKEYE SALMON

Eggs are laid in gravel sediment nests dug by females. Alevin hatch with yolk-sacs still attached but a few days later become free-swimming fry. They develop into parr, which can remain in fresh water for several years before entering salt water as silvery smolt and developing into large sea-going adults. Both sexes return to fresh water with red bodies and green heads.

fry
parr
smolt
sea-going adult
breeding adult (male)
egg
alevin

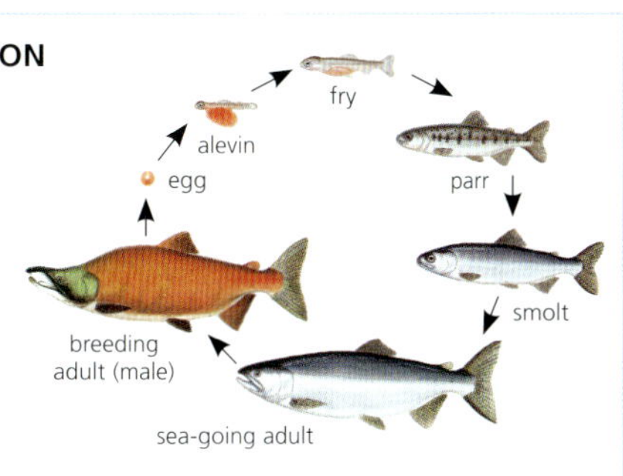

Lake trout This freshwater species feeds on other fishes, crustaceans, and aquatic insects. Females have been crossed in hatcheries with males of the brook trout (*Salvelinus fontinalis*) to produce fast-growing hybrids known as splakes.

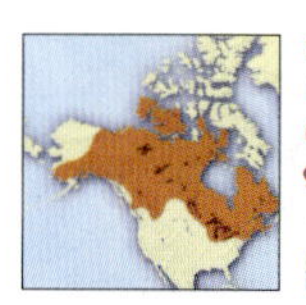

- Up to 4 ft (1.2 m)
- Up to 70 lb (32 kg)
- Oviparous
- Male & female
- Common

N. North America

Grayling Found in unpolluted streams, lakes, and occasionally brackish water throughout northern Europe, graylings spawn en masse during the northern spring when males dig nests into which females bury fertilized eggs.

- Up to 23½ in (60 cm)
- Up to 6½ lb (3 kg)
- Oviparous, hiders
- Male & female
- Common

N. Europe

Huchen The huchen is one of the largest freshwater fishes in the world. Adults are territorial predators that eat other fishes as well as frogs, reptiles, birds, and even small mammals.

- Up to 5 ft (1.5 m)
- Up to 46 lb (21 kg)
- Oviparous, hiders
- Male & female
- Endangered

Danube River basin (Europe)

DRAGONFISHES AND ALLIES

CLASS	Actinopterygii
SUPERORDER	Stenopterygii
ORDERS	2
FAMILIES	5
SPECIES	415

Although diverse and widespread, the stomiiforms are rarely encountered by people because of their mid- to deep-water lifestyle. All are predators that are adapted to habitats defined by high pressure, limited light, and low productivity. Most have long teeth and big mouths that can cope with large, infrequently encountered meals. And all but one species are equipped with light organs—photophores. Dotted along their sides and underbellies, these serve as camouflage from predators below by masking their appearance against weak light from above. Most also use them for hunting; suspended from a chin barbel or fin ray they lure prey like moths to a flame.

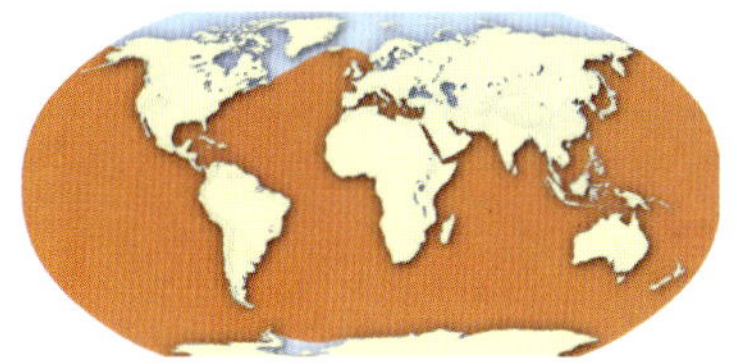

Deep-sea distribution The stomiiforms are found in the deep, open waters of all major temperate and subtropical oceans. Some extend into polar waters. They are also found over a large depth range, migrating upward at night and returning down during the day.

DEEP-SEA LIFE

Stomiiforms typically have big heads, long bodies, and dark coloration, although some are translucent or silvery. Common names such as lightfishes, bristle-mouths, loosejaws, snaggletooths, and viperfishes suggest their often bizarre appearance.

Hermaphroditism is common, an adaptation to a habitat where members of the same species may be infrequently encountered.

Eggs and larvae are buoyant, floating within plankton in surface currents, but young descend as juveniles to pursue deep-sea life. As adults, many stomiiforms rest by day in deep water but migrate at dusk to shallower depths, where food in the form of small fishes and invertebrates is more plentiful.

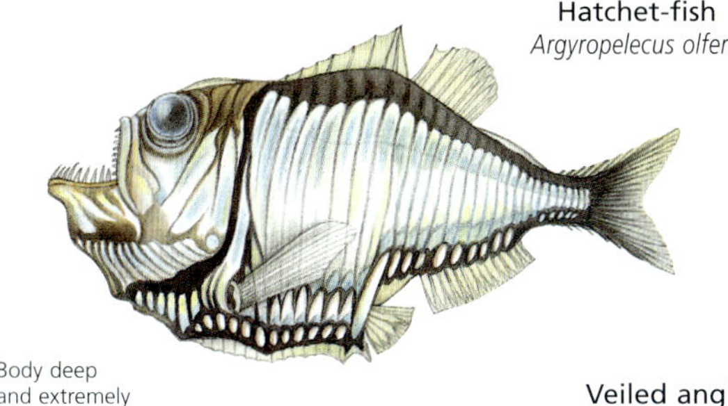

Hatchet-fish
Argyropelecus olfersi

Body deep and extremely compressed; abdominal keel-like structure

Veiled anglemouth
Cyclothone microdon

Dense stellate pigment over head, body, and fins

Scaly dragonfish
Stomias boa

Pacific viperfish
Chauliodus macouni

Deep questions Because deep-sea habitats can be as inaccessible as outer space, very little is known about the population structure, size, or breeding behavior of the stomiiforms. It is possible that one genus, *Cyclothone*, includes more individuals than any other vertebrate genus: billions of these tiny, mesopelagic fishes live in the oceans. It is also possible that some are disappearing as a result of oceanic pollution before their existence has even been recorded.

LIZARDFISHES AND ALLIES

CLASS	Actinopterygii
SUPERORDER	Cyclosquamata
ORDER	Aulopiformes
FAMILIES	13
SPECIES	229

This marine group, the aulopiforms, is represented in both coastal habitats and great depths. It includes the lizardfishes, Bombay ducks, greeneyes, ipnopids, lancetfishes, daggertooths, and telescopefishes. They exhibit an unusual mix of primitive and advanced features and are of particular interest to scientists for various reasons, including their range of peculiar eye modifications. Many deep-sea species are notable for their mode of reproduction: they are bisexual synchronous hermaphrodites, which means that they function as both sexes at the same time and may even be capable of self-fertilization. The aulopiforms are also known for the often extreme metamorphoses they pass through as they change from larvae into juveniles.

Cryptic coloration The shallow-living lizardfishes comprise 34 species of cigar-shaped fishes with large toothy mouths. The largest species is about 24 inches (60 cm) long, but most are smaller. The patterns and colors of most lizardfishes blends in well with bottom sediments and the lizardfishes' benthic habitat.

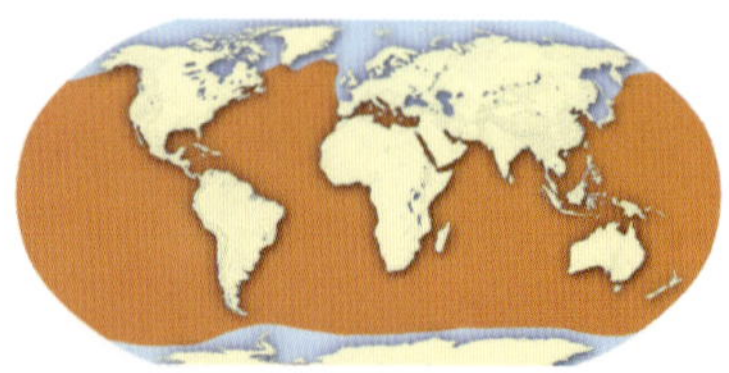

Wide distribution Lizardfishes occur in all warm seas. Greeneyes are found worldwide in tropical to warm-temperate waters. Most Bombay ducks have an Indo-Pacific distribution. The lancetfishes range widely in Atlantic and Pacific mid-waters.

ASSORTED ASSEMBLAGE

Lizardfishes, Bombay ducks, and greeneyes are inshore bottom-dwelling ambush predators. Typically well-camouflaged, they often sit propped up at the front end by their pectoral fins.

The ipnopids dwell at great depths and usually have flat heads, pencil-like bodies, and poorly developed eyes.

Attaining lengths of nearly 7 feet (2.1 m), lancetfishes are among the largest of all deep-sea predatory fishes. They have huge mouths and large, dagger-like teeth, plus a reputation for destroying undersea cables. They are sometimes seen floundering in the surf.

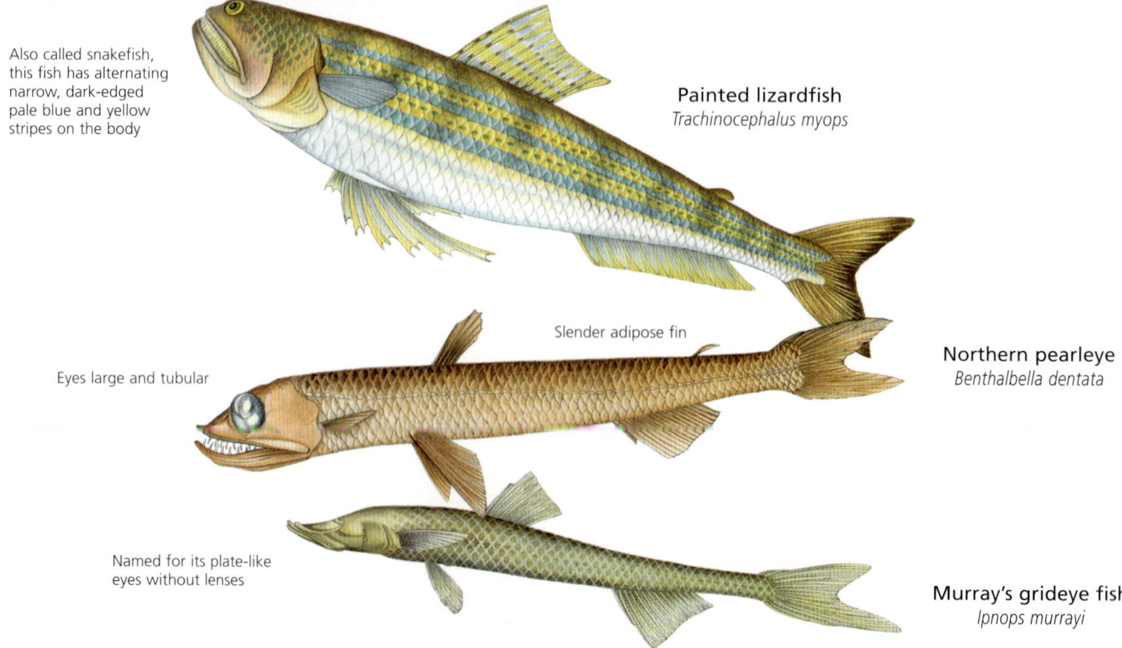

Also called snakefish, this fish has alternating narrow, dark-edged pale blue and yellow stripes on the body

Painted lizardfish
Trachinocephalus myops

Slender adipose fin

Eyes large and tubular

Northern pearleye
Benthalbella dentata

Named for its plate-like eyes without lenses

Murray's grideye fish
Ipnops murrayi

LANTERNFISHES

CLASS	Actinopterygii
SUPERORDER	Scopelomorpha
ORDER	Myctophiformes
FAMILIES	2
SPECIES	251

Of all the deep, open-water marine fishes, the lanternfishes are the most widely distributed, diverse, and abundant. Because these small plankton-eaters form large dense schools preyed upon by sea birds (particularly penguins), marine mammals, and a huge range of carnivorous fishes, they play a crucial role in virtually all marine ecosystems. They are fished to a relatively small extent for fish meal and oil but, with a global mass estimated at a staggering 660 million tons (600 million tonnes), they are otherwise regarded as a largely underused commercial resource. Together with their less abundant allies the neoscopelids, lanternfishes make up the group known as the myctophiforms.

Global resource Lanternfishes can be found in all open oceanic waters. Species distributions are related to both ocean currents and the physical and biological characteristics of the water. Many exist at great depths by day but will migrate upward to feed in surface waters by night.

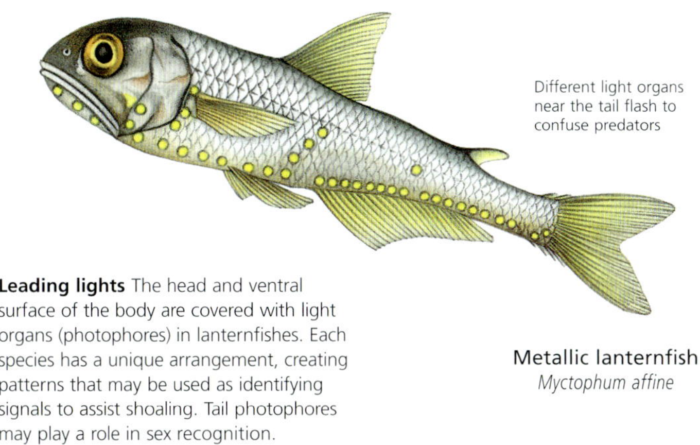

Different light organs near the tail flash to confuse predators

Leading lights The head and ventral surface of the body are covered with light organs (photophores) in lanternfishes. Each species has a unique arrangement, creating patterns that may be used as identifying signals to assist shoaling. Tail photophores may play a role in sex recognition.

Metallic lanternfish
Myctophum affine

BEARDFISHES

CLASS	Actinopterygii
SUPERORDER	Polymixiomorpha
ORDER	Polymixiiformes
FAMILY	Polymixiidae
SPECIES	10

Just 10 species in one genus make up this small but puzzling group of deep-water marine fishes, which derives its common name from the pair of sensory barbels they all have hanging from the chin. Like the more advanced teleosts (the spiny-rayed fishes), they have true fin spines (as opposed to soft rays): between four and six in the dorsal fin and four in the anal fin. However, along with this modern feature they also exhibit several primitive and unique characteristics, creating a perplexing array of attributes that has fuelled much debate about which other fish groups are their closest relatives.

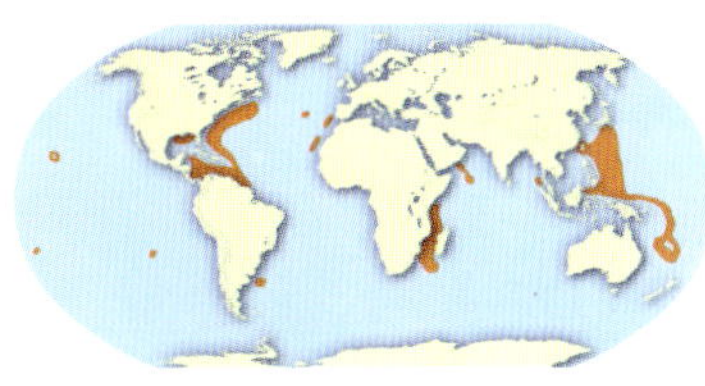

Bottom dwellers Beardfishes are known from tropical and subtropical oceanic waters. They live on the outer continental shelf and upper slope at various depths between 65 and 2,500 feet (20 and 760 m), usually on the bottom.

Dorsal fin has five spines and 34–37 soft rays

Stout beardfish
Polymixia nobilis

Rare catch Beardfish are mostly encountered as bycatch taken by commercial trawlers working the deep tropical and subtropical waters of the Indo-Pacific.

Opahs and Allies

SUBDIVISION	Euteleostei
SUPERORDER	Lampridiomorpha
ORDER	Lampridiformes
FAMILIES	7
SPECIES	23

Twenty-three species of deep-sea fishes, known as the lampridiforms, make up this group. Outwardly, their external appearance is extremely variable although most are either large and moon-shaped or elongated and snake-like. They all, however, share four unique and characteristic features, all but one of which relates to an unusual and specialized jaw arrangement that gives them highly protrusible mouths. They usually lack scales, tend to have dorsal fins that run the length of the body, and most have pelvic fins that are well forward on the body. Many have bodies adorned in radiant hues with bright red fins. This group probably arose some 65 million years ago.

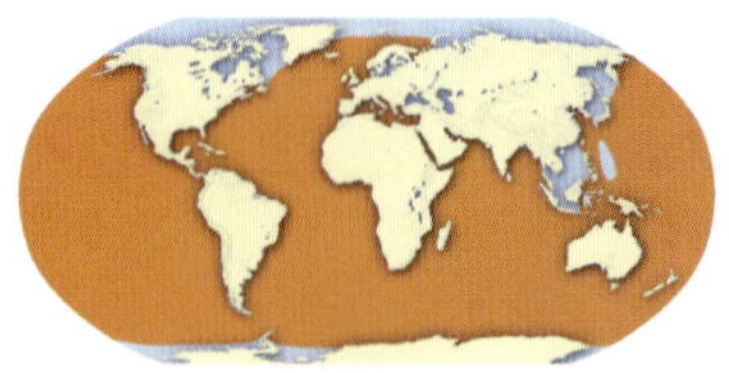

Widespread enigmas Opahs, ribbonfishes, and tube-eyes have worldwide distributions in warm waters. Velifers are found only in the Indian and Pacific Oceans. The open-ocean, deep-water lifestyles of lampridiforms mean that they are rarely encountered by people.

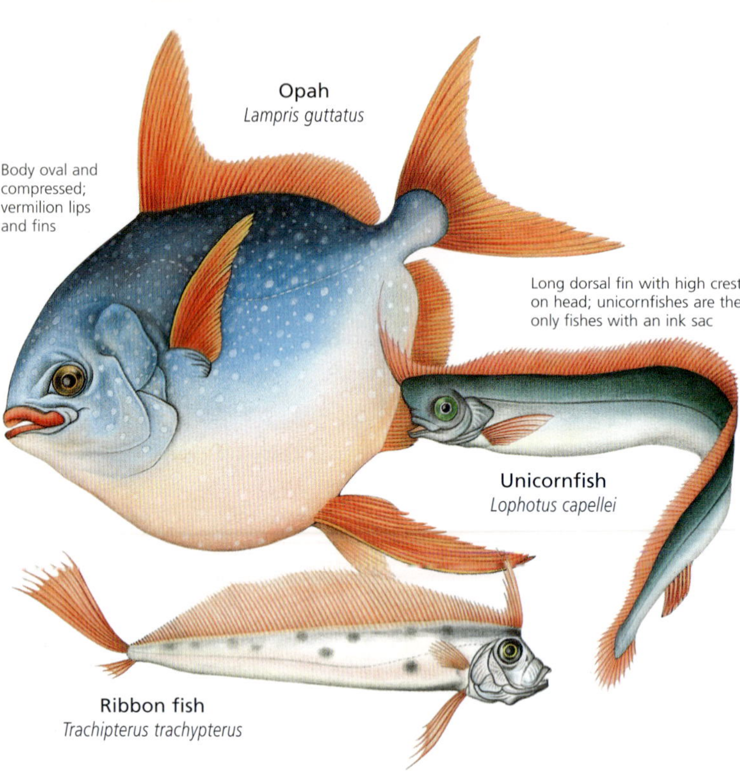

Opah
Lampris guttatus

Body oval and compressed; vermilion lips and fins

Long dorsal fin with high crest on head; unicornfishes are the only fishes with an ink sac

Unicornfish
Lophotus capellei

Ribbon fish
Trachipterus trachypterus

STRANGE LIFESTYLES

Lampridiforms are usually found at depths of between 330 and 3,300 feet (100 and 1,000 m). The lampridiforms include the crestfish, opahs, ribbonfishes, tube-eyes, and velifers, some of which exhibit unusual feeding behaviors. The tube-eye *Stylephorus chordatus*, for example, rises hundreds of feet (meters) daily from depths of about 2,600 feet (800 m) to feed on tiny crustaceans in a head-up, tail-down position.

The ribbon fish *Trachipterus trachypterus* is thought to adopt a similarly vertical position to feed on other fishes and squid.

All lampridiforms produce large eggs, up to ¼ inch (6 mm) in diameter. Most are bright shades of red, which may protect against ultraviolet rays penetrating the surface waters in which they float for up to a month before hatching. Unlike the eggs of most bony fishes, which produce feeble larvae requiring a rich yolk sac for food, lampridiform embryos develop early and are vigorous swimmers.

Mythical serpent Accounts of sea monsters probably stem from sightings of the oarfish, the world's longest fish. There has been at least one report of a specimen measuring 56 feet (17 m) long but most probably attain lengths of around 26 feet (8 m). These fish usually inhabit depths up to 660 feet (200 m) but have been seen at the surface and washed up on beaches.

CODS, ANGLERFISHES, AND ALLIES

SUBDIVISION Euteleostei
SUPERORDER Paracanthopterygii
ORDERS 5
FAMILIES 37
SPECIES 1,382

Several small but significant features of the skeleton indicate relatedness between species in this group. There are also similarities in their habitat preferences. Most, for example, are bottom dwellers although some species (particularly those that are commercially important, such as haddock, hake, and cod) form large pelagic schools. All but about 20 species comprising the group are marine. They tend to be active at night or live in dark habitats such as underwater caves or the deep sea. An unusual feature of some is that, by using special muscles located on the swim bladder, they can produce sounds. These may be important in courtship and communicating distress.

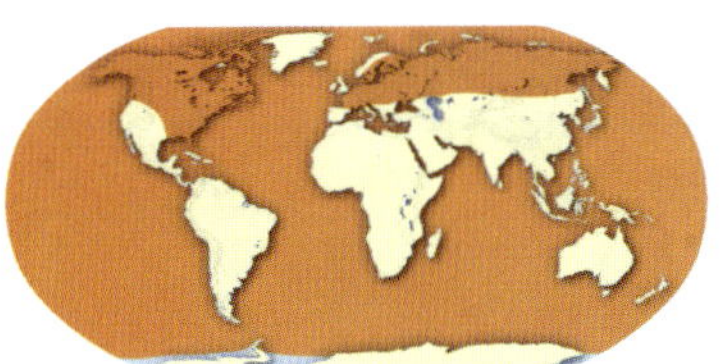

Distribution Cods and anglerfishes are found in all oceans. The cods are exclusively marine with the exception of one species—the burbot—which is found in fresh water. Troutperches and allies only live in fresh water and are restricted to North America. Toadfishes live along tropical coastlines.

CRITICAL FOOD FISHES

Some of the most significant fisheries worldwide exploit species within the family Gadidae, which includes the Alaska pollock and Atlantic cod and annually accounts for well over 10 percent of the world's total fish catch.

These fishes produce enormous numbers of eggs, among the most of any fish. A large Alaskan pollock, for example, may spawn 15 million eggs in a year. Because, however, gadiform fishes are also long-lived and late to mature, they are highly susceptible to overexploitation and some fisheries are now in trouble.

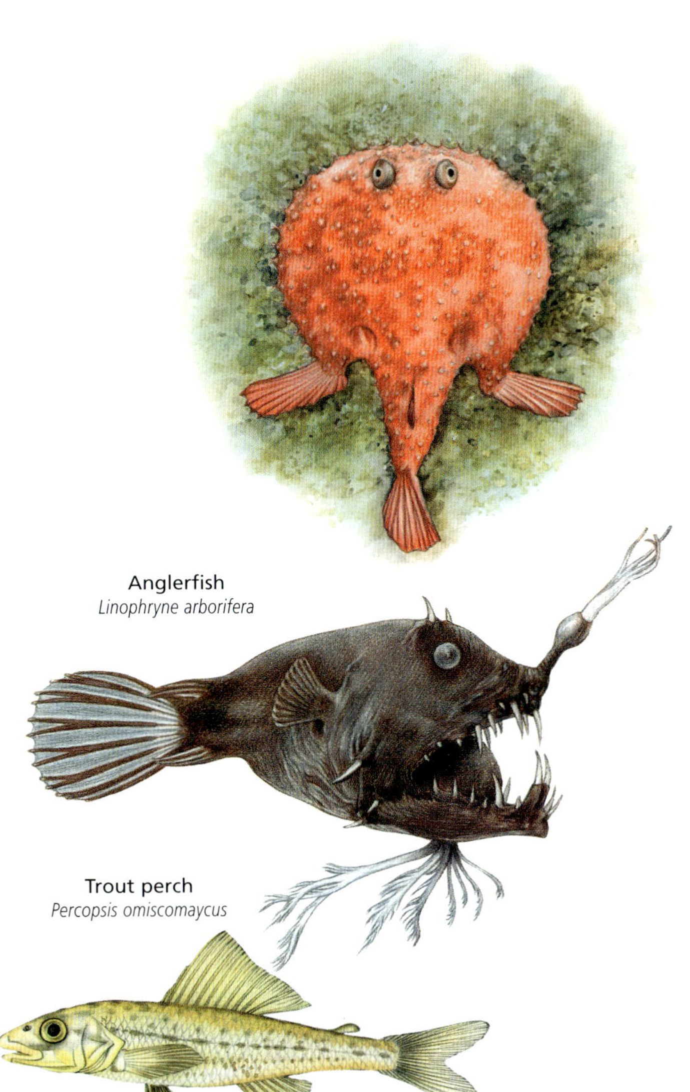

Anglerfish
Linophryne arborifera

Trout perch
Percopsis omiscomaycus

Starry handfish The expanded head of this dorsoventrally flattened bottom-dweller forms a flat rounded disc. It uses its hand-like pectoral fins, which project sideways from the back of this disc, to move across the substrate. An anglerfish, it carries a fleshy lure in a bony recess on the snout.

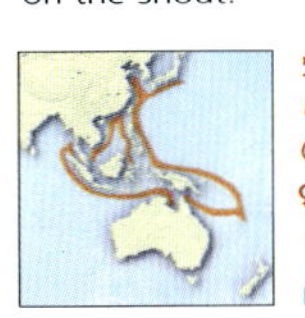

Indo-West Pacific

- Up to 12 in (30 cm)
- Up to 2 lb (900 g)
- Oviparous
- Male & female
- Common

Trout perch A native to freshwater streams and lakes in the eastern United States, the trout perch seeks rocky cover by day but ventures out by night to make short feeding migrations to shallower waters.

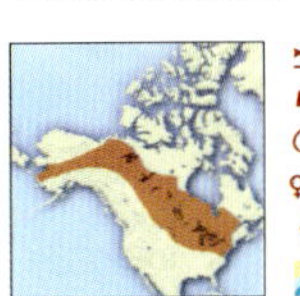

N. North America

- Up to 8 in (20 cm)
- Up to 6 oz (170 g)
- Oviparous
- Male & female
- Locally common

Blue antimora
Antimora rostrata

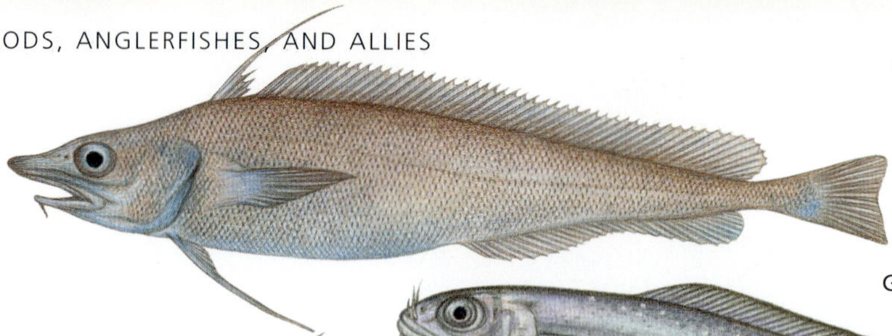

Goatsbeard brotula
Brotula multibarbata

Dorsal and anal fins long and joined with caudal fin

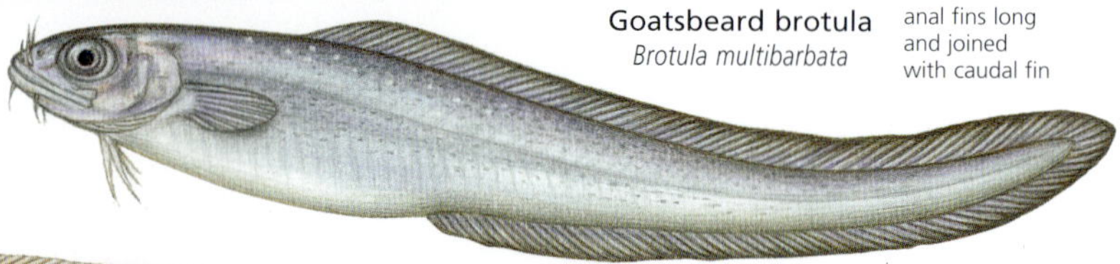

Banded whiptail
Coelorhynchus fasciatus

Longfin codling
Laemonema longipes

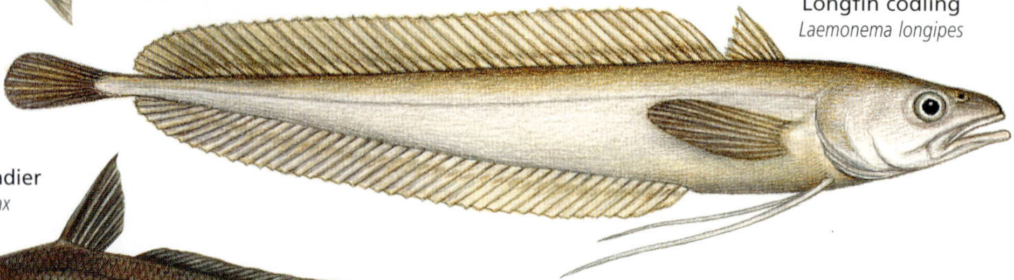

Onion-eye grenadier
Macrourus berglax

Fished commercially; has tough scales that have been known to dull knives and blades in processing plants

Tusk
Brosme brosme

Burbot
Lota lota

The only completely freshwater gadiform

European hake
Merluccius merluccius

Two dorsal fins; second dorsal and anal fins long-based with shallow notch

Red hake
Urophycis chuss

First dorsal fin ray elongated; pelvic fins long

SEXUAL PARASITES

Ceratoid anglerfishes dwell in low-density populations in the high-pressure darkness of oceanic depths. To increase chances of the sexes meeting, males of some species have evolved into "parasitic testes." As adults, they live permanently attached to and derive nourishment from their considerably larger female counterparts. In the triplewart seadevil (*Cryptopsaras couesii*), females have been found with up to four males attached.

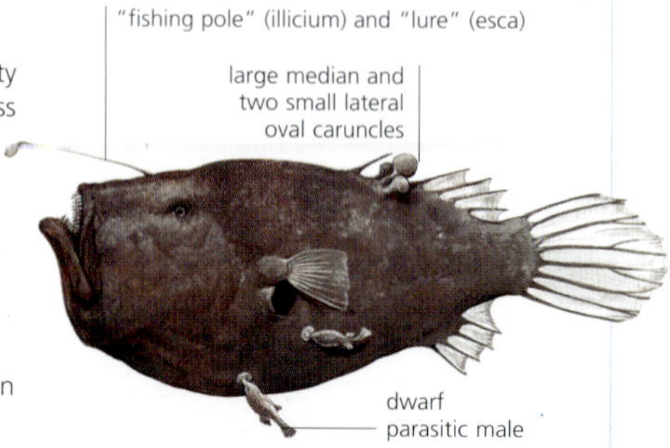

"fishing pole" (illicium) and "lure" (esca)

large median and two small lateral oval caruncles

dwarf parasitic male

European hake After years of commercial exploitation, there is widespread concern the European hake has been overfished. With a maximum weight recorded for the species about 25 pounds (11.5 kg), fish over 11 pounds (5 kg) are now rarely caught.

- Up to 4½ ft (1.4 m)
- Up to 25 lb (11.5 kg)
- Oviparous
- Male & female
- Common

E. North Atlantic, Mediterranean & Black seas

Haddock
Melanogrammus aeglefinus

Plainfin midshipman
Porichthys notatus

Tadpole fish
Raniceps raninus

Stout-bodied with a broad depressed head; minute first dorsal fin and small chin barbel

American angler
Lophius americanus

Angler (monkfish)
Lophius piscatorius

Three dorsal fins

Bib
Trisopterus luscus

Atlantic cod
Gadus morhua

Two anal fins

Haddock Both male and female haddock make "knocking" sounds when threatened. Males become particularly raucous during spawning; it is likely that their noises are significant in the reproductive process.

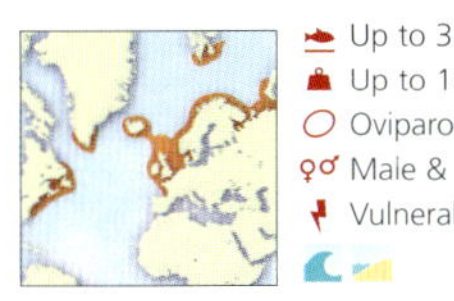

- Up to 35½ in (90 cm)
- Up to 18 lb (8.2 kg)
- Oviparous
- Male & female
- Vulnerable

North Atlantic to Spitzbergen

Angler This eastern Atlantic species of anglerfish has an enormous mouth and lies in wait for prey, well-camouflaged among sand and sediment on the ocean floor. It usually feeds on other fishes but has been known to also take seabirds.

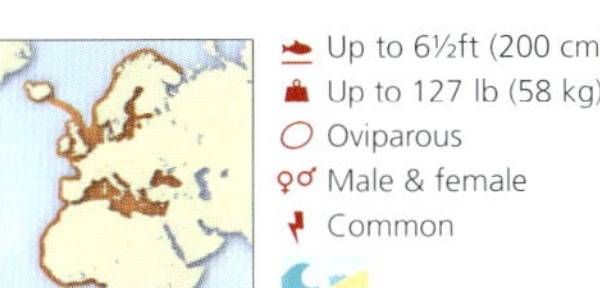

- Up to 6½ft (200 cm)
- Up to 127 lb (58 kg)
- Oviparous
- Male & female
- Common

E. North Atlantic

Atlantic cod A cold-water species exploited in large commercial fisheries in the north Atlantic, this cod is usually caught up to weights of about 25 pounds (11.5 kg) although it can grow to four times that size.

- Up to 5 ft (1.5 m)
- Up to 100 lb (45 kg)
- Oviparous
- Male & female
- Vulnerable

North Atlantic to Spitzbergen

SPINY-RAYED FISHES

CLASS	Actinopterygii
SUPERORDER	Acanthopterygii
ORDERS	15
FAMILIES	269
SPECIES	13,262

This, the largest single group of fishes, comprises the most advanced and recently evolved teleosts. Known also as acanthopterygians, the spiny-rayed fishes include more than 13,000 species in over 250 families. Common characteristics include a highly mobile and protrusible upper jaw, which has facilitated a wide range of feeding strategies, ctenoid scales that in some cases have become lost or further developed into hardened plates, and hard spines in the fins. As a group, these fishes have a ubiquitous distribution but are particularly abundant in coastal marine waters. They display an enormous breadth of specialized reproductive, behavioral, and anatomical adaptations that has seen them exploit niches not available to other fishes.

Extensive distribution The most ubiquitous of all fish groups, the spiny-rays are found in almost every aquatic environment from fresh water, through brackish to salt water, and from coastal shallows to the ocean's depths. Morphological and behavioral adaptations allow them to live in a variety of habitats, from frozen seas to ponds that dry up.

Mutual beneficiaries Living among poisonous sea anemone tentacles, clownfish escape predators, while the anemones benefit by feeding on the fishes' food scraps. A mucus layer covering clownfish keeps them safe from the poisonous stings that kill most other fishes.

ADVANCED ADAPTATIONS

Almost every variation on the basic fish body plan can be found among the spiny-rayed fishes. The flatfishes, for example, depart dramatically from the normal symmetry seen in most fishes. The wrasses and parrot-fishes have a modified pharyngeal apparatus that acts as a second set of jaws in the throat, facilitating specialized feeding.

The cyprinodontiforms include guppies, swordtails, and other resilient freshwater aquarium favorites. The flyingfishes are remarkably adapted for above-water gliding.

Gobies are mostly small fishes, such as the amphibious mudskippers, in which the pelvic fins are fused to form a cup-shaped disk.

In many triggerfishes and their relatives, scales have developed into protective body armor. Damselfishes, including clownfishes, often display highly territorial behavior. Cichlids take parental care to its extreme. The groupers are among the most robust of all marine predators, and billfishes such as marlin and sailfish are among the fastest swimming.

The scorpionfishes include some of the deadliest fishes to humans. The drums and croakers make distinctive noises when threatened and the butterflyfishes and angelfishes are among the most exquisitely colored aquatic creatures.

Brook silverside
Labidesthes sicculus

Goldie River rainbowfish
Melanotaenia goldiei

Sharpchin flyingfish
Fodiator acutus

Flat needlefish
Ablennes hians

Forktail rainbowfish
Pseudomugil furcatus

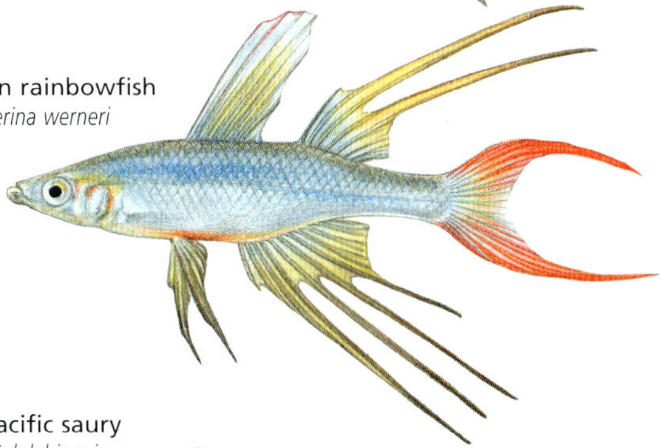

Threadfin rainbowfish
Iriatherina werneri

Pacific saury
Cololabis saira

Japanese medaka
Oryzias latipes

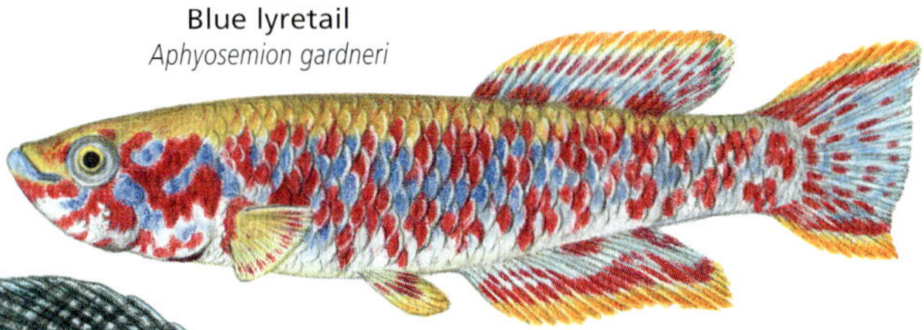

Blue lyretail
Aphyosemion gardneri

Golden pheasant panchax
Callopanchax occidentalis

Blue notho
Nothobranchius patrizii

Saberfin killie
Terranatos dolichopterus

Flagfish
Jordanella floridae

Brook silverside A native of sub-tropical North American freshwater lakes and rivers, this small translucent fish is usually found in schools near the water's surface, feeding on tiny crustaceans, insect larvae, and small flying adult insects.

Up to 5 in (13 cm)
Up to 4 oz (115 g)
Oviparous
Male & female
Common

S.E. North America

Sharpchin flyingfish Members of the genus *Fodiator* are among the weakest "flyers" of the flyingfish. Nevertheless, this species is capable of gliding over water on enlarged wing-like pectoral fins for 165 feet (50 m) or more.

Up to 9½ in (24 cm)
Up to 8 oz (225 g)
Oviparous
Male & female
Common

E. Pacific & E. Atlantic

Golden pheasant panchax Adults of this shortlived freshwater species die after spawning, at the onset of the dry season, leaving fertilized eggs to survive in mud for as long as 3 months. They hatch with the onset of the rainy season.

Up to 3 in (7.5 cm)
Up to 1 oz (30 g)
Oviparous, hiders
Male & female
Common

W. Africa

Striped foureyed fish
Anableps anableps

Guppy
Poecilia reticulata

Western mosquitofish
Gambusia affinis

Splendid alfonsino
Beryx splendens

Lateral line extends to the deeply forked caudal fin

Splitfin flashlightfish
Anomalops katoptron

Pineconefish
Monocentris japonica

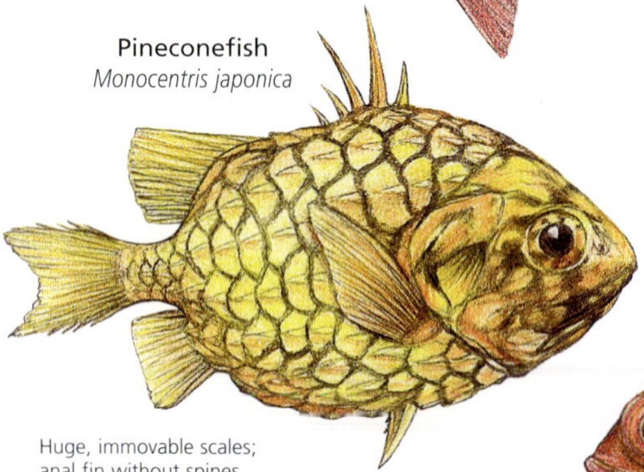

Huge, immovable scales; anal fin without spines

Crown squirrelfish
Sargocentron diadema

Alternating broad red and narrow silvery-white stripes; anal fin with four spines

Velvet whalefish
Barbourisia rufa

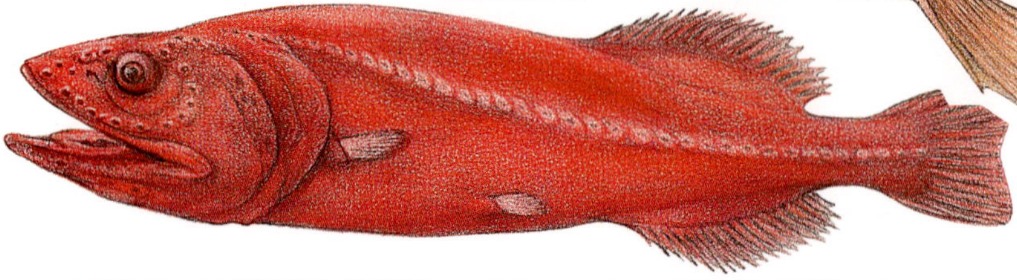

Crown squirrelfish Squirrelfish larvae have a long pelagic life and are often found way out to sea, while most adults live in shallow water around tropical reefs. This species has a short venomous cheek spine. Squirrelfishes are active at night, sheltering in caves and under ledges during the day. They feed on small fishes and invertebrates.

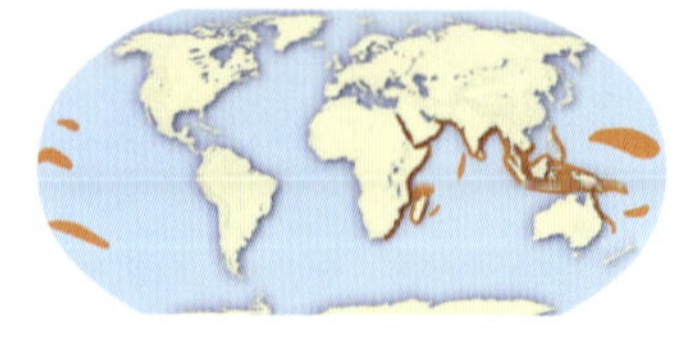

Up to 6½ in (17 cm)
Up to 8 oz (225 g)
Oviparous
Male & female
Common
Indo-Pacific, Red Sea, Oceania

Guppy The guppy belongs to one of the few bony fish families (Poeciliidae) with members that give birth to live young. Originally from South America, this popular aquarium fish has been introduced to warm freshwater lakes around the world.

Up to 2 in (5 cm)
Up to ¾ oz (20 g)
Viviparous; live bearers
Male & female
Common
N.E. South America, Barbados, Trinidad

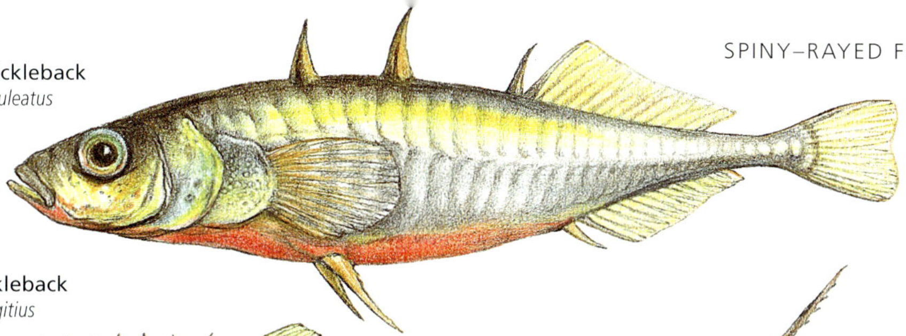

Three-spined stickleback
Gasterosteus aculeatus

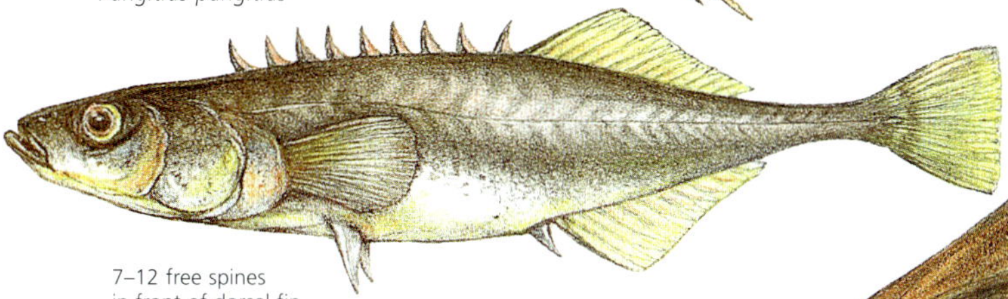

Ninespine stickleback
Pungitius pungitius

7–12 free spines in front of dorsal fin

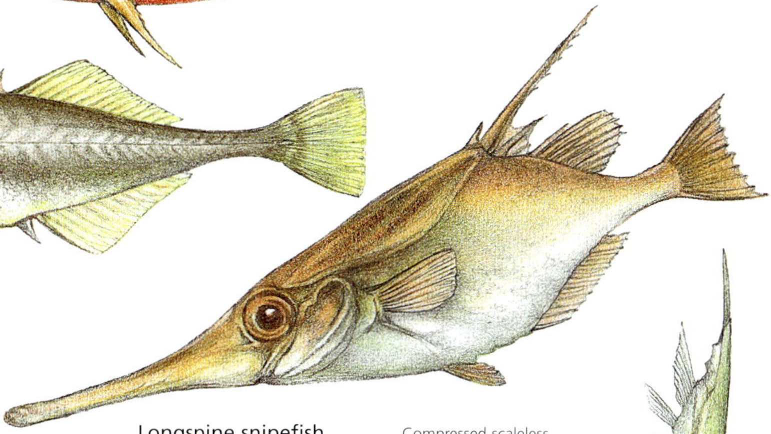

Longspine snipefish
Macrorhamphosus scolopax

Compressed scaleless body; mouth at end of long tubular snout

Shrimpfish
Aeoliscus strigatus

Indo-Pacific boarfish
Antigonia rubescens

3 anal spines with 24–48 soft rays; body highly compressed

John Dory
Zeus faber

Massive head with large, highly protrusible jaws

Shrimpfish With a semi-transparent, shrimp-like body, this fish congregates in synchronized schools in a head-down, tail-up vertical position. It lives among sea urchins and staghorn coral branches and feeds on zooplankton.

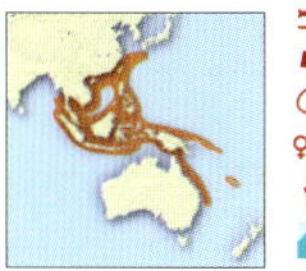
Indo-West Pacific

Up to 6 in (15 cm)
Up to 4 oz (115 g)
Oviparous
Male & female
Common

DEVOTED DADS

By cementing plants together with sticky secretions from their kidneys, male sticklebacks build elaborate nests for the females to lay their eggs in. After fertilizing the eggs, the males usually drive the females away but continue to maintain and defend the nests while attending to the eggs, fanning oxygenated water across them with their pectoral fins.

Harlequin ghost pipefish
Solenostomus paradoxus

Head set at an angle to stout body; prehensile tail

Leafy seadragon
Phycodurus eques

Seahorse
Hippocampus ramulosus

Lesser spiny eel
Macrognathus aculeatus

Ringed pipefish
Doryrhamphus dactyliophorus

East Atlantic red gurnard
Chelidonichthys cuculus

Large and blunt head, with bones forming a helmet

Flying gurnard
Dactylopterus volitans

Produces noise using muscles attached to the swimbladder

Swamp eel
Monopterus albus

Harlequin ghost pipefish Pipefishes and seahorses have a series of bony plates beneath their skin. As a result, they cannot swim by flexing the body like most other fishes but use rapid fanning movements of the fins instead.

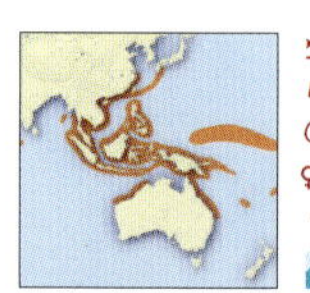

- Up to 4¾ in (12 cm)
- Up to 1 oz (30 g)
- Oviparous, brooder
- Male & female
- Uncommon

Indo-West Pacific

Leafy seadragon Like their close relatives the seahorses, male leafy seadragons incubate eggs. Females lay up to 250 onto a patch of spongy tissue on the tail's underside, where they develop for about 6 weeks before hatching.

- Up to 16 in (40 cm)
- Up to 8 oz (225 g)
- Oviparous, brooder
- Male & female
- Data deficient

S. Australia

Swamp eel This highly adaptable, air-breathing, freshwater carnivore that resembles an eel can survive out of water for long periods. It has been labeled a potential "ecological night-mare" in some areas to which it has been introduced.

- Up to 18 in (46 cm)
- Up to 24½ oz (700 g)
- Oviparous, guarders
- Hermaphrodite
- Common

S.E. Asia, Australia

Redfish
Sebastes marinus

Eggs develop inside the female's oviduct; over several months, the female releases up to 20,000 larvae, each about ⅓ inch (8 mm) long

Stonefish
Synanceia verrucosa

Radial firefish
Pterois radiata

Fins have very long spines and rays; dorsal spines are venomous

Sablefish
Anoplopoma fimbria

Baikal yellowfin
Cottocomephorus grewingkii

Spiny dorsal fin; head is also spiny

Shorthorn sculpin
Myoxocephalus scorpius

Endemic to Lake Baikal and its tributaries; huge pectoral fins

Lumpsucker
Cyclopterus lumpus

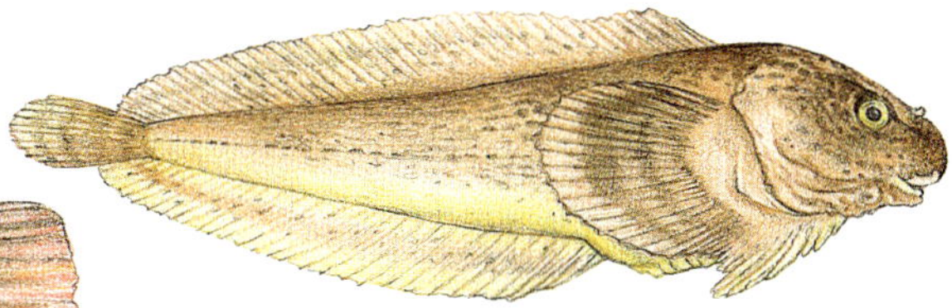

Striped seasnail
Liparis liparis

Stonefish Spines on the dorsal fin of this stonefish function like syringes to deliver agonizing stings and the most toxic venom produced by any fish. Deaths have occurred among people who have mistakenly walked on stonefish on reef flats.

- Up to 14 in (36 cm)
- Up to 4½ lb (2 kg)
- Oviparous
- Male & female
- Common

Indo-Pacific & Red Sea

Sablefish Named because of its sleek black to dark-green skin, the sablefish is an important commercial species in Alaska and to a lesser extent in Canada. It is a deep-water, long-lived fish, with reports of individuals older than 90 years.

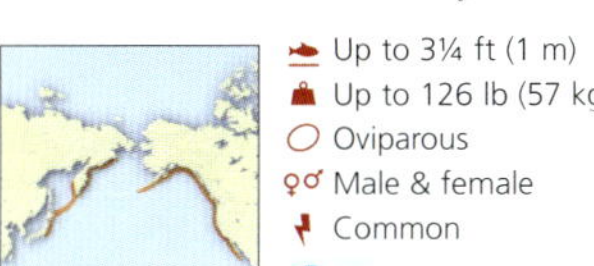

- Up to 3¼ ft (1 m)
- Up to 126 lb (57 kg)
- Oviparous
- Male & female
- Common

North Pacific

Shorthorn sculpin Renowned for eating "anything and everything," the shorthorn sculpin has a voracious appetite. Its large mouth can wrap around prey half its size and its stomach readily stretches around any large object it consumes.

- Up to 24 in (60 cm)
- Up to 2 lb (900 g)
- Oviparous
- Male & female
- Common

North Atlantic & Arctic oceans

Striped bass
Morone saxatilis

Large-mouthed predator with 7 or 8 black stripes on side

Barramundi perch
Lates calcarifer

A diadromous species, found in coastal waters, rivers, estauries, and lagoons

Comet
Calloplesiops altivelis

Sea goldie
Pseudanthias squamipinnis

Jewfish
Epinephelus itajara

Can attain lengths of over 8 feet (2.5 m); weighs around 800 pounds (365 kg)

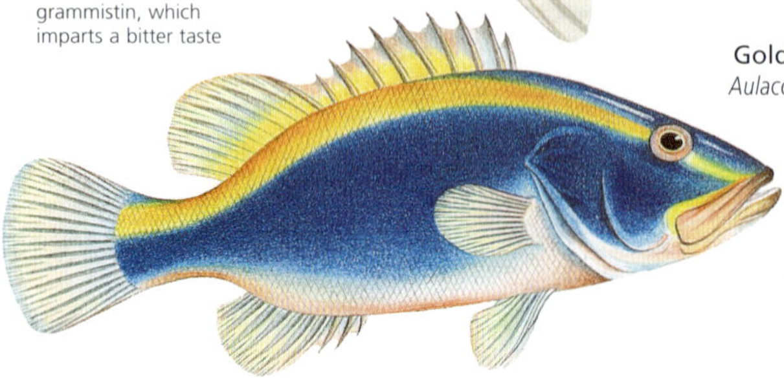

Skin contains grammistin, which imparts a bitter taste

Goldribbon soapfish
Aulacocephalus temminckii

Royal gramma
Gramma loreto

INGENIOUS MIMICRY

The comet deters predators by hiding in holes exposing only its tail, which mimics the fierce-looking head of the identically patterned turkey moray eel (*Gymnothorax meleagris*). An ocellus at the base of the dorsal fin looks like an eye. A gap between the anal and tail fins forms a "mouth."

Butterfly effect *This type of so-called "Batesian mimicry" is similar to that adopted by many butterflies.*

Barramundi perch Most barramundi are protandrous hermaphrodites: they begin life as males, reach sexual maturity after about 3 years, and become females after about 5 years. As a result, larger individuals are inevitably female.

Indo-West Pacific

- Up to 6 ft (1.8 m)
- Up to 132 lb (60 kg)
- Oviparous
- Hermaphrodite
- Common

Very large mouth, the upper jaw extending well past the eye

Largemouth bass
Micropterus salmoides

Eurasian perch
Perca fluviatilis

Pumpkinseed
Lepomis gibbosus

Broad vertical bars on body and white tips on caudal, second dorsal, and anal fins

Pilotfish
Naucrates ductor

Common name is derived from comb of elongate dorsal fin spines

Lookdown
Selene vomer

A jack with an extremely compressed body and a very steep front to the head

Roosterfish
Nematistius pectoralis

Eurasian perch Females of this species lay their eggs, several tens of thousands at a time, connected in long, white sticky mucous ribbons. These can be up to a meter in length and are left draped over submerged rocks and vegetation.

N. Eurasia

- Up to 20 in (51 cm)
- Up to 10½ lb (4.7 kg)
- Oviparous
- Male & female
- Common

Pumpkinseed As occurs within most species of the freshwater North American sunfish family, the male pumpkinseed is a nest-builder. Eggs and sperm are deposited as the male and female swim around the nest in a circular pattern.

E. North America; introd. widely

- Up to 16 in (40 cm)
- Up to 22 oz (630 g)
- Oviparous, guarders
- Male & female
- Common

Roosterfish Nematistiidae is a jack-like fish and, like many species in the jack family (Carangidae), is a popular gamefish that fights strongly when hooked. Its distinctive, elongate, backward-curving dorsal spines look like a cockscomb.

E. Pacific

- Up to 4 ft (1.2 m)
- Up to 100 lb (45 kg)
- Oviparous
- Male & female
- Common

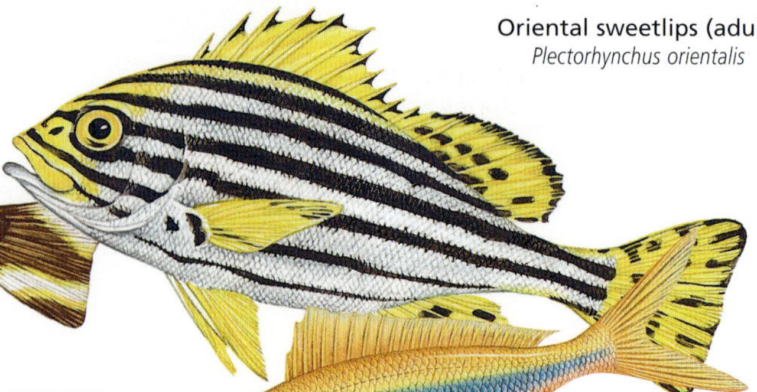

Oriental sweetlips (adult)
Plectorhynchus orientalis

Oriental sweetlips (juvenile)
Plectorhynchus orientalis

Yellowback fusilier
Caesio xanthonata

Emperor snapper
Lutjanus sebae

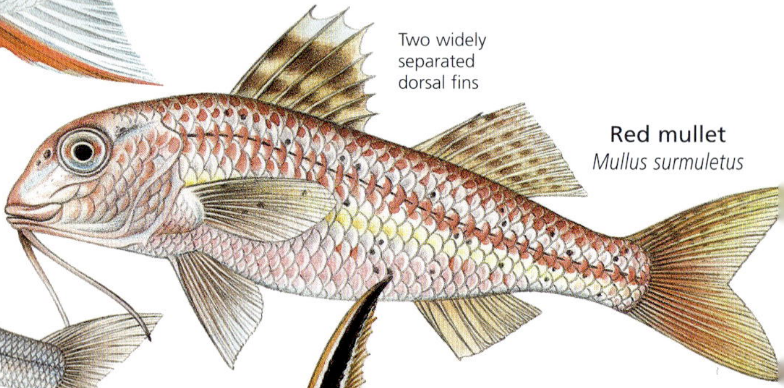

Two widely separated dorsal fins

Red mullet
Mullus surmuletus

Picarel
Spicara smaris

Jack-knifefish
Equetus lanceolatus

Mouth large and inferior

Red drum
Sciaenops ocellatus

Scales on cheeks and operculum

Twoband bream
Diplodus vulgaris

Oriental sweetlips The loose rubbery lips of sweetlips are adapted for "vacuuming" invertebrates from sandy bottoms. They feed so vigorously that silt sweeps back through their gill arches to form a cloud in the water behind.

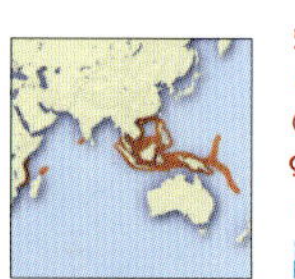

Up to 20 in (51 cm)
Up to 4 lb (1.8 kg)
Oviparous
Male & female
Common

Indo-West Pacific

Red mullet This species is not a true mullet but a goatfish (family Mullidae). It has the long sensory chin barbels characteristic of its family, and uses these to probe bottom sediments for small invertebrates.

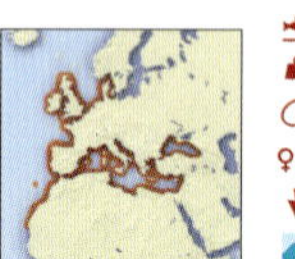

Up to 16 in (40 cm)
Up to 35 oz (1 kg)
Oviparous
Male & female
Common

E. North Atlantic & Mediterranean Sea

Emperor snapper Juvenile emperor snapper live in shallow tropical waters, often in close association with sea urchins. Large adults, which dwell at greater depths, are sometimes implicated in cases of ciguatera poisoning.

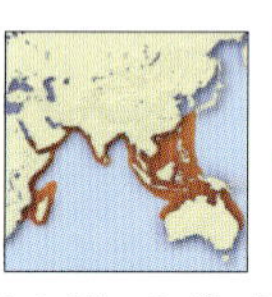

Up to 3¼ ft (1 m)
Up to 35½ lb (16 kg)
Oviparous
Male & female
Common

Indo-West Pacific, Red Sea

Amazon leaf-fish This species from South America looks and behaves like a floating dead leaf. Colored shades of brown with a twig-like chin barbel, it moves by means of transparent fins, lunging at unsuspecting prey with its enormous mouth.

- Up to 3 in (7.5 cm)
- Up to 1 oz (30 g)
- Oviparous, guarder
- Male & female
- Common

N. South America

TAKING AIM

Archerfishes fire water jets from the mouth to knock insects and other prey from overhanging branches. Rapid oral cavity compressions force water through a tube formed by the tongue pressing against the specially grooved palate. The archerfish *Toxotes jaculatrix* has a shooting range of about 5 feet (1.5 m). They will also shoot at swarms of flying insects hovering above the surface.

Clown anemonefish
Amphiprion ocellaris

Clown anemonefish are protected from harm from the sting cells of their anemone host by a distinctive undulating swimming behavior and a protective coat of mucus

Bluestreak cleaner wrasse
Labroides dimidiatus

Rainbow krib
Pelvicachromis pulcher

Small cichlid that spawns in caves and guards its eggs

Freshwater angelfish
Pterophyllum scalare

Large cichlid that feeds exclusively on fish

Peacock cichlid
Cichla ocellaris

Larvae feed on a white mucus that the parents secrete on their skin

Clown wrasse
Coris aygula

Blue discus
Symphysodon aequifasciatus

Juveniles have false eyes, shaded by orange

ORAL NURSERIES

All of the cichlids show some form of parental care, the most advanced being continuous mouth-brooding. Usually the female scoops her eggs up into her mouth quickly after laying them, nuzzles the male near his genital opening, and then draws his ejaculated sperm into her mouth where she may brood embryos and young for more than 3 weeks. Toward the end of that period, the young are released at intervals to forage.

Freshwater angelfish Eggs laid onto the submerged leaf of an aquatic plant are guarded and fanned constantly by the parents for the 3 days until hatching. Fry are gathered into a dense pack each evening for ease of protection.

- Up to 3 in (7.5 cm)
- Up to 8 oz (225 g)
- Oviparous, guarder
- Male & female
- Common

N.E. South America

Yellowhead jawfish
Opistognathus aurifrons

Constructs elaborate burrows; male broods the eggs orally

Viviparous blenny
Zoarces viviparus

Greater weever
Trachinus draco

Lies buried in sand; spines of the first dorsal fin and gill cover have venom

Atlantic mudskipper
Periophthalmus barbarus

Atlantic stargazer
Uranoscopus scaber

False cleanerfish
Aspidontus taeniatus

Lesser sandeel
Ammodytes tobianus

Burrows in sand but also forms huge schools in mid-water

Redtail surgeonfish
Acanthurus achilles

Named for the sharp spine or spines they possess on the caudal peduncle

Moorish idol
Zanclus cornutus

Atlantic stargazer A well-camouflaged bottom-dweller, the Atlantic stargazer has a small, worm-like appendage on its lower lip that it wriggles to attract prey. Defenses include a venomous spine and electric organs behind the eyes.

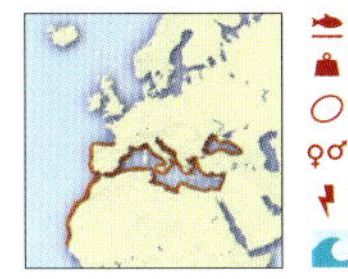

- Up to 16 in (40 cm)
- Up to 33 oz (940 g)
- Oviparous
- Male & female
- Locally common

E. Atlantic; Mediterranean & Black seas

AMPHIBIOUS FISHES

Mudskippers use their muscular tail and pectoral fins to "skip" over mud at low tide and even climb trees with the aid of a pelvic fin "suction cup." There are more than 30 species of mudskipper, found mainly in the muddy, intertidal mangrove forests of Southeast Asia and Africa. They take in oxygen through their moist skin, which is richly endowed with blood vessels.

Land lovers *The males of some mudskipper species display their masculine prowess and define territories by leaping about in the mud.*

Swordfish
Xiphias gladius

Upper jaw extended into long slender bill (rostrum)

Blue marlin
Makaira nigricans

Indo-Pacific sailfish
Istiophorus platypterus

First dorsal fin sail-like; pelvic fins very long

Wahoo
Acanthocybium solandri

Large mouth with strong, serrated triangular teeth

Skipjack tuna
Katsuwonus pelamis

Caudal fin deeply forked; upper lobe shorter than lower; no pelvic fins

Man-of-war fish
Nomeus gronovii

Silver pomfret
Pampus argenteus

Lives in association with Portuguese man-of-war

MIGRATING EYES

Larval flatfishes look like other young fishes to begin with but soon begin leaning sideways as one eye (left in some species, right in others) migrates across to take up position beside the other. Simultaneously, the front of the skull twists to bring the jaws into an oblique sideways position.

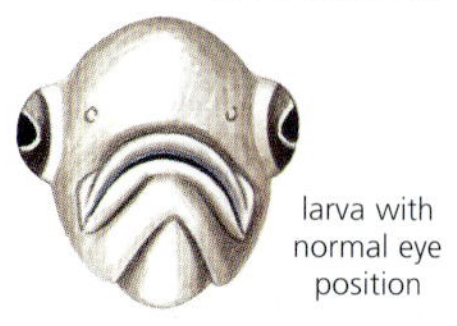

larva with normal eye position

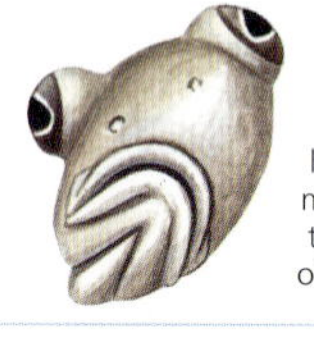

left eye migrates to top of head

adult eyes both on the right side

Indo-Pacific sailfish No other fish is known to swim faster than the sailfish, which has been clocked in excess of 68 mph (110 km/h). It uses its rapier-like beak to stun and maim prey, which it then scoops up with toothless jaws.

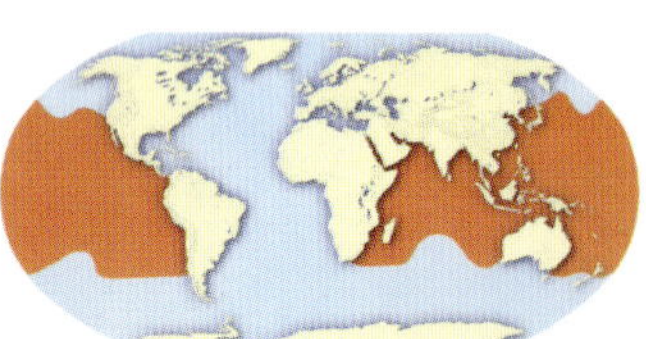

Ornate ctenopoma
Microctenopoma ansorgii

Siamese fighting fish
Betta splendens

Pearl gourami
Trichogaster leerii

Gourami
Osphronemus goramy

Can breathe moist air and survive long periods out of water

Lips have horny teeth; filter feeder and grazer on benthic algae

Kissing gourami
Helostoma temminckii

...cking lips used ...r kissing other ...hes, plants and ...her objects

Brill
Scophthalmus rhombus

Climbing perch
Anabas testudineus

Zebra sole
Zebrias zebra

Distinctive stripes give this sole its common name

Very large flatfish that is an active predator of other fishes

Atlantic halibut
Hippoglossus hippoglossus

Summer flounder
Paralichthys dentatus

Siamese fighting fish The males of this freshwater species create egg nests by blowing bubbles around leaves, which they then guard aggressively. The male-male fighting for which the species is famous involves threat displays and fin-nipping.

- Up to 2.5 in (6.6 cm)
- Up to 1 oz (30 g)
- Oviparous, guarder
- Male & female
- Common

Mekong basin (Asia)

Climbing perch With an accessory air-breathing organ associated with the gills, this fish is well-adapted to life in oxygen-depleted environments. When ponds dry up it will "walk," using its fins, in search of water and reportedly even climb low trees.

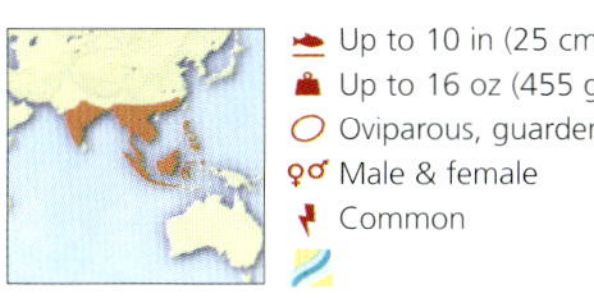

- Up to 10 in (25 cm)
- Up to 16 oz (455 g)
- Oviparous, guarder
- Male & female
- Common

S.E. Asia

Brill Like all other flatfishes, this marine species relies extensively on camouflage to hide it from both predators and prey and can change its body color to match the surrounding seabed environment.

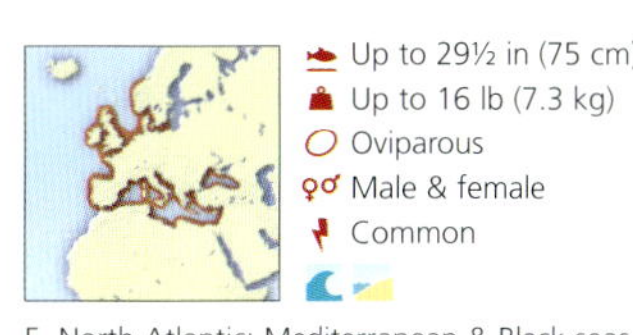

- Up to 29½ in (75 cm)
- Up to 16 lb (7.3 kg)
- Oviparous
- Male & female
- Common

E. North Atlantic; Mediterranean & Black seas

Clown triggerfish
Balistoides conspicillum

First dorsal spine can be locked erect by the small second spine; if depressed like a trigger it unlocks the first spine

Blackbar triggerfish
Rhinecanthus aculeatus

Sleeps on its side and makes a whirring noise if alarmed

Prickly leatherjacket
Chaetodermis penicilligerus

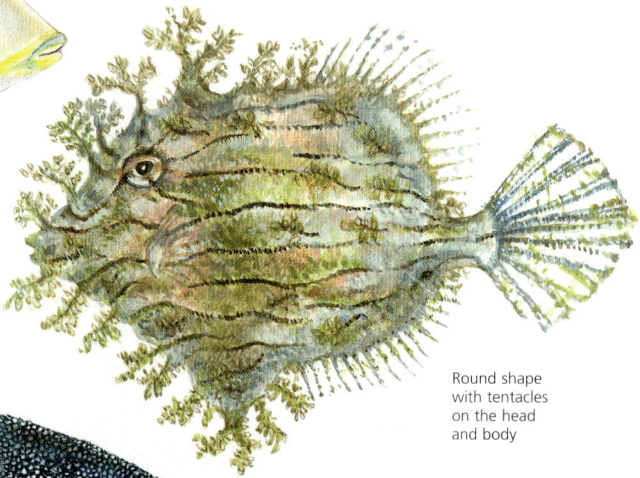

Round shape with tentacles on the head and body

Guineafowl puffer
Arothron meleagris

Body of this puffer is covered by prickles

Thornback cowfish
Lactoria fornasini

Harlequin filefish
Oxymonacanthus longirostris

Guineafowl puffer
This species is typical of pufferfishes in that it is poisonous to eat and has its teeth incorporated into the jawbone to form a parrot-like beak. This enables it to feed mainly on the soft-bodied polyps in the tips of branching coral species.

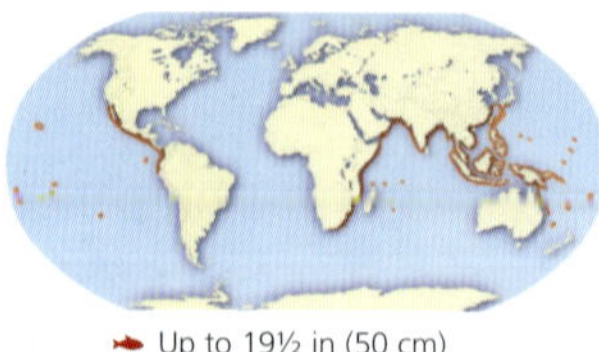

- Up to 19½ in (50 cm)
- Up to 4 lb (1.8 kg)
- Oviparous
- Male & female
- Common

Indo-Pacific: E. Africa to Americas

Thornback cowfish
Like their close relatives the boxfishes, cowfishes are encased in a protective outer armor. They also have a unique form of locomotion—octraciform swimming—which involves a sculling motion of the tail fin and no body flexing. This does not provide speed, but allows almost motionless hovering.

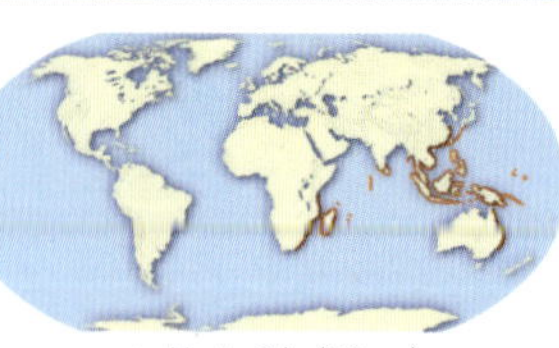

- Up to 9 in (23 cm)
- Up to 1 lb (460 g)
- Oviparous
- Male & female
- Common

Indo-West Pacific: E. Africa to Hawaii

Freshwater pufferfish
Tetraodon mbu

Four fused teeth but separated by a median suture in each jaw

Sexually dimorphic species with a body encased in a bony carapace

White-spotted boxfish
Ostracion meleagris

Narrowbanded batfish
Platax orbicularis

Compressed, deep-bodied fish; ocular band with a series of dark vermiculations

Ocean sunfish
Mola mola

Balloonfish
Diodon holacanthus

Threetooth puffer
Triodon macropterus

Large skin flap on belly and a black spot on the middle of its side

Freshwater pufferfish When they are threatened, pufferfishes pump air or water into a distensible stomach sac to greatly increase the size of their bodies and appear more formidable. This also forces spines and scales to protrude.

Central Africa

- Up to 26½ in (67 cm)
- Up to 15 lb (6.8 kg)
- Oviparous
- Male & female
- Common

PUFFERFISH POISON

Some pufferfishes are highly venomous, particularly those of the Indo-Pacific. Tetrodotoxin, which is found throughout the body but concentrated in the liver, ovaries, and gut, is a powerful nerve poison considerably stronger than cyanide. These fishes also release this toxin into the surrounding water when frightened by a potential predator.

Dicey dining *The specially prepared poison-free flesh of certain pufferfish species is highly prized, particularly in Japan and Korea, even though their organs are imbued with deadly tetrodotoxin.*

INVERTEBRATES

INVERTEBRATES
PHYLA > 30
CLASSES > 90
ORDERS > 370
SPECIES > 1.3 million

Constituting more than 95 percent of all known animal species, invertebrates are not distinguished by a single positive characteristic. Instead the group is defined by what its members lack: they have no backbone, no bones, and no cartilage. As a term of classification, invertebrate is commonly used but has little scientific validity. Unlike vertebrates, which belong to a single phylum, invertebrates are a collection of more than 30 phyla, some of which are more closely related to vertebrates than to each other. Invertebrates encompass such diverse forms as porous sponges, floating jellyfishes, parasitic flatworms, jet-propelled squids, hard-cased crabs, venomous spiders, and fluttering butterflies.

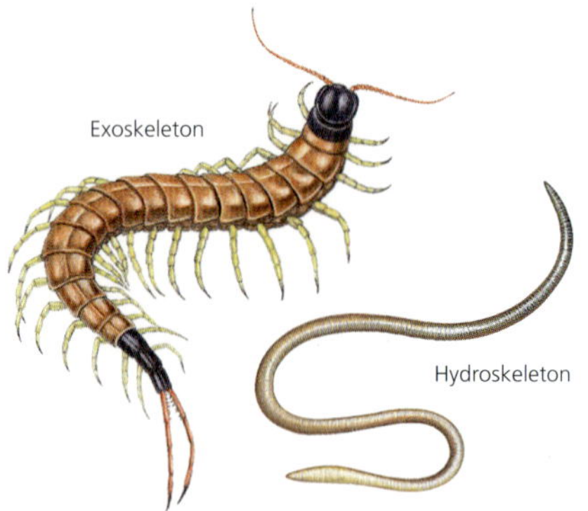

Inside and outside An internal fluid-filled hydroskeleton supports the body of worms and many other invertebrates, but is only viable in moist environments or water. An external exoskeleton helped arthropods such as centipedes to colonize land.

SOFT-BODIED CREATURES

The first animals to evolve were invertebrates, but their soft bodies left no traces. Although tracks and burrows appear in the fossil record about 1 billion years ago, the oldest fossilized animal remains date back to about 600 million years ago, near the end of the Precambrian period. Known as Ediacaran fauna, these include forms that resemble sponges, jellyfishes, soft corals, segmented worms, and echinoderms. Beginning roughly 60 million years later, the Cambrian period is associated with the explosion of invertebrate life. By its conclusion, about 500 million years ago, all of today's invertebrate phyla seem to have appeared.

Although their greatest diversity occurs in the sea, invertebrates are now found in virtually all land and water habitats. Most species are small, and some are microscopic: many rotifers are less than $\frac{1}{25,000}$ inch (0.001 mm) long. However, a few invertebrates reach staggering sizes: the elusive giant squid can be up to 59 feet (18 m) long and weigh up to 1,980 pounds (900 kg).

There are two basic invertebrate body plans. Species with radial symmetry, such as jellyfish and sea anemones, have a circular body plan and a central mouth. Others, such as worms and insects, have bilateral symmetry, with a distinct head and right and left sides.

Invertebrates lack bones, but are supported by some sort of skeleton. While many appear soft, they are held together by protein fibers, a feature of all animals. Many worms have a hydroskeleton, with fluid held under pressure inside the body cavity. Sponges and echinoderms have endoskeletons, with hard elements inside the tissues. Most mollusks and all arthropods have an exoskeleton, a hard, external casing. A mollusk's exoskeleton is a hard shell, while an arthropod's tough exoskeleton is jointed and flexible.

Most invertebrates reproduce sexually, laying vast numbers of fertilized eggs that are usually left to hatch on their own. Some species, however, develop from unfertilized eggs, and others reproduce by fragmenting or budding, with parts of their own body becoming the offspring. Invertebrate young are often quite unlike their parents and must go through metamorphosis to transform into adults.

Butterfly feeding As adults, Eastern Tiger swallowtails (*Papilio glaucus*) (above) feed on the nectar of many different species of flower.

Hunting spider Spiders are predacious invertebrates. They use venom to paralyze prey.

Modes of travel Sea anemones (opposite) are sessile as adults, remaining fixed in one spot and waving their tentacles to catch food as it comes within range. Many other invertebrates, however, are highly mobile and include swimming, burrowing, creeping, running, and flying forms.

INVERTEBRATE CHORDATES

PHYLUM	Chordata
SUBPHYLA	2
CLASSES	4
ORDERS	9
FAMILIES	47
SPECIES	> 2,000

The phylum Chordata is made up of three subphyla. The largest of these is Vertebrata, which includes all the vertebrates (mammals, birds, reptiles, amphibians, and fishes). The other two subphyla are marine invertebrate groups: Urochordata, which contains about 2,000 species of sea squirts and their relatives, and Cephalochordata, which has about 30 species of lancelets. These invertebrate chordates do not have a backbone made of vertebrae, but they do have a flexible skeletal rod known as a notochord. The notochord is present in vertebrate embryos, but is resorbed and replaced by the backbone. Vertebrates appear to have evolved from invertebrate chordates.

Ink-pot sea squirt
Polycarpa aurata,
subphylum Urochordata

Pyrosome
Pyrosoma atlanticum,
subphylum Urochordata

A pyrosome is a colony of tunicates that form a tube, which is propelled by the water expelled as they filter-feed

Common lancelet
Branchiostoma lanceolatum,
subphylum Cephalochordata

Lancelets are commercially harvested as food in Asia

Sea potato
Halocynthia papillosa,
subphylum Urochordata

Lightbulb sea squirt
Clavelina lepadiformis,
subphylum Urochordata

When a sea squirt suddenly contracts and closes its inhalant opening, a jet of water shoots out of the exhalant opening

Star ascidian
Botryllus schlosseri,
subphylum Urochordata

Star ascidians live in colonies made up of petal-like individuals around a communal exhalant opening

FIXED AND FREE

Also known as tunicates, sea squirts hatch from eggs as a tadpole-shaped larva with a notochord in its tail. After dispersing to a new location, the tadpole usually attaches itself to the seafloor, resorbs the tail and notochord, and moves its mouth to its free end. This sessile sea squirt is a bag-like creature. Water enters through an inhalant opening and leaves via an exhalant opening. On the way, the water passes through a perforated pharynx (throat), which uses mucus to filter out particles of food. Sea squirts can be solitary or colonial. Most are fixed as adults, but a small number remain free-living throughout their lives. They are almost all hermaphrodites that release eggs and sperm into the water to produce their young.

Superficially resembling little eels, lancelets (or amphioxus) can swim well, but usually stay partly buried in sand or gravel in shallow waters, with only the head protruding. They filter-feed using the same method as sea squirts, with water entering one opening, passing through the pharnx, and exiting via a second opening. The sexes are separate and the eggs are fertilized externally.

Sponges

PHYLUM	Porifera
CLASSES	3
ORDERS	18
FAMILIES	80
SPECIES	about 9,000

More than 2,000 years ago, Aristotle considered sponges to be animals, but his claim remained without proof until 1765. In the interim, most scientists believed sponges, with their lack of movement and often branching form, to be plants. Sponges are unique in the animal kingdom. They do not possess a nervous system, muscles, or stomach. Their cells do not form tissues or organs, but are specialized for particular functions, such as food collection, digestion, defense, or skeleton formation. Able to migrate throughout a sponge, the cells can transform from one type to another, which allows the sponge to completely regenerate from a fragment or even from individual cells.

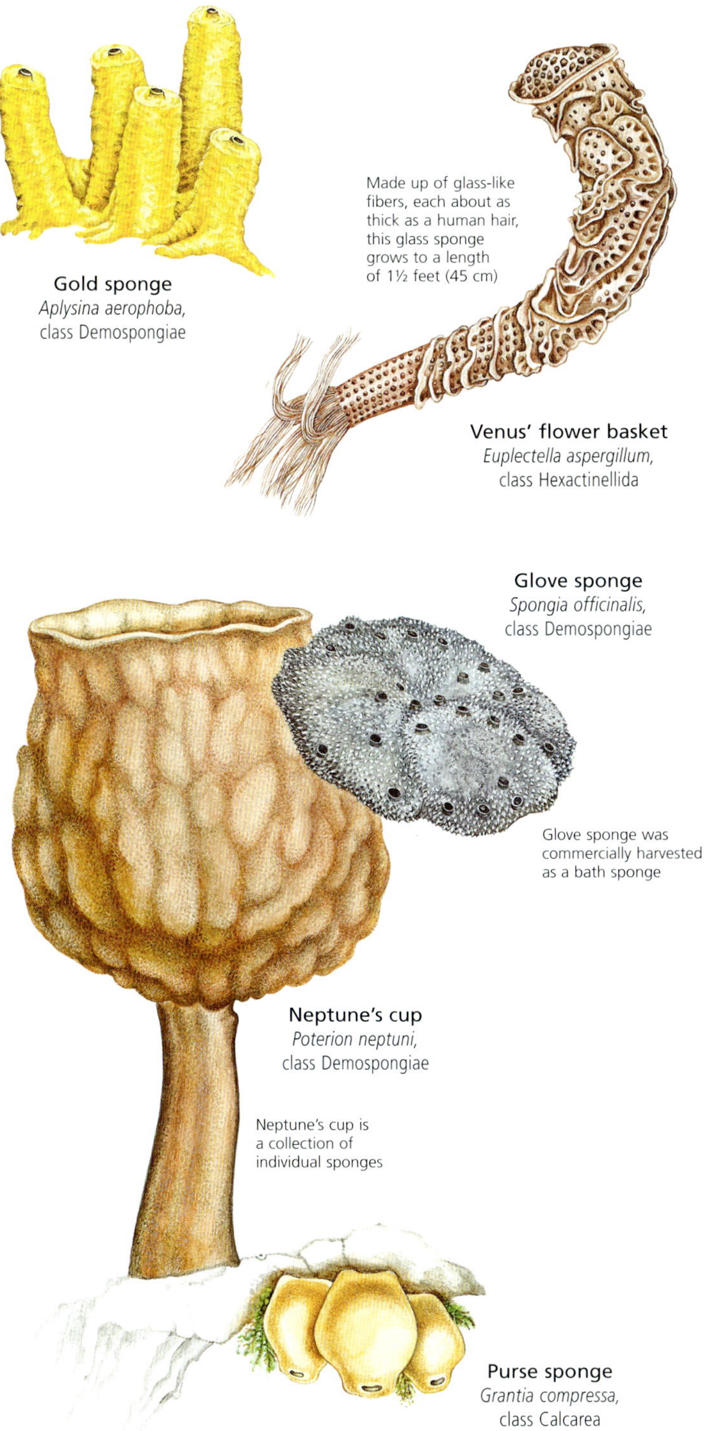

Gold sponge
Aplysina aerophoba, class Demospongiae

Made up of glass-like fibers, each about as thick as a human hair, this glass sponge grows to a length of 1½ feet (45 cm)

Venus' flower basket
Euplectella aspergillum, class Hexactinellida

Glove sponge
Spongia officinalis, class Demospongiae

Glove sponge was commercially harvested as a bath sponge

Neptune's cup
Poterion neptuni, class Demospongiae

Neptune's cup is a collection of individual sponges

Purse sponge
Grantia compressa, class Calcarea

POROUS FILTER-FEEDERS

Ranging in length from less than ½ inch to 6½ feet (1 cm to 2 m), sponges may be shaped like trees, bushes, vases, barrels, balls, cushions, or carpets, or they may simply form a shapeless mass. They are found in every marine habitat, from the shallows to the depths, and a small number of species have colonized freshwater lakes and rivers.

The skeletons of sponges are made of minerals, protein, or both. Species in the class Calcarea have skeletons of calcium carbonate and tend to be small and drab. Those in the class Hexactinellida are known as glass sponges and have skeletons made of silica. More than 90 percent of sponge species belong to the class Demospongiae. They have skeletons of silica and/or protein.

Sponges usually feed by filtering microorganisms from the water. After entering via tiny pores called ostia, the water travels through a system of canals, and is expelled via a large opening called the osculum. The collar cells that line the sponge's interior beat their whip-like flagella to maintain a constant current. A few carnivorous sponges use hook-like filaments to capture crustaceans.

Most sponges are hermaphrodites. Sperm are released into the water and carried to other sponges to fertilize their eggs. The larvae are free-swimming for a short period before attaching to a surface and developing into an adult sponge.

CNIDARIANS

PHYLUM	Cnidaria
CLASSES	4
ORDERS	27
FAMILIES	236
SPECIES	about 9,000

The phylum Cnidaria is a diverse assortment of mostly marine invertebrates, including sea anemones, corals, jellyfishes, and hydroids. All members are carnivores and use cells containing stinging nematocysts to subdue prey and deter predators. The cnidarian body is organized around a gastrovascular cavity that digests food and acts as a hydroskeleton. Food enters and waste leaves via a single opening, the mouth, which is often surrounded by tentacles. Cnidarians occur in two forms: polyp and medusa. Polyps are cylindrical and attached to a surface, with the mouth and tentacles at the free end. Medusae are free-swimming and umbrella-shaped, with the mouth and tentacles hanging down.

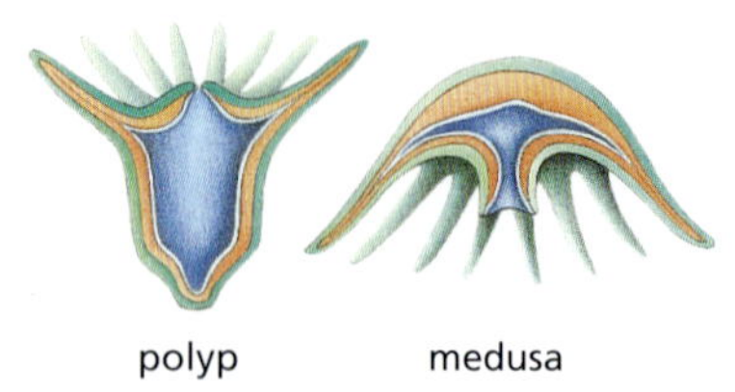

Jelly sandwich Between a cnidarian's two cell layers, the outer ectoderm and the inner endoderm, there is a jelly-like layer known as the mesoglea (shown as orange). In polyps, the mesoglea is thin, but in medusae, it forms the bulk of the animal.

POLYPS AND MEDUSAE

A small number of cnidarians live in fresh water, but most are marine, occupying all latitudes and levels of the ocean and reaching their greatest numbers in shallow tropical waters. They feed primarily on the fishes and crustaceans that swim past. Found mainly on the tentacles, a cnidarian's nematocysts (stingers) hold a coiled, barbed thread that can be ejected to spear and paralyze the prey. The tentacles then move the food to the mouth.

While corals and sea anemones exist only as polyps, many other cnidarians alternate between polyp and medusa during their life-cycle. Usually, the polyps asexually produce medusae, and the medusae sexually produce larvae that become polyps. Polyps and some medusae may grow buds or divide to create more of their own kind asexually. If the offspring detach from the parent they become clones, but if they remain attached they form colonies, such as the vast colonies that make up coral reefs. In some colonies, each member has a particular role and form, and may be specialized for feeding, defense, reproduction, or locomotion, for example.

Oaten pipes hydroid
Tubularia indivisa, class Hydrozoa

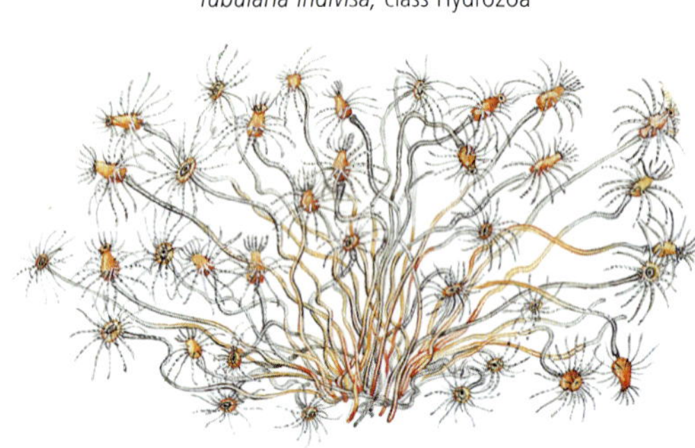

Green hydra
Chlorohydra viridis, class Hydrozoa

Freshwater jellyfish
Craspedacusta sowerbyi, class Hydrozoa

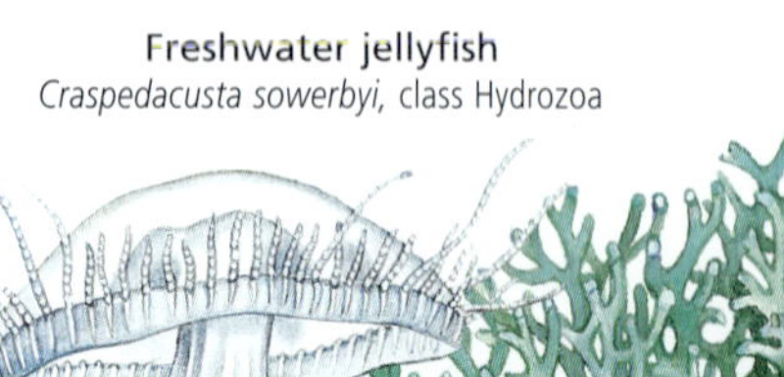

Fire coral
Millepora dichotoma, class Hydrozoa

Colony made up of tightly packed branches resembling feathers

Often mistaken for a true coral by divers, fire coral has a calcified exoskeleton that can scrape the skin, and nematocysts that can cause a burning or stinging rash

Yellow feathers
Gymnangium montagui, class Hydrozoa

Upside-down jellyfish
Cassiopeia andromeda, class Scyphozoa

Attaches to sandy ocean floor

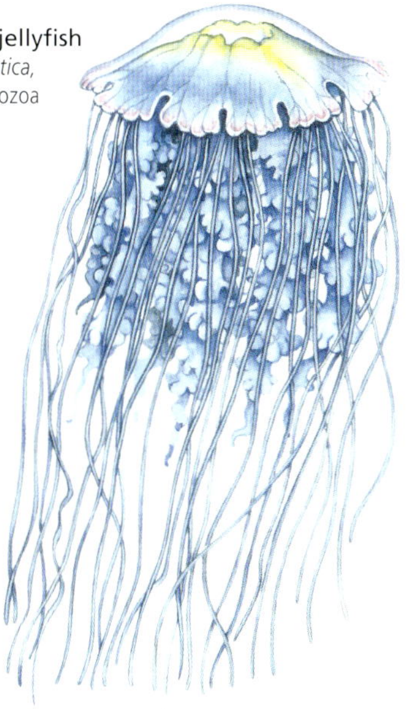

Lion's mane jellyfish
Cyanea arctica, class Scyphozoa

Portuguese man-of-war
Physalia physalis, class Hydrozoa

The Portuguese man-of-war is a free-floating colony, with a modified medusa forming the gas-filled float, and some polyps delivering a potent sting, some digesting food, and others reproducing

Hula skirt siphonophore
Physophora hydrostatica, class Hydrozoa

Dead man's fingers
Alcyonium digitatum, class Anthozoa

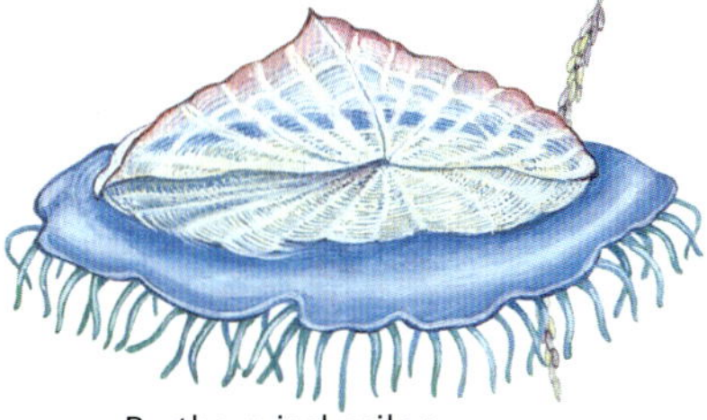

Sea wasp
Chironex fleckeri, class Cubozoa

A sea wasp contains enough venom to kill 60 people

By-the-wind sailor
Velella velella, class Hydrozoa

Phosphorescent sea pen
Pennatula phosphorea, class Anthozoa

Stalked jellyfish
Haliclystus auricula, class Scyphozoa

The stalked jellyfish exists as a polyp attached to algae and does not occur as a medusa

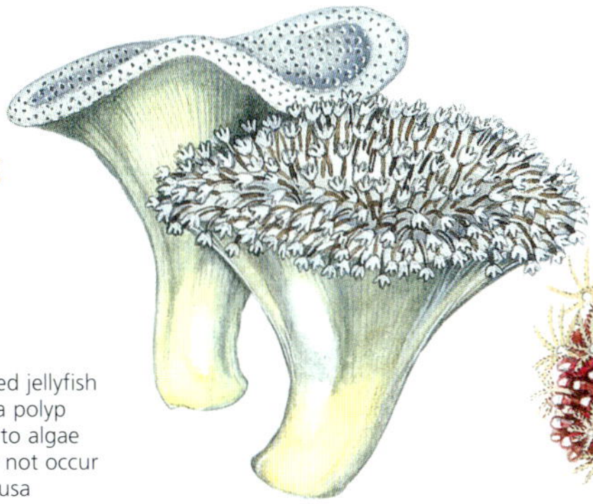

Formosan soft coral
Sarcophyton glaucum, class Anthozoa

Organ-pipe coral
Tubipora musica, class Anthozoa

YOUNG CNIDARIANS

When cnidarians sexually reproduce, they produce a microscopic planula larva, which either swims using its beating cilia (tiny hairs) or crawls along the bottom. After a time, the planula transforms into a polyp by attaching its front end to a surface and developing tentacles at its free end.

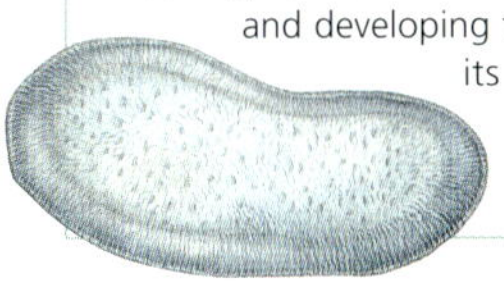

CLASS SCYPHOZOA

True jellyfishes belong to the class Scyphozoa. Generally, they spend most of their lives as medusae, but these medusae produce larvae that settle on the seabed as minute polyps. The polyps divide horizontally and break off to become medusae. Most jellyfishes are free-swimming. They squirt out jets of water that provide weak propulsion.

Species 200

Worldwide; marine

Common jelly
Found worldwide in coastal waters, the moon jellyfish (Aurelia aurita) is often seen washed up onto beaches.

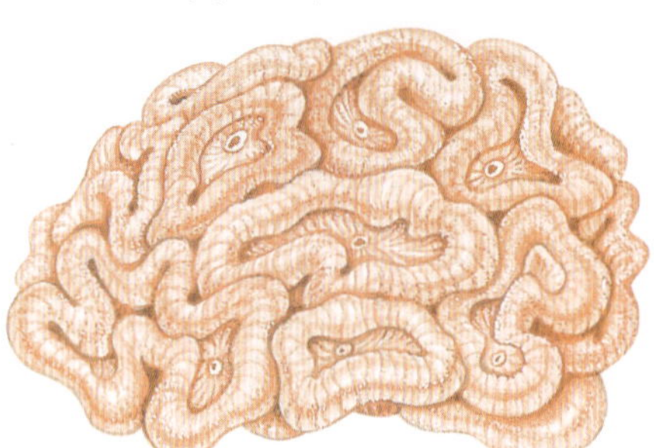

Red brain coral
Lobophyllia hemprichii, class Anthozoa

Individual polyps of brain coral fuse into wrinkled rows

Clubbed finger coral
Porites porites, class Anthozoa

Bubble coral
Plerogyra sinuosa, class Anthozoa

Black coral
Antipathes furcata, class Anthozoa

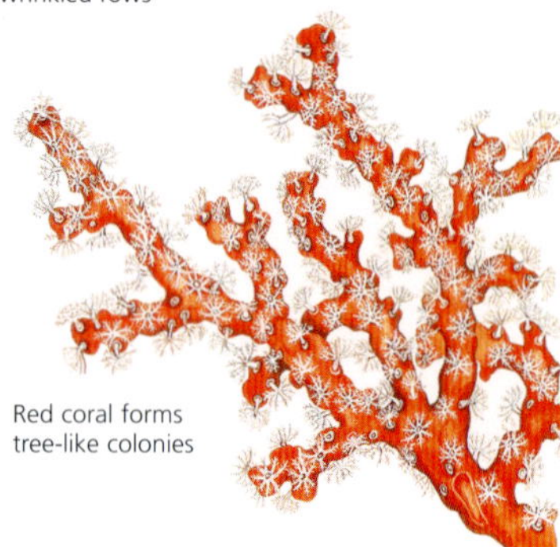

Red coral forms tree-like colonies

Red coral
Corallium rubrum, class Anthozoa

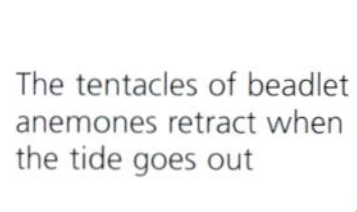

The tentacles of beadlet anemones retract when the tide goes out

Beadlet anemone
Actinia equina, class Anthozoa

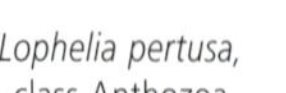

Lophelia pertusa, class Anthozoa

Lophelia pertusa is a deep-water coral that forms reefs in the cold North Atlantic

West Indian sea fan
Gorgonia flabellum, class Anthozoa

Mushroom coral
Fungia fungites, class Anthozoa

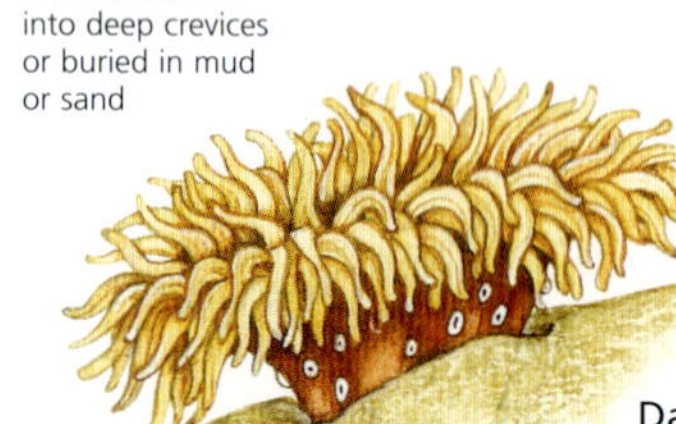

Stalk is inserted into deep crevices or buried in mud or sand

Daisy anemone
Cereus pedunculatus, class Anthozoa

Caryophyllia smithi, class Anthozoa

CLASS ANTHOZOA

Anthozoans spend their adult lives as attached polyps and most are colonial. They include the reef-building true corals, the naked sea anemones, the leathery soft corals, the many-branched sea fans and red corals, the sea pens, and the black corals. True corals secrete an exoskeleton; others have an endoskeleton.

Species 6,500

Worldwide; marine

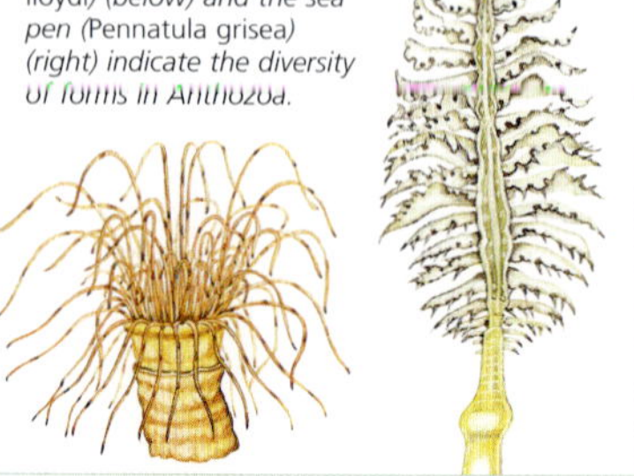

Different shapes *The tube anemone (*Cerianthus lloydi*) (below) and the sea pen (*Pennatula grisea*) (right) indicate the diversity of forms in Anthozoa.*

GREAT BARRIER REEF

Stretching more than 1,400 miles (2,240 km) along the northeastern coast of Australia, the Great Barrier Reef is a collection of coral reefs that together form the largest natural feature on Earth.

Flatworms

PHYLUM	Platyhelminthes
CLASSES	4
ORDERS	35
FAMILIES	360
SPECIES	13,000

The flatworms that make up the phylum Platyhelminthes range from microscopic free-living species, to tapeworms up to 100 feet (30 m) long that live inside humans and other vertebrates. The simplest animals possessing bilateral symmetry, flatworms have no body cavity and lack respiratory and circulatory systems. A few parasitic species also do without a digestive system. In most species, the gut has a single opening to take in food and expel waste. The indistinct head contains a brain and many of the sense organs, including ocelli (simple eyes) that can perceive light and dark, and receptors that can detect chemicals, balance, gravity, and water movement.

DIFFERENT LIFESTYLES

Flatworms are divided into four classes. The first of these, Turbelleria, is predominantly free-living. Most species are marine, but many are found in lakes, ponds, and rivers, and a few can even tolerate both fresh and salt water. There are also a number of species that live on land in moist habitats. Turbellarians usually prey on invertebrates. They move along slime trails produced by special glands, beating cilia (tiny hairs) to propel themselves.

The other three flatworm classes are all parasitic during at least part of their often complex life-cycle. Members of Monogenea are small flukes. At the rear end, they have an opisthaptor, a bulb bearing suckers and/or hooks that attach to a fish's gills or a frog's bladder. Most flukes, however, belong to Trematoda. The adults are parasites of vertebrates, using suckers to attach to the gut and other organs. The class Cestoda contains the tapeworms, internal parasites that are long, flat colonies made up of individuals called proglottids. Almost all flatworms are hermaphrodites, and tapeworms have a set of male and female sex organs in each proglottid.

A PARASITIC LIFE

While the flukes of Monogenea have a single host in their lifetime, most parasitic flatworms make use of different hosts at different stages in their life-cycle.

Eggs of the Chinese liver fluke
Opisthorchis sinensis *(class Trematoda) eggs are eaten by aquatic snails. The final-stage larvae emerge to attach to fish, which are eaten by humans. Once the flukes mature into adults, their eggs are passed out in the human host's feces.*

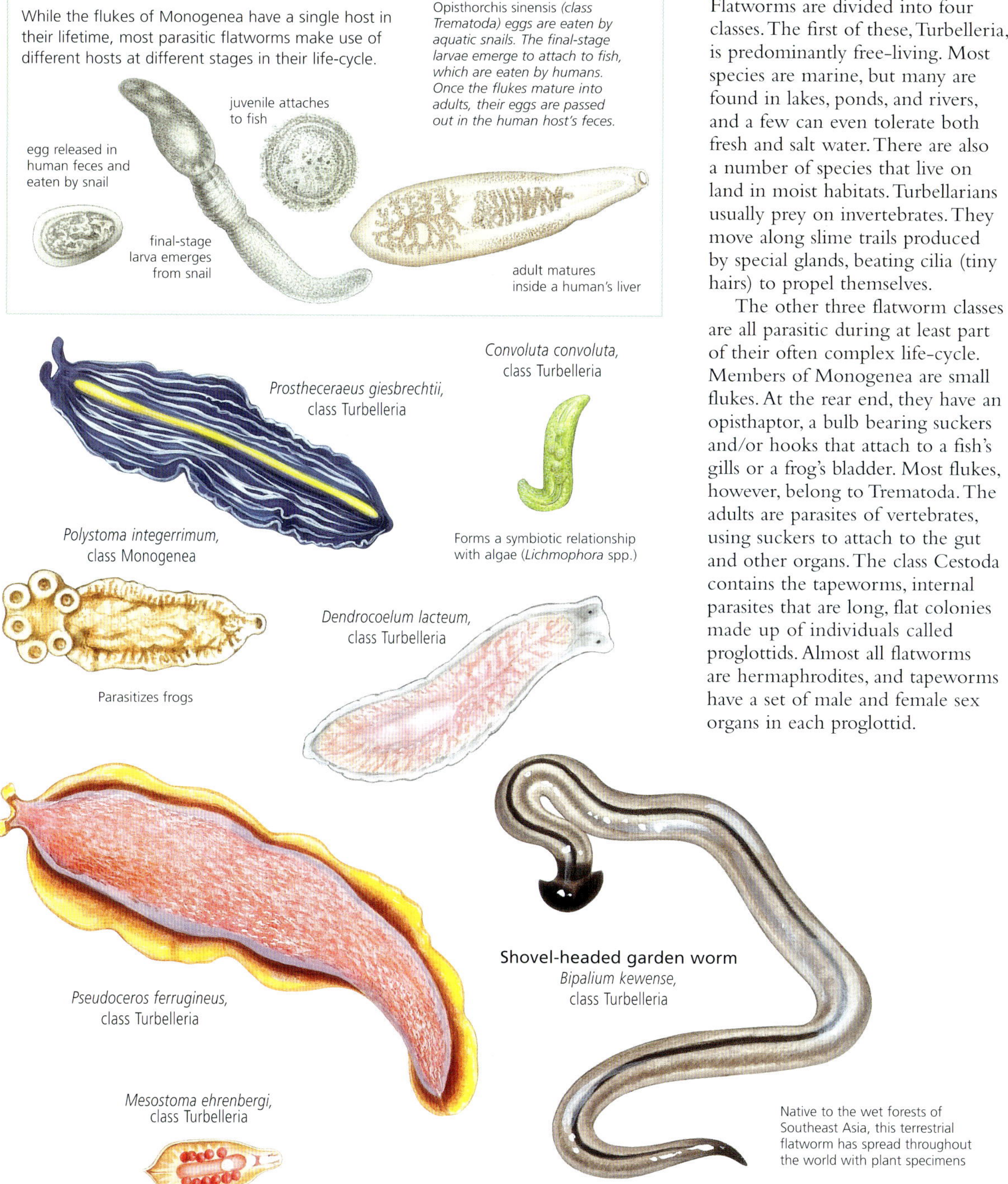

Prostheceraeus giesbrechtii, class Turbelleria

Convoluta convoluta, class Turbelleria

Forms a symbiotic relationship with algae (*Lichmophora* spp.)

Polystoma integerrimum, class Monogenea

Parasitizes frogs

Dendrocoelum lacteum, class Turbelleria

Pseudoceros ferrugineus, class Turbelleria

Shovel-headed garden worm
Bipalium kewense, class Turbelleria

Native to the wet forests of Southeast Asia, this terrestrial flatworm has spread throughout the world with plant specimens

Mesostoma ehrenbergi, class Turbelleria

ROUNDWORMS

PHYLUM	Nematoda
CLASSES	4
ORDERS	20
FAMILIES	185
SPECIES	20,000

Although a species of roundworm that parasitizes sperm whales can grow to a length of 43 feet (13 m), most members of the phylum Nematoda are microscopic. Also known as nematodes, roundworms are among the most abundant of all animals—one rotting apple was found to contain about 90,000 individual roundworms. Free-living roundworms occur in virtually every aquatic and terrestrial habitat, and those living in soil play a crucial role in recycling detritus. Parasitic roundworms are found in most groups of plants and animals. Common roundworms, hookworms, pinworms, threadworms, and other roundworms infest more than half the world's human population.

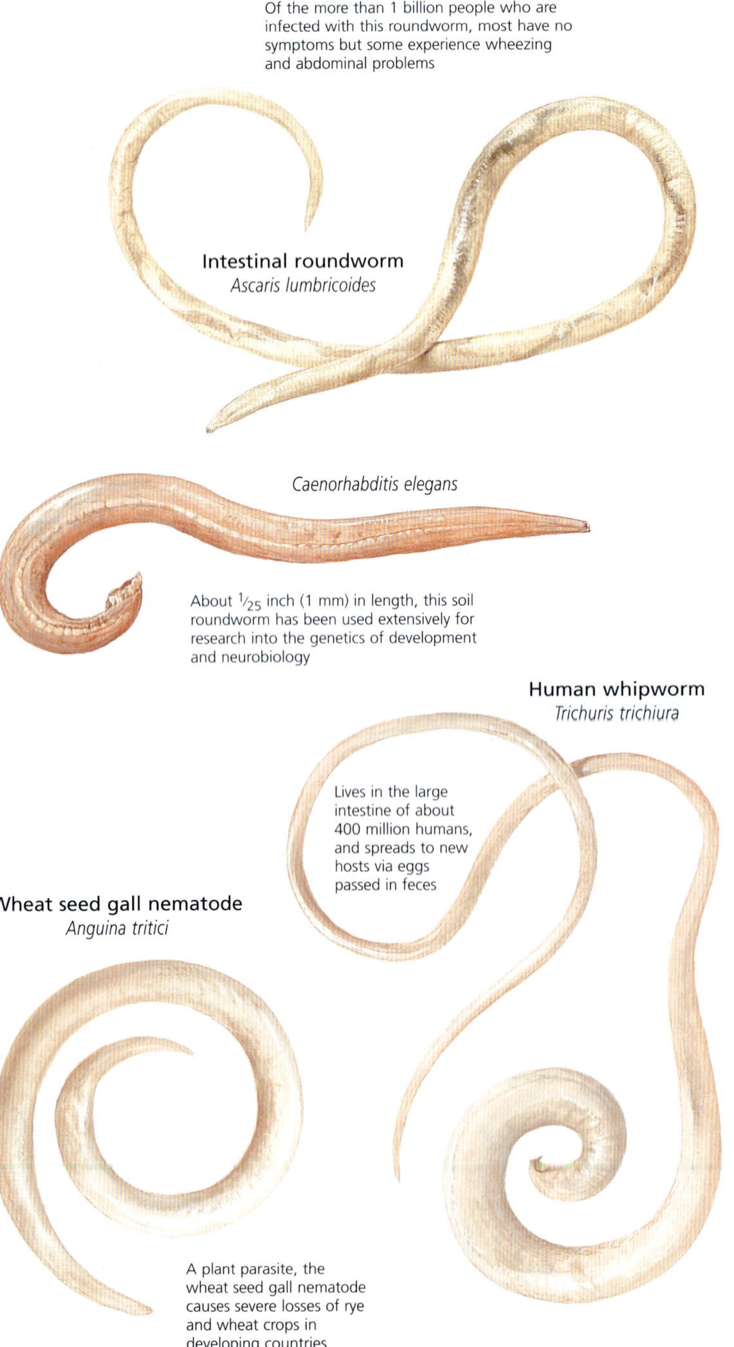

Of the more than 1 billion people who are infected with this roundworm, most have no symptoms but some experience wheezing and abdominal problems

Intestinal roundworm
Ascaris lumbricoides

Caenorhabditis elegans

About 1/25 inch (1 mm) in length, this soil roundworm has been used extensively for research into the genetics of development and neurobiology

Human whipworm
Trichuris trichiura

Lives in the large intestine of about 400 million humans, and spreads to new hosts via eggs passed in feces

Wheat seed gall nematode
Anguina tritici

A plant parasite, the wheat seed gall nematode causes severe losses of rye and wheat crops in developing countries

SIMPLE AND RESILIENT

Roundworms have long, slender, cylindrical bodies and often resemble tiny threads, a shape that allows them to live in tiny spaces between grains of soil, for example. They are bilaterally symmetrical and often tapered at both ends. The epidermis (skin) secretes a tough but flexible outer cuticle. As in arthropods such as insects, roundworms must molt as they grow. Most shed the cuticle four times before reaching maturity.

When a roundworm moves, it may appear to be thrashing aimlessly back and forth. Between the gut and the body wall, a body cavity known as a pseudocoel contains fluid under pressure. As the worm contracts its muscles, which run only lengthwise down its body, this pressure causes the body to flex from side to side—a method of locomotion that works well against soil particles or in water films.

After food enters a roundworm's mouth, it is pumped through the simple gut by a muscular pharynx (throat). Waste is released through an anus at the rear of the worm.

If they encounter unfavorable conditions, such as extreme heat or cold or drought, roundworms can enter a death-like state, known cryptobiosis, for months or even years. When conditions improve, they resume their life processes.

Although some roundworms are hermaphrodites, the sexes are more often separate. During copulation, the male holds onto the female with his hooked rear end. The young resemble miniature adults.

Mollusks

PHYLUM	Mollusca
CLASSES	7
ORDERS	35
FAMILIES	232
SPECIES	75,000

Abundant and adaptable, mollusks have diversified to fill most ecological niches. They are predominantly marine and are found in every level of the ocean, but have also colonized fresh water and land throughout the world. The diversity of their habitats is reflected in an immense variety of forms, from jet-propelled squid to creeping snails and fixed clams. Features shared by most mollusks include a well-developed head; a body cavity containing the internal organs; a specialized skin called the mantle that covers the body and secretes a shell of calcium carbonate; an intucking of the mantle to form a mantle cavity that houses the gills; and a muscular, often mucus-secreting foot.

Nudibranch
Chromodoris kuniei, class Gastropoda

Raises and lowers mantle as it crawls across the seafloor to highlight its warning coloration

Land slugs breathe air through an opening just behind the head that leads to the mantle cavity

Land slug
Bielzia coerulans, class Gastropoda

Mediterranean bonnet
Phalium granulatum undulatum, class Gastropoda

Atlantic thorny oyster
Spondylus americanus, class Bivalvia

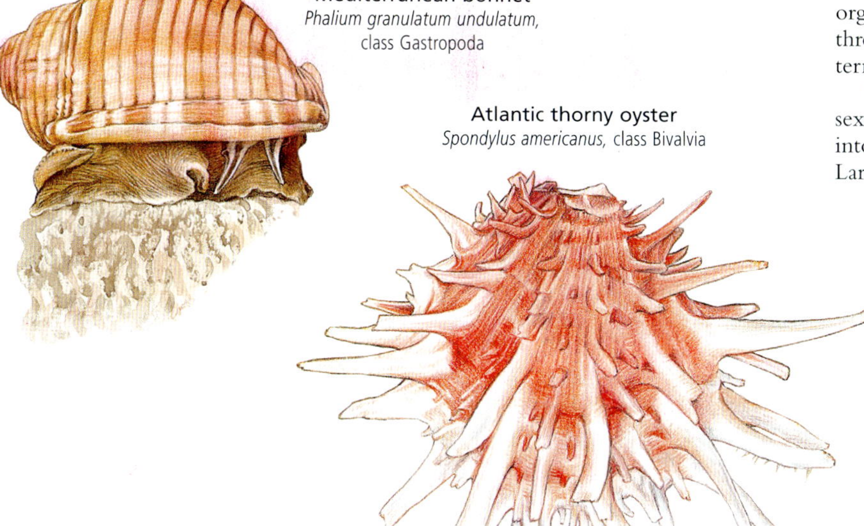

VARIED DESIGN

Mollusks are split into eight classes. The Aplacophora is a small class of worm-like mollusks that lack a shell. About 20 species that possess a low, rounded shell constitute the Monoplacophora. The chitons of Polyplacophora are protected by eight plates and creep on a sucker-like foot. Scaphopoda contains the burrowing tusk shells. The clams, oysters, mussels, and other bivalves of the class Bivalvia have a hinged, two-part shell and a tiny head. Gastropods such as snails and slugs (Gastropoda) sometimes have a spiral shell, but many have no shell at all. Octopuses, squids, and other cephalopods (Cephalopoda) have mobile arms and often lack a shell. They include the largest and most intelligent of all invertebrates.

Feeding habits among mollusks are varied. Species may feed on detritus, scrape algae from rocks, eat leaves, filter tiny organisms from the water, or actively pursue prey such as crustaceans and fishes. Food is taken in through the mouth, often with the help of a rasping, toothed organ called a radula, then passes through a complex digestive system terminating in an anus.

Most mollusks have separate sexes; some release eggs and sperm into the sea, while others copulate. Larvae tend to be free-swimming.

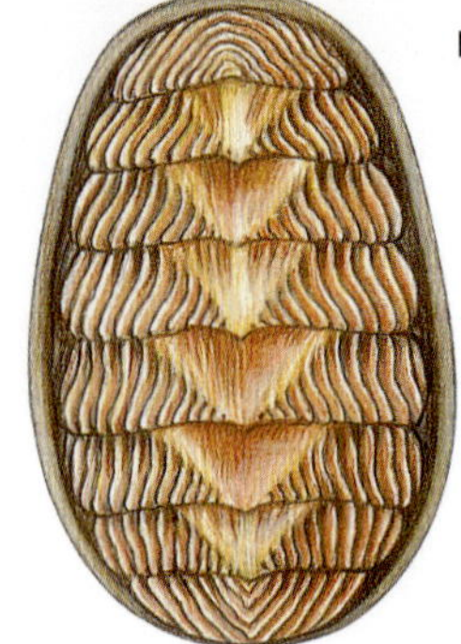

Mottled red chiton
Tonicella marmorea, class Polyplacophora

Browses on sedentary invertebrates such as sponges and bryozoans

Glistenworm
Chaetoderma canadense, class Aplacophora

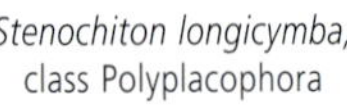

Stenochiton longicymba, class Polyplacophora

Limpets living higher on a rocky shore tend to have taller shells than those living lower down

Common limpet
Patella vulgata, class Gastropoda

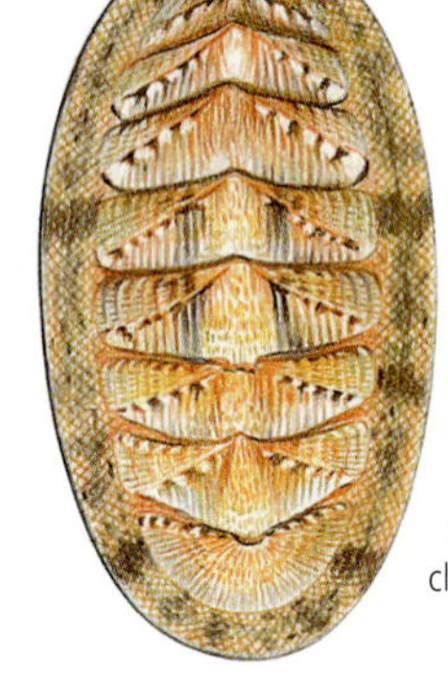

Chiton olivaceus, class Polyplacophora

Formosan tusk
Pictodentalium formosum, class Scaphopoda

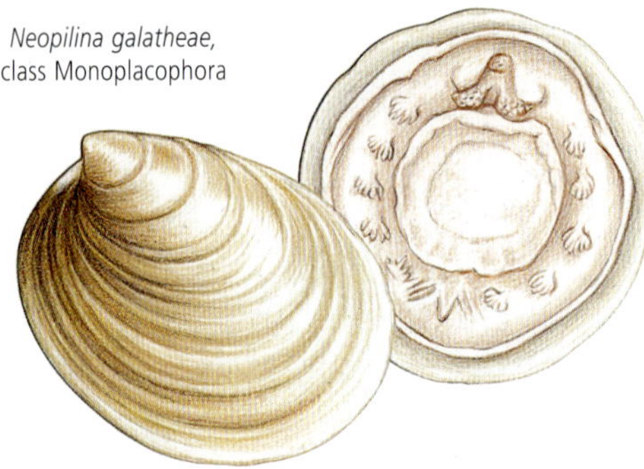

Neopilina galatheae, class Monoplacophora

West Indian green chiton
Chiton tuberculatus, class Polyplacophora

One of the most common chitons in the Caribbean

Antalis tarentinum, class Scaphopoda

Tusk shells range from 3/16 inch to 6 inches (4 mm to 15 cm) in length

Elephant tusk
Dentalium elephantinum, class Scaphopoda

Unlike the jade-green *Dentalium elephantinum*, most tusk shells are white or yelllowish

CLASS MONOPLACOPHORA

Known only from fossil shells until 1952, this rare class is made up of small, limpet-like mollusks living at great depths, from roughly 660 feet (200 m) to ocean trenches at 23,000 feet (7,000 m).

Species 20

Atlantic, Pacific, Indian Ocean; on seafloor

CLASS POLYPLACOPHORA

Ranging from 1/10 inch to 16 inches (3 mm to 40 cm) in length, the chitons in this class are long and flat and have a shell of eight overlapping plates. They creep slowly on their broad, flat foot.

Species 500

Worldwide; intertidal zone to deeper seafloor

CLASS SCAPHOPODA

This class contains the tusk shells, burrowing mollusks with long, tubular shells that are open at both ends. A muscular foot and a small head bearing sticky, thread-like tentacles protrude from the large end of the shell.

Species 500

Worldwide; sandy or muddy seafloor

Spiny cockle
Acanthocardia aculeata, class Bivalvia

Corrugations strengthen cockle shell

Common piddock
Pholas dactylus, class Bivalvia

Bores into soft rock, clay, and peat

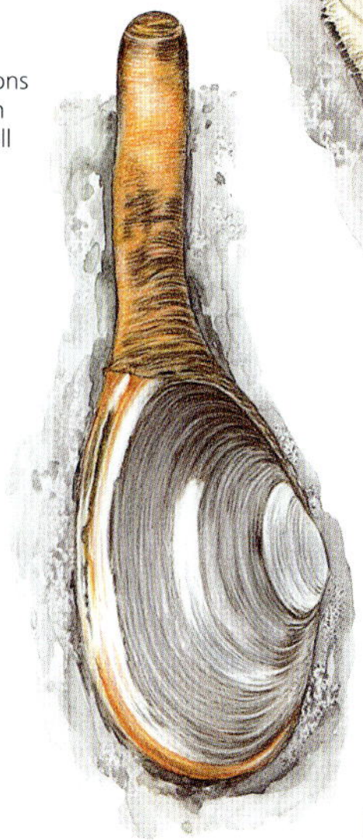

About 4 inches (10 cm) in length and burrows up to 20 inches (50 cm) into the sand

Sand gaper
Mya arenaria, class Bivalvia

Burrows in sand, mud, or gravel

Grooved razor clam
Solen vagina, class Bivalvia

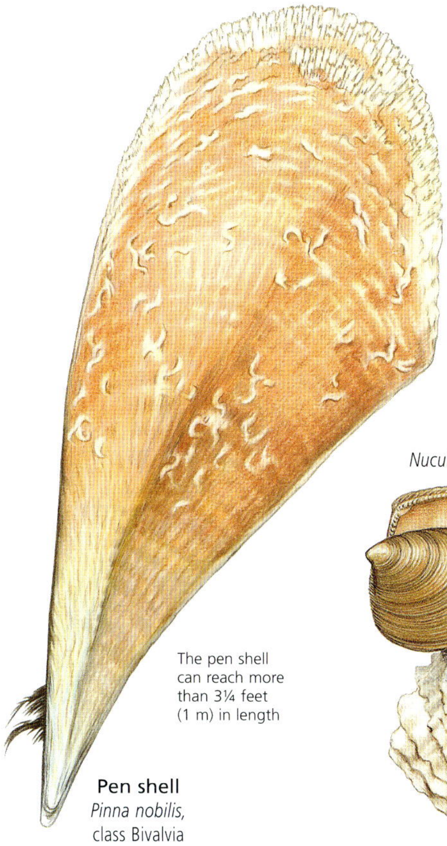

The pen shell can reach more than 3¼ feet (1 m) in length

Pen shell
Pinna nobilis, class Bivalvia

Nut shell
Nucula nucleus, class Bivalvia

Flat oyster
Ostrea edulis, class Bivalvia

Common mussel
Mytilus edulis, class Bivalvia

Shape, size, and color of common mussel's shell vary according to the local conditions

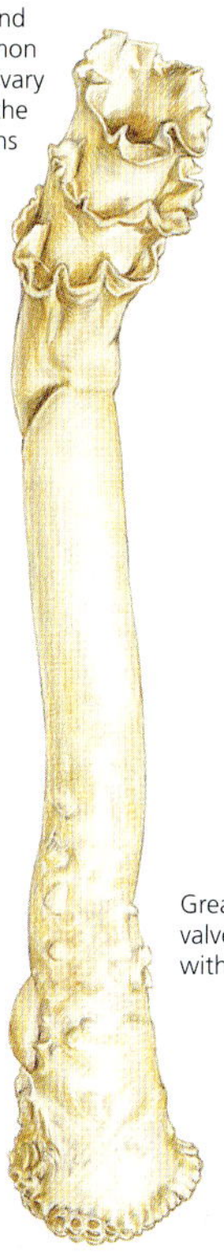

Watering-pot shells secrete a calcareous tube

File clam
Lima hians, class Bivalvia

Edge of mantle bears many small tentacles

Swan mussel
Anodonta cygnaea, class Bivalvia

Freshwater pearl mussel
Margaritana margaritifera, class Bivalvia

Greatly reduced valves have fused with tube

Watering-pot shell
Penicillus javanus, class Bivalvia

INSIDE A BIVALVE

Clams, oysters, mussels, and other bivalves live inside a hinged two-valve shell. A bivalve's mantle forms a sheet of tissue lining the two valves. The adductor muscles are useD to close the shell, and the blade-like foot is often used for burrowing. Bivalves use their gills to filter particles from the water and their palps to sort them by size.

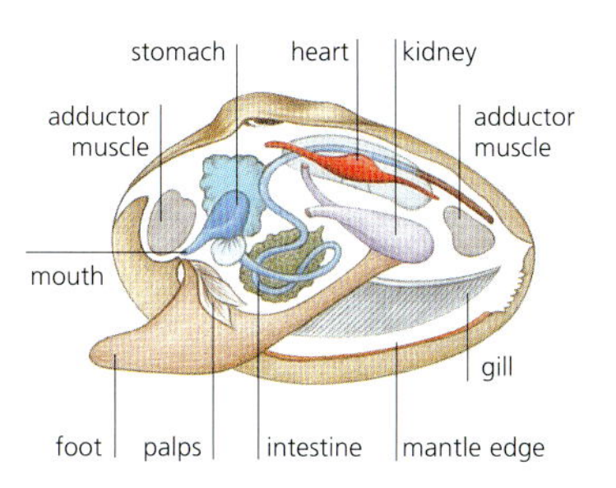

GIANT CLAM

The largest bivalve lives on the tropical coral reefs of the Indian and Pacific oceans. Weighing up to 700 pounds (320 kg), giant clams (*Tridacna gigas*) remain embedded in the sediment, filtering plankton from the water. The bulk of their nutrition, however, comes from the algae that they host.

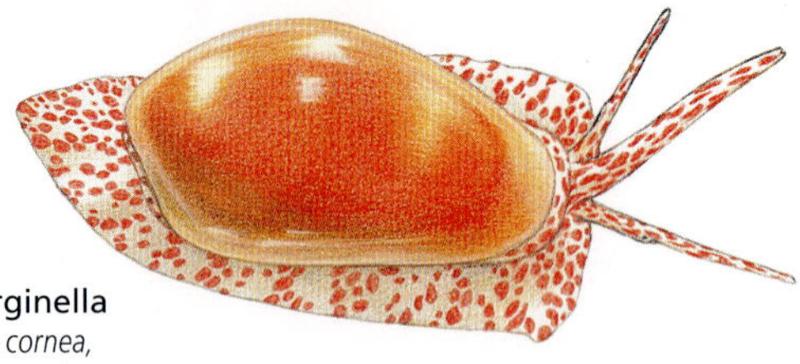

Plain marginella
Marginella cornea,
class Gastropoda

Phyllidia ocellata,
class Gastropoda

Color of *Phyllidia ocellata* is variable, but it can be identified by its knobby tubercles and white-ringed black spots

Purple bubble-raft snail
Janthina janthina, class Gastropoda

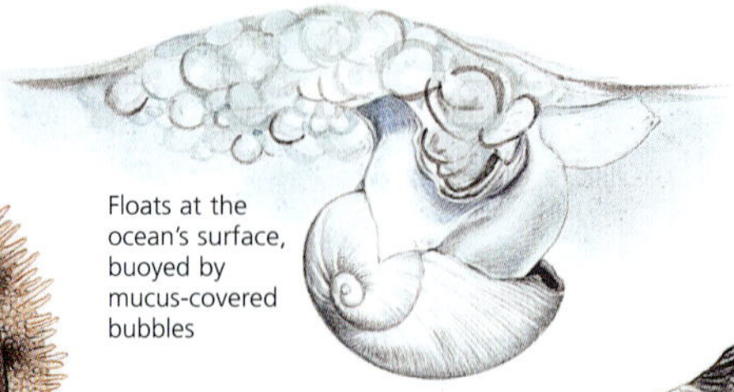

Floats at the ocean's surface, buoyed by mucus-covered bubbles

European edible abalone
Haliotis tuberculata,
class Gastropoda

Crown conch
Melongena corona, class Gastropoda

The mucus of *Murex brandarius* was used by the ancient Greeks to produce a dye called Tyrian purple, which was so rare that it was used only for the clothes of royalty

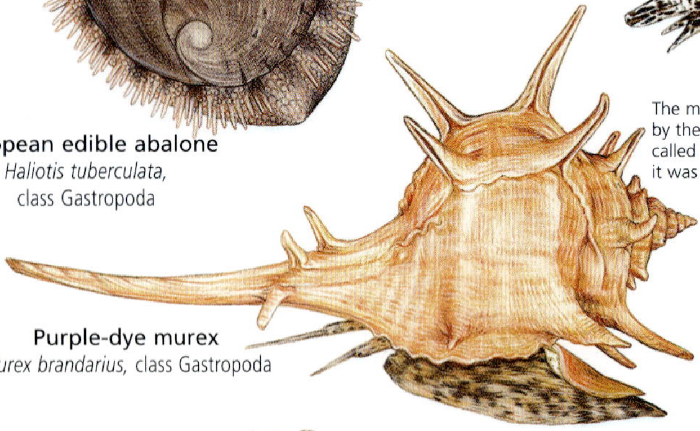

Purple-dye murex
Murex brandarius, class Gastropoda

Trumpet triton
Charonia tritonis,
class Gastropoda

Lettuce sea slug
Elysia crispata,
class Gastropoda

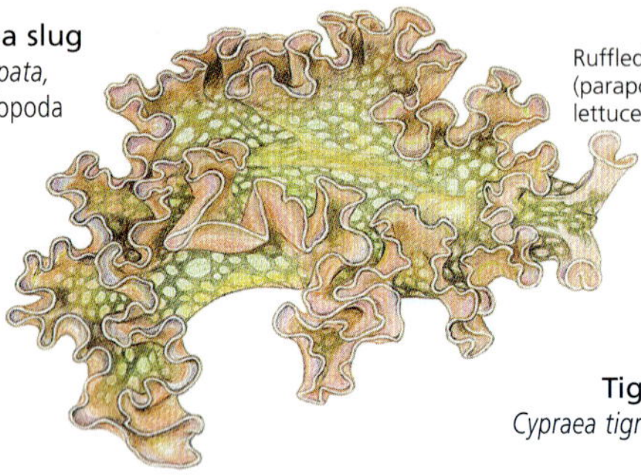

Ruffled flaps of tissue (parapodia) resemble lettuce leaves

Tiger cowry
Cypraea tigris, class Gastropoda

Floats upside down and feeds on Portuguese man-of-war, ingesting the host's nematocysts (stinging cells) to use in its own defense

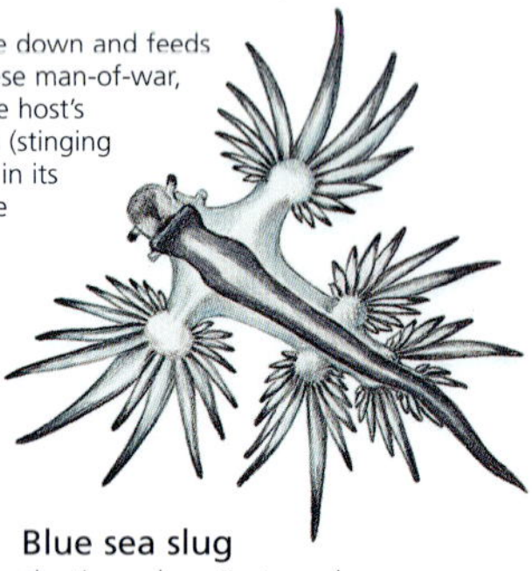

Blue sea slug
Glaucus atlanticus, class Gastropoda

Common egg cowrie
Ovula ovum, class Gastropoda

Queen conch
Strombus gigas, class Gastropoda

CLASS GASTROPODA

This is the largest class of mollusks and includes snails, slugs, limpets, and nudibranchs. Some gastropods have lost their shells, but most have a spiral shell that contains a twisted body mass. Gastropods usually have a distinct head with eyes and tentacles. The head retractS into the shell. The muscular foot is used for creeping, swimming, or burrowing.

Species 60,000

Worldwide

Gastropod cannibal *The banded tulip (Fasciolaria hunteria) can be very aggressive and will eat other members of its species.*

THE GASTROPOD SHELL

While gastropod shells display immense variety in their details, the typical shell is a conical spire. The mantle adds new material to the inner and outer lips of the shell at slightly different rates, resulting in the spiral form.

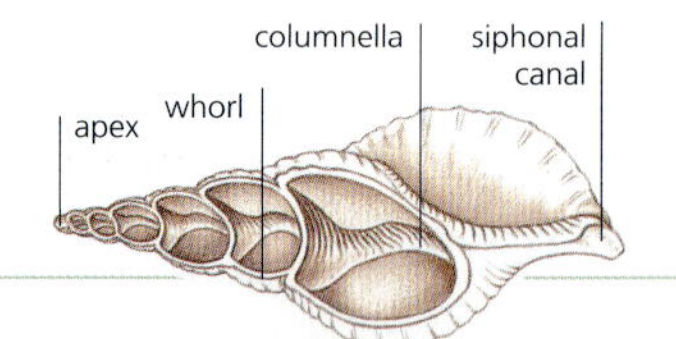

Gray garden slug
Deroceras reticulatus, class Gastropoda

Chocolate arion
Arion rufus, class Gastropoda

Tree slug
Lehmania marginata, class Gastropoda

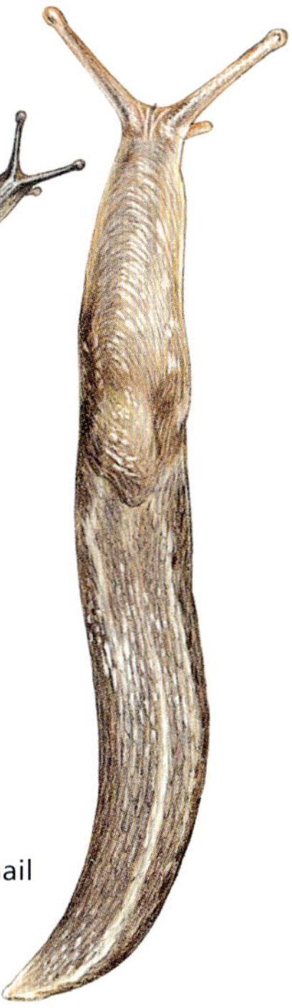

Eared pond snail
Lymnaea auricularia, class Gastropoda

The eared pond snail lives in overgrown, stagnant waters

Wormslug
Boettgerilla pallens, class Gastropoda

Daudebardia rufa, class Gastropoda

River limpet
Ancylus fluviatilis, class Gastropoda

Manus Island tree snail
Papustyla pulcherrima, class Gastropoda

The giant tiger snail's shell can grow to 1 foot (30 cm) in length, making it one of the largest snails

White-lipped garden snail
Cepaea hortensis, class Gastropoda

Escargot
Helix pomatia, class Gastropoda

Giant tiger snail
Achatina achatina, class Gastropoda

The escargot has been eaten by humans since prehistoric times

LIFE ON LAND

Several times during the course of evolution, gastropods have adapted to life on land. There are now roughly 20,000 land snail species. In most, gills have been lost and replaced by a lung. Some species, however, have both a gill chamber and a lung. Out of water, the shell has become lightweight and now protects against desiccation as well as against predators. The mantle and the mucus that cover the body also help to keep the snail moist.

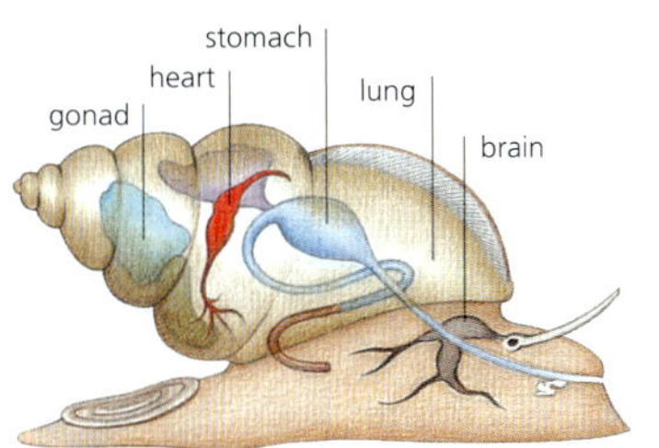

Air breather *Most land snails lack gills. Instead, the mantle wall has become filled with blood vessels and acts as a lung.*

MATING HERMAPHRODITES

Land snails are hermaphrodites but do not self-fertilize. Before mating, two snails will usually circle each other, touch tentacles, intertwine bodies, and bite each other. In the copulation that follows, sperm are exchanged so that the eggs of both snails can be fertilized.

All squids have eight arms and two long tentacles, all with suckers

Long-finned squid
Loligo vulgaris, class Cephalopoda

Like many cephalopods, cuttlefishes react to danger by immediately changing color to camouflage themselves, and may also squirt a black ink called sepia to confuse a predator

Common cuttlefish
Sepia officinalis, class Cephalopoda

To escape predators, flying squids can shoot out of the water and glide for a distance

Flying squid
Ommastrephes sagittatus, class Cephalopoda

The ram's horn squid is the only surviving member of the family Spirulidae

Lesser cuttlefish (dwarf bobtail)
Sepiola rondeleti, class Cephalopoda

Chambered nautiluses have about 90 sticky tentacles without suckers

The internal shell, or pen, of the ram's horn squid acts as a buoyancy device

Ram's horn squid
Spirula spirula, class Cephalopoda

Abraliopsis morisii, class Cephalopoda

Has light-producing organs at the tip of its two tentacles and on the underside of its body

Common nautilus
Nautilus pompilius, class Cephalopoda

Class Cephalopoda The cephalopod foot lies close to the head and has been modified to form arms, tentacles, and a funnel. Nautiluses retain a large external shell; in squids and cuttlefishes the shell has been reduced and covered by tissue; octopuses have lost their shell altogether.

Species 600

Worldwide; marine

THE CHAMBERED NAUTILUS

The most primitive of the cephalopods, chambered nautiluses live in the last chamber of their large shell. The other chambers are filled with gas, and the nautilus controls its buoyancy by adding fluid to or removing fluid from them.

outer shell
shell chambers regulate depth
"pinhole" eye
beak
arms
funnel, where water is expelled for propulsion
gills
gonads

REPRODUCING CEPHALOPODS

To attract a mate, a male cephalopod will perform some form of courtship. Some cuttlefish and squid species, for example, display a particular pattern on their body to impress females and deter rival males. A male cephalopod delivers his sperm in packages called spermatophores. He transfers these into the female's mantle cavity during copulation using a modified arm known as a hectocotylus. In the tiny male paper nautilus, which is much smaller than a female, the hectocotylus breaks away from his body and crawls into the female. Male cephalopods usually die soon after mating.

Egg laying *Female cephalopods usually attach their eggs to rocks or seaweed. The eggs may be laid singly, in bunches like grapes, or in clusters protected by a hardened gelatinous covering.*

JET PROPULSION

Most cephalopods swim by expelling jets of water from the mantle cavity through the funnel. The highly mobile funnel can be pointed forward or backward, allowing the cephalopod to swim in either direction. Combined with their streamlined shape, this jet propulsion allows squids to attain speeds of up to 25 miles per hour (40 km/h).

SEGMENTED WORMS

PHYLUM	Annelida
CLASSES	2
ORDERS	21
FAMILIES	130
SPECIES	12,000

Also known as annelids, segmented worms have a head and a long body made up of segments that look like rings from the outside. Each segment has its own fluid-filled body cavity that acts as a hydroskeleton and contains a separate set of excretory, locomotory, and respiratory organs. The segments are united, however, by common digestive, circulatory, and nervous systems. Segmented worms crawl or swim by wriggling from side to side, or burrow by waves of contractions that pass down the length of the body. Stiff bristles known as chetae protrude from each segment and provide traction. While many species are highly active, others live in burrows or tubes.

Red feather duster
Spirographis spallanzanii,
class Polychaeta

Instantly withdraws its feathery crown if disturbed

In the soil Long and thin with a small head, earthworms are shaped for burrowing. With up to 650 individuals per square yard (800 per sq. m), earthworms play a crucial role in aerating and fertilizing the soil.

IN WATER AND ON LAND

Segmented worms are found in all levels of the ocean, in freshwater lakes and rivers, and on land. They include filter-feeders, predators, and blood-suckers, as well as deposit-feeders such as earthworms, which eat sediment to extract nutrients.

Commonly called bristleworms, the species in the class Polychaeta are almost entirely marine and usually swim or crawl using paddle-like appendages known as parapodia. The parapodia are often lost in tube-living forms, which have a modified head region for food-gathering. Typically, the sexes are separate and release eggs and sperm into the water for fertilization.

The class Clitellata includes many terrestrial and freshwater species such as earthworms and leeches, as well as some marine species. Most are hermaphrodites that exchange sperm by copulation. They all bear a clitellum, a ring of glandular skin that secretes a cocoon for the eggs.

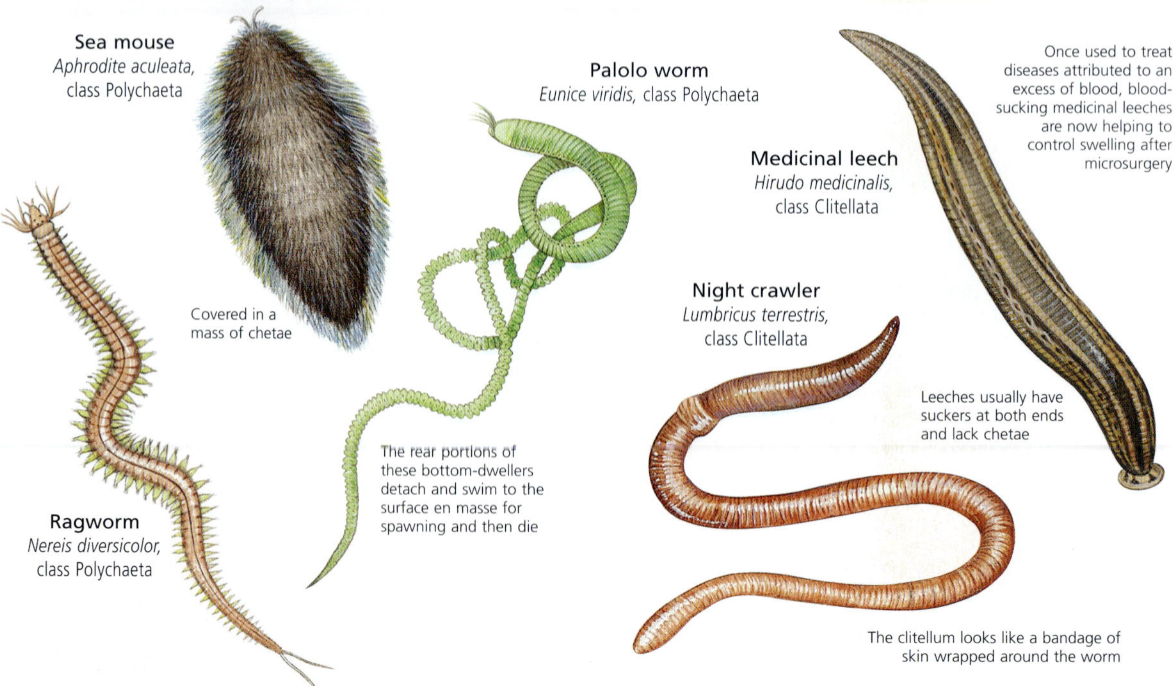

Sea mouse
Aphrodite aculeata,
class Polychaeta

Covered in a mass of chetae

Ragworm
Nereis diversicolor,
class Polychaeta

Palolo worm
Eunice viridis, class Polychaeta

The rear portions of these bottom-dwellers detach and swim to the surface en masse for spawning and then die

Medicinal leech
Hirudo medicinalis,
class Clitellata

Once used to treat diseases attributed to an excess of blood, blood-sucking medicinal leeches are now helping to control swelling after microsurgery

Leeches usually have suckers at both ends and lack chetae

Night crawler
Lumbricus terrestris,
class Clitellata

The clitellum looks like a bandage of skin wrapped around the worm

METAMORPHOSIS

While some newly hatched invertebrates resemble little adults, most look quite distinct from their parents and have a very different lifestyle. To become adults, these young must go through a transformation known as metamorphosis. Corals, clams, many crustaceans, and most other invertebrates metamorphose into adults after a brief larval stage. For most insects, the larval stage is prolonged. In some insect groups, such as crickets and bugs, metamorphosis is incomplete. The hatched form, called a nymph, is structured like the adult, but lacks wings and full sex organs until it emerges from the final molt. Other insects, such as butterflies and bees, have complete metamorphosis. The young, called larvae, differ so much from their parents that they must go through a pupal stage, in which their bodies break down and are built up again as adults.

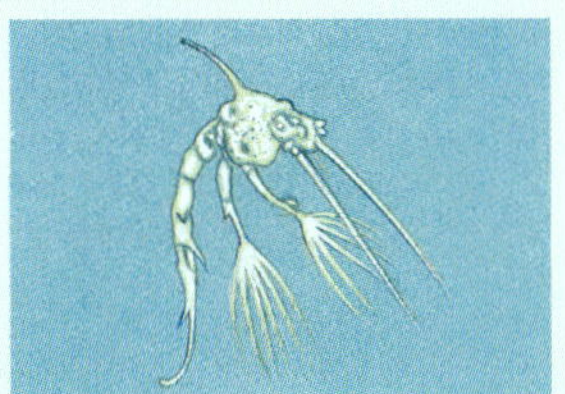

Zoea larva The ten-legged decapods—which account for about one-quarter of all crustacean species—often hatch as a zoea, a free-swimming, spiny larva. The zoea metamorphoses into a postlarval megalopa, with adult appendages.

Egg to nymph to adult A dragonfly develops through incomplete metamorphosis. **1.** Male and female mate. **2.** Female lays eggs in water or in stem of water plant. **3.** Aquatic nymph hatches. **4.** Nymph feeds on tadpoles and worms and grows through a series of molts. **5.** Nymph emerges from the water for its final molt. **6.** Adult bursts from the nymphal skin. **7.** Adult rests while its wings dry out. **8.** Feeding on flying insects, adult flies to a new location to find a mate.

ARTHROPODS

PHYLUM	Arthropoda
CLASSES	22
ORDERS	110
FAMILIES	2,120
SPECIES	1.1 million

The insects, spiders, crustaceans, centipedes, and other invertebrates in the phylum Arthropoda account for three-quarters of all known animal species, and millions more remain to be discovered. Arthropods have adapted to fill virtually every ecological niche on land, in fresh water, and in the oceans. Consequently, their anatomy and lifestyles are extraordinarily diverse, but they share a number of defining features. The name arthropod means "jointed feet," and arthropods do have jointed appendages and segmented bodies. They are distinguished from other invertebrates by their tough but flexible exoskeleton, which provides both protection and support.

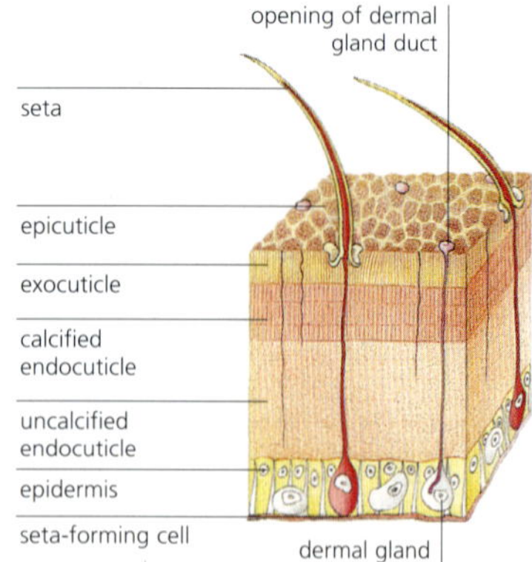

Arthropod armor Secreted by epidermal cells, the arthropod exoskeleton has thin layers of wax and protein, known as the epicuticle, overlying various layers of chitin and protein. Tanning of the protein in the exocuticle hardens the exoskeleton.

Insects everywhere Insects account for about 90 percent of all arthropod species. Ubiquitous on land and in fresh water, but rare in the ocean, insects include predators such as the praying mantis (below), and nectar-feeders such as butterflies.

WAYS OF BREATHING

The exoskeleton is usually too impermeable to allow gas exchange. While some tiny arthropods can breathe directly through the body wall, most have developed specialized structures. Aquatic arthropods tend to have gills. Derived from gills are book lungs, found in many arachnids. Terrestrial arthropods often rely on minute air ducts known as tracheae, which deliver air to all parts of the body.

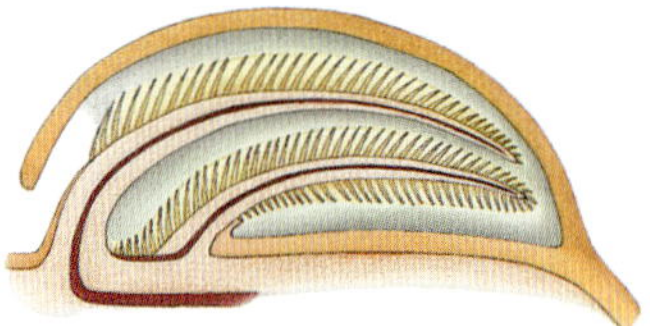

Gills *Gills collect oxygen from water and help to maintain the arthropod's salt balance. In crustaceans, the gills are external and often positioned on the legs. Horseshoe crabs have unique structures called book gills, with flaps like the pages of a book.*

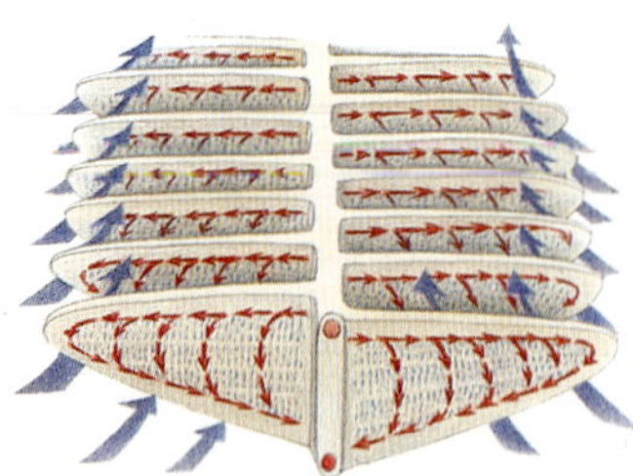

Book lungs *Most likely derived from the book gills found on horseshoe crabs, book lungs are found in many arachnids. While blood circulates within the hollow, flat plates of a book lung, air circulates through the space between the plates.*

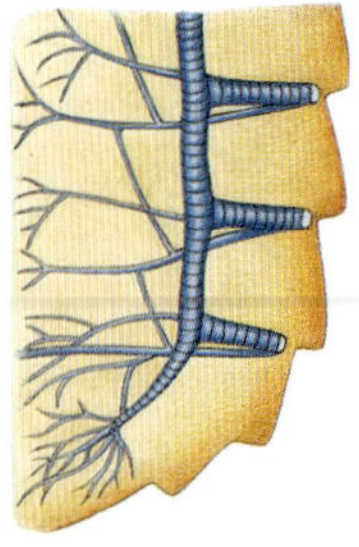

Tracheae *Found in insects, arachnids, and centipedes and millipedes, tracheae collect air through openings in the exoskeleton called spiracles, and transport it to the tissues or blood.*

Legs and bodies Tarantulas and other spiders have eight legs and a body divided into two parts: the cephalothorax (fused head and thorax) and the abdomen. Insects have six legs and a three-part body made up of a head, thorax, and abdomen.

Deadly bites Few arthropods deliver a bite that is fatal to humans. Far more deadly are the diseases transmitted by blood-sucking arthropods such as mosquitoes, which transfer malaria and other illnesses from one vertebrate host to another.

EARLY ARRIVALS

The first arthropods appeared in the oceans about 530 million years ago and included crustaceans; ancestors of horseshoe crabs; and the now-extinct trilobites. These early species tended to have numerous body segments, each bearing a similar pair of appendages. Over time, the appendages became specialized for particular tasks, such as locomotion, food collection, sensory perception, and copulation. Furthermore, the segments became arranged into distinct regions of the body called tagma, such as the head, thorax, and abdomen of insects.

Arthropods were the first animals to leave the sea and the first to take to the sky. Scorpions had ventured onto land by 350 million years ago and were soon followed by the first terrestrial insects. Before long, winged insects had emerged. The success of arthropods on land rests in part on their waxy exoskeleton, which stops them from drying out. Jointed legs also make arthropods highly mobile, thus enhancing their ability to find food and mates, evade predators, and colonize new places. In insects, the development of wings increases the advantage.

While many arthropods are themselves predators, almost all are prey to vertebrates or to other invertebrates. The compound eyes found on many arthropod species are made up of multiple lenses and are extremely effective at detecting motion. Simple eyes, with a single lens, detect light intensity. Hairs (known as setae), projections, and slits on the antennae, mouthparts, and legs detect subtle vibrations. A cerebral ganglion (brain) connects via a nerve cord to ganglia (clusters of nerve cells) in each body segment. In arthropods' open circulatory system, a heart pumps hemolymph (blood) to bathe the organs.

The sexes are separate in most arthropod species, and reproduction usually involves the male's sperm fertilizing the female's eggs. Often, the sperm is transferred in packages known as spermatophores, which the female can use when she is ready. The life stages of arthropods vary greatly. Some species are like small adults when they hatch, while others add segments as they grow. Many insect larvae look nothing like their parents and must go through a pupal stage to transform into a winged adult.

ARACHNIDS

PHYLUM	Arthropoda
SUBPHYLUM	Chelicerata
CLASS	Arachnida
ORDERS	17
FAMILIES	450
SPECIES	80,000

Including some of the most feared and fascinating of all invertebrates, the class Arachnida encompasses spiders, scorpions, harvestmen (or daddy-longlegs), and mites and ticks, along with several lesser-known groups. Apart from some families of aquatic mites and a few species of water spiders, all arachnids are terrestrial, and the majority are predators of other invertebrates. Many spiders use silk webs to snare prey, and both scorpions and spiders inject their prey with venom to paralyze or kill it. Most arachnids are unable to swallow solid food and must squirt digestive enzymes into the prey and then suck the liquefied meal into their mouth.

Night vision Most arachnid species are nocturnal and their simple eyes detect only variations in light. Like most spiders, wolf spiders have eight eyes.

Arachnid anatomy A spider displays many of the typical arachnid features, including a two-part body, eight jointed legs, a pair of chelicerae, a pair of pedipalps, and a number of simple eyes. It also has venom glands in its chelicerae, and silk glands at its rear, both of which play an important role in hunting and defense.

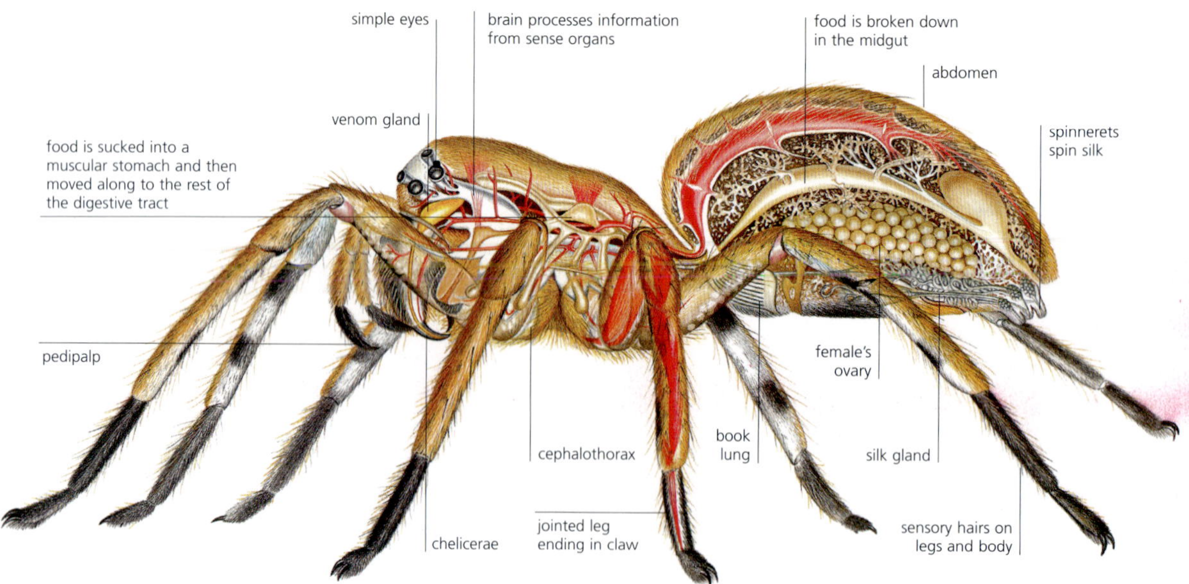

EIGHT-LEGGED PREDATORS

Arachnids have a segmented body, a tough but flexible exoskeleton, and jointed limbs. Mites and ticks have rounded bodies made up of a single region, but other arachnids have two-part bodies, with a cephalothorax holding the eyes, mouthparts, and limbs, and an abdomen containing many of the internal organs.

Arachnids have eight legs, unlike insects, which have six. They also have two pairs of appendages near the mouth. The first pair, known as chelicerae, may be pincer-like or can form fangs that inject venom. They are used to subdue prey. The second pair, the pedipalps, are often used to sense the surroundings. Scorpions and some other arachnids use pedipalps to seize prey, while male spiders use them to transfer sperm to females.

To find prey and avoid danger, arachnids rely on the fine sensory hairs on their body and legs. They also have a number of simple eyes. Fine slits in the cuticle may detect odors, gravity, or vibrations.

Arachnids breathe using book lungs or a tracheal system, or both. Book lungs probably developed from gills and are made up of stacked leaves of tissue. The tracheal system takes air in through tiny pores called spiracles and distributes it to tissues via a network of tubes.

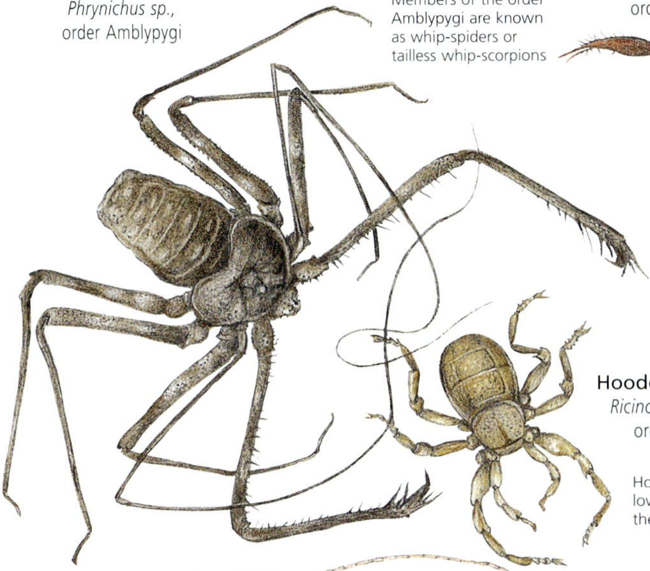

Whip-spider
Phrynichus sp.,
order Amblypygi

Members of the order Amblypygi are known as whip-spiders or tailless whip-scorpions

Book scorpion
Chelifer cancroides,
order Pseudoscorpiones

Book scorpions are often found living in books and furniture

European buthid
Buthus occitanus,
order Scorpiones

Hooded tick spider
Ricinoides sjoestedti,
order Ricinulei

Hood can be lowered over the mouth

Extremely long front legs act as tactile organs

Fat-tailed scorpion
Androctonus australis,
order Scorpiones

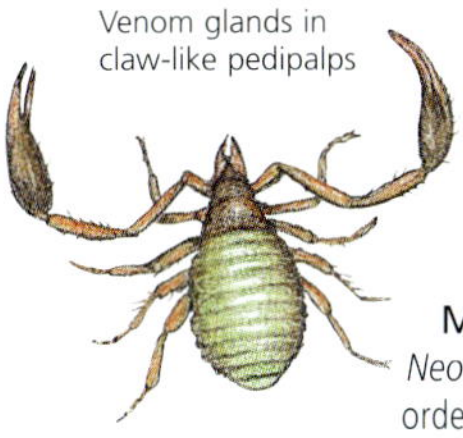

Venom glands in claw-like pedipalps

Moss neobisid
Neobisium carcinoides,
order Pseudoscorpiones

Whip-spider
Charinus milloti, order Amblypygi

The first pair of legs on a whip-scorpion is used as feelers

Whip-scorpion
Thelyphonus caudatus, order Uropygi

Short-tailed whip-scorpion
Schizomus crassicaudatus,
order Schizomida

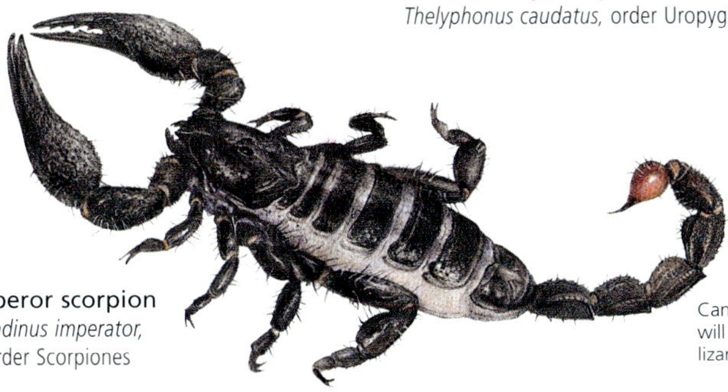

Emperor scorpion
Pandinus imperator,
order Scorpiones

Can weigh more than 2 ounces (60 g) and will hunt small vertebrates such as mice and lizards, crushing them in its large pedipalps

ORDER SCORPIONES

Scorpions are distinguished from other arachnids by two large clawed pedipalps and a segmented abdomen. The tip of the tail holds a prominent stinger, which is sometimes used to subdue prey but more often plays a defensive role. Scorpions hunt by night.

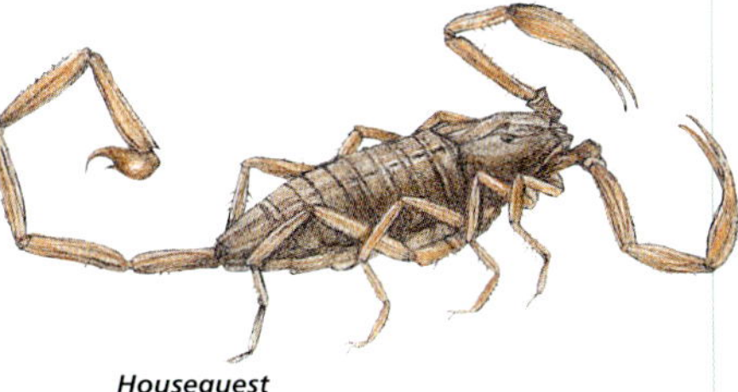

Houseguest
*The slenderbrown scorpion (*Centruroides *gracilis) usually lives in tropical forests, but may dwell in houses in regions where it has been introduced.*

Species 1,400

Warmer regions; under rocks & bark

PARENTAL CARE

Most female arachnids lay their eggs in soil or another safe location and leave them to hatch on their own. In *Euscorpius carpathicus* and other scorpions, however, the mother produces live young. She catches them in her first or first and second pairs of legs. They crawl onto the mother's back and are carried around until after their first molt, about 3 to 14 days later.

EYES OF THE SPIDER

Most spiders are nocturnal and tend to rely more on touch than on vision. Day-active species, however, generally have excellent eyesight at close range. Spiders usually have four pairs of simple eyes grouped in a characteristic pattern according to the family.

Night hunter *The woodlouse-eating spider (family Dysderidae) has six tiny eyes instead of the usual eight and relies on touch to uncover its prey.*

Seeing in the dark *The two huge eyes of the ogre-faced spider (family Deinopidae) enable it to see prey in near-total darkness.*

Wide field *The huntsman spider (family Heteropodidae) is an active hunter. Its wideset eyes provide all-round vision.*

Close range *The diurnal crab spider (family Thomisidae) depends on its keen close-up vision as it ambushes insect prey.*

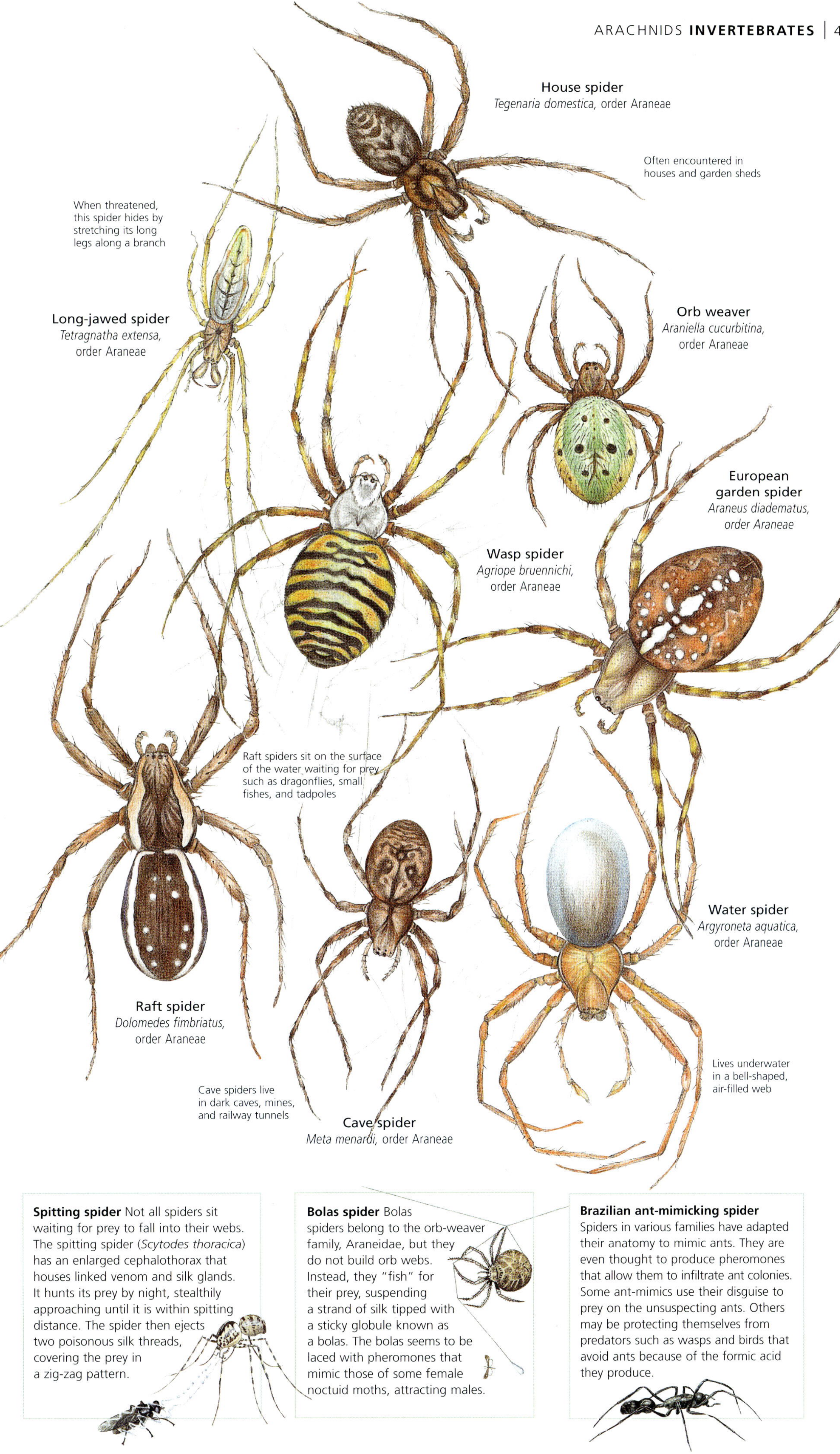

Spitting spider Not all spiders sit waiting for prey to fall into their webs. The spitting spider (*Scytodes thoracica*) has an enlarged cephalothorax that houses linked venom and silk glands. It hunts its prey by night, stealthily approaching until it is within spitting distance. The spider then ejects two poisonous silk threads, covering the prey in a zig-zag pattern.

Bolas spider Bolas spiders belong to the orb-weaver family, Araneidae, but they do not build orb webs. Instead, they "fish" for their prey, suspending a strand of silk tipped with a sticky globule known as a bolas. The bolas seems to be laced with pheromones that mimic those of some female noctuid moths, attracting males.

Brazilian ant-mimicking spider
Spiders in various families have adapted their anatomy to mimic ants. They are even thought to produce pheromones that allow them to infiltrate ant colonies. Some ant-mimics use their disguise to prey on the unsuspecting ants. Others may be protecting themselves from predators such as wasps and birds that avoid ants because of the formic acid they produce.

SILK AND WEBS

While only some species build webs, almost all spiders produce silk. Made of the protein fibroin, spider silk is as strong as nylon thread but has greater elasticity. A spider may have up to eight silk glands in its abdomen, each yielding a different kind of silk. One kind of silk is used for the dragline a spider usually spins after itself like a mountain climber's safety line. Other types are used for cocooning fertilized eggs or wrapping up captured prey. Male spiders may use silk to hold their packages of sperm. Some spiderlings use threads of silk like balloons to carry them aloft so they can disperse. The best known use of silk is among the webspinning spiders, which build a remarkable variety of webs to snare their prey.

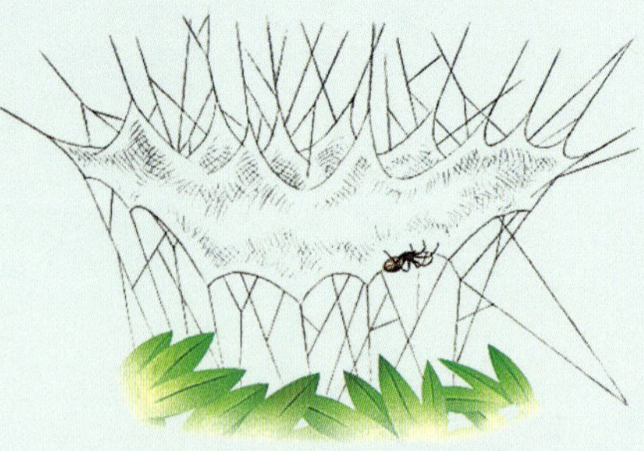

Hammock web *The hammock web spun by a money spider over low bushes may look messy, but any insect caught in its fine lattice weave usually falls into another web suspended below.*

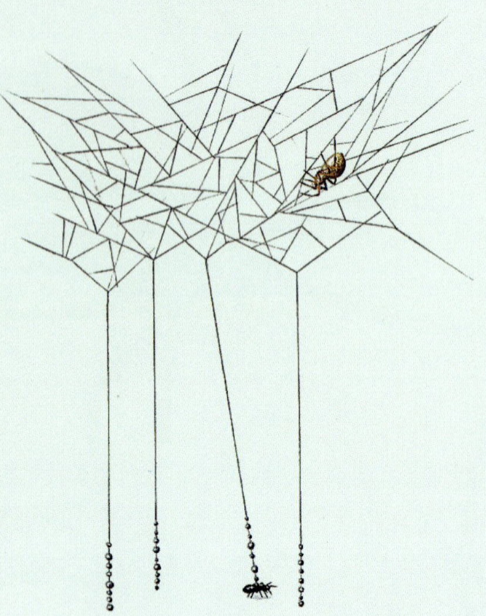

Scaffold web *A scaffold web has stretched traplines with sticky ends attached to the ground. If a crawling insect walks into a trapline, the line snaps back, with the victim dangling in the air.*

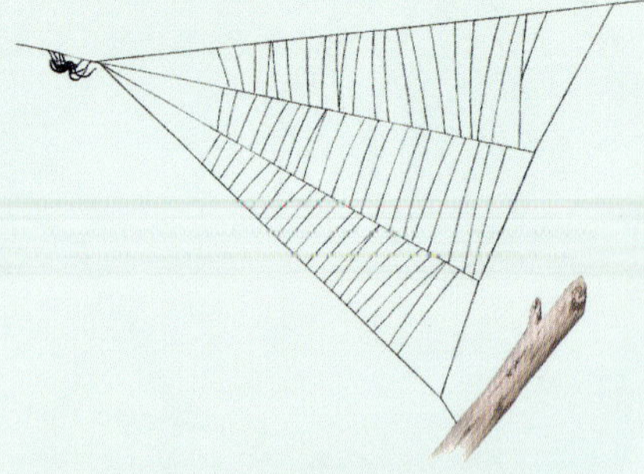

Triangle web *A spider holds a triangular web in its front legs while anchored by a thread of silk to the twig behind it. When an insect hits the web, the spider releases its grip and the web collapses, entangling the prey.*

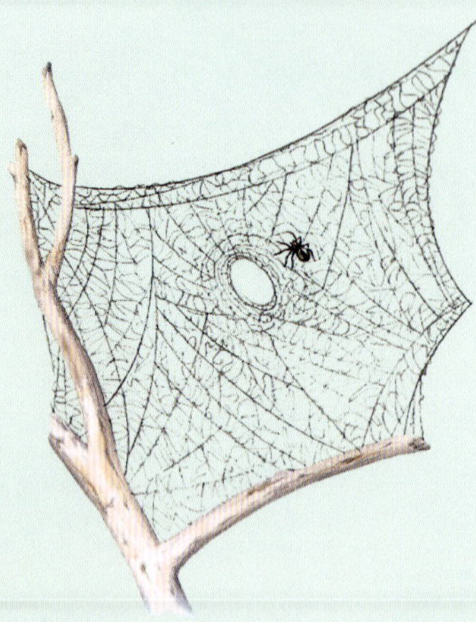

Lace-sheet web *The trap spun by a lace-web weaver is made of fine, woolly silk. It may not be sticky, but insects soon get tangled up in its many fibers, and remain trapped until the spider is ready for food.*

Orb web *The most elaborate of all webs, the orb web covers a large area using a minimum of silk. An outer framework supports a continuous spiral and a series of spokes. The web is suspended between tree branches to capture flying insects.*

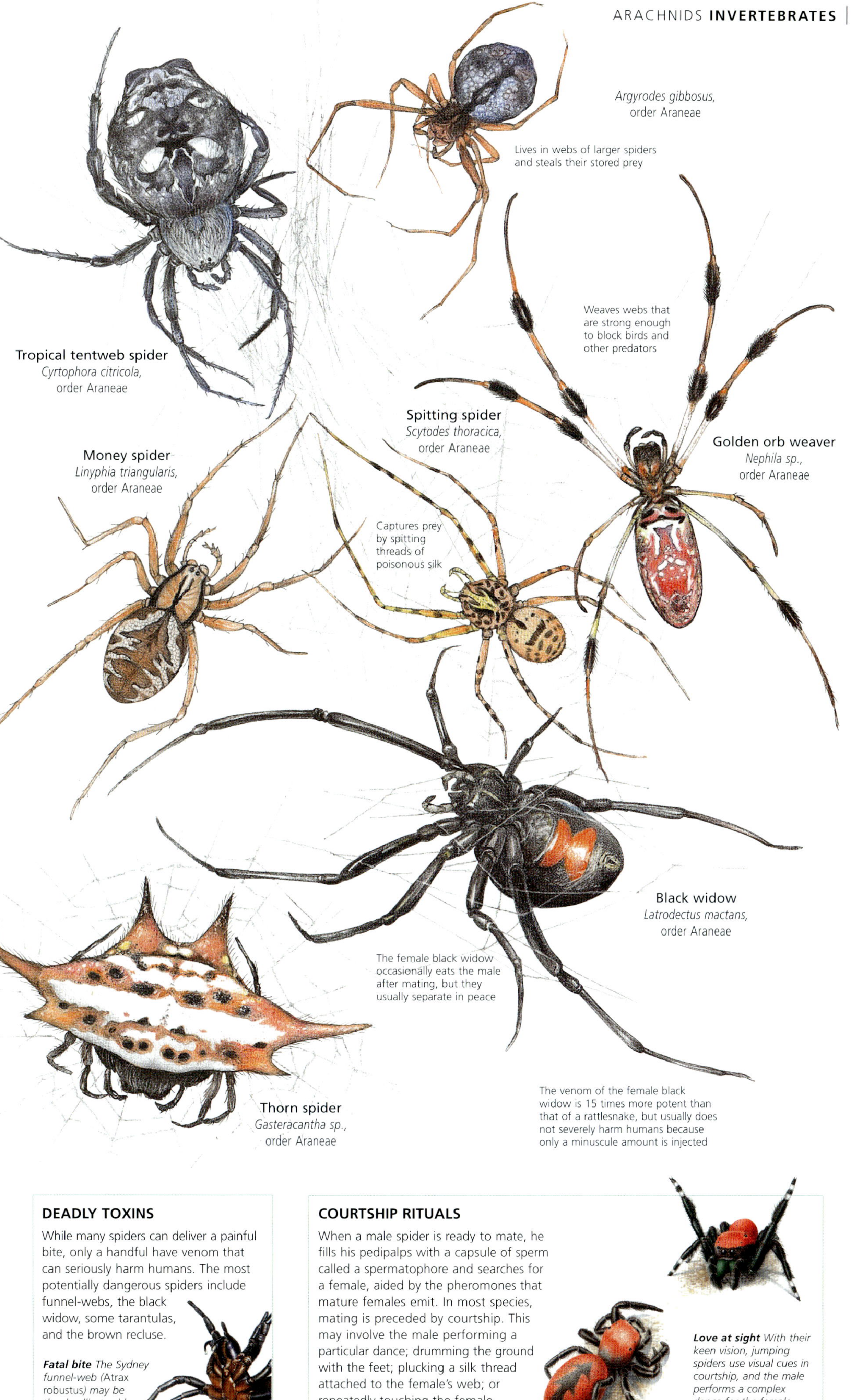

DEADLY TOXINS

While many spiders can deliver a painful bite, only a handful have venom that can seriously harm humans. The most potentially dangerous spiders include funnel-webs, the black widow, some tarantulas, and the brown recluse.

***Fatal bite** The Sydney funnel-web (Atrax robustus) may be the deadliest spider.*

COURTSHIP RITUALS

When a male spider is ready to mate, he fills his pedipalps with a capsule of sperm called a spermatophore and searches for a female, aided by the pheromones that mature females emit. In most species, mating is preceded by courtship. This may involve the male performing a particular dance; drumming the ground with the feet; plucking a silk thread attached to the female's web; or repeatedly touching the female.

***Love at sight** With their keen vision, jumping spiders use visual cues in courtship, and the male performs a complex dance for the female.*

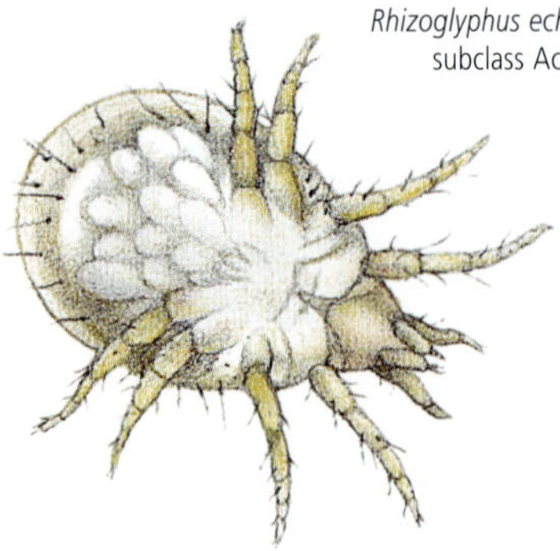

Bulb mite
Rhizoglyphus echinopus, subclass Acari

Harvestman
Opilio parietinus, order Opiliones

Harvestmen have scent glands that release a pungent odor when the arachnid is threatened

Harvestman
Phalangium opilio, order Opiliones

Trogulus tricarinatus, order Opiliones

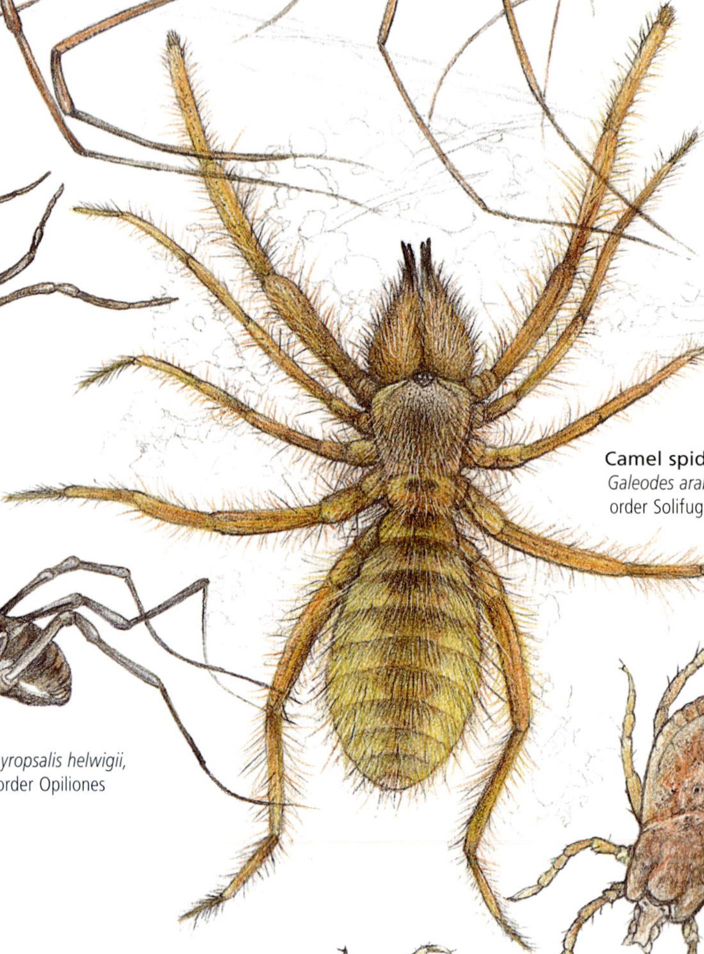

One of the fastest arthropods, the camel spider runs on only three pairs of legs; it uses the first pair as feelers to detect prey such as lizards and mice

Camel spider
Galeodes arabs, order Solifugae

Ischyropsalis helwigii, order Opiliones

Phthiracarus sp., subclass Acari

Pigeon tick
Argas reflexus, subclass Acari

The pigeon tick can spread relapsing fever and other diseases

Sheep tick
Dermacentor marginatus, subclass Acari

SUBCLASS ACARI

Classified in seven orders in the subclass Acari, mites and ticks are the most varied and abundant arachnids. They flourish in almost every kind of habitat, from polar caps to deserts, thermal springs to ocean trenches. With some species small enough to live inside a human hair follicle, however, mites are so tiny that they are rarely noticed. Ticks tend to be larger, but rarely exceed ½ inch (1 cm) in length.

Species 30,000

Worldwide

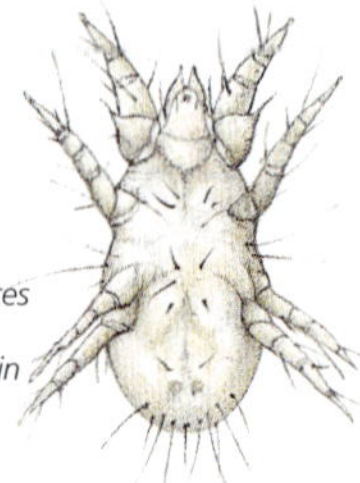

Dust to dust *Flour mites (Acarus siro) are found living in the dust of grain crops, in animal cages, and in food stores.*

DISEASE SPREADERS

While most members of Acari are free-living inhabitants of soil, leaf litter, and water, some are parasites of other animals or plants. Those mites and ticks that feed on vertebrates often transmit bacteria that cause potentially fatal diseases in humans.

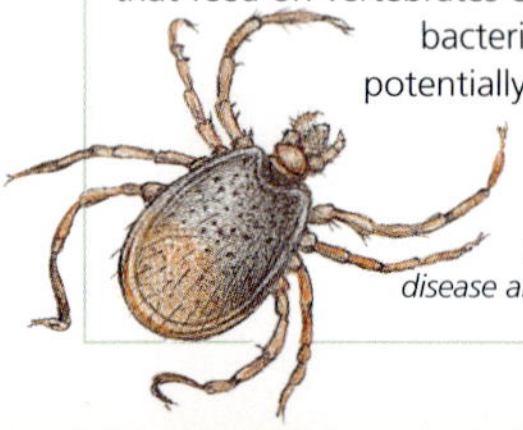

Sick tick *Some ticks spread Lyme disease and other illnesses.*

Ctenoglyphus palmifer, subclass Acari

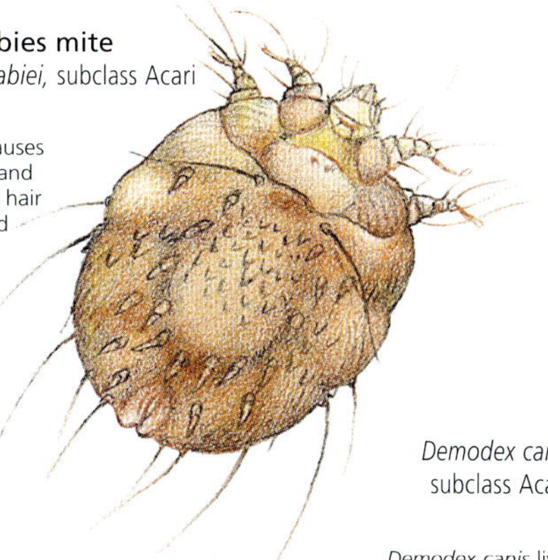

Scabies mite
Sarcoptes scabiei, subclass Acari

The scabies mite causes mange in humans and animals, leading to hair loss, dermatitis, and weight loss

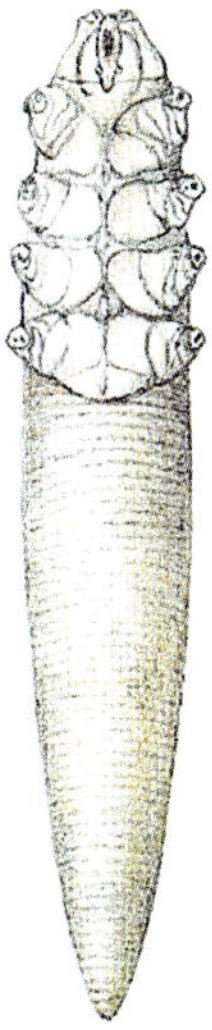

Demodex canis, subclass Acari

Demodex canis lives in the hair follicles and sebaceous glands of most dogs and can cause mange

Pear bud mite (leaf blister mite)
Eriophyes piri, subclass Acari

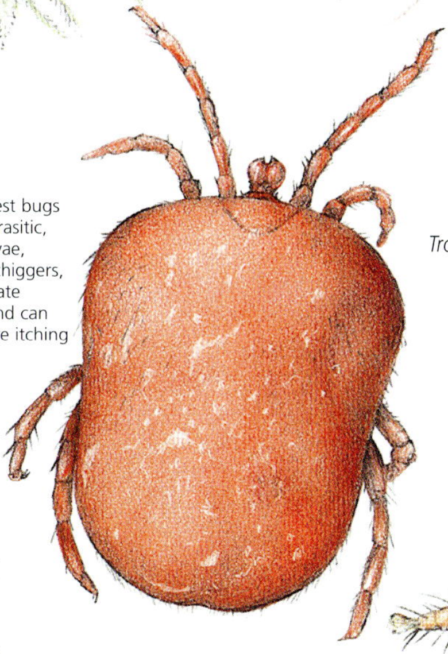

Adult harvest bugs are non-parasitic, but the larvae, known as chiggers, are vertebrate parasites and can cause severe itching

Harvest bug
Trombidium holosericeum, subclass Acari

The harvest bug is covered in tiny red hairs that create the appearance of velvet

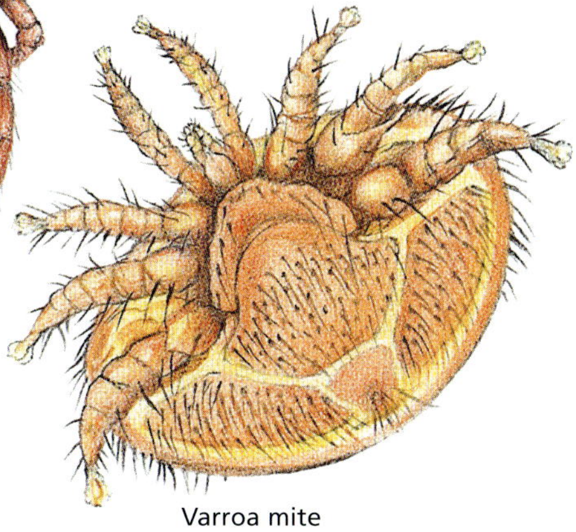

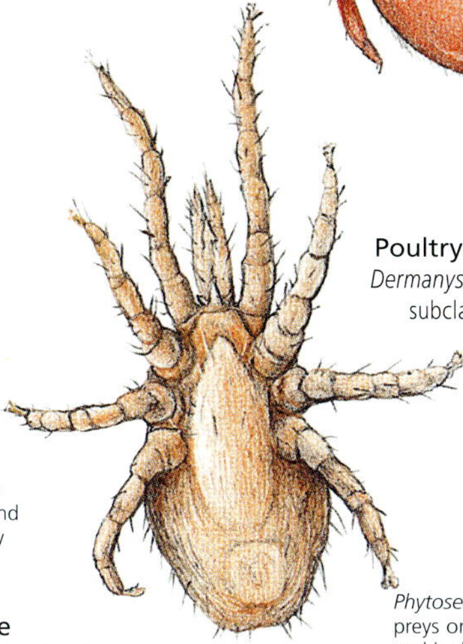

Poultry red mite
Dermanyssus gallinae, subclass Acari

Varroa mite
Varroa jacobsoni, subclass Acari

The straw itch mite is a parasite of insect larvae found in stored grain and straw; it can cause itchy dermatitis in humans

Straw itch mite
Pyemotes ventricosus, subclass Acari

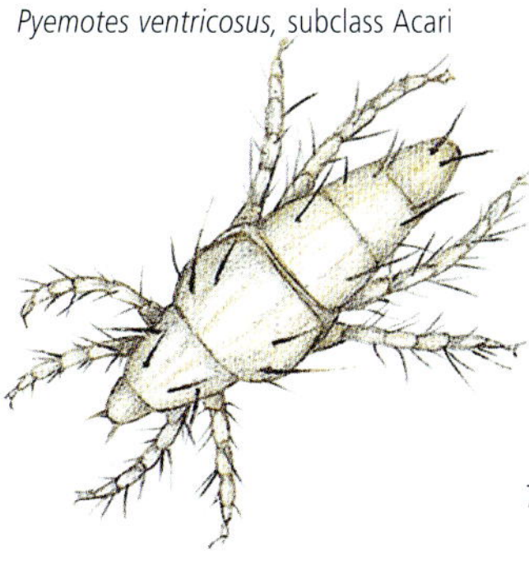

Phytoseiulus persimilis preys on spider mites and is deliberately used to control them in greenhouses

Spider mite
Tetranychus telarius, subclass Acari

Spider mites feed on the underside of leaves, causing them to yellow and die

Phytoseiulus persimilis, subclass Acari

AQUATIC MITES

Some mites have adapted to a life in water. They are found in every kind of freshwater habitat, from puddles to deep lakes, hot springs to raging rivers. Some have also colonized the ocean, where they live on mudflats, coral reefs, and the seafloor, absent only from open waters.

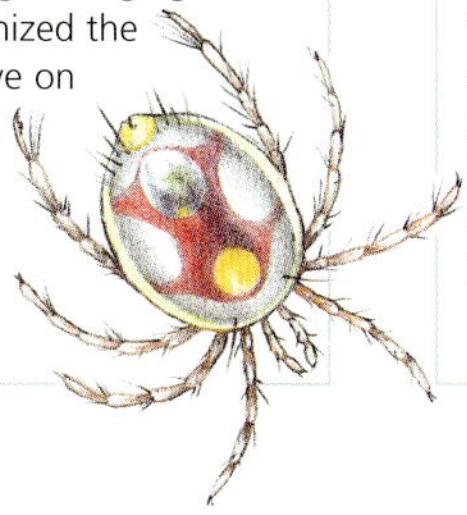

HONEYBEE THREAT

A honeybee colony can be severely harmed or even killed by the varroa mite (*Varroa jacobsoni*). An adult female mite moves into a honeybee's brood cell to lay her eggs. Both the mite and her hatched offspring then feed on the bee larva as it matures. The offspring mate while still in the brood cell. The males then die, but the females emerge attached to the young adult bee.

Snug fit *The varroa mite usually inserts itself between the honeybee's body segments, making it difficult to see.*

Horseshoe crabs

PHYLUM	Arthropoda
SUBPHYLUM	Chelicerata
CLASS	Merostomata
ORDER	Xiphosura
FAMILY	Limulidae
SPECIES	4

More closely related to arachnids than to crustaceans, the horseshoe crabs of the class Merostomata have changed little in the past 200 million years. The four living species, found off the east coasts of North America and Asia, are marine. Their body is made up of a cephalothorax and abdomen, each protected by a hard shell. The horseshoe-shaped cephalothorax bears the pincer-like chelicerae, which are used for seizing worms and other prey from the muddy seafloor, and five pairs of legs, used for both walking and handling food. The formidable tail spine, or telson, does not act as a weapon but helps the horseshoe crab to right itself and to plow through mud.

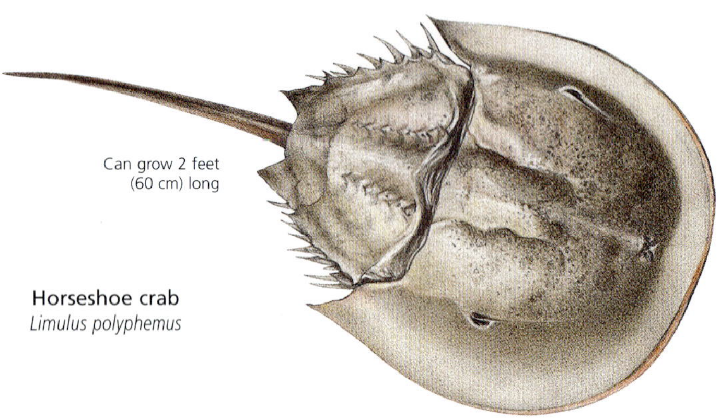

Can grow 2 feet (60 cm) long

Horseshoe crab
Limulus polyphemus

SEA AND SHORE

Horseshoe crabs spend most of their time on the seafloor at depths of about 100 feet (30 m), breathing through book gills on the abdomen. They push through mud, and can also swim on their back and walk.

In spring, horseshoe crabs come ashore to breed. A male will cling onto a female's back as she makes her way over the tidal zone. The female scoops out holes in the sand and lays up to 300 eggs in each, and the male then covers the eggs with sperm to fertilize them. The larvae reach adulthood after about 16 molts over 9 to 12 years.

Sea spiders

PHYLUM	Arthropoda
SUBPHYLUM	Chelicerata
CLASS	Pycnogonida
ORDER	Pantopoda
FAMILIES	9
SPECIES	1,000

With their long legs, sea spiders bear a superficial resemblance to spiders, but they evolved separately and have many unique features. A sea spider's body is greatly reduced. The small head region, known as the cephalon, holds a long proboscis with the mouth; four simple eyes on a stalk; and, usually, two chelicerae (to grasp prey) and two pedipalps (used as sensors). It also bears the first pair of walking legs anda pair of appendages called ovigers, used for grooming and egg-carrying. A segmented trunk holds another three pairs of legs, and leads to a tiny abdomen at the rear. Because the abdomen is so small, the digestive and reproductive organs extend into the legs.

BOTTOM-DWELLERS

Sea spiders live in every ocean, from warm, shallow waters to icy depths of 23,000 feet (7,000 m). Most are small, but some deep-sea species have a leg-span of up to 28 inches (70 cm). Although some sea spiders can swim, most are bottom-dwellers and feed on soft-bodied invertebrates such as sponges and corals.

To breed, a female sea spider releases her eggs into the water and the male covers them with sperm. The male then carries the fertilized eggs on his ovigers until they hatch.

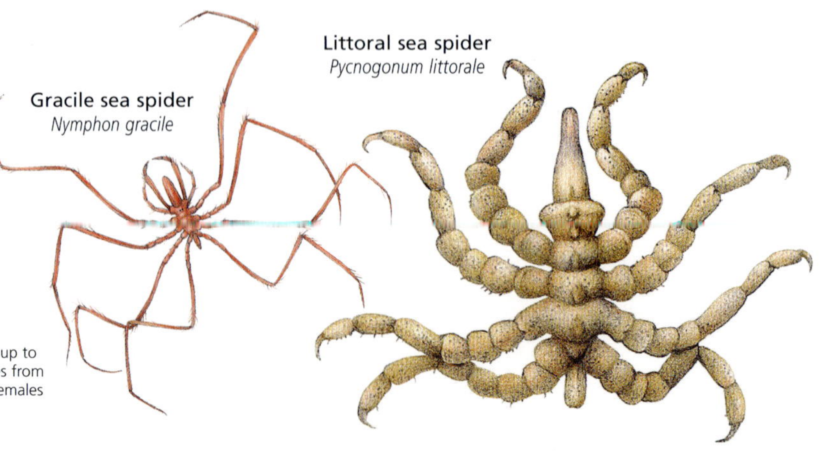

Gracile sea spider
Nymphon gracile

Male can carry up to four egg masses from four different females on his ovigers

Littoral sea spider
Pycnogonum littorale

Myriapods

PHYLUM Arthropoda	
SUBPHYLUM Myriapoda	
CLASSES 4	
ORDERS 20	
FAMILIES 140	
SPECIES 13,500	

Centipedes (class Chilopoda), millipedes (class Diplopoda), symphylids (class Symphyla), and pauropods (class Pauropoda) are all myriapods with long, segmented bodies; simple eyes; a pair of jointed antennae; and numerous pairs of legs. Most centipedes are carnivores and hunt through leaf litter for other small invertebrates. They paralyze prey with a pair of venomous fangs on the underside of the head that can deliver a nasty bite even to humans. Millipedes are plant-feeders and do not bite, but many react to danger by curling into a coil and emitting a toxic substance. Resembling tiny centipedes, symphylids and pauropods live in leaf litter and soil and feed on decaying plant matter.

Tropical species Centipedes are most diverse in tropical forests. Their many legs can be short and hooked, or long and slender. The final pair may be antenna-like or pincer-like.

MANY LEGS

Centipedes have flattened, worm-like bodies, with one pair of legs attached to each body segment except the last. They have at least 15 and as many as 191 pairs of legs. The largest centipede, *Scolopendra gigantea* of the American tropics, can grow to 11 inches (28 cm) long. It is strong enough to prey on mice, frogs, and other small vertebrates.

A millipede's body is rounded and made up of doubled segments known as diplosomites, most of which bear double pairs of legs. Millipedes range from $\frac{1}{12}$ inch to 11 inches (2 mm to 28 cm) long, and have up to 200 pairs of legs.

Measuring no more than ½ inch (1 cm) in length, symphylids have 12 pairs of legs. Pauropods are even smaller, at less than $\frac{1}{12}$ inch (2 mm) long, and bear nine pairs of legs.

Most myriapods are nocturnal, live in moist forests, and tend to be hidden in leaf litter or soil or under rocks or logs. Some species, however, are found in grasslands or deserts. Millipedes usually lay their eggs in a soil nest, while some centipedes curl around their eggs and young to protect them.

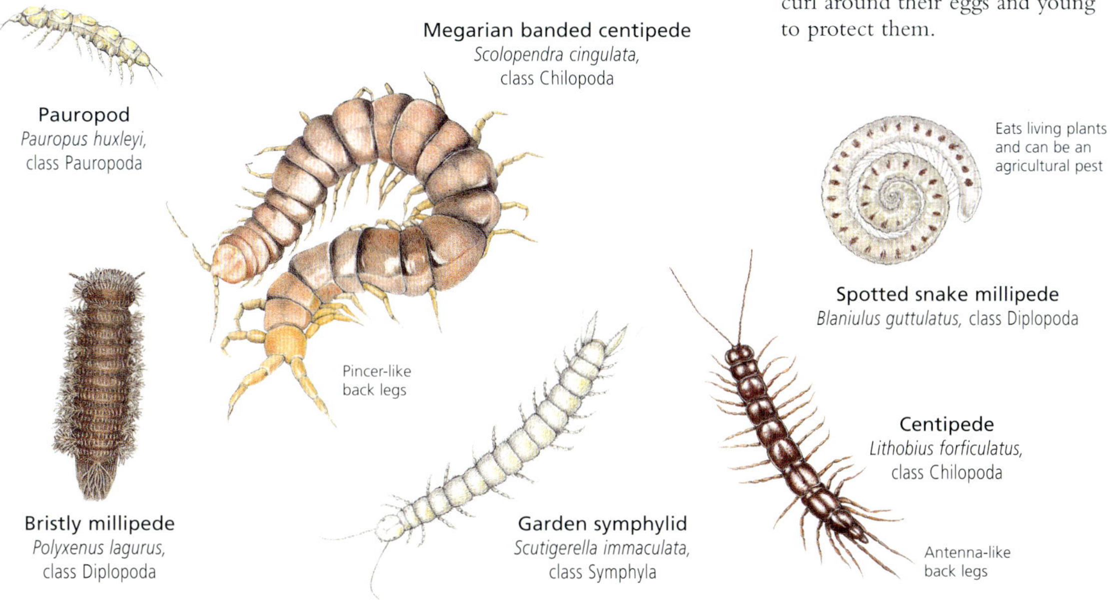

Pauropod *Pauropus huxleyi*, class Pauropoda

Megarian banded centipede *Scolopendra cingulata*, class Chilopoda

Spotted snake millipede *Blaniulus guttulatus*, class Diplopoda

Centipede *Lithobius forficulatus*, class Chilopoda

Bristly millipede *Polyxenus lagurus*, class Diplopoda

Garden symphylid *Scutigerella immaculata*, class Symphyla

Crustaceans

PHYLUM	Arthropoda
SUBPHYLUM	Crustacea
CLASSES	11
ORDERS	37
FAMILIES	540
SPECIES	42,000

From water fleas less than 1⁄100 inch (0.25 mm) long, to giant spider crabs with legs spanning 12 feet (3.7 m), crustaceans are an extraordinarily diverse group. Although some species have adopted a terrestrial lifestyle, it is in aquatic environments that crustaceans have flourished, gradually evolving to exploit every marine and freshwater niche. Free-swimming planktonic species, such as krill, form the basis of vast aquatic food webs. Crabs and other bottom-dwellers may burrow into or crawl over the sediment. Barnacles remain cemented in place, filtering food from the water. There are also parasitic crustaceans, some of which exist in their adult form as a collection of cells inside the host.

Attractive claw In many crustaceans, the first pair of limbs has been modified to form claws known as chelipeds. As a male fiddler crab (*Uca* sp.) matures, the right cheliped grows disproportionately, until it accounts for 65 percent of the crab's total weight. The crab waves this huge claw to attract mates and intimidate rival males. It may also use it to make sounds that entice females into its burrow.

CRUSTACEAN ANATOMY

Like other arthropods, crustaceans have a segmented body, jointed legs, and an exoskeleton, and grow by molting. The exoskeleton can be thin and flexible, as in water fleas, or rigid and calcified, as in crabs. The body usually has three regions—head, thorax, and abdomen—but in many of the larger species, the head and thorax form a cephalothorax, which is protected by a shield-like carapace. The abdomen often has a tail-like extension called a telson.

A typical crustacean's head holds two pairs of antennae; a pair of compound eyes, often on stalks; and three pairs of biting mouthparts. The thorax and sometimes the abdomen carry the limbs, each of which usually has two branches. In many species, particular limbs have become specialized for walking, swimming, food collection, or defense. Crabs, for example, have modified their first pair into claws called chelipeds.

While some female crustaceans lay eggs in water, many brood eggs on their body. The eggs of some species hatch into free-swimming nauplius larvae, with two pairs of antennae, a pair of mandibles, and a single simple eye. In most species, the eggs hatch at a more advanced stage or even as miniature adults.

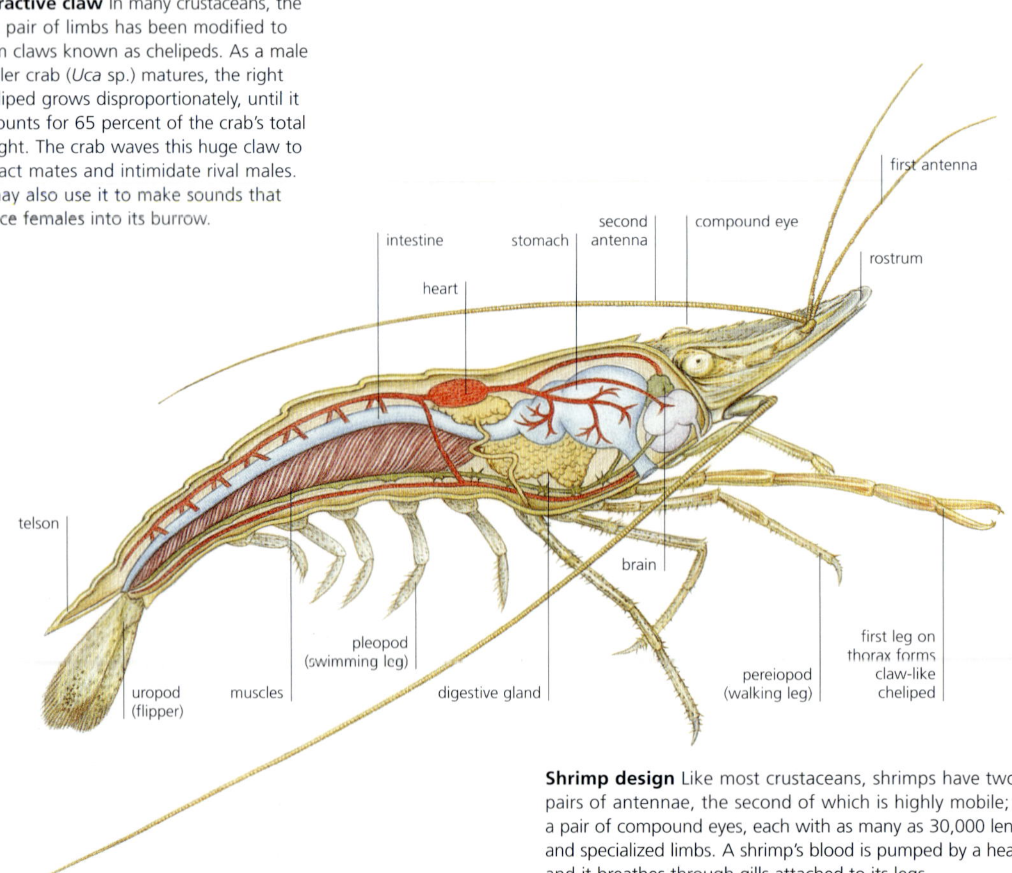

Shrimp design Like most crustaceans, shrimps have two pairs of antennae, the second of which is highly mobile; a pair of compound eyes, each with as many as 30,000 lenses; and specialized limbs. A shrimp's blood is pumped by a heart, and it breathes through gills attached to its legs.

Water flea
Bosmina longirostris,
class Branchiopoda

Marketed as pets called "sea monkeys," brine shrimp can hatch from dried eggs years after they are laid

Brine shrimp
Artemia salina,
class Branchiopoda

Giant water flea
Leptodora kindtii,
class Branchiopoda

Water flea
Daphnia pulex,
class Branchiopoda

Water flea
Polyphemus pediculus,
class Branchiopoda

Water fleas swim jerkily using branched antennae

Copepod
Cyclops strenuus,
class Copepoda

Copepod
Eudiaptomus vulgaris,
class Copepoda

Copepod
Ergasilus sieboldi,
class Copepoda

Cephalocarid
Hutchinsoniella macracantha,
class Cephalocarida

This is one of nine species in Cephalocarida, a class of primitive crustaceans that feed on seafloor detritus

Unchanged for 220 million years, the tadpole shrimp is one of the oldest known living animal species

The clam shrimp lives inside a bivalve shell

Tadpole shrimp
Triops cancriformis,
class Branchiopoda

The clam shrimp swims using its legs and antennae

In male fairy shrimps, the second pair of antennae is enlarged to clasp the female during mating

Clam shrimp
Limnadia lenticularis,
class Branchiopoda

Fairy shrimp
Chirocephalus grubei, class Branchiopoda

RECENT DISCOVERY

When *Speleonectes lucayensis* was discovered in a sea cave in the Bahamas in 1981, it prompted the description of a new class of crustacean, Remipedia. Since then, several more species, all found in Caribbean or Australian caves connected to the sea, have been added to the class. These primitive forms have long, worm-like bodies with up to 32 segments, each bearing a pair of legs. They are carnivorous.

Swimming blind *Like other members of Remipedia,* Speleonectes lucayensis *lacks eyes. It swims on its back using its many legs as oars.*

Class Copepoda Found in vast numbers in the oceans, copepods are a crucial link in the marine food web. The body tends to be cylindrical. They lack compound eyes, but have a single simple eye retained from the nauplius larval stage. Some species are parasitic on fish and other aquatic animals.

Species 8,500

Worldwide; mainly marine

Freshwater fish louse
Argulus foliaceus, class Branchiura

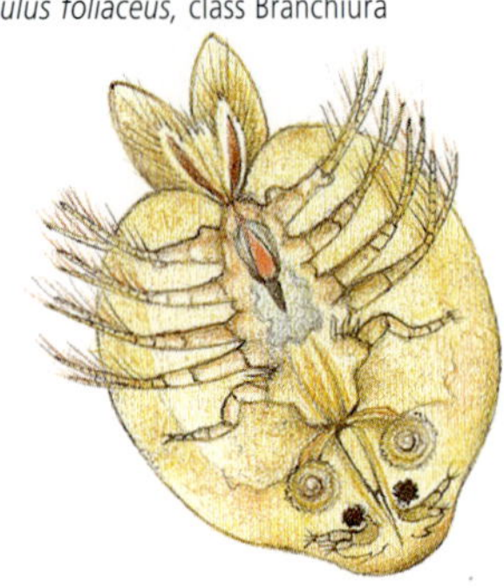

Adult fish lice parasitize freshwater fish, feeding on the skin or in the gills of the host and causing tissue damage

Acorn barnacle
Balanus tintinnabulum, class Cirripedia

Acorn barnacles use their feathery cirri to sweep plankton from the water, but can retract them behind a trapdoor when the tide is out

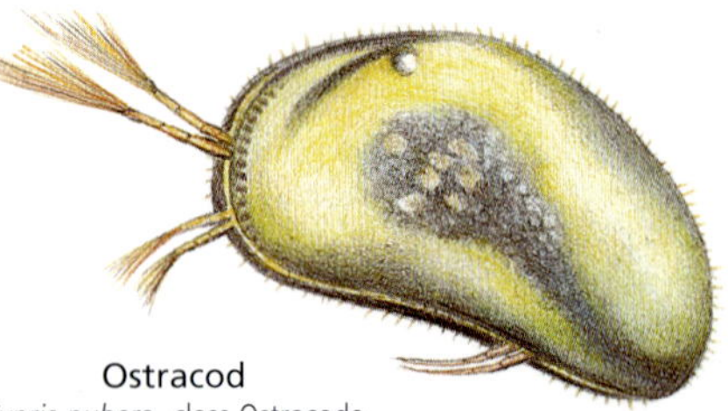

Ostracod
Cypris pubera, class Ostracoda

Ostracod
Candona suburbana, class Ostracoda

Barnacle
Ascothorax ophioctenis, class Cirripedia

Primitive barnacle that parasitizes brittle stars

Like other members of the small class Mystacocarida, *Derocheilocaris remanei* lives among sand grains in the intertidal zone

Mystarocarid
Derocheilocaris remanei, class Mystacocarida

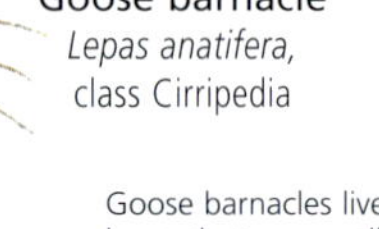

Goose barnacle
Lepas anatifera, class Cirripedia

Goose barnacles live in large clusters, usually in deep water, attached by flexible stalks to surfaces such as rock and wood

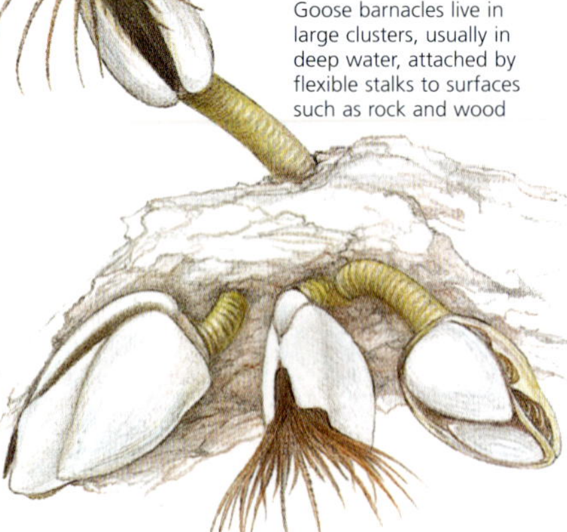

Conchoderma auritum attaches to the skin of a whale, but does not feed on the host's body, filtering food from the water instead

Barnacle
Conchoderma auritum, class Cirripedia

Class Ostracoda Known as mussel shrimps or seed shrimps, the members of this class are distinguished by their carapace, which takes the form of a hinged, bivalve shell. Only the antennae and end bristles of the legs emerge from the shell.

Species 6,000

Worldwide; in all types of water bodies

TOTAL INVASION

The parasite *Sacculina carcini* can only be recognized as a barnacle by its free-swimming nauplius larvae. When a female larva has matured, it invades a host, usually the green crab (*Carcinus maenas*). The larva metamorphoses into a needle-like form that injects cells into the host. The cells form a root-like system, called an interna, throughout the crab's body. The interna eventually develops a reproductive body outside the crab.

Parastygocaris andeni,
class Malacostraca,
order Stygocaridacea

Thermosbaena mirabilis,
class Malacostraca,
order Thermosbaenacea

Amphipod
Caprella anatifera,
class Malacostraca,
order Amphipoda

Diastylis rathkei,
class Malacostraca,
order Cumacea

Opossum shrimp
Mysis relicta,
class Malacostraca, order Mysidacea

All stomatopods have large raptorial appendages for spearing or smashing their aquatic prey

Mantis shrimp
Odontodactylus scyllarus,
class Malacostraca, order Stomatopoda

Amphipod
Gammarus fossarum,
class Malacostraca, order Amphipoda

Bathynella natans,
class Malacostraca,
order Bathynellacea

Bathynella natans is a subterranean crustacean that lives in gravel near lakes

Common pill wood louse
Armadillidium vulgare,
class Malacostraca,
order Isopoda

Armadillidium vulgare is a fully terrestrial species that lives and breeds on land, brooding its eggs in a fluid-filled pouch

Tasmanian mountain shrimp
Anaspides tasmaniae,
class Malacostraca,
order Anaspidacea

Having changed little in the past 250 million years, the Tasmanian mountain shrimp is considered a "living fossil"

Giant isopod
Bathynomus giganteus,
class Malacostraca, order Isopoda

Aquatic sowbug
Asellus aquaticus,
class Malacostraca, order Isopoda

Nebalia bipes,
class Malacostraca,
order Leptostraca

Class Malacostraca By far the largest class of crustaceans, Malacostraca contains 13 orders. Almost all species have six head segments, eight thoracic segments, and six abdominal segments. Except for the first head segment, each segment usually bears two appendages.

Species 25,000

Worldwide; mainly aquatic

Familiar crustaceans *The class Malacostraca includes Daum's reef lobster (*Enoplometopus daumi*), which joins other lobsters, crabs, and shrimps in the order Decapoda.*

PREDACIOUS EYES

The Stomatopoda is a highly specialized order in the class Malacostraca. Known as mantis shrimp, its members actively prey on fish, crabs, and mollusks. They have complex compound eyes.

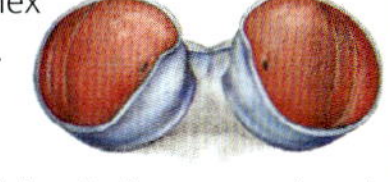

Stomatopod vision *Each compound eye is divided by a band that provides color vision and contrast. The rest of the eye provides monochromatic vision and perspective.*

Emperor shrimp
Periclimenes imperator,
class Malacostraca,
order Decapoda

The emperor shrimp lives on hosts such as sea snails and sea cucumbers for protection and transport

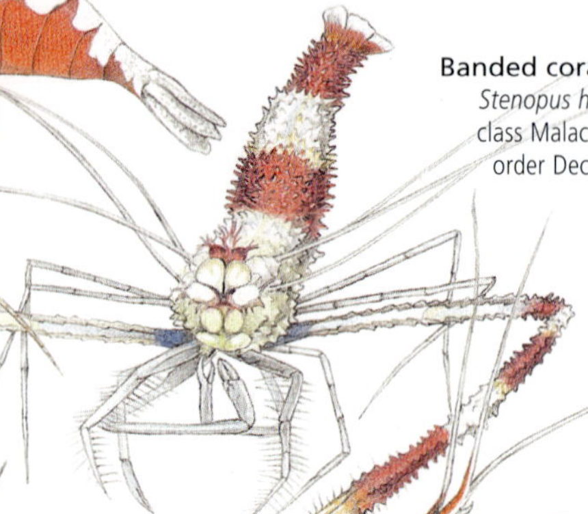

Banded coral shrimp
Stenopus hispidus,
class Malacostraca,
order Decapoda

Randall's pistol shrimp
Alphaeus randalli,
class Malacostraca,
order Decapoda

Brown shrimp
Crangon crangon,
class Malacostraca,
order Decapoda

The cleaner shrimp feeds on parasites that it cleans off passing fish

Cleaner shrimp
Lysmata amboinensis,
class Malacostraca,
order Decapoda

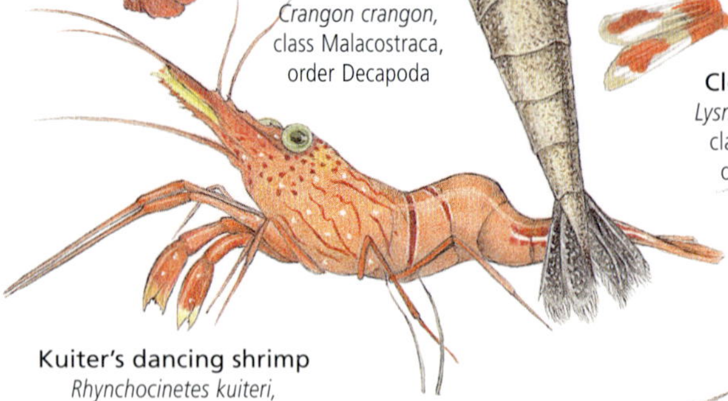

Kuiter's dancing shrimp
Rhynchocinetes kuiteri,
class Malacostraca,
order Decapoda

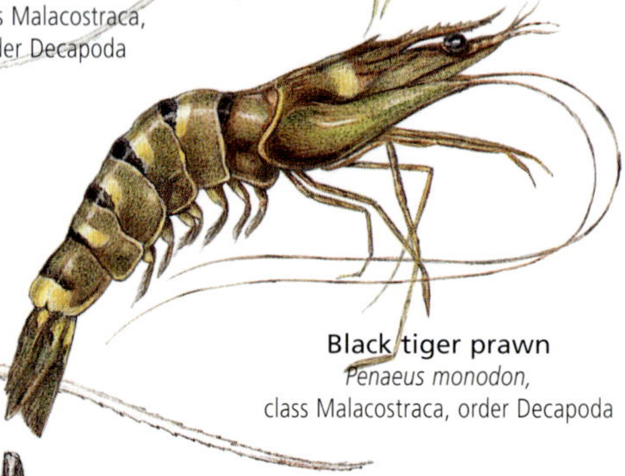

Black tiger prawn
Penaeus monodon,
class Malacostraca, order Decapoda

Painted spiny lobster
Panulirus versicolor,
class Malacostraca,
order Decapoda

The Atlantic lobster is such an important fishery species that it is often known as the "king of seafood"

Spiny lobster
Palinurus vulgaris,
class Malacostraca,
order Decapoda

Atlantic lobster (American lobster)
Homarus americanus,
class Malacostraca,
order Decapoda

Spiny lobsters lack the large claws of true lobsters

Order Decapoda Named after their five pairs of thoracic legs, decapods account for about one-quarter of all crustacean species. They can be divided into swimmers, which include shrimps, and crawlers, which include lobsters, crayfish, and crabs.

Species 8,000

Worldwide; mainly marine

CRUCIAL KRILL

The small order Euphausiacea comprises about 85 species of krill, planktonic creatures with a shrimp-like body. They are a crucial link in marine food webs, particularly in the Antarctic, where the southern krill is the keystone species. Krill usually spend the day in the ocean depths, but gather in vast swarms at the surface to feed at night. In turn, krill are the staple food of many seabirds, fishes, squids, baleen whales, and seals.

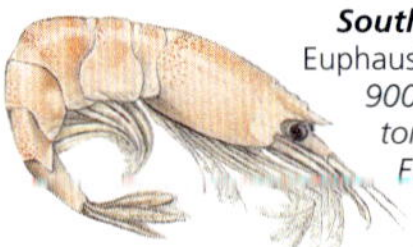

Southern krill *The biomass of* Euphausia superba *may be about 900 million tons (800 million tonnes)—more than that of Earth's human population*

Northern krill *The dominant krill species in the North Atlantic is the northern krill (*Meganyctiphanes norvegica*).*

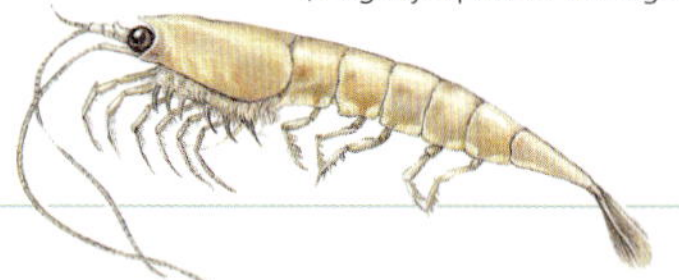

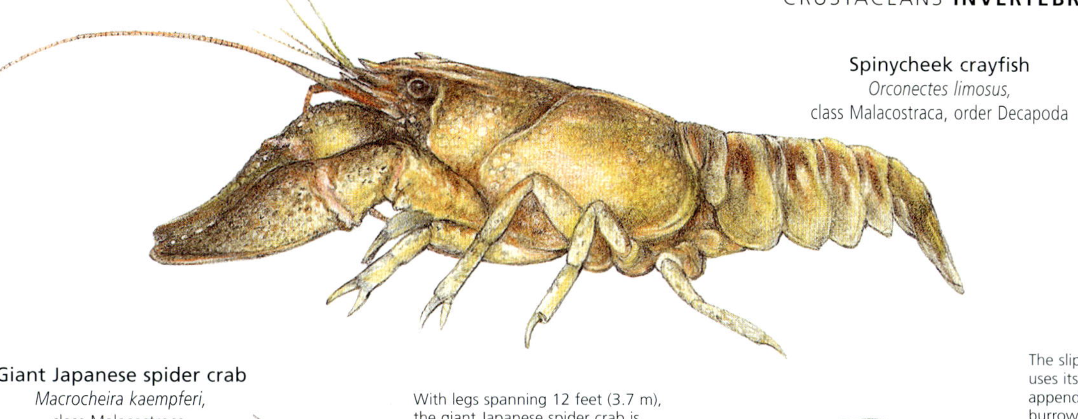

Spinycheek crayfish
Orconectes limosus,
class Malacostraca, order Decapoda

Giant Japanese spider crab
Macrocheira kaempferi,
class Malacostraca,
order Decapoda

With legs spanning 12 feet (3.7 m), the giant Japanese spider crab is the largest of all arthropods

The slipper lobster uses its spade-like appendages to burrow into mud, sand, or gravel

Slipper lobster
Scyllarus arctus,
class Malacostraca, order Decapoda

European green crab
Carcinus maenas,
class Malacostraca,
order Decapoda

Horn-eyed ghost crab
Ocypode ceratophthalma,
class Malacostraca, order Decapoda

Chinese mitten crab
Eriocheir sinensis,
class Malacostraca,
order Decapoda

The Chinese mitten crab's common name was inspired by dense patches of hair on its chelipeds

An Asian species, the Chinese mitten crab has been introduced in Europe and North America, where it is having a severe impact on native species

Alaskan king crab
Paralithodes camtschatica,
class Malacostraca,
order Decapoda

CRUSTACEAN MIGRATIONS

Land crabs have adopted a terrestrial lifestyle, but some return to water to breed. As adults, the red crabs (*Gecarcoidea natalis*) of Christmas Island live in the humid, inland rain forest. When it is time to lay eggs, the island's entire population of 40 millioncrabs undertakes a perilous week-long journey back to the coast.

Males arrive first and dig the nesting burrows. After mating, the females brood their eggs for about 2 weeks, then release them in the ocean. The few aquatic larvae that survive will return to shore after a month and molt into young terrestrial crabs that slowly move inland.

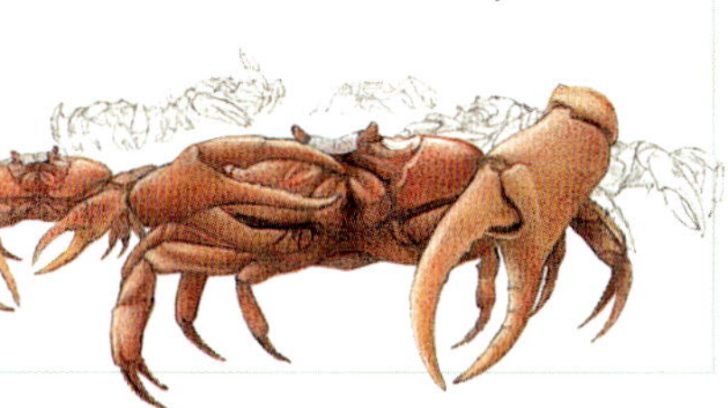

PORTABLE HOMES

To protect their soft abdomens, hermit crabs (family Paguridae) and most land hermit crabs (family Coenobitidae) live in the abandoned shells of gastropods. As the crabs grow, they must find larger shells to move into.

INSECTS

PHYLUM	Arthropoda
SUBPHYLUM	Hexapoda
CLASS	Insecta
ORDERS	29
FAMILIES	949
SPECIES	>1 million

By most measures, insects are the most successful animals ever to have lived on Earth. With about a million described species, they account for more than half of all known animal species, and many more insects are still to be discovered, with estimates of the total number of insect species ranging from 2 million to 30 million. In terms of sheer numbers, insects overwhelm all other animal forms: some scientists believe that ants and termites alone make up 20 percent of the world's animal biomass. Some form of insect has managed to colonize the hottest deserts and the coldest polar zones, as well as virtually every land and freshwater habitat in between, and a handful live in the sea.

Pollinators Pollen sticks to a honeybee as it drinks nectar from a flower. When the bee visits the next flower, some pollen will be dislodged. Most flowering plants rely on insects for pollination.

Purpose-built mouthparts The varied feeding habits of insects are reflected by the shape of their mouthparts. Mosquitoes have beaks that can pierce the skin of prey and suck up the body fluids, while aphids have piercing mouthparts for gathering plant juices. Moths have a long, coiled proboscis, that can be unfurled to collect nectar.

SUCCESS STORY

Insects owe their incredible success to various aspects of their biology. Their tough, flexible exoskeleton provides protection without overly restricting movement. It is covered in a waxy coating that minimizes moisture loss, enabling insects to survive in dry conditions.

As the first creatures and the only invertebrates ever to develop powered flight, insects were able to find both food and mates, evade their predators, and colonize new areas with great efficiency. In most species, the wings can be folded over the body when at rest, so that the insect can exploit confined spaces inside bark, dung, leaf litter, or soil, for example.

Insects breathe through spiracles, small openings in the sides of their body that can be closed to prevent moisture loss. Rather than being transported around the body in the blood, oxygen is distributed directly into the body tissues by a series of tiny pipes known as tracheae. This method of gas diffusion works well only over short distances, which may be the main reason insects have remained small. Most insects are no more than an inch or two (a few centimeters) long. Their small size allowed insects to take advantage of a wide range of microhabitats, a major factor in their great diversity.

Sensory organs are scattered over an insect's body. The head usually has two compound eyes as well as three simple eyes, or ocelli. The two antennae can detect scent,

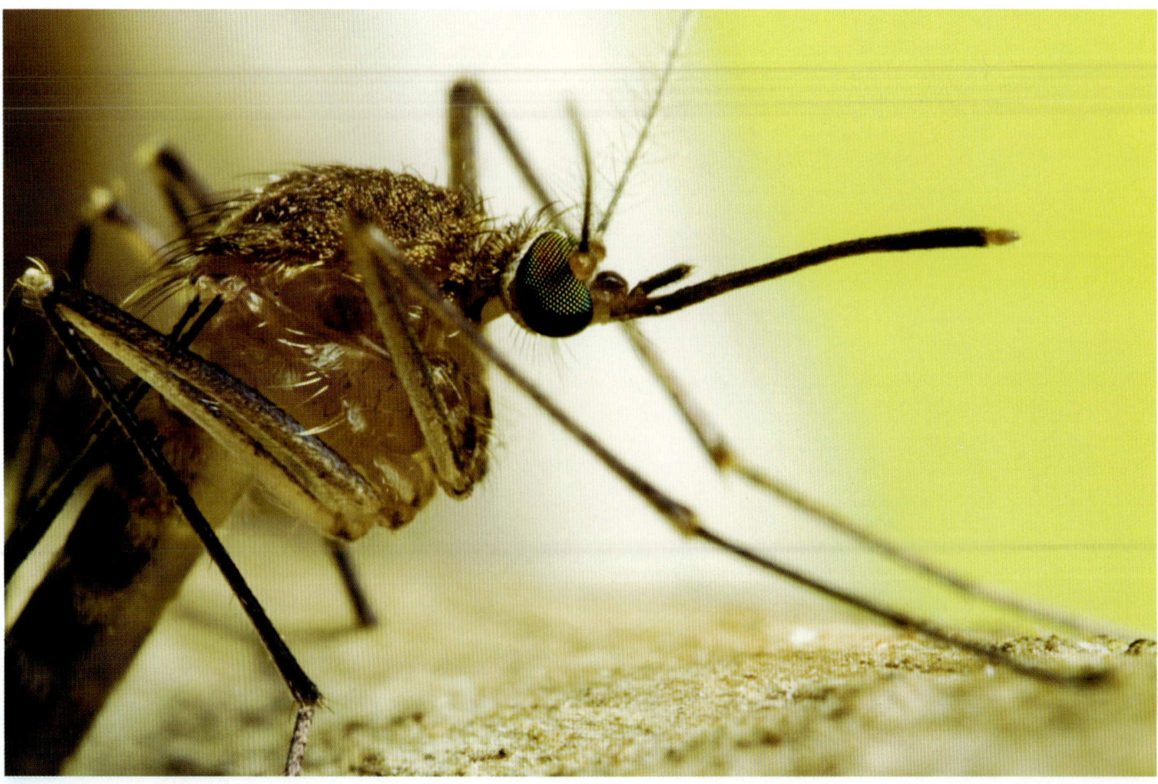

On a roll Male and female dung beetles fashion animal droppings into a ball, before the female lays one or more eggs inside it.

taste, touch, and sound. Hearing organs can also be on the body or legs. Other sensors detect changes in air pressure, humidity, and temperature.

The life-cycles of many insects allow for very rapid reproduction, enabling populations to respond quickly to favorable conditions and recover from catastrophes. Most insects use internal fertilization, but the female is able to store the male's sperm and use it over time. Some insects, such as aphids, can reproduce without fertilization to rapidly increase their numbers, but also use sexual reproduction to ensure genetic diversity. Insect eggs are protected by a shell-like membrane called the chorion that prevents them from drying out, another factor in the successful colonization of dry habitats. While some insects resemble small versions of their parents when they hatch, most do not and must go through some form of metamorphosis. The juveniles and adults often occupy different ecological niches and eat different foods, which helps to avoid competition between them.

Because some species can cause human discomfort, spread disease, and destroy food stores and crops, insects are often regarded as pests. This ignores the fact that the vast majority of insects cause little harm and play a crucial role in the world's environments. About three-quarters of all flowering plants rely on insects for pollination, and many animals depend on insects for food.

Insect anatomy While insects display a staggering diversity of shapes and forms, they follow the same basic body plan. Like other arthropods, insects have segmented bodies made up of a head, thorax, and abdomen; segmented limbs; and a tough exoskeleton. They are distinguished by having three pairs of legs and, usually, two sets of wings, a single pair of antennae, and a pair of compound eyes.

INSECT FLIGHT

While some insects are wingless, the vast majority are winged as adults. Beetles and grasshoppers fly with a slow wing beat of 4–20 beats per second (bps) and limited maneuverability. Bees and true flies, with wing beats of about 190 bps, can hover and dart. Some midges have wing beats of 1,000 bps. The fastest sustained fliers are dragonflies, which can maintain speeds of 31 miles per hour (50 km/h).

Folding wings *The hindwings of a mantis fold up like fans when not in use. This protects the wings and allows the mantis to squeeze into tight spaces.*

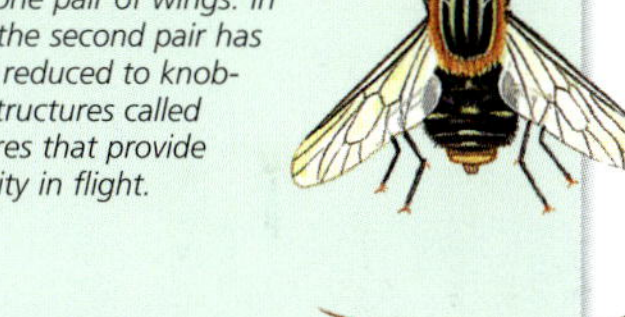

Stabilizers *True flies appear to have only one pair of wings. In fact, the second pair has been reduced to knob-like structures called halteres that provide stability in flight.*

Plumed wings *In thrips and plume moths, the wings look like tiny feathers and are made up of fine hairs supported by a midrib.*

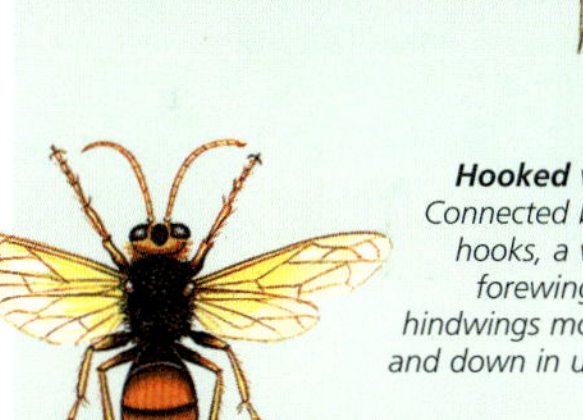

Hooked wings *Connected by tiny hooks, a wasp's forewings and hindwings move up and down in unison.*

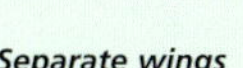

Separate wings *A dragonfly's forewings and hindwings can beat in unison or independently. As in early flying insects, the wings do not fold back over the body.*

Protective wings *In lady beetles and other beetles, the forewings have been modified to form protective cases known as elytra. The hindwings are used for flying.*

DRAGONFLIES & DAMSELFLIES

PHYLUM	Arthropoda
SUBPHYLUM	Hexapoda
CLASS	Insecta
ORDER	Odonata
FAMILIES	30
SPECIES	5,500

The first members of the order Odonata emerged more than 300 million years ago—100 million years before dinosaurs appeared—and included the largest insect that has ever lived, a dragonfly with a wingspan of 28 inches (70 cm). Today's dragonflies and damselflies are much smaller, with wingspans from ¾ inch to 7½ inches (18 mm to 19 cm). Often seen flying near water, these voracious aerial predators are most abundant in the tropics, but can be found worldwide except for the polar zones. Damselflies have a fluttery flight and usually rest with the wings near the body, while dragonflies are strong, agile flyers and always rest with the wings held out from the body.

Sharp sight A dragonfly's large compound eyes may be made up of 30,000 lenses, providing superb vision and a wide field of view. Strong, biting mouthparts indicate the insect's carnivorous diet.

AQUATIC BEGINNINGS

A dragonfly or damselfly spends most of its life as a wingless aquatic nymph. The nymph breathes via gills and feeds on other insect larvae, tadpoles, and small fishes. It has a specialized lower mouthpart known as a mask that usually rests under its head but can shoot out to snatch prey in its pincers. Over a period of a few weeks to 8 years, depending on the species, the nymph may molt up to 17 times. For the final molt, it climbs out of the water and sheds its nymphal skin to reveal an adult with prominent eyes, sharp mouthparts, two pairs of transparent wings, a sloping thorax, and a long, slender abdomen.

A newly emerged adult dragonfly or damselfly flies away from water and starts to feed, snatching flying insects on the wing. To find mates, dragonflies will gather along a body of water, and males may engage in aerial contests. After mating, the male usually guards the female as she lays her eggs in the water. Most adults live for only several weeks.

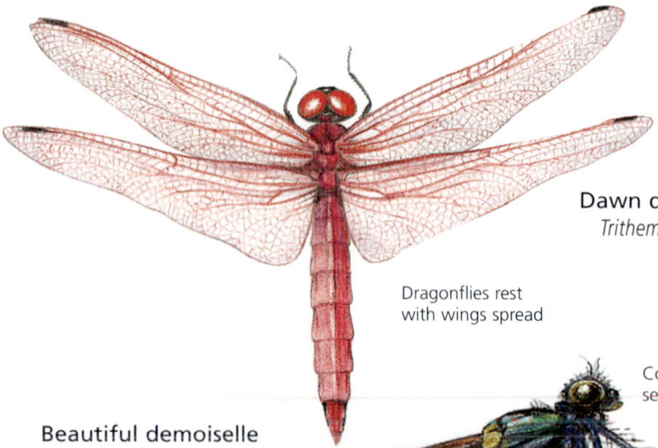

Dawn dropwing
Trithemis aurora

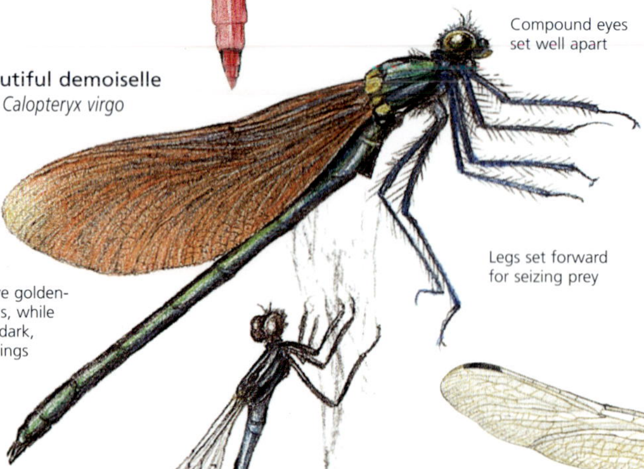

Dragonflies rest with wings spread

Beautiful demoiselle
Calopteryx virgo

Compound eyes set well apart

Legs set forward for seizing prey

Females have golden-brown wings, while males have dark, iridescent wings

Damselflies rest with wings together

Azure damselfly
Coenagrion puella

Enormous compound eyes set close together

Blue dasher
Pachydiplax longipennis

Mantids

PHYLUM	Arthropoda
SUBPHYLUM	Hexapoda
CLASS	Insecta
ORDER	Mantodea
FAMILIES	8
SPECIES	2,000

Waiting to ambush prey, a mantid will sit perfectly still, holding its large forelegs folded up before it—a posture that inspired the common name of praying mantis for some species. With lightning-quick reflexes, the raptorial forelegs will shoot out to snatch insect prey. Most mantids are medium-sized, measuring about 2 inches (5 cm) long, but some tropical species reach lengths of 10 inches (25 cm). These giant mantids may add small birds and reptiles to their primarily insectivorous diet. While the majority of mantids live in the tropics or subtropics, some are found in the warm temperate areas of southern Europe, North America, South Africa, and Australia.

Precision hunting With its forward-facing eyes supplying binocular vision, a praying mantis uses its spiky, hooked forelimbs to impale passing prey. The mantis's strong mandibles quickly devour the meal.

SILENT HUNTERS

Mantids are masters at hiding from both predators and prey. They are the only insects that can turn their head without moving other parts of the body, allowing them to silently observe potential victims. Most species also have cryptic coloration that helps them blend in with the grass, leaves, twigs, or flowers of their environment. An alarmed mantid will adopt a threat posture, raising and rustling its wings and rearing up to show off its vivid warning coloration. Mantids that are targeted by bats at night possess a single "ear" on their thorax that can detect a bat's ultrasonic signals.

In some species, the female mantid will devour the male during copulation. Females mate only once, but from that single mating, they can produce up to 20 ootheca, or egg cases, each containing from 30 to 300 eggs. The young emerge from the ootheca as highly active nymphs, looking like miniature, wingless versions of their parents. Ready to hunt, they may even eat each other. The survivors disperse. After a series of molts, the nymphs mature and develop wings and adult coloration.

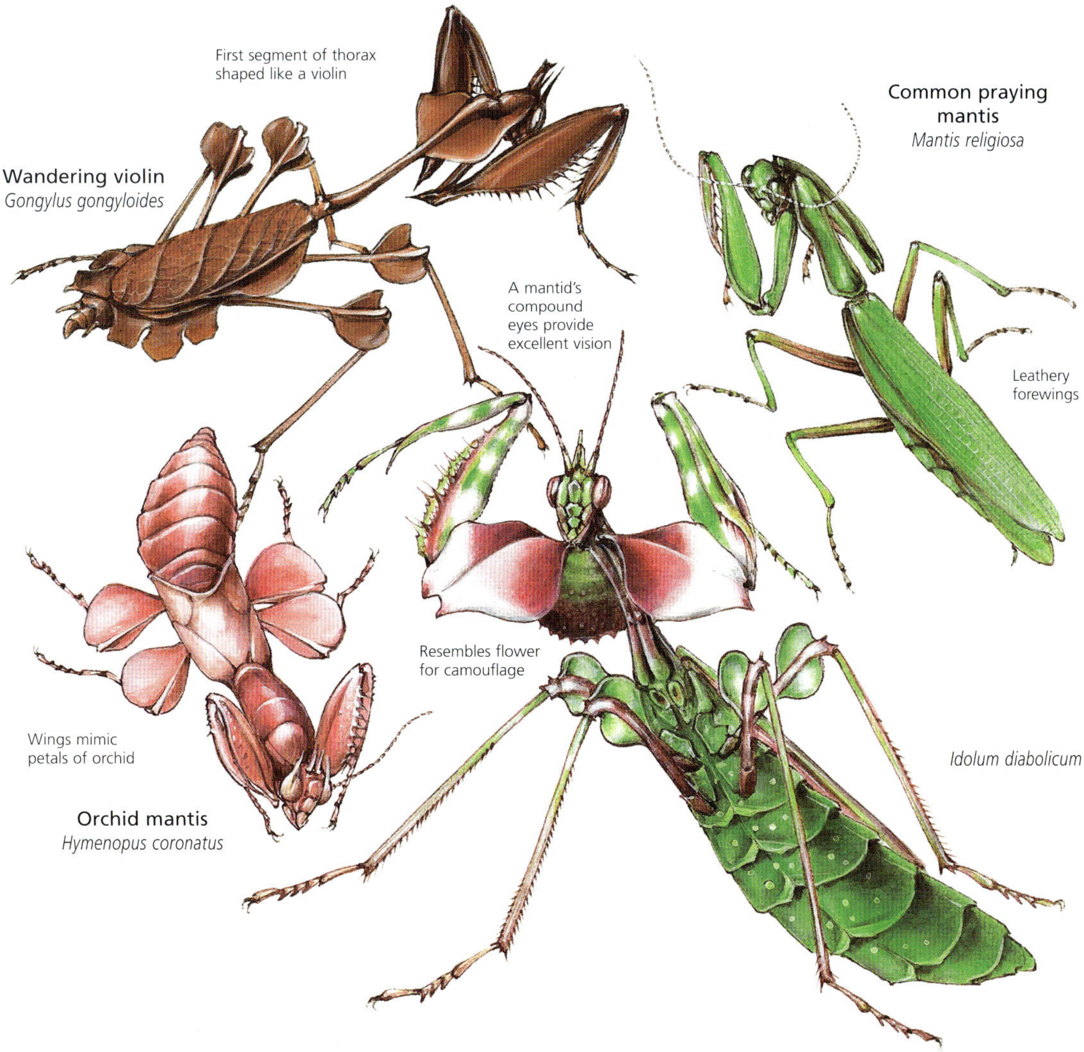

First segment of thorax shaped like a violin

Wandering violin
Gongylus gongyloides

Common praying mantis
Mantis religiosa

A mantid's compound eyes provide excellent vision

Leathery forewings

Resembles flower for camouflage

Wings mimic petals of orchid

Orchid mantis
Hymenopus coronatus

Idolum diabolicum

Cockroaches

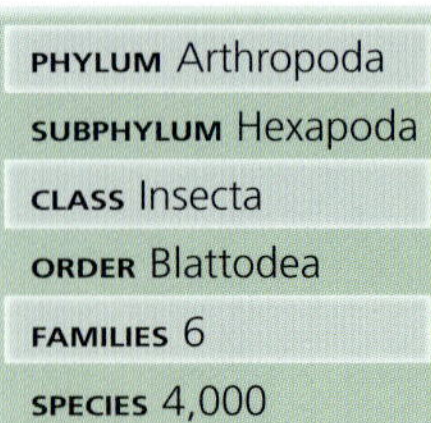

PHYLUM Arthropoda
SUBPHYLUM Hexapoda
CLASS Insecta
ORDER Blattodea
FAMILIES 6
SPECIES 4,000

As scavenging insects willing to feed on almost any plant or animal product, including stored food, trash, paper, and clothing, some cockroaches thrive in human environments and are widely regarded as repulsive pests. In fact, less than 1 percent of all cockroach species are pests—the remainder perform an important ecological role by recycling leaf litter and animal excrement in forests and other habitats. While most cockroaches are found in warm tropical zones, about 25 species have spread worldwide, accidentally transported on ships. Cockroaches are among the most primitive of all living insects, with an anatomy that has changed little in more than 300 million years.

Safe eggs Female cockroaches deposit their eggs in an ootheca, or egg case. The nymphs emerge soft and white, but soon harden and turn brown. They eat the ootheca as their first food.

SENSITIVE INSECTS

A range of sensors helps cockroaches to detect minute changes in their surroundings. Their long antennae can find minuscule amounts of food and moisture. Sensors on the legs and abdomen can pick up tiny air movements, prompting the insect to flee danger in a split second. With its oval, flattened shape, a cockroach can scuttle into tiny crevices. Not all cockroaches have wings, but in those that do the forewings are usually hardened and opaque, and the hindwings are translucent.

Mating between cockroaches may be initiated when a female emits a pheromone to attract males. Females usually lay 14–32 eggs in a hardened case known as an ootheca. The ootheca may then be left to hatch alone; carried at the base of the mother's abdomen; or incubated inside the mother's body so that she produces live young. Nymphs molt up to 13 times before emerging as adults. A cockroach usually lives for 2–4 years.

Giant cockroach
Blaberus giganteus

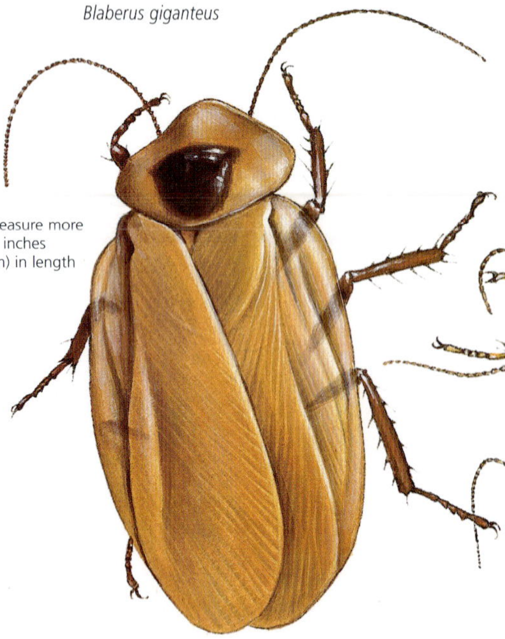

Can measure more than 3 inches (7.5 cm) in length

Green banana roach
Panchlora nivea

Madagascan hissing cockroach
Gromphadorina portentosa

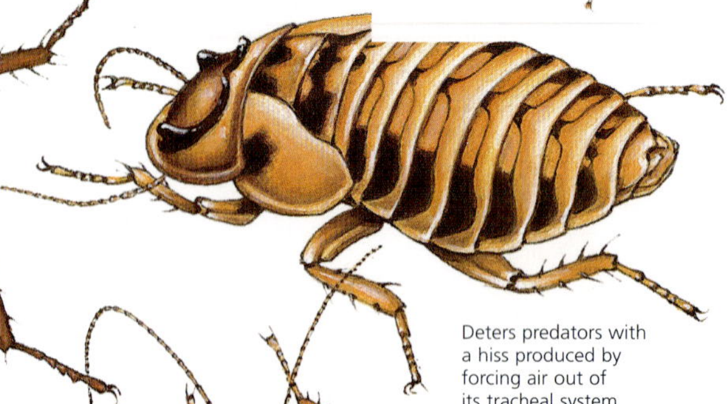

Deters predators with a hiss produced by forcing air out of its tracheal system

Attaphila fungicola

Smallest known cockroach, at $\frac{3}{16}$ inch (4 mm) long; lives in nests of leafcutter ants and feeds on fungus

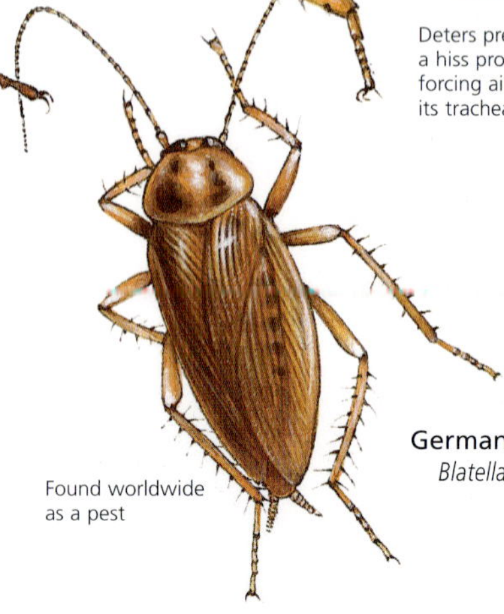

German cockroach
Blatella germanica

Found worldwide as a pest

TERMITES

PHYLUM	Arthropoda
SUBPHYLUM	Hexapoda
CLASS	Insecta
ORDER	Isoptera
FAMILIES	7
SPECIES	2,750

Termites may be the world's most monogamous animals. A king and queen mate for life, producing thousands or millions of offspring that operate together as a highly structured colony. These small insects are often referred to as white ants, but although their social system and anatomy resemble those of ants, they evolved independently and are most closely related to cockroaches. Found throughout much of the world, termites are most abundant in tropical rain forests, where there can be as many as 25,000 individuals per square mile (10,000 per sq km). By feeding on dead wood, they recycle nutrients in their natural habitats but can severely damage buildings in urban environments.

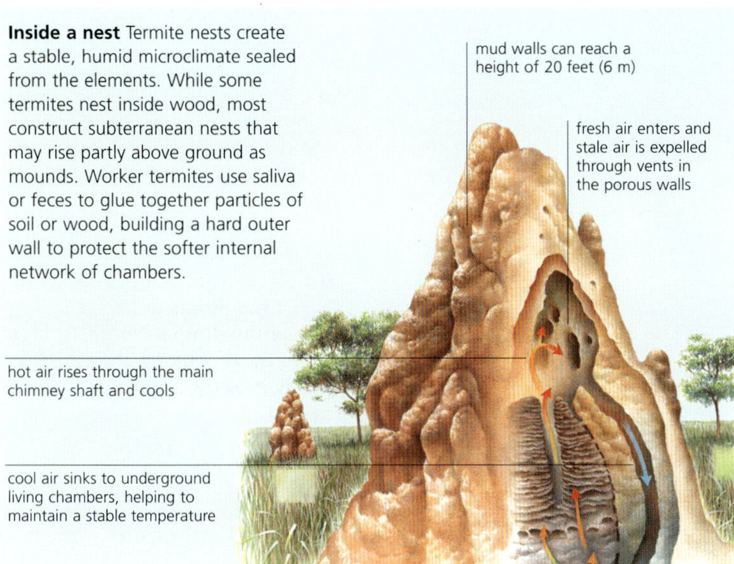

Inside a nest Termite nests create a stable, humid microclimate sealed from the elements. While some termites nest inside wood, most construct subterranean nests that may rise partly above ground as mounds. Worker termites use saliva or feces to glue together particles of soil or wood, building a hard outer wall to protect the softer internal network of chambers.

CASTE SOCIETY

A mature termite colony has three castes: reproductives, workers, and soldiers. The main reproductives are the king and queen, who produce all the colony's other members and can live for 25 years. There may also be secondary reproductives ready to take over if the king or queen dies.

Workers and soldiers can be male or female. They are wingless, usually lack eyes and mature reproductive organs, and may live for 5 years. The pale and soft-bodied workers feed all other colony members and maintain the nest. Soldiers defend the colony, usually using powerful mandibles. Termite nymphs can develop into any caste, depending on the needs of the colony.

At a particular time each year, a swarm of winged males and females will emerge from a termite nest and disperse. They then shed their wings and form pairs, becoming the kings and queens of new colonies.

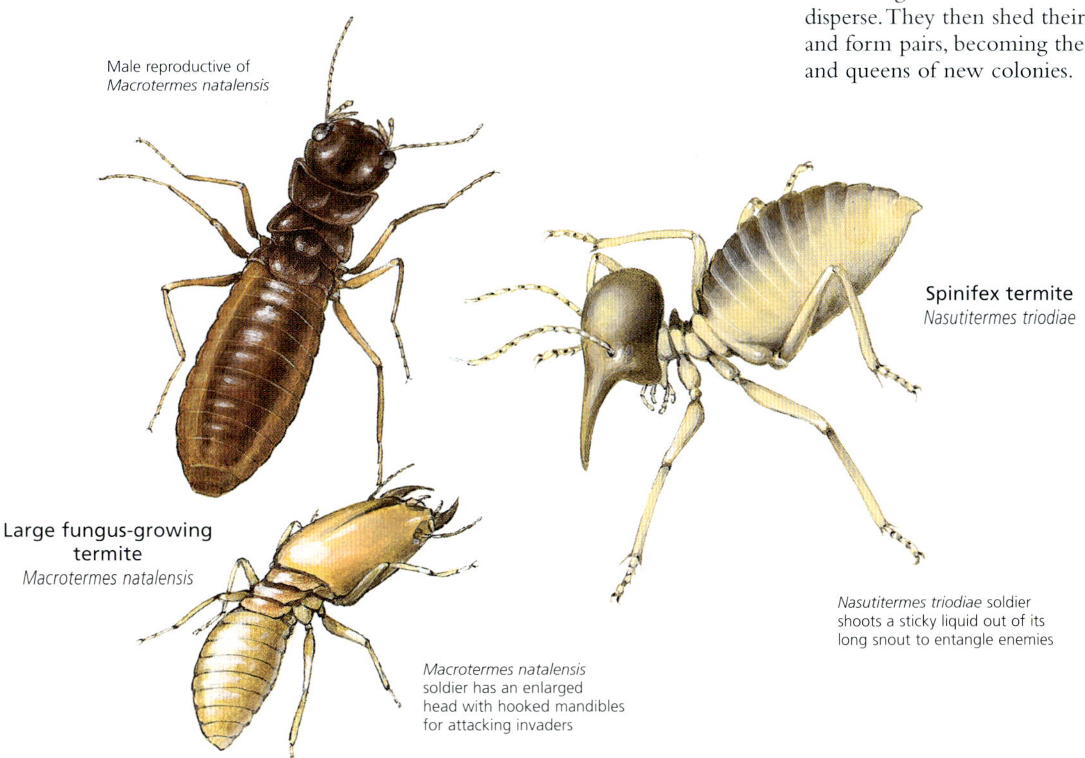

Male reproductive of *Macrotermes natalensis*

Large fungus-growing termite
Macrotermes natalensis

Macrotermes natalensis soldier has an enlarged head with hooked mandibles for attacking invaders

Spinifex termite
Nasutitermes triodiae

Nasutitermes triodiae soldier shoots a sticky liquid out of its long snout to entangle enemies

CRICKETS & GRASSHOPPERS

PHYLUM	Arthropoda
SUBPHYLUM	Hexapoda
CLASS	Insecta
ORDER	Orthoptera
FAMILIES	28
SPECIES	> 20,000

Renowned for their songs and leaping ability, the members of the order Orthoptera include grasshoppers, locusts, crickets, katydids (or bush-crickets), and their kin. They are distinguished by elongated hindlegs that allow them to jump. Most species are winged, with slender, toughened forewings protecting the membranous, fan-shaped hindwings. While most orthopterans are ground-dwellers in grasslands and forests, some are arboreal, burrowing, or semiaquatic, and species can be found in deserts, caves, bogs and marshes, and seashores. Grasshoppers, groundhoppers, and a few katydids are herbivores, while most other orthopterans are omnivorous.

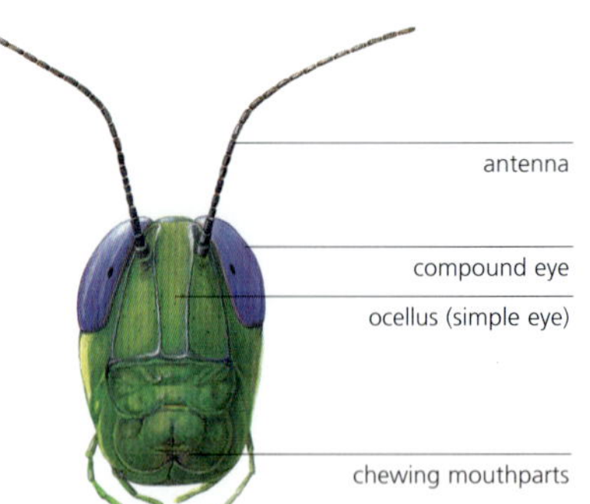

Information collector An orthopteran's head features two fine, sensitive antennae, two large compound eyes, and mouthparts adapted for chewing vegetation.

SINGING INSECTS

Most male orthopterans can make sounds by rubbing two parts of the body together, a technique known as stridulation. Crickets and long-horned grasshoppers move a scraper on one forewing along a row of teeth on the other forewing, while short-horned grasshoppers rub a ridged surface on the hindleg against a forewing. The sounds can be heard by organs located on the legs or abdomen.

There are three kinds of song: a calling song, to attract females from a distance; a courtship song, to entice a nearby female to mate; and a battle call, to deter rival males. The songs, some of which are too high-pitched to be heard by human ears, are unique to each species. Cricket songs tend to be affected by the weather, with the rate of chirps increasing with the temperature.

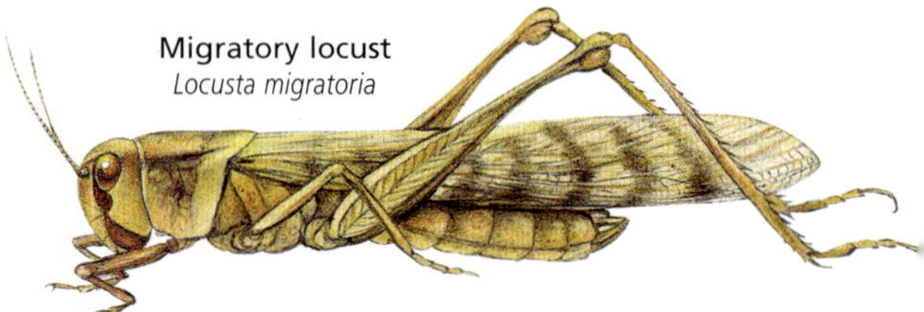

Migratory locust
Locusta migratoria

In response to overcrowding, many millions of locusts may form swarms that devastate crops as they migrate

Great green bush-cricket
Tettigonia viridissima

Tropical leaf katydid
Siliquofera grandis

Wings mimic leaves

European mole cricket
Gryllotalpa gryllotalpa

Enlarged forelegs for digging

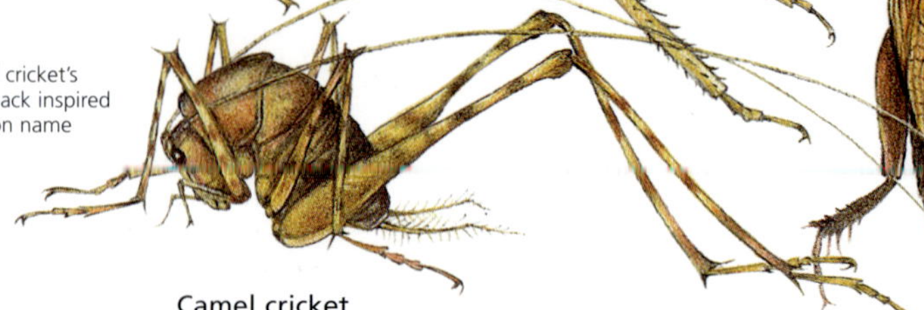

The camel cricket's humped back inspired its common name

Camel cricket
Tachycines asynamorus

Blue-winged grasshopper
Oedipoda coerulescens

Bugs

PHYLUM	Arthropoda
SUBPHYLUM	Hexapoda
CLASS	Insecta
ORDER	Hemiptera
FAMILIES	134
SPECIES	> 80,000

Ranging from minute wingless aphids to giant frog-catching water bugs, the members of the order Hemiptera are extremely diverse. The name Hemiptera means "half wing," a reference to the fact that many, but not all, species have forewings that are leathery at the base and membranous at the tip. The one feature that all bugs have in common is their piercing and sucking mouthparts. With these, they pierce the surface of a plant or animal and inject saliva to start digesting the food, which they then suck into their mouth. Most bug species feed on plant sap, and some of these are significant agricultural pests. Other bugs suck the blood of vertebrates, or prey on other insects.

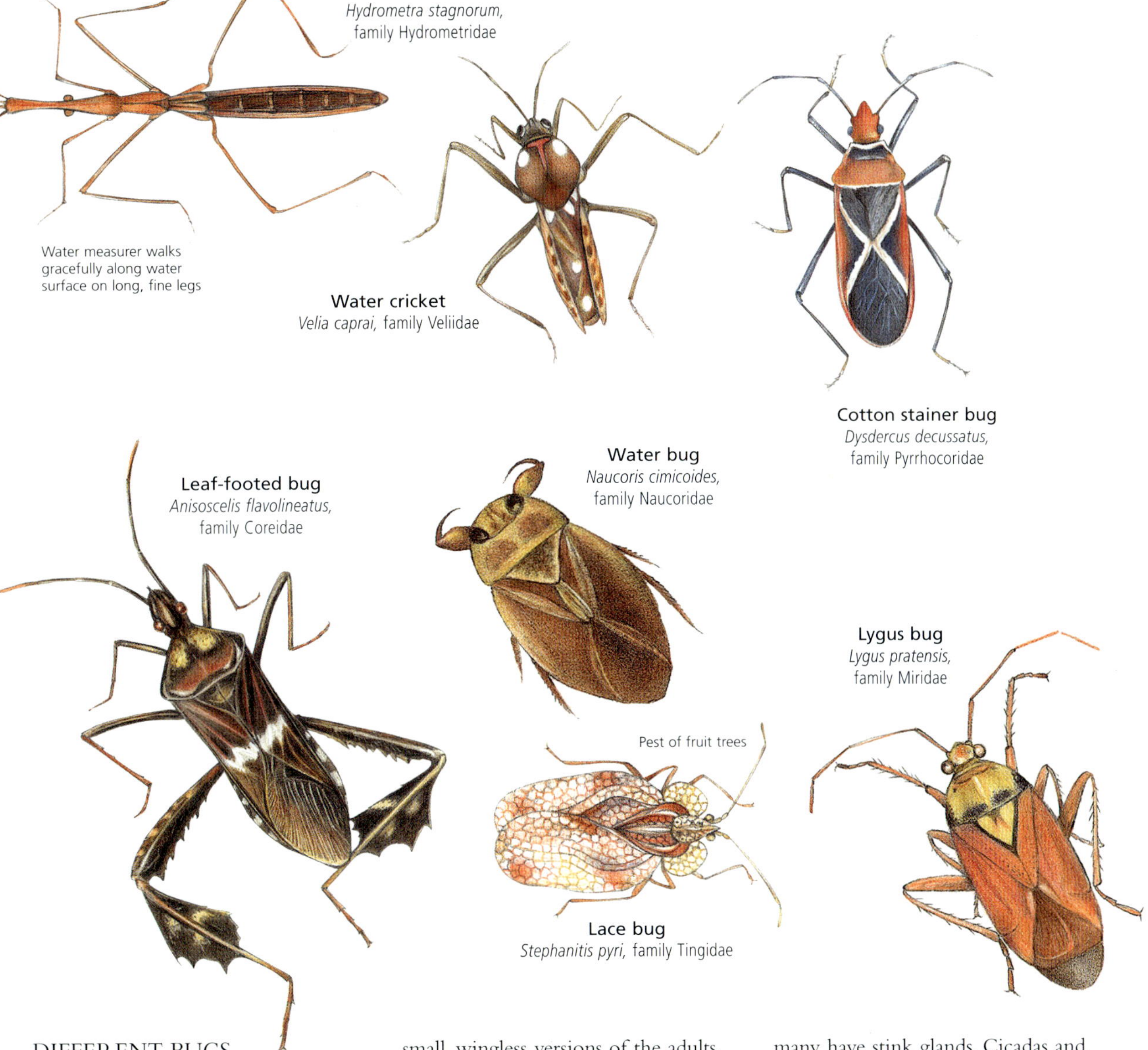

Water measurer
Hydrometra stagnorum, family Hydrometridae

Water measurer walks gracefully along water surface on long, fine legs

Water cricket
Velia caprai, family Veliidae

Cotton stainer bug
Dysdercus decussatus, family Pyrrhocoridae

Leaf-footed bug
Anisoscelis flavolineatus, family Coreidae

Water bug
Naucoris cimicoides, family Naucoridae

Lygus bug
Lygus pratensis, family Miridae

Pest of fruit trees

Lace bug
Stephanitis pyri, family Tingidae

DIFFERENT BUGS

Distributed throughout the world, bugs can be found in almost all terrestrial habitats. Some species have specialized for an aquatic life—these include sea skaters (*Halobates* sp.), the only insects on the open ocean. Bugs range from $\frac{1}{25}$ inch to 4½ inches (1 mm to 11 cm) in length. They undergo incomplete metamorphosis, with nymphs that usually resemble small, wingless versions of the adults. In cicadas and some other species, however, the burrowing nymphs and the adults look quite different.

The order Hemiptera is divided into three suborders. In the true bugs (suborder Heteroptera), the forewings are toughened at the base and sit flat, concealing membranous hindwings. True bugs can flex their head and mouthparts forward, and many have stink glands. Cicadas and hoppers (Auchenorrhyncha) hold their uniform forewings over the abdomen like a tent. The head and mouthparts point down and back. Aphids, scale insects, mealy bugs, whiteflies, and their kin all belong to the suborder Sternorrhyncha. Most have soft bodies kept moist by a covering of wax or froth. Wings are often absent or reduced in adults.

Red cicada
Tibicen haematodes,
family Cicadidae

Male cicada vibrates tymbals on abdomen to produce songs

The hindwings of the peanut-headed lanternfly bear large spots that deter predators by mimicking the eyes of a larger animal

Peanut-headed lanternfly
Laternaria phosphorea,
family Fulgoridae

Peanut-shaped head

Lanternfly
Lanternaria candelaria,
family Fulgoridae

Treehopper
Hemikyptha punctata,
family Membracidae

Treehopper
Bocydium globulare, family Membracidae

Green shield bug
Palomena prasina,
family Pentatomidae

Bed bug
Cimex lectuarius,
family Cimicidae

Bed bugs suck blood from vertebrates such as birds, bats, and humans

Squash bug
Coreus marginatus,
family Coreidae

Large milkweed bug
Oncopeltus fasciatus,
family Lygaeidae

Assassin bug
Rhinocoris irracundus,
family Reduviidae

Family Reduviidae This family contains the assassin bugs, predators that usually target other insects, but sometimes suck the blood of vertebrates. The bugs swing their curved proboscis forward to puncture the prey and inject paralyzing saliva, then suck up the body fluids.

Kissing bug *The conenose bug (*Triatroma sanguisuga*) is a blood-sucker and sometimes bites sleeping humans near the mouth.*

PERIODICAL CICADAS

The noisiest insects in the world, male cicadas (family Cicadidae) produce their sound with structures on the abdomen known as tymbals. While many cicadas have a life-cycle of up to 8 years, most of which is spent underground as a nymph, the periodical cicadas of the *Magicicada* genus are marked by their synchronized development. After up to 18 years underground, an area's entire population will mature and emerge at the same time.

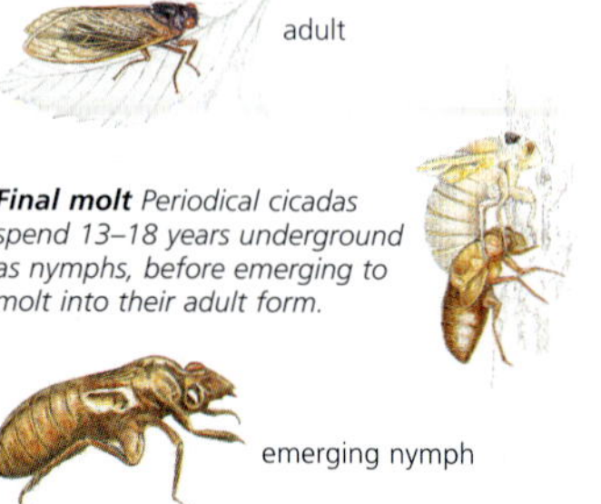

Final molt *Periodical cicadas spend 13–18 years underground as nymphs, before emerging to molt into their adult form.*

European fruit lecanium
Parthenolecanium corni, family Coccidae, superfamily Coccoidea

Greenhouse whitefly
Trialeurodes vaporarium, family Aleyrodidae

Woolly apple aphid
Eriosoma lanigerum, family Aphididae

Long-tailed mealy bug
Pseudococcus longispinus, family Pseudococcidae, superfamily Coccoidea

Nettle ensign scale
Orthezia urticae, family Ortheziidae, superfamily Coccoidea

Apple leaf sucker
Psylla mali, family Psyllidae

Flata rubra, family Flatidae

Red and black froghopper (spittlebug)
Cercopis sanguinolenta, family Cercopidae

Treehopper
Heteronotus reticulatus, family Membracidae

Euscelis plebejus, family Cicadellidae

Treehopper
Oeda inflata, family Membracidae

APHIDS AS FOOD

Aphids provide food for many other insects. They are prey for many lady beetles (family Coccinellidae), hover flies (family Syrphidae), and lacewings (order Neuroptera). They also have a happier relationship with some species of ants, which protect them from predators and the elements. In return, the ants stroke, or "milk," the aphids to feed on honeydew, the sweet waste product of the aphids' plant-sap diet.

Aphid predator *In its lifespan of 1–2 months, a lady beetle can consume more than 2,000 aphids.*

Family Aphididae Aphids are tiny, soft-bodied plant-suckers that cause great damage to crops. They are able to rapidly increase their numbers by producing wingless or winged females from unfertilized eggs.

Cabbage lover *Native to Europe, the cabbage aphid (*Bravicoryne brassicae*) has spread to many parts of the world. It feeds on both wild and cultivated plants.*

WATER INSECTS

Only about 3 percent, or 30,000, insect species are truly aquatic for at least part of their life-cycle. Water insects have had to adapt to the low oxygen content of water, as well as to the difficulty of moving about in still water or staying put in flowing water. Some insects, such as water striders, live only on the surface. Many, such as backswimmers, breathe air but carry it with them when they swim underwater. Others, such as dragonfly nymphs, spend all their time submerged, using gills to absorb oxygen from the water. The vast majority of aquatic insects live in fresh water, and only 300 or so species live in saltwater habitats, possibly because crustaceans evolved in the sea first and offer too much competition.

Aquatic locomotion Aquatic insects have evolved various strategies for moving around. Water scorpions (family Nepidae) can swim but often crawl along the bottom of a pond and spend much of their time hanging from pond weeds, waiting to ambush prey. They breathe air through a tube that sticks out above the surface. Water boatmen (family Corixidae) have long, hairy hindlegs that can "row" powerfully through the water. Water striders (family Gerridae) have fine hairs on their feet that allow them to make small jumps along the surface.

Bubble breather Predaceous diving beetles (family Dytiscidae) spend virtually their entire lives in water. The adults come to the surface to collect air, which they store in a bubble under their elytra (forewings). They swim using their hindlegs, which are fringed in thick hairs.

water scorpion hangs from pond weeds as it attacks a tadpole

water strider uses surface tension to "walk" on water

water boatman swims by using its legs as oars

BEETLES

PHYLUM	Arthropoda
SUBPHYLUM	Hexapoda
CLASS	Insecta
ORDER	Coleoptera
FAMILIES	166
SPECIES	> 370,000

When asked what his study of the natural world had revealed about its Creator, the scientist J. B. S. Haldane replied, "an inordinate fondness for beetles." Of all known animal species, about one in four is a beetle. Members of the order Coleoptera have colonized almost all of Earth's habitats, from Arctic tundra and exposed mountaintops to deserts, grasslands, woodlands, and lakes. They reach their greatest diversity, however, in the lush rain forests of the tropics. Most beetles are distinguished by their hardened, leathery forewings, known as elytra, which protect the membranous, flying hindwings and allow the insect to live in tight spaces under bark or in leaf litter.

Competing males The huge, branched mandibles of male stag beetles resemble a stag's antlers and perform a similar function, being used in competitions with rival males to win the right to mate.

DIVERSE FORMS

From lady beetles to diving beetles, scarab beetles to fireflies, the order Coleoptera encompasses enormous diversity. The feather-winged beetle (*Nanosella fungi*) is just $\frac{1}{100}$ inch (0.25 mm) long, while the South American longhorn beetle (*Titanus giganteus*) can exceed 6½ inches (16 cm) in length. Adult beetles can be oval and flattened, long and slender, or squat and domed. While a few beetles are strong fliers, most are clumsy in flight, and some lack wings and cannot fly at all.

The mouthparts of beetles are all adapted for biting, but they have been put to many purposes. Plant-eating beetles may eat roots, stems, leaves, flowers, fruit, seeds, orwood. Carnivorous beetles usually attack invertebrate prey. A key ecological role is filled by scavenging beetles, which recycle dead animals and plants, excrement and other waste.

Most beetles live on the ground, but some have specialized to live in trees, in water, and underground—a few even make their home in the nests of ants and termites.

Beetles develop by complete metamorphosis, with the larvae looking markedly different from the parents and going through a non-feeding, pupal stage before becoming adults.

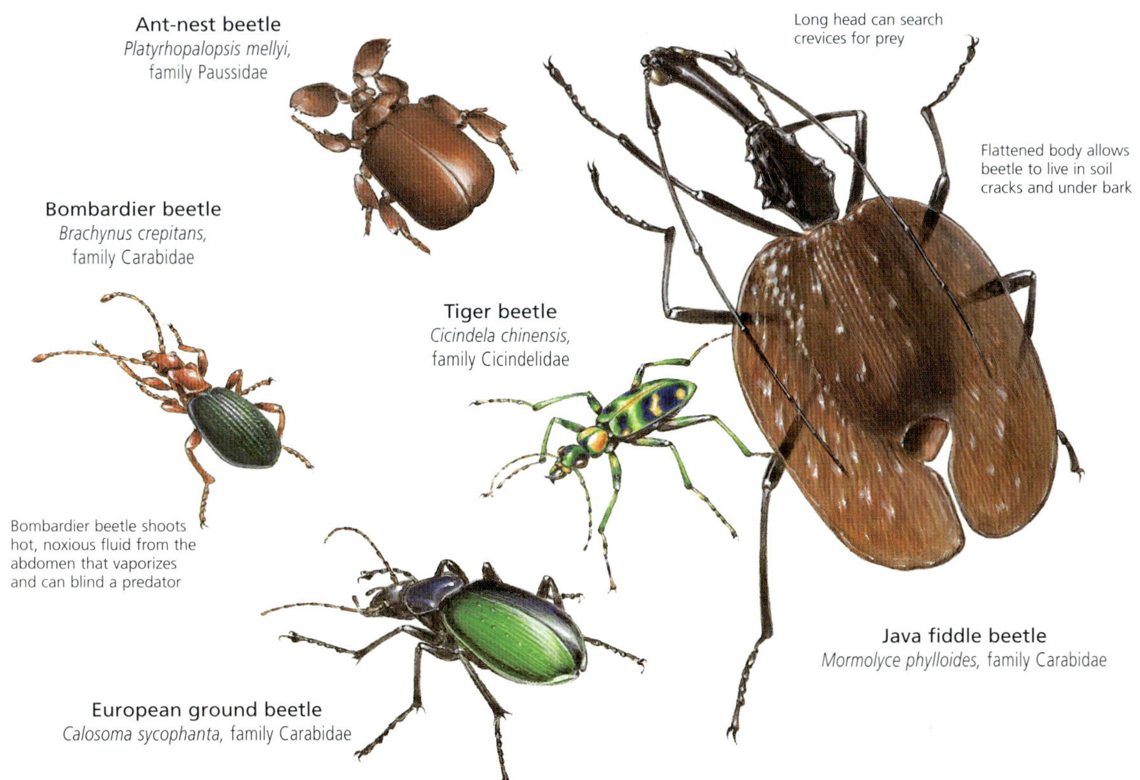

Seven-spotted lady beetle
Coccinella septempunctata,
family Coccinellidae

Larva of *Coccinella septempunctata*

Vivid color warns predators of lady beetle's foul taste

European splendor beetle
Anthaxia hungarica,
family Buprestidae

Golden rove beetle
Emus hirtus,
family Staphylinidae

Golden rove beetle is covered in fine hairs

Flesh-eating beetle
Phosphuga atrata,
family Silphidae

Common carpet beetle
Anthrenus scrophulariae,
family Dermestidae

Feeds on carpet, wool, and other animal products

Water scavenger beetle
Sphaeridium scarabaeoides,
family Hydrophilidae

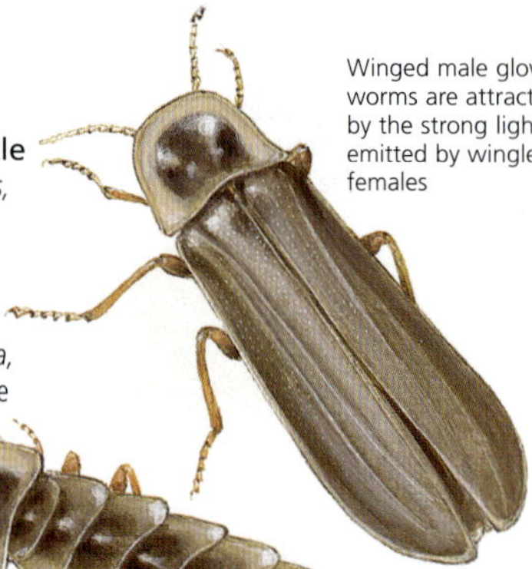

Winged male glow worms are attracted by the strong light emitted by wingless females

Glow worm
Lampyris noctiluca,
family Lampyridae

Larva of *Lampyris noctiluca*

European burying beetle
Necrophorus vespillo,
family Silphidae

Buries corpses of small vertebrates as food for its young

Fuscous soldier beetle
Cantharis fusca,
family Cantharidae

Beneficial insect that feeds on aphids and caterpillars

Bee beetle
Trichodes apiarius,
family Cleridae

Parasitizes bee hives

Family Buprestidae With spectacular colors and a metallic sheen, the jewel beetles in this family include some of the world's most attractive insects. They have a long oval body and are good fliers. Many adults feed on nectar, while larvae tend to be wood-borers.

Shiny wings *The metallic wood-boring beetle* (Euchroma gigantea) *is used in jewelry.*

Family Coccinellidae The lady beetles of this family are among the best known of all insects. They are usually brightly colored and spotted, with a compact shape and short antennae. Most lady beetles prey on aphids and other pests, and are desired guests in gardens.

Lines and spots *The convergent lady beetle* (Hippodamia convergens) *usually has 13 spots.*

Family Lampyridae This family is made up of fireflies. All firefly larvae glow, producing light via a chemical reaction while giving off little heat. Their luminescence may be a warning to predators of their unpleasant taste. Most adult fireflies also produce light to attract mates.

Aggressive mimic *Some female fireflies mimic flash patterns of other species to lure males as prey.*

Goliath beetle
Goliathus meleagris, family Scarabaeidae

Golden spider beetle
Niptus hololeucus, family Ptinidae

Spanish fly
Lytta vesicatoria, family Meloidae

Black blister beetle
Meloe violaceus, family Meloidae

Rhinoceros dung beetle
Oxysternon conspicillatum, family Scarabaeidae

Cigarette beetle (tobacco beetle)
Lasioderma serricorne, family Anobiidae

Hercules beetle
Dynastes hercules, family Scarabaeidae

Fire beetle
Pyrophorus noctilucus, family Elateridae

Green scarab
Eudicella gralli, family Scarabaeidae

BEETLE LIFECYCLE

All beetles develop through complete metamorphosis. In most species, eggs are fertilized sexually. The eggs hatch into a larva that has chewing mouthparts but otherwise bears little resemblance to the adult. The larva molts several times, growing larger until it is ready to pupate. Enclosed in a cocoon, the larva transforms into a pupa with adult features. It is soft and pale when it emerges but soon hardens and colors.

Family Elateridae This family comprises the click beetles. If a click beetle ends up lying upside down, it will straighten its back and propel itself into the air to land right way up. As it does so, a hinge-like structure on the elytra makes a loud click. Adults feed on foliage, while the larvae live in soil and eat roots and bulbs.

Species 9,000

Worldwide; near plants and in soil

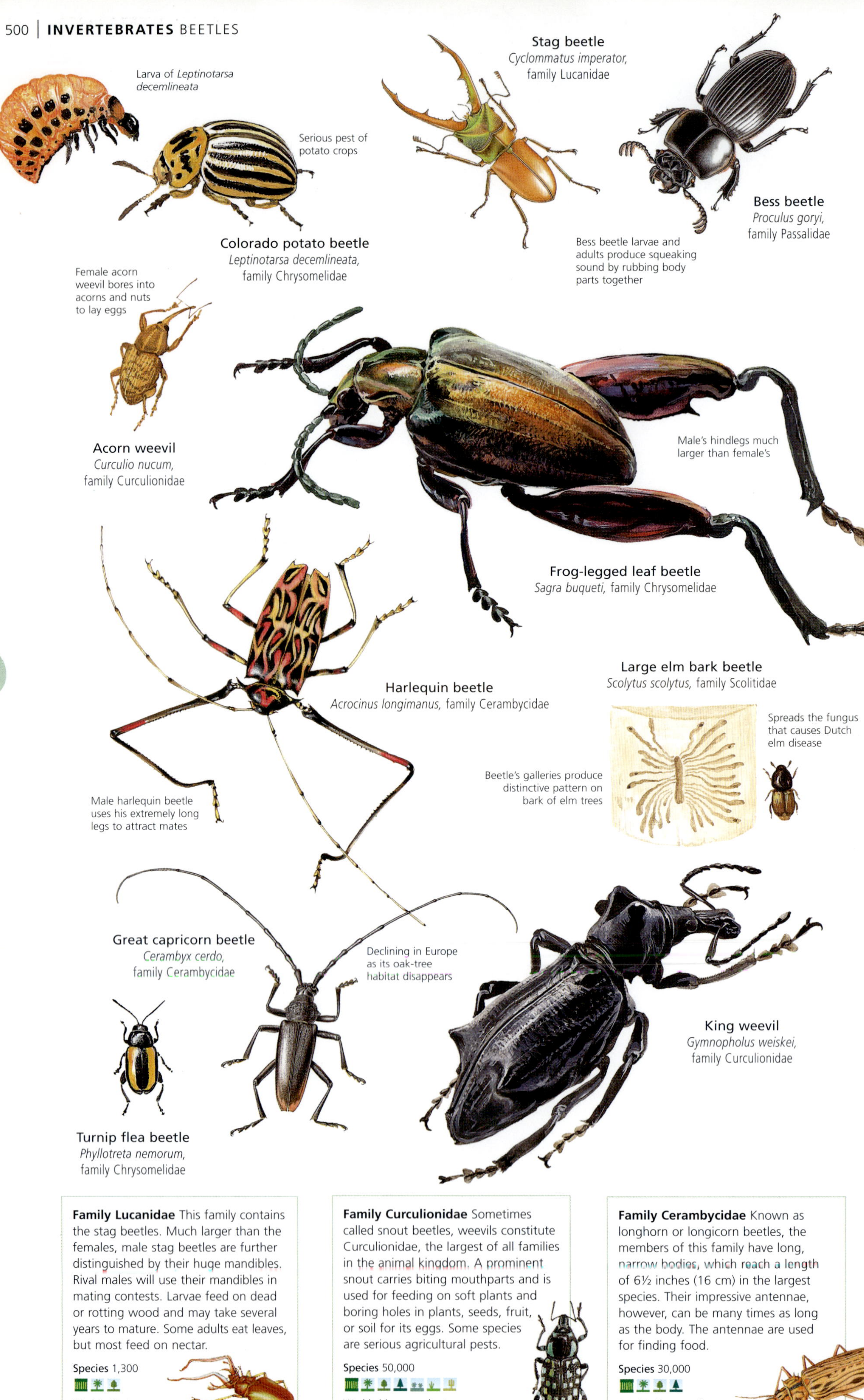

Family Lucanidae This family contains the stag beetles. Much larger than the females, male stag beetles are further distinguished by their huge mandibles. Rival males will use their mandibles in mating contests. Larvae feed on dead or rotting wood and may take several years to mature. Some adults eat leaves, but most feed on nectar.

Species 1,300

Worldwide; in trees

Family Curculionidae Sometimes called snout beetles, weevils constitute Curculionidae, the largest of all families in the animal kingdom. A prominent snout carries biting mouthparts and is used for feeding on soft plants and boring holes in plants, seeds, fruit, or soil for its eggs. Some species are serious agricultural pests.

Species 50,000

Worldwide; near plants

Family Cerambycidae Known as longhorn or longicorn beetles, the members of this family have long, narrow bodies, which reach a length of 6½ inches (16 cm) in the largest species. Their impressive antennae, however, can be many times as long as the body. The antennae are used for finding food.

Species 30,000

Worldwide; near flowers and sap

Flies

PHYLUM	Arthropoda
SUBPHYLUM	Hexapoda
CLASS	Insecta
ORDER	Diptera
FAMILIES	130
SPECIES	120,000

Houseflies and mosquitoes are the most ubiquitous members of the order Diptera, which also includes gnats, midges, blowflies, fruit flies, crane flies, horseflies, hover flies, and other true flies. While most insects fly with four wings, dipterans are generally distinguished by their single pair of functional wings. The hindwings are reduced to halteres, tiny clubbed stalks that vibrate up and down in time with the forewings, helping the fly to balance during flight. Some species have lost their wings altogether and are flightless. Although often associated with moist environments and decaying matter, flies live almost everywhere in the world except Antarctica.

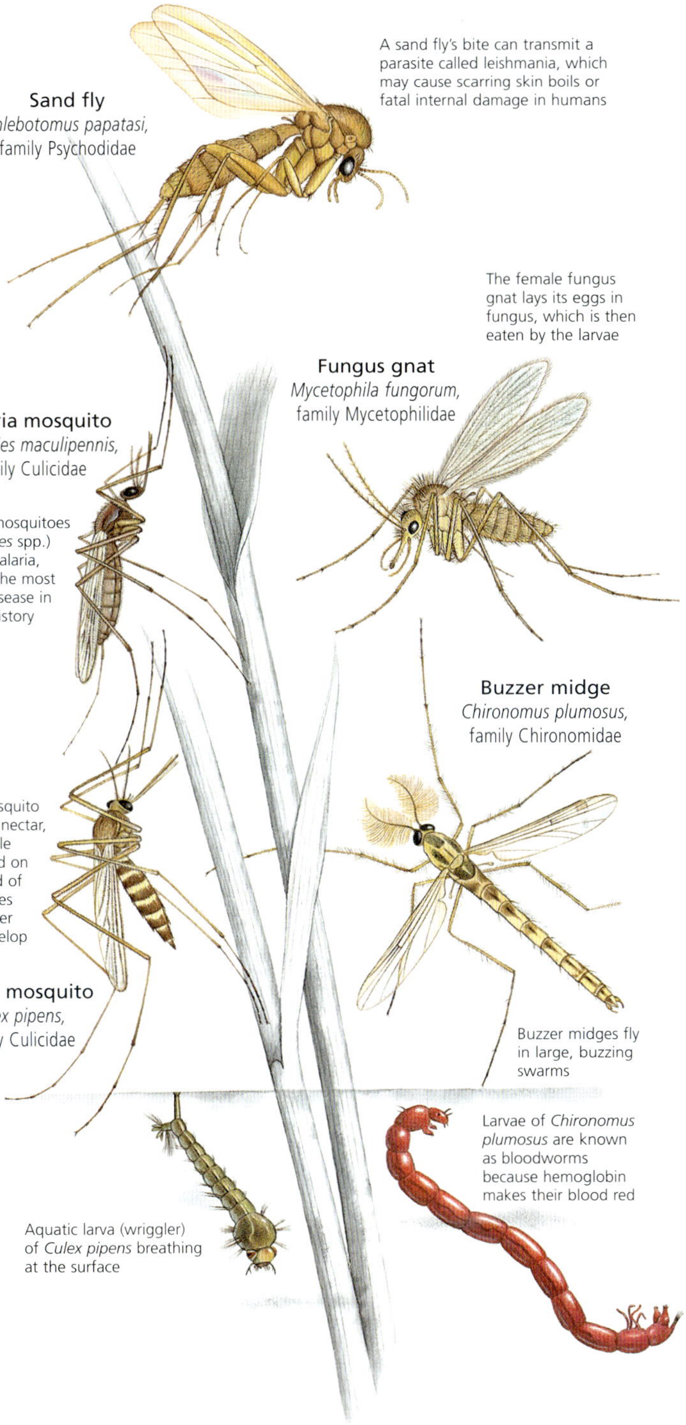

Sand fly
Phlebotomus papatasi, family Psychodidae

A sand fly's bite can transmit a parasite called leishmania, which may cause scarring skin boils or fatal internal damage in humans

Fungus gnat
Mycetophila fungorum, family Mycetophilidae

The female fungus gnat lays its eggs in fungus, which is then eaten by the larvae

Malaria mosquito
Anopheles maculipennis, family Culicidae

Malaria mosquitoes (*Anopheles* spp.) spread malaria, possibly the most deadly disease in human history

Buzzer midge
Chironomus plumosus, family Chironomidae

Buzzer midges fly in large, buzzing swarms

Male mosquito feeds on nectar, but female must feed on the blood of vertebrates to help her eggs develop

House mosquito
Culex pipens, family Culicidae

Aquatic larva (wriggler) of *Culex pipens* breathing at the surface

Larvae of *Chironomus plumosus* are known as bloodworms because hemoglobin makes their blood red

STICKY SUCKERS

Flies are small insects, ranging from midges $\frac{1}{25}$ inch (1 mm) in length, to robber flies about 3 inches (7 cm) long. Their feet have sticky pads with tiny claws, allowing the insects to walk on smooth surfaces, even upside down along a ceiling.

With mouthparts designed for sucking, flies can consume only liquid food. Houseflies have fleshy pads on their mouth that can sponge up a meal. Mosquitoes have piercing mouthparts for feeding on nectar and blood. Robber flies use their mouthparts to stab insect prey.

Flies begin life as eggs, from which larvae hatch. Often known as maggots, the larvae usually have pale, soft bodies and lack true legs. After several molts, they pupate into the adult form.

Their feeding habits, abundance, and wide distribution have made flies responsible for the spread of deadly diseases such as malaria and sleeping sickness. They also play key ecological roles—as pollinators, as decomposers of organic matter, and as links in the food chain.

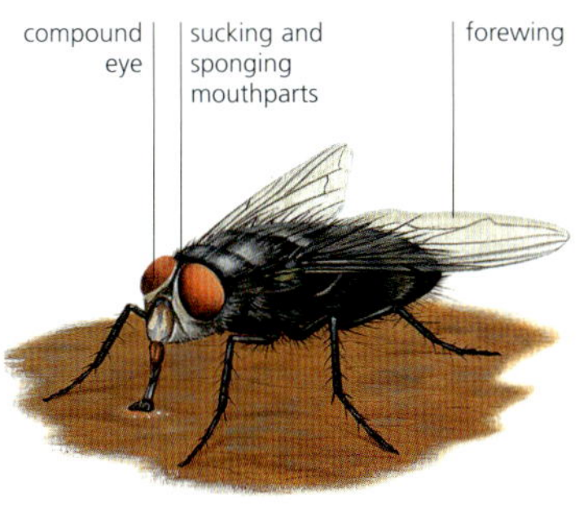

Eyes and wings Mostly active by day, flies rely on their large compound eyes, each with up to 4,000 lenses. The first segment of the thorax is enlarged to hold the huge muscles that propel the forewings.

Bumble fly (hover fly)
Volucella bombylans, family Syrphidae

Resembles bumblebee but has larger eyes and smaller antennae

Hanging fly
Empis tesselata, family Empididae

Male offers prey to female to entice her to mate

Horsefly
Tabanus bovinus, family Tabanidae

American fruit fly
Rhagoletis pomonella, family Tephritidae

Robber fly
Laphria flava, family Asilidae

Robber fly stabs insect prey with its proboscis, injects paralyzing saliva, then sucks up body fluids

Greater bee fly
Bombylius major, family Bombyliidae

Parasite of ground-nesting bees

Soldier fly resembles wasp in appearance and behavior but does not sting

Soldier fly
Stratiomys chamaeleon, family Stratiomyidae

Drone fly
Eristalis tenax, family Syrphidae

Hover fly
Syrphus ribesii, family Syrphidae

Eyes on the end of long stalks may enhance stereoscopic vision or may simply advertise a male's genetic health to prospective mates

Gout fly larvae live inside the stems of cereal crops, causing a gouted appearance

Stalk-eyed fly
Diopsis tenuipes, family Diopsidae

Cheese skipper
Piophila casei, family Piophilidae

Pest in cheeses and cured meats

Gout fly
Chlorops pumilionis, family Chloropidae

FLIGHTLESS FLIES

In several families of flies, some species have lost their wings and do not fly. The family Tipulidae (crane flies) includes the flightless snow fly (*Chionea* sp.). Able to tolerate temperatures of 20°F (–7°C), it avoids the many predators that appear in warmer months.

Pollinating pest *The narcissus fly (*Merodon equestris*), considered a pest because its larvae eat plant bulbs, but, like other hover flies, the adults are important pollinators.*

Family Syrphidae The hover flies of this family will hover in one spot, then dart forward or sideways before hovering again. The larvae of many hover-fly species eat aphids.

Species 6,000

Worldwide; on flowers

Wingless parasite *The bee louse (*Braula caeca*), a wingless fly, lays its eggs in beehives so the larvae can feed on the honey.*

Family Tabanidae Male stout flies feed on nectar and sap, while the females are bloodsuckers because they need protein to develop their eggs. Females use their blade-like mouthparts to slice the skin of mammals so they can suck the blood.

Species 4,000

Worldwide; near mammals

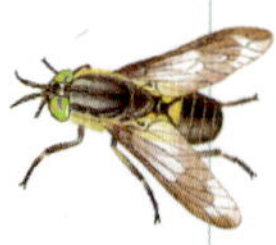

Common name *Species in the genus Tabanus are usually called horseflies.*

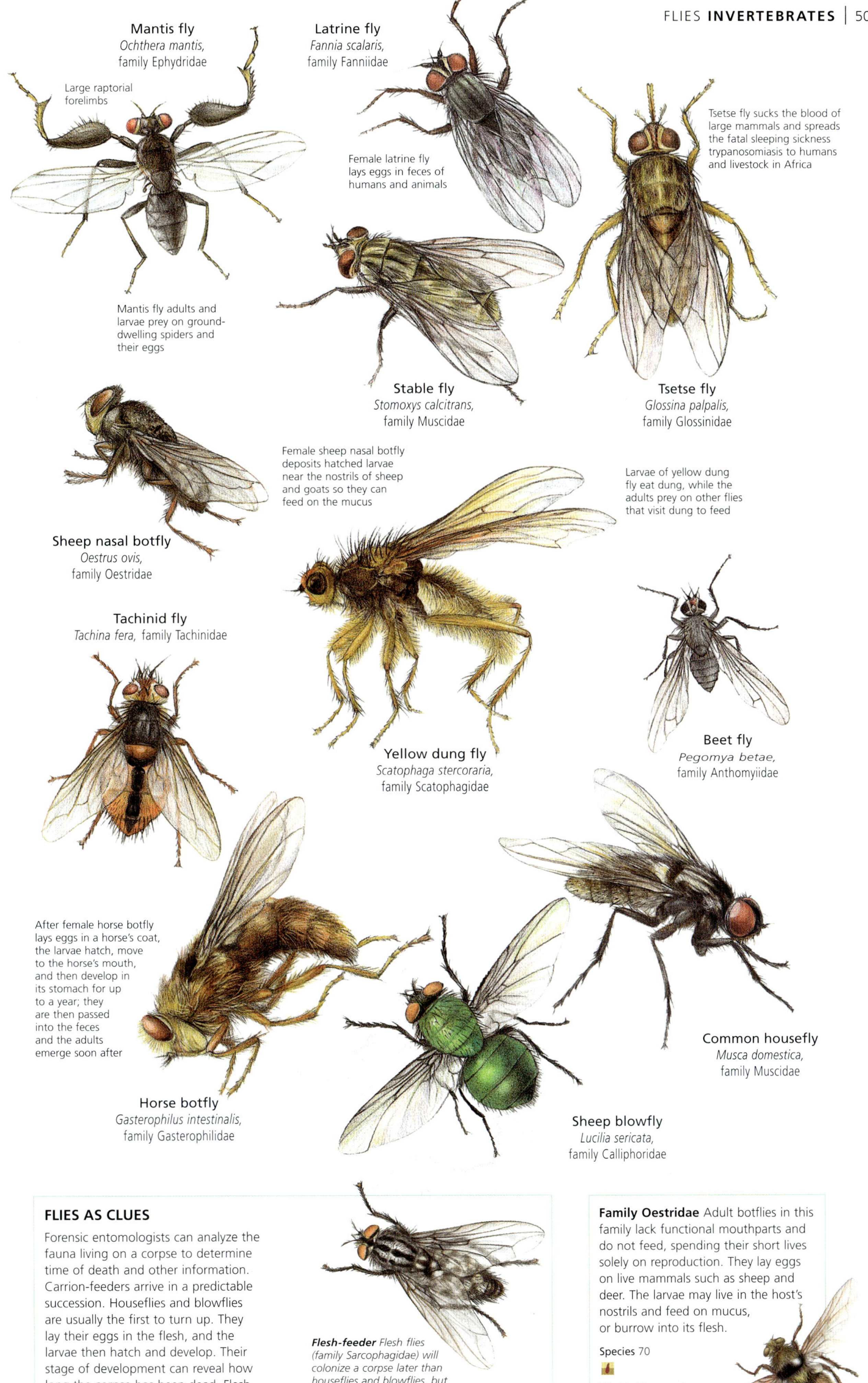

Mantis fly
Ochthera mantis, family Ephydridae

Latrine fly
Fannia scalaris, family Fanniidae

Stable fly
Stomoxys calcitrans, family Muscidae

Tsetse fly
Glossina palpalis, family Glossinidae

Sheep nasal botfly
Oestrus ovis, family Oestridae

Tachinid fly
Tachina fera, family Tachinidae

Yellow dung fly
Scatophaga stercoraria, family Scatophagidae

Beet fly
Pegomya betae, family Anthomyiidae

Common housefly
Musca domestica, family Muscidae

Horse botfly
Gasterophilus intestinalis, family Gasterophilidae

Sheep blowfly
Lucilia sericata, family Calliphoridae

FLIES AS CLUES

Forensic entomologists can analyze the fauna living on a corpse to determine time of death and other information. Carrion-feeders arrive in a predictable succession. Houseflies and blowflies are usually the first to turn up. They lay their eggs in the flesh, and the larvae then hatch and develop. Their stage of development can reveal how long the corpse has been dead. Flesh flies arrive later.

Flesh-feeder *Flesh flies (family Sarcophagidae) will colonize a corpse later than houseflies and blowflies, but catch up by depositing live larvae rather than eggs.*

Family Oestridae Adult botflies in this family lack functional mouthparts and do not feed, spending their short lives solely on reproduction. They lay eggs on live mammals such as sheep and deer. The larvae may live in the host's nostrils and feed on mucus, or burrow into its flesh.

Species 70

Worldwide; near sheep, goats & deer

BUTTERFLIES AND MOTHS

PHYLUM	Arthropoda
SUBPHYLUM	Hexapoda
CLASS	Insecta
ORDER	Lepidoptera
FAMILIES	131
SPECIES	165,000

With their delicate fluttering flight and often intricate wing patterns, butterflies and moths are the most studied and admired of all insects. They belong to the order Lepidoptera, which means "scale wing" in Greek. Their four broad wings are covered by tiny, overlapping scales—hollow, flattened hairs that create the bright colors and iridescence of diurnal species. Almost all butterflies and moths are plant-feeders. Their larvae, known as caterpillars, have biting mouthparts for chewing plants, which can make them pests. Most adults have a long proboscis for sucking flower nectar and are important pollinators, though a few lack mouthparts and do not feed at all.

Nectar-feeder To drink nectar from flowers, butterflies and moths extend a long proboscis, which sits coiled like a watchspring when not in use. Many flowering plants evolved with colors and shapes to attract these pollinating insects.

DAY AND NIGHT FLYERS

Absent only from the polar ice caps and the oceans, butterflies and moths are most abundant in the tropics but are found virtually anywhere that land plants grow. Most are highly specialized for feeding on particular flowering species.

As adults, butterflies and moths have slim bodies, broad wings, long antennae, and two large compound eyes. Their wingspans range from less than ¼ inch to 12 inches (6 mm to 30 cm). More than 85 percent of species in Lepidoptera are moths, most of which fly at night and have dull wings that are coupled by a spine-like structure known as a frenulum. Butterflies fly by day, display vibrant colors, have clubbed antennae, and lack the frenulum. Some moth species, however, are also diurnal and can be very brightly colored.

Caterpillars are covered in hairs, and walk using three pairs of true legs and several false legs. When ready to pupate, they usually enclose themselves in a cocoon or chrysalis, using silk from modified salivary glands. The entire life-cycle takes from a few weeks to a few years.

Larvae of European pine shoot moth feed on pine shoots, causing considerable damage in pine forests

European pine shoot moth
Rhyacionia buoliana, family Tortricidae

Bird-cherry ermine
Yponomeuta evonymella, family Yponomeutidae

Larvae of bird-cherry ermine feed on leaves of bird-cherry trees

Meal moth
Pyralis farinalis, family Pyralidae

Carpet moth (tapestry moth)
Trichophaga tapetzella, family Tineidae

Larvae of carpet moth eat clothes and other textiles

Moths usually rest with their wings apart or held like a roof over the abdomen, while most butterflies rest with their wings together in a vertical position

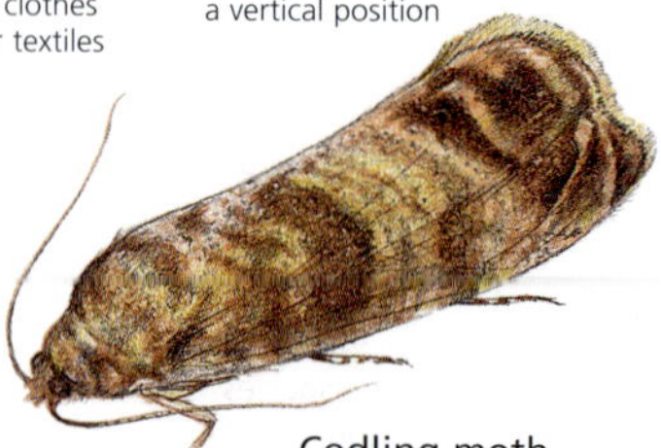

Longhorn moth
Adela reamurella, family Adelidae

Codling moth
Cydia pomonella, family Tortricidae

Spirit moth
Canephora hirsuta, family Psychidae

Family Hepialidae The moths in this primitive family are known as ghost moths or swift moths. The adults lack functional mouthparts and do not feed, spending all their time on reproduction. As they fly, females release enormous quantities of eggs; some species deposit up to 30,000 eggs.

Species 500

Worldwide except C.W. Africa & Madagascar

Family Sesiidae The clear-winged moths in this family lose most of their scales so that their wings are largely transparent. Along with yellow and brown stripes on the abdomen, the clear wings help many species mimic stinging insects such as bees, wasps, and hornets.

Species 1,000

Worldwide; near flowers

Family Hesperiidae The members of this family are known as skippers or darters because they rapidly flit from flower to flower. They are considered to be an intermediate form between butterflies and moths. They lack the wing-coupling structure found in most moth species, often resting with wings held vertically.

Species 3,000

Worldwide except New Zealand

Family Papilionidae The swallowtails of this family are named after the tail-like extensions on the hindwings of most species. All swallowtail caterpillars have an osmetrium, an organ on their head that can release a foul odor.

Species 600

Worldwide; near flowers

Vivid warning *Bright wing markings warn predators that swallowtails have an unpleasant taste.*

Family Noctuidae Noctuid or owlet moths make up the largest family in the order Lepidoptera. These night flyers have hearing organs on their thorax that can pick up echolocation signals emitted by bats, their main predators.

Species 35,000

Worldwide; on plants

Camouflage *The mottled wing coloration of many noctuid moths helps them blend in with their woodland habitat.*

Family Geometridae Called geometer, looper, or inchworm moths, members of this family tend to be small with slender bodies. The larvae, which often resemble twigs when resting, lack at least one middle pair of false legs.

Species 20,000

Worldwide; on foliage

Leaf mimic *The wings of the Southeast Asian geometer moth (*Sarcinodes restitutaria*) resemble leaves.*

COCOONS AND CHRYSALISES

When they are ready to transform into adults, caterpillars stop eating and find a safe spot. Some pupate underground without any further protection, but most moths spin a cocoon of silk, while almost all butterflies create a chrysalis, a hard shell formed from the caterpillar's skin and secured by silk to a branch.

Sticky cocoon *The larvae of bagworm moths (*Thyridopteryx ephemeraeformis*) live in cases of silk and plant debris, which harden into cocoons when they are ready to pupate.*

Transforming chrysalis *The pale green chrysalis of the monarch butterfly becomes transparent when the pupa is ready to emerge.*

Family Nymphalidae Known as brush-footed butterflies, the species in this family have very small and hairy forelegs, and walk using only the middle and back legs. The underside of the wings tends to be drab, while the upperside usually features strongly contrasting colors.

Species 5,200

Worldwide; near flowers

BEES, WASPS, ANTS, & SAWFLIES

PHYLUM	Arthropoda
SUBPHYLUM	Hexapoda
CLASS	Insecta
ORDER	Hymenoptera
FAMILIES	91
SPECIES	198,000

The order Hymenoptera is named after the Greek words for "membrane wing," and most of its species have two pairs of transparent wings. Although many bees, wasps, and sawflies are solitary creatures, some species of bees and all species of ants live in highly structured societies, whose thousands or millions of members belong to different castes and perform specific tasks. Many of the world's most beneficial insects are hymenopterids. Bees and some wasps are the chief pollinators of both crops and wild plants, while many parasitic wasps play a crucial role in controlling the populations of other insects. Honeybees have been domesticated for their honey and wax.

Pine sawfly
Diprion pini,
family Diprionidae

Lays eggs inside pine needles

Larva of *Arge ochropus*

Rose sawfly
Arge ochropus,
family Argidae

ANATOMY AND LIFE-CYCLE

Generally small to medium-sized, hymenopterids include the parasitic *Megaphragma caribea*, which, at just $^{3}/_{500}$ inch (0.17 mm) long, is one of the smallest insects. Hymenopterids' hindwings are attached to the larger forewings by tiny hooks so that they beat together. The more primitive sawflies and their kin lack the slim waist that separates the thorax and abdomen of bees, wasps, and ants.

Hymenopterids have mouthparts designed for biting or for biting and sucking. Sawflies, gall wasps, and some ants and bees are herbivorous, while most other species in the order are predatory or parasitic. In many species, the female's ovipositor (egg-laying organ) has become modified—sawflies use it for sawing into plant stems, where they lay their eggs; while bees, wasps, and ants often use it for piercing or stinging predators or prey.

Female hymenopterids may lay their eggs in soil, in plants, in nests or hives, or in living insect hosts. In many species, the female determines whether the eggs are fertilized, with unfertilized eggs producing male larvae, and fertilized eggs producing females. In solitary species, the larvae hatch near a food source and develop independently, but in social species they rely on constant care from the adults. Development is by complete metamorphosis, and larvae usually pupate in a cocoon.

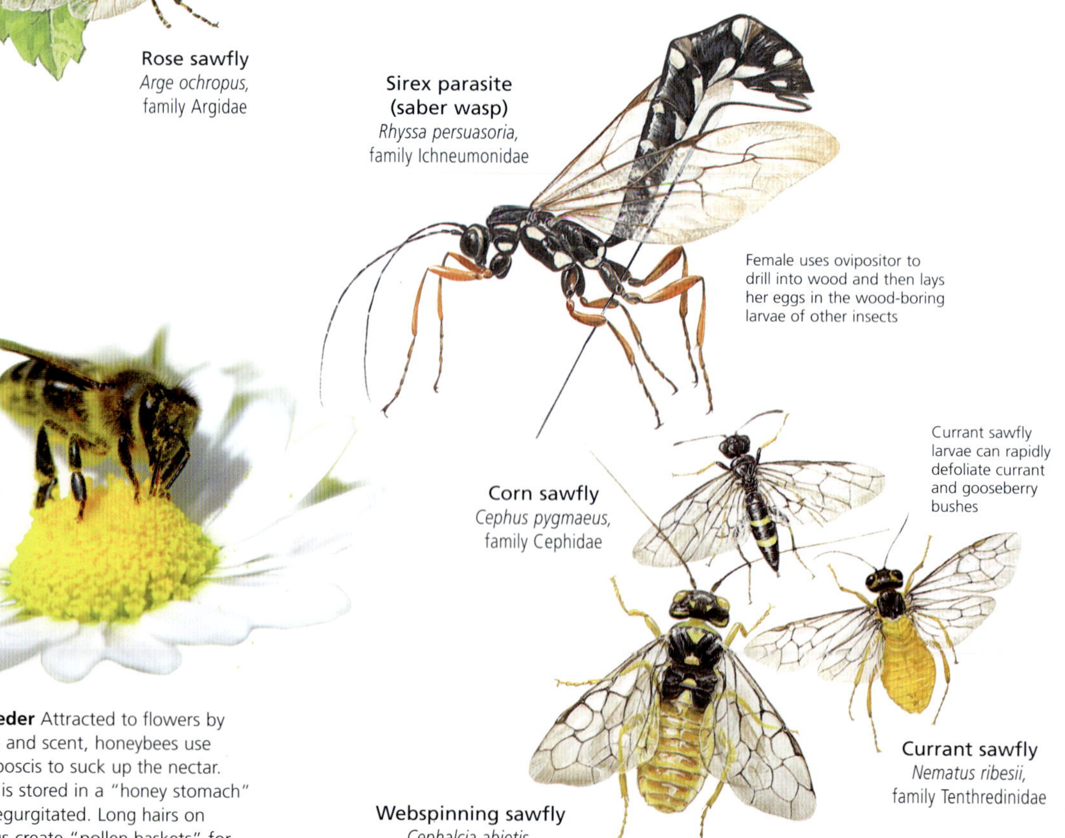

Sirex parasite (saber wasp)
Rhyssa persuasoria,
family Ichneumonidae

Female uses ovipositor to drill into wood and then lays her eggs in the wood-boring larvae of other insects

Flower feeder Attracted to flowers by their colors and scent, honeybees use a long proboscis to suck up the nectar. The nectar is stored in a "honey stomach" and later regurgitated. Long hairs on the hindlegs create "pollen baskets" for carrying the pollen back to the hive.

Corn sawfly
Cephus pygmaeus,
family Cephidae

Currant sawfly larvae can rapidly defoliate currant and gooseberry bushes

Currant sawfly
Nematus ribesii,
family Tenthredinidae

Webspinning sawfly
Cephalcia abietis,
family Pamphiliidae

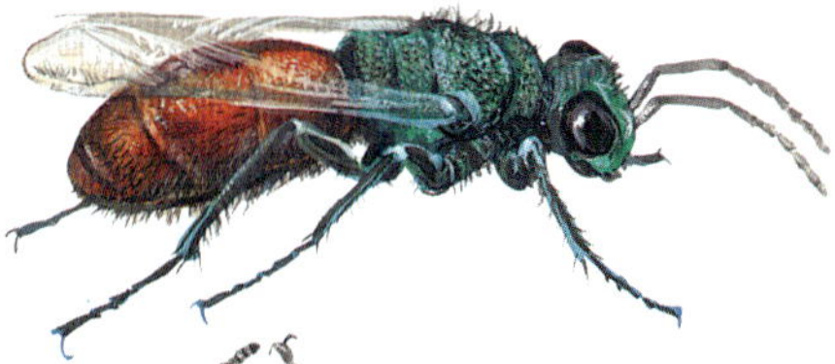

Ruby-tailed wasp (common cuckoo wasp)
Chrysis ignita,
family Chrysididae

American pelecinid wasp
Pelecinus polyturator,
family Pelecinidae

Velvet wasp
Megascolia maculata,
family Scoliidae

Oak apple gall wasp
Biorrhiza pallida,
family Cynipidae

Alternates an all-female generation with a generation of males and females

Clover seed chalcid
Bruchophagus gibbus,
family Eurytomidae

Woolly aphid parasite
Aphelinus mali,
family Aphelinidae

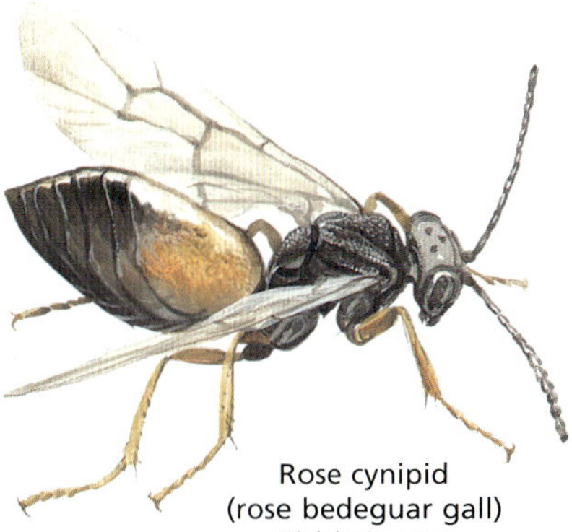

Rose cynipid (rose bedeguar gall)
Diplolepis rosae,
family Cynipidae

Torymid wasp
Torymus bedeguaris,
family Torymidae

Yellow ophion (red wasp)
Ophion luteus,
family Ichneumonidae

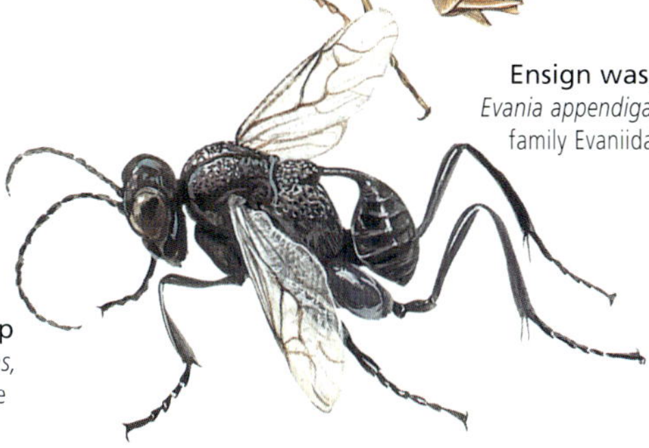

Torymid wasp parasitizes gall-forming insects

Ensign wasp
Evania appendigaster,
family Evaniidae

Eucharitid wasp
Eucharis adscendens,
family Eucharitidae

Family Chrysididae Known as cuckoo wasps, ruby-tailed wasps, or jewel wasps, the species in this family lay their eggs in the nests of bees, wasps, or other insects. The cuckoo wasp larvae then feed on the nest's larvae or the food that has been provided for them to eat.

Species 3,000

Worldwide; near flowers & insect hosts

Family Scoliidae Adult scoliid wasps feed on flower nectar and pollen, but the larvae are parasites. The adult female will burrow into soil to find the grubs of scarab beetles. She then paralyzes each grub with a sting and lays an egg into it. When the wasp larva hatches, it will feed on the paralyzed grub.

Species 8,000

Worldwide; near scarab beetles

Family Cynipidae This family contains the gall wasps. Most are inconspicuous insects less than ¼ inch (5 mm) long. Females lay their eggs on trees, where the larvae hatch and secrete saliva that forces the plant to form a protective gall around them. The larvae feed on the gall until they emerge as adults.

Species 1,400

Mainly Northern Hemisphere; on trees

HIVE OF ACTIVITY

The social life of the honeybee (genus *Apis*) is among the most complex in the animal world. A queen bee can lay up to 1,500 eggs a day. Fertilized eggs develop into workers or queens, and unfertilized eggs become drones. The workers feed the hatched larvae before they are capped inside cells to pupate. When a hive reaches its optimum size, the old queen flies off to found a new colony, followed by thousands of workers. The first new queen to emerge will rule the old nest.

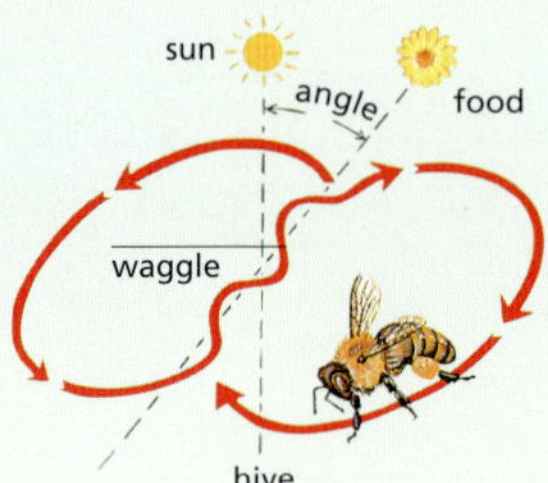

Dance of the bees A worker bee uses a figure-eight dance to convey information about a new food source. The rate of tail-waggling indicates the distance to the food source, while the direction is signaled by the angle of the dance to the sun.

Inside a beehive All the drones and workers in a hive (below) are the offspring of the queen, whose only task is to lay eggs. Drones are male bees, and their only job is to mate with new queens. Worker bees gather food, maintain the hive, and feed the queen and her young. Glands in their head produce royal jelly for the larvae. Workers also make honey by regurgitating nectar and spreading it in cells to dehydrate, then capping it with wax.

Bee young Worker bees bring food to the larvae as they mature inside water-resistant waxen cells. These larvae (above) are transforming into pupae, and will emerge as adults.

Being queen *A queen bee is the largest bee in the hive. She can live for up to 5 years.*

Hard workers *Workers are females that cannot breed. They live for only a few weeks.*

Larva cells *Most larvae become workers. They are first fed royal jelly, and later pollen and honey. Queen larvae are fed only royal jelly. Drones develop from unfertilized eggs.*

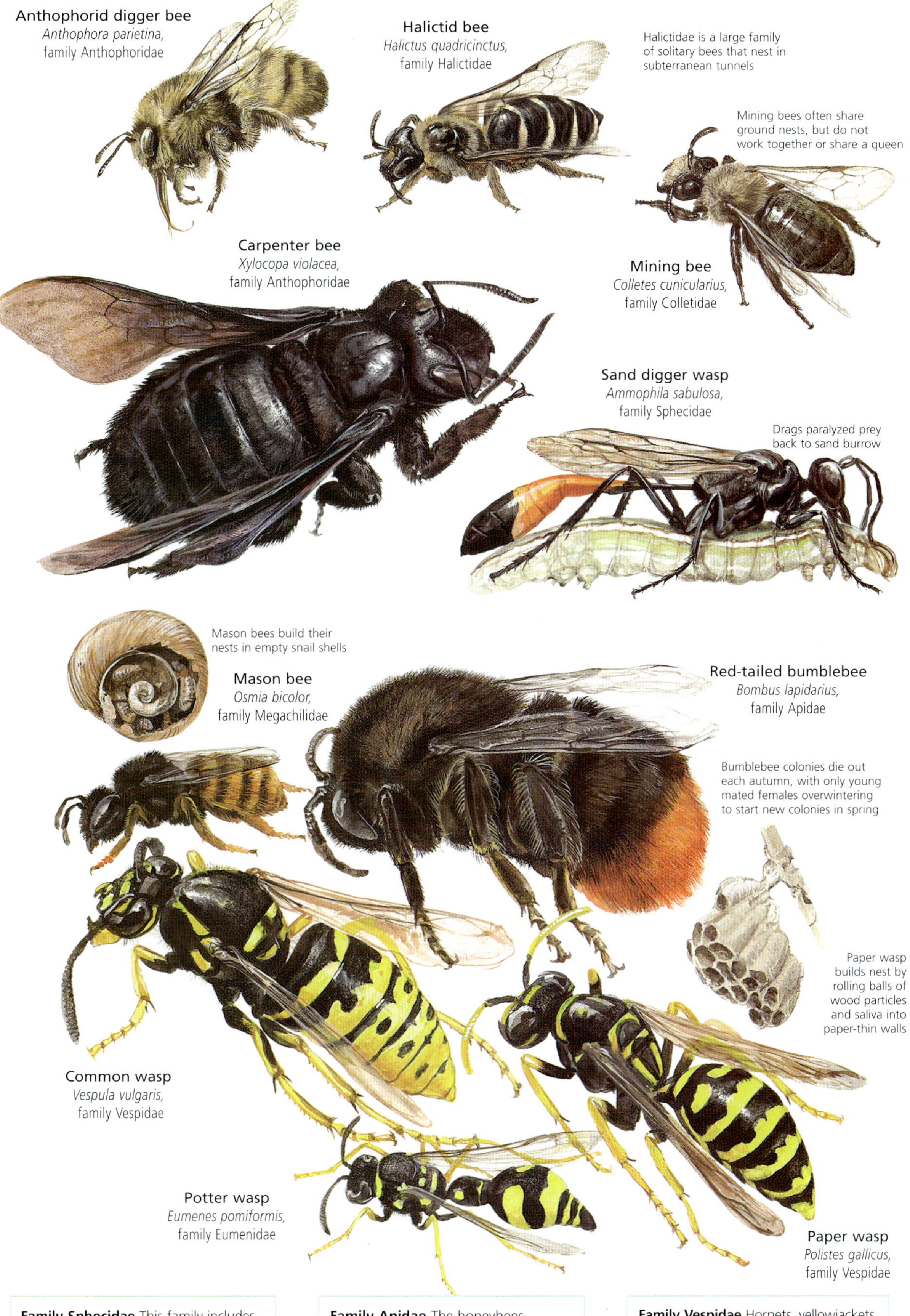

Family Sphecidae This family includes species known as digger wasps, sand wasps, mud-dauber wasps, and thread-waisted wasps. A female sphecid wasp paralyzes arthropod prey with her sting and deposits it in her nest.

Species 8,000

Worldwide; often in sandy soil

Solitary hunter *The female bee wolf will carry paralyzed honeybees back to its sandy burrow.*

Family Apidae The honeybees, bumblebees, and stingless bees in this family live in communal nests centered on a queen. Workers forage for nectar and pollen, which they bring back to the nest to feed the queen and her larvae.

Species 1,000

Worldwide; near flowers

Fierce bee *A colony of Southeast Asia's giant honeybees (*Apis dorsata*) is capable of stinging a person to death.*

Family Vespidae Hornets, yellowjackets, and paper wasps all use chewed wood and saliva to build their papery nests. In most species, the colony dies out in late autumn, leaving only a few mated queens to overwinter and establish new colonies in spring.

Species 4,000

Worldwide

Self-sacrifice *To defend its colony, a worker* Brachygastra lecheguana *will sting an attacker and can die when the sting detaches.*

Honey ant
Myrmecocystus hortideorum, family Formicidae

Honey ants called repletes store food in their swollen abdomen and hang from the ceiling of the colony's nest

Myrmica laevinodis, family Formicidae

Myrmica laevinodis feeds on honeydew produced by aphids in return for protecting them from parasites

Red wood ant
Formica rufa, family Formicidae

Formica rufa worker

Formica rufa queen

Formica rufa male

Leafcutter workers carry leaf fragments up to three times their own weight

Pharaoh ant
Monomorium pharaonis, family Formicidae

Carpenter ant
Camponotus herculeanus, family Formicidae

Leafcutter ant
Atta sexdens, family Formicidae

Slave-making ant
Polyergus rufescens, family Formicidae

Giant hunting ant
Dinoponera grandis, family Formicidae

Slave-making ant raids nests of closely related species to steal pupae, which then emerge in its own nest and become its workers

Bull ant
Myrmecia forficata, family Formicidae

Weaver ant
Oecophylla smaragdina, family Formicidae

ANT SOCIETY

While ant societies vary in complexity, they all contain three basic castes—queen, worker, and male. A new colony forms when winged females and males disperse from a nest and mate. The males die, while the females lose their wings and start to lay eggs as queens.

Male *Produced from unfertilized eggs, males are short-lived, their only function being to inseminate a virgin queen.*

Soldier *In some species, large workers with enormous heads act as soldiers. Their role is to defend the colony from invaders.*

Queen *After establishing the colony, the queen's sole task is to lay eggs. She has a large thorax and abdomen.*

Worker *The vast majority of a colony is made up of female workers, who gather food, tend the queen and her eggs and larvae, and maintain the nest.*

OTHER INSECTS

PHYLUM	Arthropoda
SUBPHYLUM	Hexapoda
CLASSES	Insecta
ORDERS	19*
FAMILIES	218*
SPECIES	> 3,000*

The remaining 19 orders of Insecta are a diverse assortment. They include primitive, wingless forms such as silverfish and bristletails; mammal parasites such as fleas and lice; insect parasites such as strepsipterans (order Strepsiptera); crop pests such as thrips; and insects that spend most of their lives in water, such as mayflies, caddisflies, dobsonflies, and stoneflies. Stick insects and leaf insects are plant-feeders; angel insects (order Zorotypidae), snakeflies, antlions, lacewings, and hanging flies prey on other invertebrates; rock crawlers (order Grylloblattodea), webspinners, earwigs, and scorpionflies recycle detritus; and some booklice species will feed on paper or stored cereals.

Snakefly
Raphidia notata, order Raphidioptera

Spiny devil walking stick
Eurycantha horrida, order Phasmatodea

Leaf insect
Phyllium siccifolium, order Phasmatodea

Silverfish
Lepisma saccharina, order Thysanura

Bristletail
Machilis helleri, order Archaeognatha

Order Thysanura This order contains the silverfish, flightless insects that may closely resemble the earliest insects. They have long antennae on their head, and three "tails" extending from the abdomen. Silverfish hatch from eggs laid in crevices near food sources.

Species 370

Worldwide; in trees, caves & buildings

House guest *This silverfish* (Lepismodes iniquilinus) *is often found in houses, where it feeds on paper, glue in book bindings, starched clothing, and dry foods.*

Order Phasmatodea The supreme mimics in this order are known as stick insects and leaf insects. They feed on foliage. Most species are wingless, and, when wings are present, they usually are fully developed and functional only in the males. Many females can produce offspring from unfertilized eggs.

Species 3,000

Worldwide, esp. in the tropics; near plants

Stick mimic *When disturbed, stick insects such as* Bacillus rossii *may freeze for hours or they may sway as if blowing in the wind.*

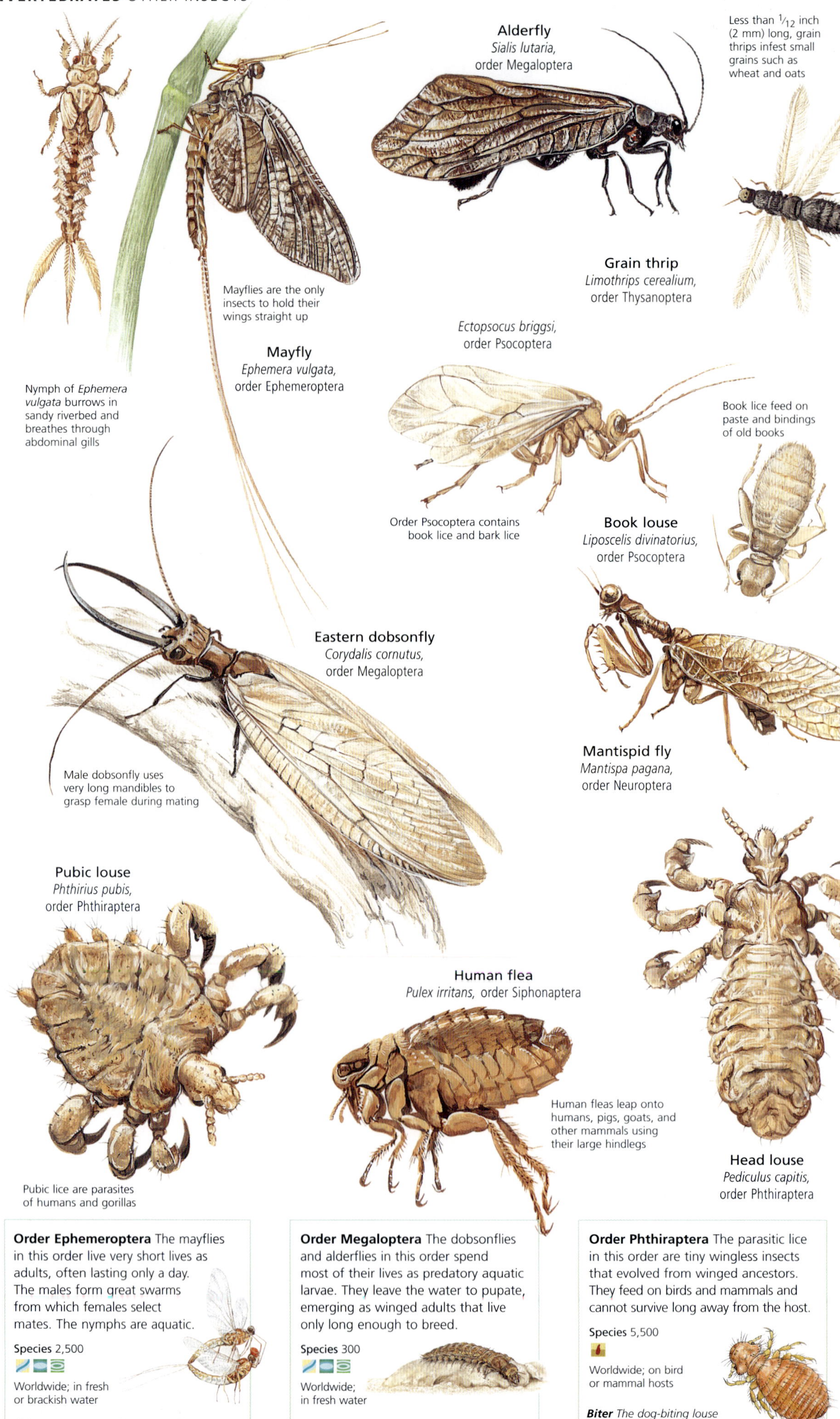

Order Ephemeroptera The mayflies in this order live very short lives as adults, often lasting only a day. The males form great swarms from which females select mates. The nymphs are aquatic.

Species 2,500

Worldwide; in fresh or brackish water

Flying romance *Mayflies mate in flight, with the male using his forelegs to grasp the female.*

Order Megaloptera The dobsonflies and alderflies in this order spend most of their lives as predatory aquatic larvae. They leave the water to pupate, emerging as winged adults that live only long enough to breed.

Species 300

Worldwide; in fresh water

Toebiter *Larva of* Archichauliodes *breathes through the leg-like gills along its abdomen.*

Order Phthiraptera The parasitic lice in this order are tiny wingless insects that evolved from winged ancestors. They feed on birds and mammals and cannot survive long away from the host.

Species 5,500

Worldwide; on bird or mammal hosts

Biter *The dog-biting louse uses its chewing mouthparts to feed domestic dogs and other canids.*

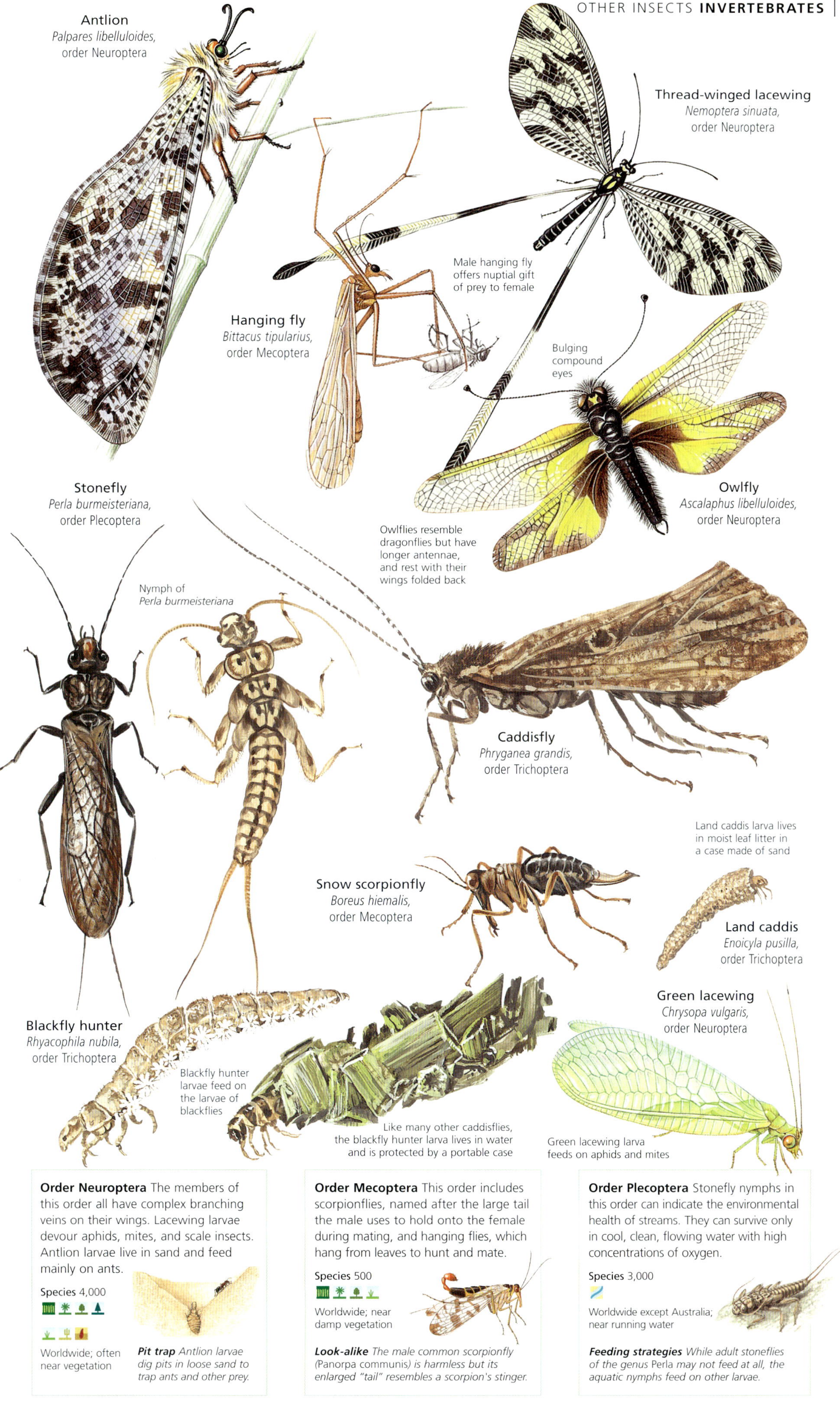

Order Neuroptera The members of this order all have complex branching veins on their wings. Lacewing larvae devour aphids, mites, and scale insects. Antlion larvae live in sand and feed mainly on ants.

Species 4,000

Worldwide; often near vegetation

Pit trap *Antlion larvae dig pits in loose sand to trap ants and other prey.*

Order Mecoptera This order includes scorpionflies, named after the large tail the male uses to hold onto the female during mating, and hanging flies, which hang from leaves to hunt and mate.

Species 500

Worldwide; near damp vegetation

Look-alike *The male common scorpionfly (*Panorpa communis*) is harmless but its enlarged "tail" resembles a scorpion's stinger.*

Order Plecoptera Stonefly nymphs in this order can indicate the environmental health of streams. They can survive only in cool, clean, flowing water with high concentrations of oxygen.

Species 3,000

Worldwide except Australia; near running water

Feeding strategies *While adult stoneflies of the genus* Perla *may not feed at all, the aquatic nymphs feed on other larvae.*

INSECT ALLIES

PHYLUM	Arthropoda
SUBPHYLUM	Hexapoda
CLASSES	3
ORDERS	5
FAMILIES	31
SPECIES	8,300

Hexapods are arthropods with six legs and a body divided into three parts—head, thorax, and abdomen. The vast majority of hexapods are insects, but there are also three classes of non-insect hexapods—the springtails (class Collembola), proturans (Protura), and diplurans (Diplura). Found in soil and leaf litter, these tiny creatures range from $\frac{1}{50}$ inch to 1¼ inches (0.5 mm to 3 cm) in length. They are distinguished from insects by their mouthparts, which are fully tucked into the head. As in the most primitive of insects, the silverfish and the bristletails, these tiny soil-dwellers evolved from a wingless ancestor and they continue to molt throughout their life.

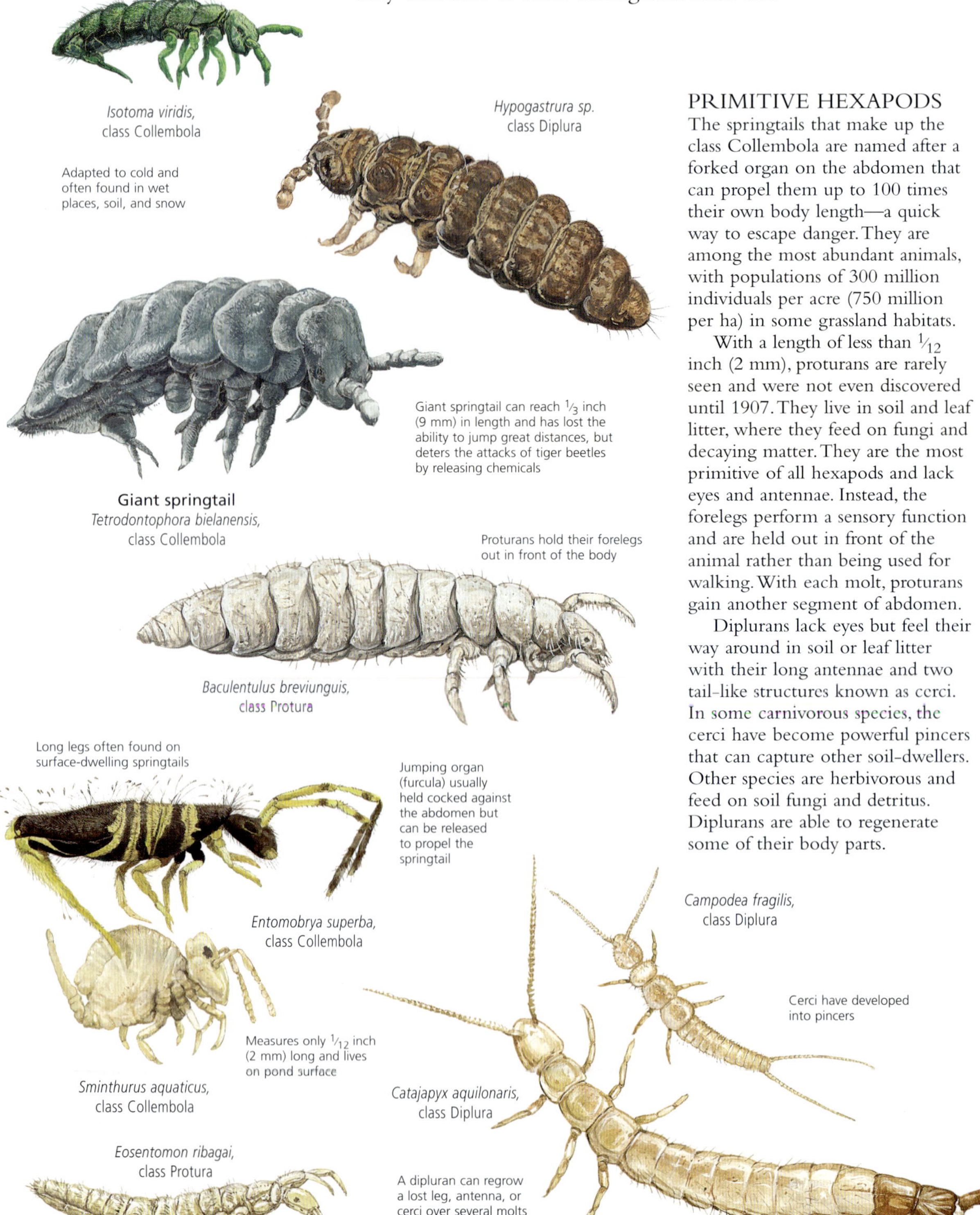

Isotoma viridis, class Collembola

Hypogastrura sp. class Diplura

Giant springtail *Tetrodontophora bielanensis*, class Collembola

Baculentulus breviunguis, class Protura

Entomobrya superba, class Collembola

Sminthurus aquaticus, class Collembola

Eosentomon ribagai, class Protura

Catajapyx aquilonaris, class Diplura

Campodea fragilis, class Diplura

PRIMITIVE HEXAPODS

The springtails that make up the class Collembola are named after a forked organ on the abdomen that can propel them up to 100 times their own body length—a quick way to escape danger. They are among the most abundant animals, with populations of 300 million individuals per acre (750 million per ha) in some grassland habitats.

With a length of less than $\frac{1}{12}$ inch (2 mm), proturans are rarely seen and were not even discovered until 1907. They live in soil and leaf litter, where they feed on fungi and decaying matter. They are the most primitive of all hexapods and lack eyes and antennae. Instead, the forelegs perform a sensory function and are held out in front of the animal rather than being used for walking. With each molt, proturans gain another segment of abdomen.

Diplurans lack eyes but feel their way around in soil or leaf litter with their long antennae and two tail-like structures known as cerci. In some carnivorous species, the cerci have become powerful pincers that can capture other soil-dwellers. Other species are herbivorous and feed on soil fungi and detritus. Diplurans are able to regenerate some of their body parts.

ECHINODERMS

PHYLUM	Echinodermata
CLASSES	5
ORDERS	36
FAMILIES	145
SPECIES	6,000

The sea stars, sea urchins, brittle stars, feather stars, and sea cucumbers that belong to the phylum Echinodermata display diverse body forms, but share some defining characteristics. Adults usually have pentamerous symmetry, with the body arranged into a five-part radial pattern around a central axis. Most internal organs are also arranged in this pattern. An internal skeleton of calcareous plates provides protection and support. The plates often bear spines or small lumps—the name echinoderm means "spiny skin." The body cavity includes a water-vascular system, a network of water vessels that controls the tube feet used for locomotion, feeding, respiration, and sensory functions.

Radial pattern In sea cucumbers, such as the red-lined sea cucumber (*Thelenota rubralineata*), the five-part symmetry is internal and the skeleton is reduced.

SPINY BOTTOM-DWELLERS

All echinoderms are marine and most are mobile bottom-dwellers, although sea lilies are fixed to the seafloor by a long stalk, and a few sea cucumbers float in the open ocean. Echinoderms are found throughout the world at all depths. Sea urchins and sea stars are very commonly seen along the seashore.

The phylum Echinodermata is divided into five classes. Members of the class Crinoidea include the sessile sea lilies as well as the mobile feather stars. They are filter-feeders and, unlike other echinoderms, have their mouth facing away from the substrate. The other echinoderm classes include predators, grazers of algae, and deposit-feeders. The sea stars of class Asteroidea and brittle stars of class Ophiuroidea have arms radiating from their body. Their skeletal plates are held together by muscles, which allows flexibility. In the class Echinoidea, sea urchins and sand dollars have a rigid skeleton of fused plates that supports a globular or flattened form, usually featuring prominent spines. Sea cucumbers (class Holothuroidea) are generally soft-bodied, with their skeletons reduced to minute spicules.

Although a few echinoderms can reproduce asexually, most have separate sexes that release eggs and sperm into the water for fertilization. The larvae are often free-swimming and metamorphose into bottom-dwelling adults. In some species, the larval stage is omitted and the newly hatched young resemble small adults.

Feather star
Tropiometra afra, class Crinoidea

Sea apple
Pseudocolochirus violaceus, class Holothuroidea

Sea apples release a toxin called holothurin that is lethal to many marine animals

Synapta maculata can grow to a length of 16½ feet (5 m)

Synaptid sea cucumber
Synapta maculata, class Holothuroidea

Common starfish
Asterias rubens, class Asteroidea

Pelagic sea cucumber
Pelagothuria natatrix, class Holothuroidea

Red pencil urchin
Heterocentrotus mammillatus, class Echinoidea

Pencil urchin often associates with the striped bumblebee shrimp (*Gnathophyllum americanum*)

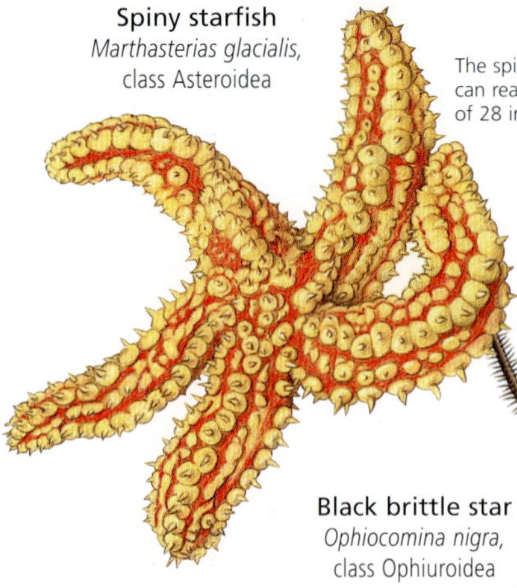

Spiny starfish
Marthasterias glacialis, class Asteroidea

The spiny starfish can reach a diameter of 28 inches (70 cm)

Sand dollar (sea biscuit)
Clypeaster humilis, class Echinoidea

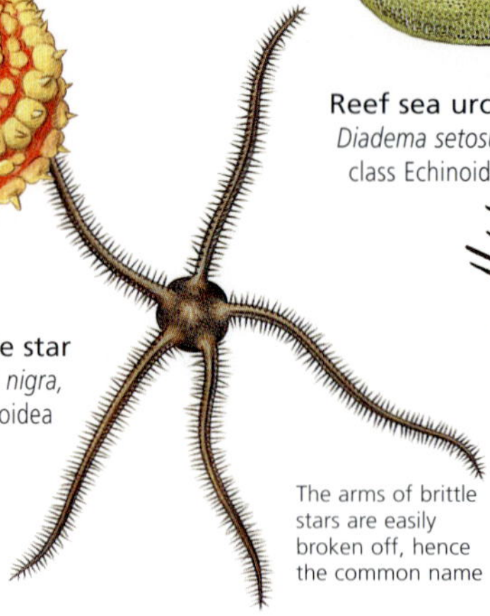

Reef sea urchin
Diadema setosum, class Echinoidea

Black brittle star
Ophiocomina nigra, class Ophiuroidea

The arms of brittle stars are easily broken off, hence the common name

Sea urchins move slowly using their tube feet and spines

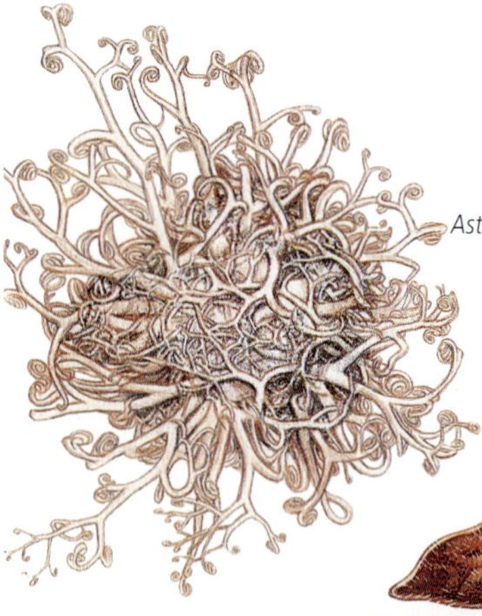

Gorgon's head
Astrospartus mediterraneus, class Crinoidea

Goose-foot starfish
Anseropoda placenta, class Asteroidea

Heart urchin (sea potato)
Echinocardium cordatum, class Echinoidea

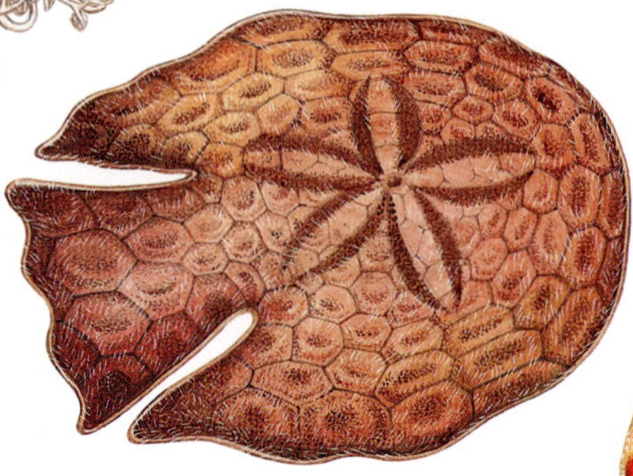

Sea pancake
Echinodiscus auritus, class Echinoidea

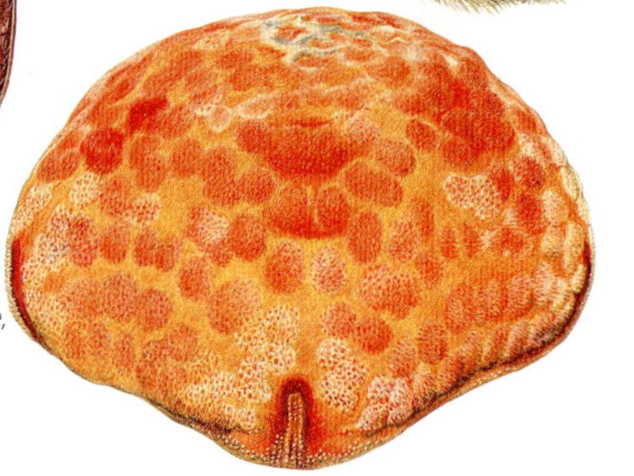

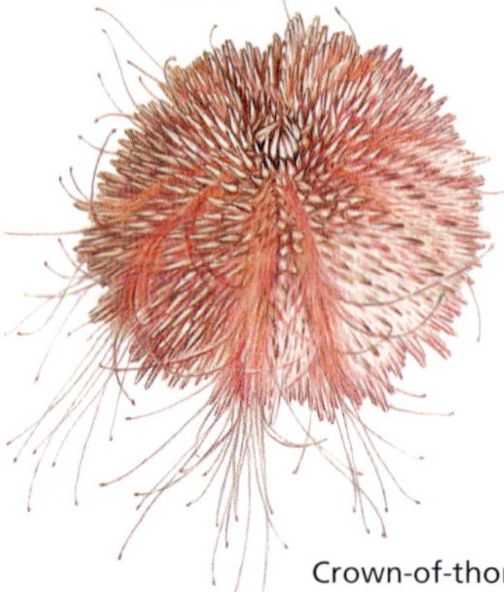

Edible sea urchin
Echinus esculentus, class Echinoidea

Cushion star
Culcita novaeguineae, class Asteroidea

Crown-of-thorns starfish
Acanthaster planci, class Asteroidea

An increase in the numbers of this coral predator has devastated large areas of the Great Barrier Reef

African red knob sea star
Protoreaster linckii, class Asteroidea

Class Asteroidea Also known as starfish, the sea stars in this class move and feed using the suckered tube feet along their arms. Most are predators that evert their stomach over sessile or slow-moving prey to begin digesting it.

Species 1,500

Worldwide; on the seafloor

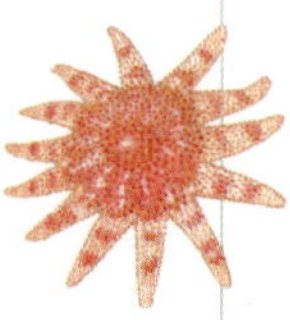

Many arms *The common sun star (Crossaster papposus) can have as many as 40 arms.*

Class Ophiuroidea In contrast to sea stars, in which the arms grade into the central disk, the brittle stars and basket stars of Ophiuroidea have sharply demarcated arms. The class includes predators, scavengers, and filter-feeders.

Species 2,000

Worldwide; on the seafloor

Surface star *The brittle star (Ophiura ophiura) usually lives on the surface of sandy or muddy sediment.*

Class Echinoidea The sea urchins in this class have a spherical body protected by a skeletal case called a test, and long, movable spines. Sand dollars are flattened and covered in smaller spines.

Species 950

Worldwide; on or in the seafloor

Burrower *A coastal species, the purple sea urchin (Paracentrotus lividus) uses its spines and teeth to burrow into soft rocks.*

Other invertebrates

MINOR INVERTEBRATE PHYLA
PHYLA 25
CLASSES > 40
ORDERS > 60
SPECIES > 12,000

The preceding pages have covered eight major invertebrate phyla, as well as two subphyla of invertebrate chordates. There are, however, another 25 phyla of invertebrates, many of which are profiled in this chapter. Although the phylum Bryozoa contains 5,000 species, most of these minor groups are small—Phoronida (horseshoe worms), for example, contains a mere 20 species. The members of these phyla also tend to be physically small and are often microscopic. Most are marine, although many are found in fresh water and some occur on land. Often overlooked, the minor invertebrate phyla all display fascinating solutions to the challenges presented by their environment.

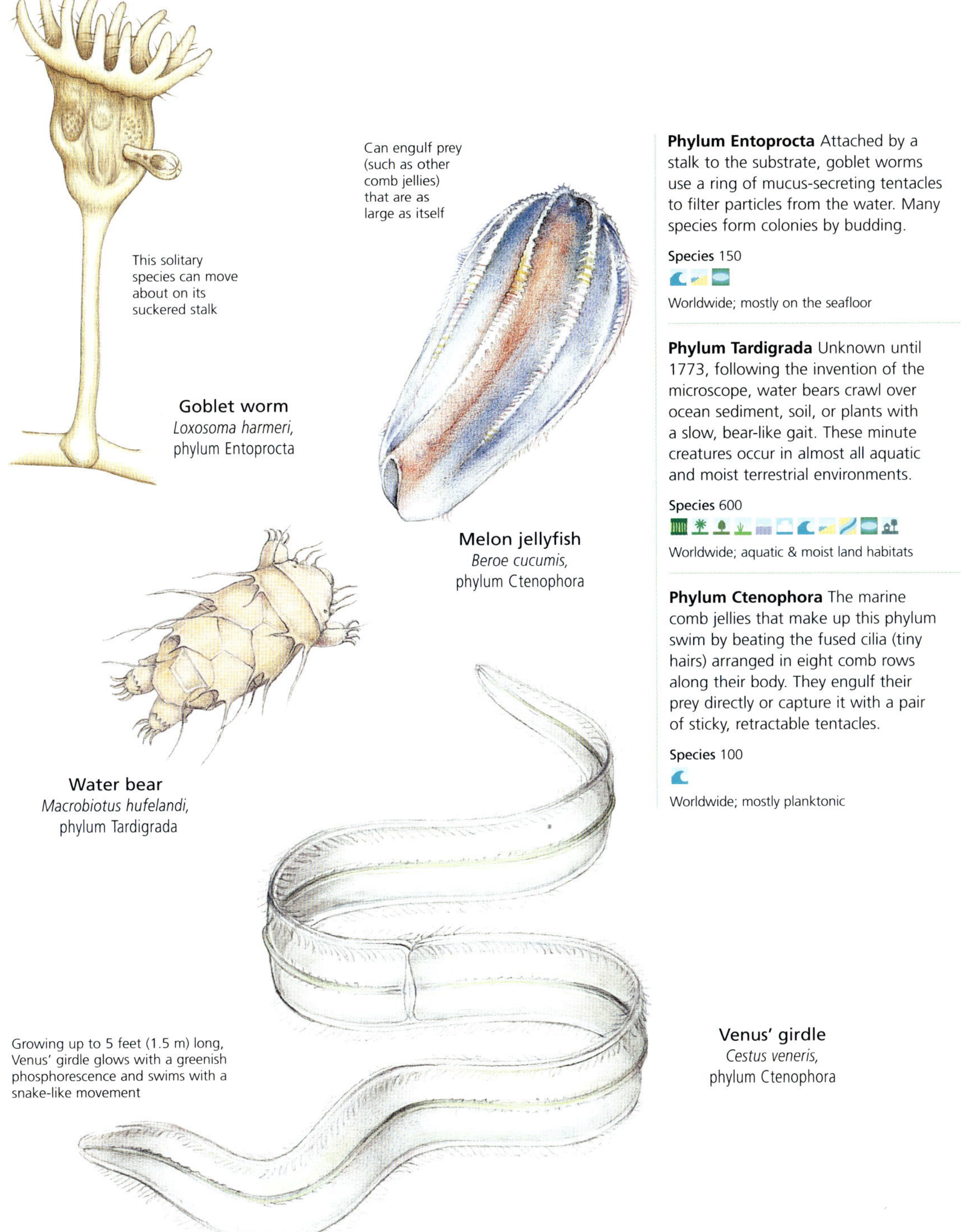

This solitary species can move about on its suckered stalk

Goblet worm
Loxosoma harmeri, phylum Entoprocta

Can engulf prey (such as other comb jellies) that are as large as itself

Melon jellyfish
Beroe cucumis, phylum Ctenophora

Water bear
Macrobiotus hufelandi, phylum Tardigrada

Growing up to 5 feet (1.5 m) long, Venus' girdle glows with a greenish phosphorescence and swims with a snake-like movement

Venus' girdle
Cestus veneris, phylum Ctenophora

Phylum Entoprocta Attached by a stalk to the substrate, goblet worms use a ring of mucus-secreting tentacles to filter particles from the water. Many species form colonies by budding.

Species 150

Worldwide; mostly on the seafloor

Phylum Tardigrada Unknown until 1773, following the invention of the microscope, water bears crawl over ocean sediment, soil, or plants with a slow, bear-like gait. These minute creatures occur in almost all aquatic and moist terrestrial environments.

Species 600

Worldwide; aquatic & moist land habitats

Phylum Ctenophora The marine comb jellies that make up this phylum swim by beating the fused cilia (tiny hairs) arranged in eight comb rows along their body. They engulf their prey directly or capture it with a pair of sticky, retractable tentacles.

Species 100

Worldwide; mostly planktonic

Phylum Rotifera Sometimes called wheel animals, rotifers are microscopic aquatic creatures. A crown of cilia around the mouth beat rapidly to collect food and propel the rotifer, creating the impression of a whirling wheel.

Species 1,800

Worldwide; in aquatic vegetation and moist terrestrial habitats

Phylum Hemichordata The members of this phylum have a three-part body made up of a proboscis at the head; a collar in the middle, which bears tentacles in some species; and a trunk containing the digestive and reproductive organs.

Species 90

Worldwide; on the seafloor

Phylum Chaetognatha The arrow worms of this phylum are voracious predators. Their head has up to 14 large grasping spines for seizing prey such as small fishes. When the spines are not in use, a hood formed from the body wall covers the head to protect them.

Species 90

Worldwide; in plankton & on the seafloor

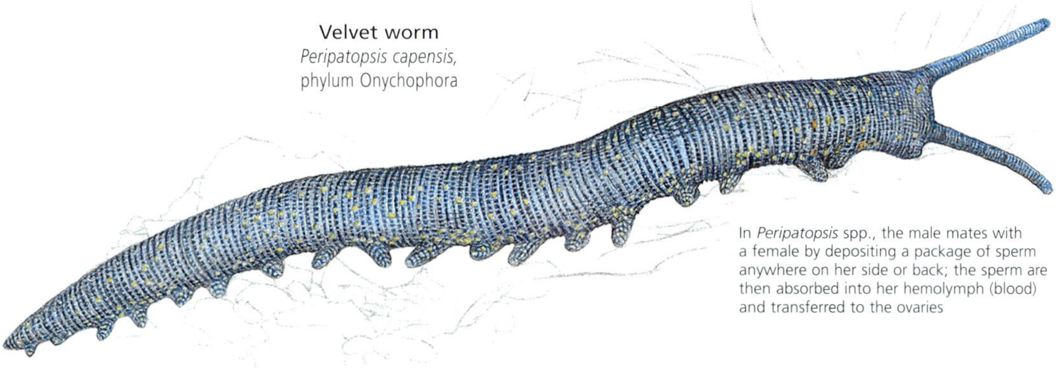

Velvet worm
Peripatopsis capensis, phylum Onychophora

In *Peripatopsis* spp., the male mates with a female by depositing a package of sperm anywhere on her side or back; the sperm are then absorbed into her hemolymph (blood) and transferred to the ovaries

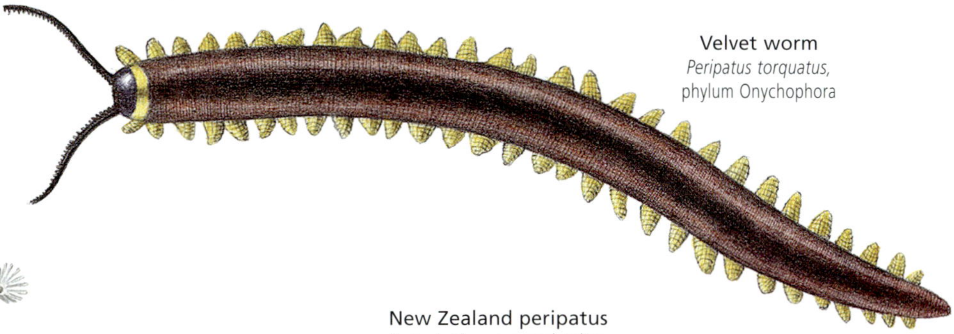

Velvet worm
Peripatus torquatus, phylum Onychophora

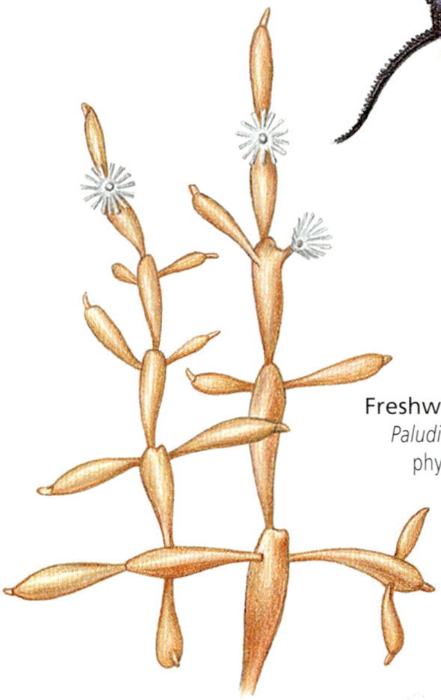

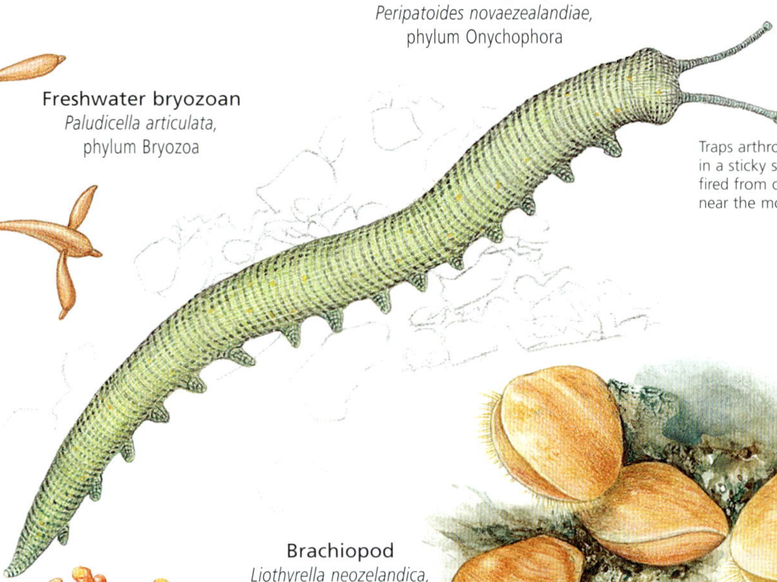

New Zealand peripatus
Peripatoides novaezealandiae, phylum Onychophora

Traps arthropod prey in a sticky substance fired from openings near the mouth

Freshwater bryozoan
Paludicella articulata, phylum Bryozoa

Brachiopod
Liothyrella neozelandica, phylum Brachiopoda

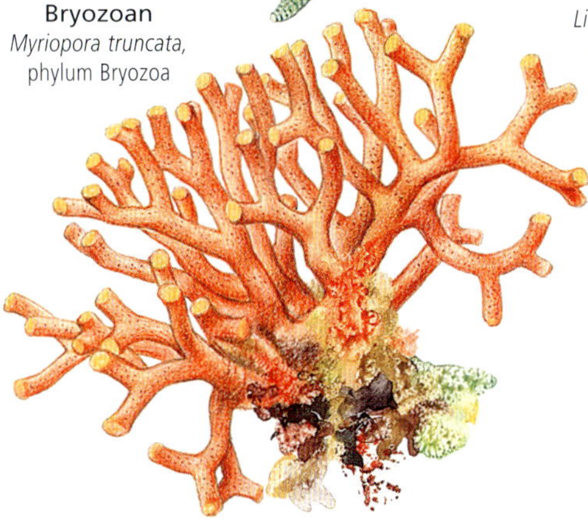

Bryozoan
Myriopora truncata, phylum Bryozoa

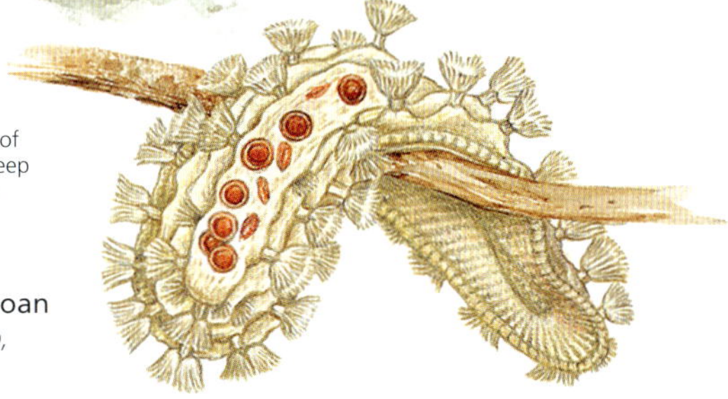

Unlike most bryozoans, which are fixed, a colony of *Cristatella mucedo* can creep on its flattened underside

Freshwater bryozoan
Cristatella mucedo, phylum Bryozoa

Phylum Onychopora Since their thin, non-waxy cuticle does little to prevent water loss, the velvet worms of this phylum are restricted to moist land habitats. To entangle prey, velvet worms squirt out sticky threads from slime glands near the mouth.

Species 70

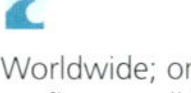

Tropical & southern temperate zones

Phylum Brachiopoda With their bivalved shells, brachiopods look like clams or mussels, but are in fact much more closely related to bryozoans. Both filter food particles from the water with a lophophore, a crown of hollow tentacles around the mouth.

Species 350

Worldwide; on the seafloor at all levels of the ocean

Phylum Bryozoa Sometimes called moss animals, bryozoans are tiny, aquatic, and colonial. Most secrete a protective covering around the body with an opening for the lophophore (feeding tentacles). Colonies are created by budding.

Species 5,000

Worldwide; on the bottom of aquatic habitats

Glossary

abdomen The part of the body containing the digestive system and the reproductive organs. In insects and spiders, the abdomen makes up the rear of the body.

adaptation A change in an animal's behavior or body that allows it to survive and breed in new conditions.

adaptive radiation A situation in which animals descended from a common ancestor evolve to exploit different ecological niches that are not being filled by other animals, as in the finches of the Galápagos Islands and the lemurs of Madagascar.

algae The simplest forms of plant life.

altricial Helpless at birth. Describes newly hatched chicks of some birds; cf precocial.

amphibious Able to live on land and in water. Amphibians (frogs, toads, salamanders, caecilians, and newts) are vertebrates that are similar to reptiles, but have moist skin and lay their eggs in water.

amplexus A breeding position of frogs and toads, in which the male holds the female with his front legs.

anadromous Living in the ocean but returning to fresh water to spawn, as in salmon; cf catadromous.

anal fin An unpaired fin on the lower surface of a fish's abdomen. It plays an important role in swimming.

annulus (pl. annuli) A marking resembling a ring, as in caecilians; or a growth ring, as on a scale of a fish, that is used in estimating age.

antenna (pl. antennae) A slender organ on an animal's head, used to sense smells and vibrations around it. Insects have two antennae.

aquaculture Cultivation of the resources of the sea or inland waters, as opposed to their exploitation.

aquatic Living all or most of the time in water; cf amphibious, terrestrial.

arachnid An arthropod with four pairs of walking legs; includes spiders, scorpions, ticks, and mites.

arboreal Living all or most of the time in trees.

arthropod An animal with jointed legs and a hard exoskeleton. Arthropods make up the largest group of animals on Earth and include insects, spiders, crustaceans, centipedes, and millipedes.

avian Of or about birds. Birds form the class Aves in the animal kingdom.

baleen The comb-like, fibrous plates found in some whales; often referred to as whalebone. The plates hang from the upper jaw and are used to sieve food from sea water.

barb A part of a bird's feather. Barbs emerge from the central shaft of a feather in a parallel arrangement, like the teeth on a comb.

barbel A slender, fleshy outgrowth on an animal's lips or near its mouth. It is equipped with sensory and chemical receptors and is used by such animals as bottom-dwelling fishes to find food.

benthic Relating to or occurring at the bottom of a body of water or the depths of the ocean; bottom dwelling (as in fishes).

biodiversity The total number of species of plants and animals in a particular location.

birds of prey Flesh-eating land birds that hunt and kill their prey. Hawks, eagles, kites, falcons, buzzards, and vultures are diurnal birds of prey; owls are nocturnal birds of prey.

blubber A thick layer of insulating fat in whales, seals, and other large marine mammals.

bony fish A fish with a bony skeleton. Other characteristics are a covering over the gills, and a swim bladder. Most bony fishes also have scales on their skin.

book lungs Lungs found in primitive insects and arachnids, consisting of "pages" of tissue through which blood and oxygen circulate.

brachiation A form of movement in which an animal (such as an ape) swings from hold to hold by its arms.

brood parasite An animal, often a bird, that tricks another species into raising its young, as in some cuckoos. A young brood parasite often kills all its nestmates so that they do not compete with it for food or care.

browser A plant-eating mammal that uses its hands or lips to pick leaves from trees and bushes (as in koalas and giraffes) or low-growing plants (as in the black rhino).

canine teeth The teeth between the incisors and molars of mammals.

carnassials Special cheek teeth with sharp, scissor-like edges used by carnivores to tear up food.

carnivore An animal that eats mainly meat. Most carnivorous mammals are predators; others are both predators and scavengers. Most carnivores eat some plant material as well as meat.

carrion The rotting flesh and other remains of dead animals.

cartilaginous fish A fish with a skeleton made of cartilage, such as a shark, ray, or chimaera.

catadromous Living in fresh water but returning to the sea to spawn, as in freshwater eels; cf anadromous.

caudal Relating to an animal's tail. In fishes, the caudal fin is the tail fin.

cephalic Of, relating to, or situated on or near the head.

cephalothorax In arachnids and crustaceans, a region of the body that combines the head and thorax. It is covered by a hard body case.

chelicera (pl. chelicerae) A pincerlike, biting mouthpart, such as a fang, as in spiders, ticks, scorpions, and mites.

chitin A hard, plastic-like substance that gives an exoskeleton its strength.

chrysalis The form taken by a butterfly in the pupal stage of its metamorphosis, or the case in which it undergoes this stage.

claspers The modified inner edges of the pelvic fins of male sharks, rays, and chimaeras, used for the transferring of sperm to the female.

cloaca An internal chamber in fishes, amphibians, reptiles, birds, and monotremes, into which the contents of the reproductive ducts and the waste ducts empty before being passed from the body.

clutch The full and completed set of eggs laid by a female bird or reptile in a single nesting attempt.

cocoon In insects and spiders, a case made of silk used to protect themselves or their eggs. In amphibians, a protective case made of mud, mucus, or a similar material, in which the animal rests during estivation.

cold-blooded See ectothermic.

complete metamorphosis A form of development in which a young insect changes shape from an egg to a larva, to a pupa, to an adult. Insects such as beetles and butterflies develop by complete metamorphosis.

compound eye An eye that is made up from many smaller eyes, each with its own lens; found in most insects, but not in spiders.

conspecific Of the same species.

convergent evolution The situation in which totally unrelated groups develop similar structures to cope with similar evolutionary pressures.

coral reef A structure made from the skeletons of coral animals, or polyps, found in warm waters.

crest In lizards, a line of large, scaly spines on the neck and back. In birds, a line of feathers on the top of the head. Lizards and many birds can raise or lower the crest as a means of communication with others.

crop A thin-walled, saclike pocket of the gullet, used by birds to store food before digestion or to feed chicks by regurgitation.

crocodilians Crocodiles, caimans, alligators, and gavials. Members of the order Crocodilia.

crustacean A mostly aquatic animal, such as a lobster, crab, or prawn, that has a hard external skeleton.

deforestation The cutting down of forest trees for timber, or to clear land for farming or building.

dewlap In a lizard, a flap of skin, sometimes brightly colored, on the throat. A dewlap usually lies close to the neck and can be extended to communicate with other lizards.

dimorphic Having two distinct forms within a species. Sexual dimorphism is the situation in which the male and female of a species differ in size and/or appearance.

dioecious Born either male or female and remaining that way through life; cf hermaphroditic.

display Behavior used by an animal to communicate with its own species, or with other animals. Displays can include postures, actions, or showing brightly colored parts of the body, and may signal threat, defense, or readiness to mate.

diurnal Active during the day. Most reptiles are diurnal because they rely on the Sun's heat to provide energy for hunting and other activities.

divergent evolution The situation in which two or more similar species become more and more dissimilar due to environmental adaptations.

DNA A molecule, found in chromosomes of a cell nucleus, that contains genes. DNA stands for deoxyribonucleic acid.

domestication The process of taming and breeding an animal for human use. Domesticated animals include pets, as well as animals used for sport, food, or work, such as sheep, horses, and dairy cattle.

dorsal fin The large fin on the back of some fishes and aquatic mammals, which helps the animal to keep its balance as it moves through the water. Some fishes have two or three dorsal fins.

dragline A strand of silk that spiders leave behind them as they move.

drone A male honeybee. Drones mate with young queens, but unlike worker bees, they do not help in collecting food or maintaining the hive.

echolocation A system of navigation that relies on sound rather than sight or touch. Dolphins, porpoises, many bats, and some birds use echolocation to tell them where their prey is, and if anything is in their way.

ecosystem A community of plants and animals and the environment to which they are adapted.

ectothermic Unable to internally regulate the body temperature, instead relying on external means such as the Sun, as in cold-blooded animals such as reptiles; cf endothermic.

egg sac A silk bag that some spiders spin around their eggs. Some egg sacs are portable, so the eggs can be carried from place to place. Others are more like a blanket, and are attached to leaves or suspended from twigs.

egg tooth A special scale on the tip of the upper lip of a hatchling lizard or snake. It is used to break a hole in the egg so that the newborn animal can escape. The egg tooth falls off within a few days of hatching.

electroreceptors Specialized organs found in some fishes (such as sharks) and mammals (such as platypuses) that detect electrical activity from the bodies of other animals. They also help the animal to navigate by detecting distortions in the electrical field of its surroundings—for example, those caused by a reef.

elytra The thickened front wings of beetles that cover and protect the back wings.

embryo An unborn animal in the earliest stages of development. An embryo may grow inside its mother's body, or in an egg outside her body.

endothermic Able to regulate the body temperature internally, as in warm-blooded animals; cf ectothermic.

estivate To spend a period of time in a state of inactivity to avoid unfavorable conditions. During times of drought, many frogs greatly reduce their overall metabolic rate and estivate underground until rain falls.

evolution Gradual change in plants and animals, over many generations, in response to their environment.

exoskeleton A hard external skeleton or body case. All arthropods are protected by an exoskeleton.

exotic A foreign or non-native species of animal or plant, often introduced into a habitat by humans.

feral A wild animal or plant, or a species that was once domesticated but has returned to its wild state.

fledgling A young bird that has recently grown its first true feathers and has just left its nest.

fry Young or small fishes.

gastroliths Stones swallowed by such animals as crocodilians, that stay in the stomach to help crush food.

gestation period The period of time during which a female animal is pregnant with her young.

gills Organs that collect oxygen from water and are used for breathing. Gills are found in many aquatic animals, including fishes, some insects, and the larvae of amphibians.

gizzard In birds, the equivalent of the stomach in mammals. Grit and stones inside the gizzard help to grind up food. Food passes from the gizzard to the intestines.

gravid Full of eggs; pregnant.

grazer An animal that eats grasses and plants that grow on the ground.

groom Of an animal, to clean, repair, and arrange the fur.

grub An insect larva, usually that of a beetle.

gullet Found in birds, the gullet is the equivalent of the esophagus in mammals. This tube passes food from the bill to the gizzard.

habitat The area in which an animal naturally lives. Many different kinds of animals live in the same environment (for example, a rain forest), but each kind lives in a different habitat within that environment. For example, some animals in a rain forest live in the trees, while others live on the ground.

harem A group of female animals that mate and live with one male.

hatchling A young animal, such as a bird or reptile, that has recently hatched from its egg.

heat-sensitive pit Sense organs in some snakes that detect tiny changes in temperature.

herbivore An animal that eats only plant material, such as leaves, bark, roots, and seeds; cf carnivore, omnivore.

hermaphrodite A plant or animal either possessing both male and female sex organs at the same time, and thus able to fertilize itself (known as simultaneous hermaphroditism), or one that starts life as one sex and later changes to the other (sequential hermaphroditism); see also protandrous, protogynous.

hibernate To remain completely inactive during the cold winter months. Some animals eat as much as they can before the winter, then curl up in a sheltered spot and fall into a very deep sleep. They live off stored fat, and slow their breathing and heartbeat to help conserve energy until spring. Insects may hibernate as eggs, larvae, pupae, or adults.

hoof The toe of a horse, antelope, deer, or related animal that is covered in thick, hard skin with sharp edges.

hybrid The offspring of parents of two different species.

incisors The front teeth of an animal, located between the canines, used for cutting food.

incomplete metamorphosis A system of development in which a young insect gradually changes shape from an egg, to a nymph, to an adult.

incubate To keep eggs in an environment, outside the female's body, in which they can develop and hatch. Most birds incubate their eggs by warming them with their body heat. The eggs of other birds and reptiles may incubate in soil, leaf litter, or a similar covering, with no further input from the parents.

insectivore An animal that eats only or mainly insects or invertebrates. Some insectivores also eat small vertebrates, such as frogs, lizards, and mice. In mammals, a member of the order Insectivora.

introduced An animal or plant species imported from another place by humans and deliberately or accidentally released into a habitat.

invertebrate An animal with no backbone. Many invertebrates are soft-bodied animals, such as worms, leeches, or octopuses, but most have an exoskeleton, or hard external skeleton, such as insects.

ivory The tusks or enlarged teeth of elephants and walrus, or objects manufactured from them.

Jacobson's organ Two small sensory pits on the top part of the mouth of snakes, lizards, and mammals. They use this organ to analyze small molecules picked up from the air or ground with their tongue.

keratin A protein found in horns, hair, scales, feathers, and fingernails.

krill Small, shrimplike crustaceans that live in huge numbers in Arctic and Antarctic waters.

lateral line A sensory canal system along the sides of a fish. It detects moving objects by registering disturbances (or pressure changes) in the water.

larva (pl. larvae) A young animal that looks completely different from its parents. An insect larva, sometimes called a grub, maggot, or caterpillar, changes into an adult by either complete or incomplete metamorphosis. In amphibians, the larval stage is the stage before metamorphosis that breathes with gills rather than lungs (for example, tadpoles).

live-bearing Giving birth to young that are fully formed.

luminous Reflecting or radiating light. Some deepsea fishes can produce their own light source by using luminous bacteria.

maggot The larva of a fly.

mammal A warm-blooded vertebrate that suckles its young with milk and has a single bone in its lower jaw. Although most mammals have hair and give birth to live young, some, such as whales and dolphins, have very little or no hair; others, the monotremes, lay eggs.

mandibles Biting jaws of an insect.

marsupial A mammal that gives birth to young that are not fully developed. These young are usually protected in a pouch (where they feed on milk) before they can move around independently.

mesopelagic Of or associated with the deep sea, from about 600 to 3,000 feet (200 to 1,000 m).

metamorphosis A way of development in which an animal's body changes shape. Many invertebrates, including insects, as well as some vertebrates, such as amphibians, metamorphose as they mature.

migration A usually seasonal journey from one habitat to another. Many animals migrate vast distances to another location to find food, or to mate and lay eggs or give birth. Some deep-sea fishes migrate vertically at night to feed.

molars A mammal's side cheek teeth, used for crushing and grinding.

mollusk An animal, such as a snail or squid, with no backbone and a soft body that is often partly or fully enclosed by a shell.

molt To shed an outer layer of the body, such as hair, skin, scales, feathers, or the exoskeleton.

monotreme A primitive mammal with many features in common with reptiles. Monotremes lay eggs and have a cloaca. They are the only mammals that lack teats, although they feed their young milk released through ducts on their belly.

morph A color or other physical variation within, or a local population of, a species.

musth In elephants, a time of high testosterone levels when the musth gland between the eye and ear secretes fluid; associated with mating and characterized by increased aggression and searching for females.

mutualism An alliance between two species that is beneficial to both, as in ox-peckers and grazing herbivores.

neoteny The retention of some immature or larval characteristics into adulthood. Some salamander species are usually or often neotenic.

niche The ecological role played by a species within an animal community.

nocturnal Active at night. Nocturnal animals have special adaptations, such as large, sensitive eyes or ears, to help them find their way in the dark. All nocturnal animals rest during the day.

nomadic Lacking a fixed territory, instead wandering from place to place in search of food and water.

nymph The young stage of an insect that develops by incomplete metamorphosis. Nymphs are often similar to adults, but do not have fully developed wings.

ocellus A kind of simple eye with a single lens. Insects usually have three ocelli on the top of their head.

omnivore An animal that eats both plant and animal food. Omnivores have teeth and a digestive system designed to process almost any kind of food.

opposable Describing a thumb that can reach around and touch all of the other fingers on the same hand, or a toe that can similarly touch all of the other toes on the same foot.

order A major group used in taxonomic classification. An order forms part of a class, and is further divided into one or more families.

osteoderm In reptiles, a lump of bone in the skin that provides protection against predators. Most crocodilians and some lizards are protected by osteoderms, as well as by thick, strong skin.

oviparous Reproducing by laying eggs. Little or no development occurs within the mother's body; instead, the embryos develop inside the egg; cf ovoviviparous, viviparous.

ovipositor A tubelike organ through which female insects lay their eggs. The stinger of bees and wasps is an modified ovipositor.

ovoviviparous Reproducing by giving birth to live young that have developed from eggs within the mother's body. The eggs may hatch as they are laid or soon after; cf oviparous, viviparous.

pair bond A partnership maintained between a male and a female animal, particularly birds, through one or several breeding attempts. Some species maintain a pair bond for life.

paleontology The scientific study of life in past geological periods.

parallel evolution The situation in which related groups living in isolation develop similar structures to cope with similar evolutionary pressures.

parasitism The situation in which an animal or plant lives and/or feeds on another living animal or plant, sometimes with harmful effects.

passerine Any species of bird belonging to the order Passeriformes. A passerine is often described as a songbird or a perching bird.

pectoral fins In fishes and aquatic mammals, the paired fins attached to each side of the animal and used for lift and control of movement. In fishes, they are usually located just behind the gills.

pedicel The narrow "waist" that connects an insect's head to its thorax, or a spider's cephalothorax to its abdomen.

pedipalp One of a pair of small, leglike organs on the head of insects and the cephalothorax of spiders and scorpions, used for feeling or for handling food. In spiders, they are also used for mating.

pelagic Swimming freely in the open ocean; not associated with the bottom; cf benthic.

pelvic fins Paired fins, located on the lower part of a fish's body.

photophores Luminous organs on some deep-sea fishes.

pheromone A chemical released by an animal that sends a signal and affects the behavior of others of the same species. Many animals use pheromones to attract mates, or to signal danger.

phytoplankton Tiny, single-celled algae that float on or near the surface of the sea.

placental mammal A mammal that does not lay eggs (as monotremes do), or give birth to underdeveloped young (as marsupials do). Instead, it nourishes its developing young inside its body with a blood-rich organ called a placenta.

plankton The plant (phytoplankton) or animal (zooplankton) organisms that float or drift in the open sea. Plankton forms an important link in the food chain.

precocial Active and self-reliant at birth. Describes newly hatched chicks of some birds, such as ducks and chickens; cf altricial.

predator An animal that lives mainly by killing and eating other animals.

preen Of a bird, to clean, repair, and arrange plumage.

prehensile Grasping or gripping. Some tree-dwelling mammals and reptiles have prehensile feet or a tail that can be used as an extra limb to help them stay safely in a tree. Elephants have a prehensile "finger" on the end of their trunk so they can pick up small pieces of food. Browsers, such as giraffes, have prehensile lips to help them grip leaves. The prehensile tongue of parrots enables them to extract kernels from shells.

proboscis In insects, a long, tubular mouthpart used for feeding. In some mammals, a proboscis is an elongated nose, snout, or trunk. That of an elephant has many functions, such as smelling, touching, and lifting.

protandrous Of sequential hermaphrodites (as in some fishes), starting life as a male and later becoming female.

protogynous Of sequential hermaphrodites (as in some fishes), starting life as a female and later becoming male.

pupa (pl. pupae) The stage during which an insect transforms from a larva to an adult.

queen A female insect that begins a social insect colony. The queen is normally the only member of the colony that lays eggs.

quill Long, sharp hair of an echidna, porcupine, anteater, and a few other mammals; used for defense.

raptor A diurnal bird of prey, such as a hawk or falcon. The term is not used to describe owls.

roost A place or site used by some animals, such as birds and bats, for sleeping. Also, the act of traveling to or gathering at such a place.

refection A practise of some rodents, in which they eat their own fecal matter to gain maximum nutrition from the food, before passing feces as dry pellets.

retractile claws The claws of cats and similar animals that are usually protected in sheaths. Such claws spring out when the animal needs them to capture prey or to fight.

rostrum A tubular, beak-like feeding organ on the head of some insects.

ruminants Hoofed animals—cattle, buffalo, bison, antelopes, gazelles, sheep, goats, and other members of the family Bovidae—with a four-chambered stomach. One of these chambers is the rum.

savanna Open grassland with scattered trees. Most savannas are found in subtropical areas that have a distinct summer wet season.

scales In reptiles, distinct thickened areas of skin that vary in size, from very small to large; in fishes, small plates that form part of the external body covering.

scavenger An animal that eats carrion—often the remains of animals killed by predators.

scutes In a turtle or tortoise, the horny plates that cover the bony shell.

sedentary Having a lifestyle that involves little movement; also used to describe animals that do not migrate.

semi-aquatic Living some of the time in water, and some of the time on land.

social Living in groups. Social animals can live in breeding pairs, sometimes together with their young, or in a colony or herd of thousands.

spawn To release eggs and sperm together directly into the water.

species A group of animals with very similar features that are able to breed together and produce fertile young.

spermatophore A container or package of sperm that is passed from male to female during mating.

spicule A minute, needle-like, siliceous or calcareous body, found in invertebrates.

spinnerets The fingerlike appendages of spiders that are connected to silk glands, found near the tip of the abdomen.

squalene A substance found in the liver oil of deep-sea sharks. It is refined and used as a high-grade machine oil in high-technology industries, as a human health and dietary supplement, and in cosmetics.

stinger A hollow structure on the tail of insects and scorpions that pierces flesh and injects venom.

stereoscopic vision Vision in which both eyes face forward, giving an animal two overlapping fields of view and thus allowing it to judge depth.

stridulate To make a sound by scraping objects together. Many insects communicate in this way, some by scraping their legs against their body.

stylet A sharp mouthpart used for piercing plants or animals.

subantarctic Of the oceans and islands just north of Antarctica.

swim bladder A gas-filled, baglike organ in the abdomen of bony fishes. It enables the fishes to remain at a particular depth in the water.

symbiosis An alliance between two species that is usually (but not always) beneficial to both. Animals form symbiotic relationships with plants, microorganisms, and other animals; cf mutualism, parasitism.

sympatric Of two or more species, occurring in the same area.

syndactylic Having fused toes on the hindfoot, as in bandicoots, kangaroos, and wombats.

tadpole The larva of a frog or toad. Tadpoles are aquatic and take in oxygen from the water through gills.

taxonomy The system of classifying living things into various groups and subgroups according to similarities in features and adaptations.

teleosts Bony fishes.

terrestrial Living all or most of the time on land; cf amphibious, aquatic.

temperate Describes an environment or region that has a warm (but not very hot) summer and a cool (but not very cold) winter. Most of the world's temperate regions are located between the tropics and the polar regions.

tentacles On marine invertebrates, long, thin structures used to feel and grasp, or to inject venom. On caecilians, sensory organs located on the sides of the head, possibly used for tasting and smelling.

thermal A column of rising air, used by birds to gain height, and on which some birds soar to save energy.

thorax The middle part of an animal's body. In insects, the thorax is divided from the head with a narrow "waist," or pedicel. In spiders, the thorax and head make up a single unit.

torpid In a sleep-like state in which bodily processes are greatly slowed. Torpor helps animals to survive difficult conditions such as cold or lack of food. Estivation and hibernation are types of torpor.

trachea A breathing tube in an animal's body. In vertebrates, there is one trachea (or windpipe), through which air passes to the lungs. Insects and some spiders have many small tracheae that spread throughout the body.

tropical Describes an environment or region near the Equator that is warm to hot all year round.

tropical forests Forests growing in tropical regions, such as central Africa, northern South America, and southeast Asia, that experience little difference in temperature throughout the year.

tundra A cold, barren area where much of the soil is frozen and the vegetation consists mainly of mosses, lichens, and other small plants adapted to withstand intense cold. Tundra is found near the Arctic Circle and on mountain tops.

tusks The very long teeth of such mammals as elephants, pigs, hippos, musk deer, walruses, and narwhals; used in fights and for self-defense.

ungulate A large, plant-eating mammal with hoofs. Ungulates include elephants, rhinoceroses, horses, deer, antelope, and wild cattle.

vertebrate An animal with a backbone. All vertebrates have an internal skeleton of cartilage or bone. Fishes, reptiles, birds, amphibians, and mammals are vertebrates.

vestigial Relating to an organ that is non-functional or atrophied.

vibrissae Specialized hairs, or whiskers, that are extremely sensitive to touch.

viviparous Reproducing by means of young that develop inside the mother's body and are born live; sometimes called placental viviparity. Most mammals and some fishes (such as sharks) are viviparous.

warm-blooded See endothermic.

worker A social insect that collects food and tends a colony's young, but which usually cannot reproduce.

zooplankton The tiny animals that, together with phytoplankton, form the plankton that drifts on or near the sea's surface. Zooplankton are eaten by some whales, fishes, and seabirds.

zygodactylous Of birds, having two of the four toes pointing forward, and the other two pointing backward.

INDEX

A

D

I

J

K

L

M

P

Q

R

S

U

V

PHOTOGRAPHS
All photographs © iStockphoto.com and Shutterstock, except p 422 and p 430 Sea Pics.

ILLUSTRATIONS
All species gallery illustrations © MagicGroup s.r.o. (Czech Republic)—www.magicgroup.cz
Pavel Dvorský, Eva Göndörová, Petr Hloušek, Pavla Hochmanová, Jan Hošek, Jaromír a Libuše Knotkovi, Milada Kudrnová, Petr Liška, Jan Maget, Vlasta Matoušová, Jiří Moravec, Pavel Procházka, Petr Rob, Přemysl Vranovský, Lenka Vybíralová.

Additional illustrations © Weldon Owen Pty Ltd by:
Susanna Addario, Alistair Barnard, Priscilla Barret, Andre Boos, Anne Bowman, Peter Bull Art Studio, Leonello Calvetti, Martin Camm, Barry Croucher, Andrew Davies/Creative Communication, Kevin Deacon, Fiammetta Dogi, Sandra Doyle, Simone End, Christer Eriksson, Alan Ewart, John Francis, Giuliano Fornari, Jon Gittoes, Mike Golding, Ray Grinaway, Gino Hasler, Phil Hood, Robert Hynes, Ian Jackson, Frits Jan Maas, David Kirshner, Frank Knight, Alex Lavroff, John Mac, Robert Mancini, Map Illustrations, James McKinnon, Karel Mauer, Ken Oliver, Erik van Ommen, Sandra Pond, Mick Posen, Tony Pyrzakowski, John Richards, Edwina Riddell, Barbara Rodanska, Trevor Ruth, Peter Schouten, Peter Scott, Chris Shields, Kevin Stead, Roger Swainston, Guy Troughton, Chris Turnbull, Trevor Weekes, Rod Westblade, Wildlife Art Ltd.

MAPS/GRAPHICS
All maps and information graphics © Weldon Owen Pty Ltd by Andrew Davies/Creative Communication, except maps pp 388–435 by Brian Johnston.

INDEX
Puddingburn Publishing Services.

The publishers wish to thank Brendan Cotter, Helen Flint, Frankfurt Zoological Society, Maria Harding, Tanzania National Parks, and Shan Wolody for their assistance in the preparation of this volume.

Read to Learn

Main Idea **Many important substances in our lives are acids or bases.**

What Are Acids and Bases?

How do you know whether something is an **acid** or a **base**? Vinegar, orange juice, and lemon juice are acids. An acid tastes sour and turns blue litmus paper red. Bases, like ammonia and baking soda, taste bitter and turn red litmus paper blue. (You should never test acids and bases by tasting them, however.)

In Lesson 5 you learned about chemical reactions. When you mixed vinegar, an acid, with baking soda, a base, a chemical reaction occurred. Acids and bases can react with each other to form water and a salt.

In a reaction acids "give away" hydrogen particles, or *hydronium ions*. Bases give off *hydroxide ions*.

This is a simplified hydronium ion. Acids release them in solution.

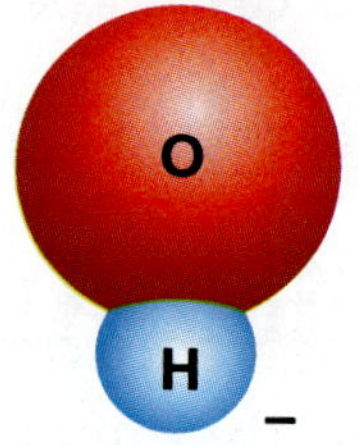

Bases release hydroxide ions, such as this, in solution.

Hydroxide ions are particles that are made up of one oxygen atom and one hydrogen atom linked together. Hydronium and hydroxide ions combine to make water. What is left of the acid and base also combines to make a new substance—a salt.

If you place a blue and a red strip of litmus paper in a glass of water, both strips will stay the same color. This tells you that water is neither acidic nor basic. A solution that is neither acidic nor basic is **neutral**. Water is a neutral substance.

When hydrochloric acid is added to a base, sodium hydroxide, a chemical reaction occurs to produce water and sodium chloride (table salt).